In Situ Bioreclamation

Applications and Investigations for Hydrocarbon and Contaminated Site Remediation

Edited by

Robert E. Hinchee

and

Robert F. Olfenbuttel

Battelle Memorial Institute
Columbus, Ohio

BUTTERWORTH–HEINEMANN
Boston London Oxford Singapore Sydney Toronto Wellington

Recognizing the importance of preserving what has been written, it is the policy of Butterworth-Heinemann to have the books it publishes printed on acid-free paper, and we exert our best efforts to that end.

LC 91-073683

ISBN 0-7506-9301-0

Butterworth-Heinemann
80 Montvale Avenue
Stoneham, Massachusetts 02180

10 9 8 7 6 5 4 3 2 1

Printed in the United States of America

Contents

Articles

Technical Notes

On-Site Bioreclamation A Companion Volume

Contents

Articles

Technical Notes

Foreword

This book is one of two that resulted from the international symposium *"In Situ* and On-Site Bioreclamation" held in San Diego in March of 1991. The two volumes, In Situ *Bioreclamation: Applications and Investigations for Hydrocarbon and Contaminated Site Remediation* and *On-Site Bioreclamation: Processes for Xenobiotic and Hydrocarbon Treatment,* consist of selected papers submitted by symposium participants. The symposium brought together more than 700 people from 20 countries in Europe, North America, Asia, and Australia. It was apparent that the interest in the biological treatment of contaminated soil, water, and gas is tremendous. It was also obvious that the state of the art varies from proven commercially available technologies for some applications, such as reactor treatment of petroleum hydrocarbons in aqueous steams, to fundamental research in others, such as genetic engineering to develop new strains of biodegradable xenobiotics. Together, these volumes represent the most complete and up-to-date set of papers published on the topic.

In response to an invitation to submit papers, we received 100 papers. The papers were sent to two peer reviewers and, following revision, 82 were accepted for publication. Two different formats are used, articles and technical notes. Articles represent a substantial technical contribution; technical notes are brief technology descriptions or reports of preliminary or less substantial studies.

Appreciation is extended to the more than 100 people who served as technical reviewers. This relatively thankless task is essential to assure quality in publishing. The peer reviewers are listed at the end of this Foreword. Additionally, Lewis Semprini of Stanford University played an invaluable role in inviting and reviewing many of the papers on chlorinated solvent degradation; as a result of his efforts this subject is very well covered.

Numerous organizations and individuals contributed to organizing and supporting both the symposium and these resulting publications. Battelle Memorial Institute underwrote the symposium and subsequent publications and assumed all of the financial risks of this effort. The following is a list of people and organizations who either served as cosponsors or organized technical sessions contributing to this effort:

Mick Arthur, Battelle Columbus
Bruce Bauman, American Petroleum Institute
Christian Bocard, Institut Francais du Pétrole, France
Rob Booth, Environment Canada
D. B. Chan, U.S. Naval Civil Engineering Laboratory
Soon Cho, Ajou University, Korea
Giovanni Ferro, D'Appolonia, Italy
Volker Franzius, Umweltbundesamt, Germany
Giancarlo Gabetto, Castalia, Italy
Ronald Hoeppel, U.S. Naval Civil Engineering Laboratory
Hidemi Iguchi, Mitsubishi Corporation, Japan
Massimo Martinelli, ENEA, Italy
Jeff Means, Battelle Columbus
Kun Mo Lee, Ajou University, Korea
Ross N. Miller, U.S. Air Force HSD/YAQE
Donna Palmer, Battelle Columbus
Gloria Patton, U.S. Department of Energy
Chongrak Polprasert, Asian Institute of Technology, Thailand
Agusto Porta, Battelle Geneve, Switzerland
Parmely "Hap" Prichard, U.S. Environmental Protection Agency
Emilio Santucci, Castalia, Italy
Paul Schaltzberg, David Taylor Research Center
Lewis Semprini, Stanford University
John Skinner, U.S. Environmental Protection Agency
Terri Stewart, Battelle Pacific Northwest Laboratories
Klass Visscher, RIVM, The Netherlands
John Wilson, U.S. Environmental Protection Agency

Cher Paul at Battelle served diligently in the monumental task of organizing and assembling both volumes. She read and edited every paper both for consistency of style and for readability. She also supervised the logistics of actually assembling the books. Others who contributed to putting the volumes together were Karl Nehring, Tracy Melchiori, Shauna Kearney, Gina Melaragno, Jennifer Eddy, Janelle McClary, Sharon Manwering, Susan Snyder, Darlene Whyte, and Lisa Becker of Battelle and Tom Bigelow of Kenyon College. Mr. Nehring provided administrative support and handled communications with the publisher. Ms. Melchiori, Ms. Kearney, and Ms. Melaragno coordinated the communications with the authors and peer reviewers. Mr. Bigelow, an editor, supported Ms. Paul. Ms. Eddy and Ms. McClary coordinated art and layout, Ms. Manwering was the text processor, and Ms. Snyder, Ms. Whyte, and Ms. Becker proofread the work.

In the rapidly evolving field of bioremediation, much of the work presented in these volumes will become dated in a few years. A consensus of participants in the 1991 symposium called for a second symposium, which will be held in 1993. The 1993 symposium will most probably also result in publication; at this time its format is undecided, but that it will be needed is undoubtable.

Rob Hinchee
May 1993

Peer Reviewers

Dan Acton, Beak Consultants Limited, Canada
Pradeep Aggarwal, Argonne National Laboratory
Robert C. Ahlert, Rutgers University
Pedro J. J. Alvarez, University of Michigan
Mick Arthur, Battelle Columbus
Erik Arvin, Technical University of Denmark

Dr. Ballerini, Institut Français du Pétrole, France
Chris Barber, CSIRO, Australia
E. R. Barenschee, Degussa Corporation, Germany
James F. Barker, University of Waterloo, Canada
Denise M. Barnes, Ecosystems Engineering
Thomas C. Beard, Battelle Columbus Division
Philip B. Bedient, Rice University
Sanjoy K. Bhattacharya, Tulane University
Christian Bocard, Institut Français du Pétrole, France
Gary Boettcher, Geraghty & Miller, Inc.
Robert C. Borden, North Carolina State University
Edward J. Bouwer, Johns Hopkins University
Hans Breukelman, DSM Research B.V., The Netherlands
Andrew Briefer, Western Geologic Resources, Inc.
Fred Brockman, Battelle Pacific Northwest Laboratory
Thomas M. Brouns, Battelle Pacific Northwest Laboratory
Richard A. Brown, Groundwater Technology, Inc.
Gunter Brox, EIMCO Process Equipment
Gaylen R. Brubaker, Remediation Technologies, Inc.
David Burris, U.S. Air Force
Timothy E. Buscheck, Chevron Corporation

Jason A. Caplan, ESE Biosciences
Peter J. Chapman, U.S. Environmental Protection Agency
Thomas H. Christensen, Technical University of Denmark
John A. Cioffi, Ecova Corporation

Gary L. Conaway, Ebasco Services, Inc.
Robert Coutant, Battelle Columbus
Craig Criddle, Michigan State University

Greg B. Davis, CSIRO, Australia
Lois Davis, Sybron Chemicals, Inc.
Wendy J. Davis-Hoover, U.S. Environmental Protection Agency
W. P. de Bruin, Agricultural University, Wageningen, The Netherlands
Barbara A. Denovan, University of Washington
Jeffrey C. Dey, Groundwater Technology Inc.
Ludo Diels, SCK-CEN/VI TO, Belgium
Philip M. Digrazia, University of Tennessee
Ellen L. Dorwin, Computer Sciences Corporation
William Doucette, Utah State University
Rob Douglass, The Traverse Group, Inc.
Douglas C. Downey, Engineering Science, Inc.
Wayne C. Downs, U.S. Environmental Protection Agency
Jean Ducreux, Institut Français du Pétrole, France
R. Ryan Dupont, Utah State University

Elizabeth A. Edwards, Standford University
Richard Egg, Texas A&M University
Mary Pat Eisman, U.S. Naval Civil Engineering Laboratory
W. Eng, University of Tennessee

Margaret Findlay, ABB Environmental Services
Paul E. Flathman, OHM Corporation
John Flyvbjerg, Technical University of Denmark
Samuel Fogel, ABB Environmental Services
Louis B. Fournier, Groundwater Technology, Inc.

William T. Frankenberger, Jr., University of California, Riverside
James K. Fredrickson, Battelle Pacific Northwest Laboratory
David L. Freedman, University of Illinois

Dr. Gatellier, Institut Français du Pétrole, France
W. Kennedy Gauger, Institute of Gas Technology
Richard M. Gersberg, San Diego State University
James M. Gossett, Cornell University
Chris Griffen, The Traverse Group, Inc.
Mirat Gurol, Drexel University

Cathleen J. Hapeman-Somich, U.S. Department of Agriculture
Joop Harmsen, The Winand Staring Centre, The Netherlands
Kirk Hatfield, University of Florida
Edward Heyse, U.S. Air Force
Duane D. Hicks, Texas A&M University
Ronald Hoeppel, U.S. Naval Civil Engineering Laboratory
Scott G. Huling, U.S. Environmental Protection Agency

Charles Imel, Ecosystems Engineering
J. Mark Inglis, Western Geologic Resources, Inc.

Bjørn K. Jensen, Technical University of Denmark
Rex Johnson, The Traverse Group, Inc.
Mark Jones, Delta Environmental Consultants, Inc.

Paul Kemp, Woodward-Clyde Consultants
Khalique A. Khan, Ensotech Inc.
J. M. Henry King, University of Tennessee
T. Kent Kirk, U.S. Department of Agriculture

Calvin A. Kodres, U.S. Naval Civil Engineering Laboratory
Sydney S. Koegler, Battelle Pacific Northwest Laboratory
William M. Korreck, The Traverse Group, Inc.
Peter Kroopniek, Groundwater Technology, Inc.
Curtis A. Kruger, Midwest Research Institute

R. Lamar, U.S. Department of Agriculture
Robert LaPoe, U.S. Air Force
Jim League, Bogart Environmental
Kun Mo Lee, Ajou University, South Korea
Michael D. Lee, Du Pont Environmental Remediation Services
Richard F. Lee, Skidaway Institute of Oceanography
Alfred P. Leuschner, ReTec
M. Tony Lieberman, ESE Biosciences
Carol D. Litchfield, Environment America
C. Deane Little, Stanford University
John Lyngkilde, Technical University of Denmark

David Major, Beak Consultants Limited, Canada
Donn L. Marrin, InterPhase Environmental, Inc.
Michael M. Martinson, Delta Environmental Consultants, Inc.
Perry L. McCarty, Stanford University
Gloria McCleary, EA Engineering, Science and Technology
Linda McConnell, Midwest Research Institute
Mike McFarland, Utah State University
Stephen P. McGrath, AFRC Institute of Arable Crops Research, United Kingdom
Jeffrey Means, Battelle Columbus
Blaine Metting, Battelle Pacific Northwest Laboratory
Barbara J. Mickelson, Delta Environmental Consultants, Inc.
Ross N. Miller, U.S. Air Force
Ali Mohagheghi, Solar Energy Research Institute

Tom Naymik, Battelle Columbus Division
Michael J. K. Nelson, Ecova Corporation
Bruce Nielsen, U.S. Air Force
Per H. Nielsen, Technical University of Denmark
Victor K. Nowicki, WMS Associates Ltd., Canada

Joseph E. Odencrantz, Lavine-Fricke Consulting Engineers
Richard A. Ogle, Applied Earth Sciences
Say Kee Ong, Battelle Columbus
Rick L. Ornstein, Battelle Pacific Northwest Laboratory
David Ostendorf, University of Massachusetts

Donna Palmer, Battelle Columbus
Anthony V. Palumbo, Oak Ridge National Laboratory
Gene F. Parkin, University of Iowa
Christopher J. Perry, Battelle Columbus
Chuck Pettigrew, U.S. Air Force
George Philippidis, Solar Energy Research Institute
Michael Piotrowski, Woodward-Clyde Consultants
Parmely "Hap" Pritchard, U.S. Environmental Protection Agency
Barb Prosen, The Traverse Group, Inc.

John F. Quensen, III, Michigan State University

Roger D. Reeves, Massey University, New Zealand
Martin Reinhard, Stanford University
H. James Reisinger, II, Integrated Environmental Technologies, Inc.
H. S. Rifai, Rice University
Christopher J. Rivard, Solar Energy Research Institute
George Robinson, Rice University

Eric K. Schmitt, ESE Biosciences Group
Gosse Schraa, Agricultural University, Wageningen, The Netherlands
Doug Selby, Las Vegas Valley Water District
Lewis Semprini, Stanford University
Patrick Sferra, U.S. Environmental Protection Agency
Daniel R. Shelton, U.S. Department of Agriculture
Ronald Sims, Utah State University
Rodney S. Skeen, Battelle Pacific Northwest Laboratory
George J. Skladany, Envirogen
Lloyd Slezak, Dayton & Knight Ltd., Canada
Larry Smith, Battelle Columbus
Jim Spain, U.S. Air Force
Sjef Staps, Grontmij N.V. Consulting Engineers
Jan Stepek, EA Engineering
Gerald W. Strandberg, Oak Ridge National Laboratory

William C. Tacon, Battelle Columbus
James M. Tiedje, Michigan State University
Francis T. Tran, Asian Institute of Technology, Thailand
R. D. Tyagi, INRS-EAU, Canada

A. J. Valsangkar, WMS Associates Ltd., Canada
Jean-Paul Vandecastelle, Institut Français du Pétrole, France
J. van Eyk, Delft Geotechnics, The Netherlands
Al Venosa, U.S. Environmental Protection Agency
Catherine Vogel, U.S. Air Force
Timothy Vogel, University of Michigan
Evangelos A. Voudrias, Georgia Institute of Technology

Martin Werner, Washington State Department of Ecology
Godage B. Wickramanayake, Environ Corporation
Barbara H. Wilson, Rice University
Nancy Winters, Washington State Department of Ecology
Will Wright, III, California Department of Health Services
Lin Wu, University of California, Davis
Robert E. Wyza, Battelle Columbus

Lily Young, New York University Medical Center

A Review of European Bioreclamation Practice

Augusto Porta[*]
Battelle Europe

Remediation technology in Europe has evolved significantly over the past few years. Although certain developments have paralleled those in the United States, Europeans have often led the way, developing technologies that were later adopted in America.

And Europeans will continue to make great strides in bioreclamation research and development. With soil rehabilitation efforts in the European Community expected to reach $10 billion by the year 2000—ground-cleaning efforts in general are expected to grow from $30 billion in the next decade to $130 billion over the long term—bioreclamation technologies will be invaluable tools in solving a myriad of pollution problems in Europe.

Progressive Countries and Companies

A growing number of European countries and companies alike are involved in bioremediation work. The Third International KfK/TNO Conference on Contaminated Soil, held December 10–14, 1990, in Karlsruhe, Federal Republic of Germany, underscored this when 24 papers on biological techniques were presented—all by European firms, mainly from The Netherlands, Denmark, and Germany. These three regions, especially, have high levels of public awareness that will continue to influence environmental legislation. Tables 1 through 4

* Battelle Europe, 7, route de Drize, 1227 Carouge, Geneva, Switzerland

TABLE 1. German organizations active in bioremediation.

Company	Location	State of Development	Soil Treatment	Groundwater Treatment	*In-Situ* Treatment	On-Site and/or Off-Site Treatment
A. Alexander KG GmbH u. Co.[a]	Berlin					
Anakat, Institut für Biotechnologie	Berlin	Full-Scale Demonstration	X		X	X
Argus Umweltbiotechnologe GmbH	Berlin	Industrial-Scale Demonstration	X		X	X
B & R Ingenieurgesellschaft für Baustoff-Recycling und Umwelttechnik mbH[a]	Düsseldorf					
b-d-s- Boden- und Deponie-Sanierungs GmbH[a]	München					
Bauer Spezialtiefbau GmbH	Schrobenhausen	Full-Scale Demonstration	X			X
Bil'inger u. Berger Bau AG[a]	Mannheim					
Biodetox Gesellschaft zur biologischen Schadstoffentsorgung GmbH	Alsen/b. Bückeburg	Industrial-Scale	X	X	X	X
Bonnenberg & Drescher	Aldenhoven	Pilot Plant	X		X	
BTB, Bartels Technik & Bau GmbH[a]	Berlin					
Degussa AG	Hanau	Research at Laboratory	X		X	
Dekon, Dekontaminations-gesellschaft mbH[a]	Dortmund					
Detlef Hegemann Engineering GmbH[a]	Bremen					
Deutsche Shell	Hamburg	Pilot Projects (Shell Bioreg)	X			X
Ecosystem-Gruppe	Meerbusch	Research	X	X	X	

(a) These companies offer microbiological processes, but detailed information has not been collected.

TABLE 1. (Continued)

Company	Location	State of Development	Soil Treatment	Groundwater Treatment	*In-Situ* Treatment	On-Site and/or Off-Site Treatment
Erd- und Grundbauinstitut der Rein-Ruhr Ingenieurgesellschaft mbH[(a)]	Dortmund					
Este GmbH	Hamburg	Full-Scale Demonstration (Shell Bioreg)	X	X		X
Fraunhofer Institut für Grenzllächen und Bioverfahrenstechnik	Stuttgart	Biodegradation of Explosives				
Frühbis GmbH u. Co. KG[(a)]	Edenkoben/ Weinstr.					
GBF, Gesellschaft für Biotechnologische Forschung MbH	Braunschweig		X		X	
GMA, Gesellschaft für Müllund Abfallbeseitigung mbH u. Co. KG[(a)]	Schortens					
Groth u. Co.	Pinneberg	Full-Scale Demonstration	X	X		X
Herost Umwelttechnik		Pilot Scale		X		X
Hochtief AG	Essen	Laboratory & Industrial Scale	X		X	X
Industrie Abwasser Technik Kurt Lissner GmbH u. Co.[(a)]	Ganderkesee					
Institut für Molekularbiologie und Analytik GmbH	Zeppelinheim	Field Demonstration Completed	X		X	
Jastram-Werke GmbH KG[(a)]	Hamburg					
Kloeckner Decotec GmbH	Duisburg	Full-Scale Demonstration	X	X	X	X
Kolsch GmbH[(a)]	Siegen					

(a) These companies offer microbiological processes, but detailed information has not been collected.

TABLE 1. (Continued)

Company	Location	State of Development	Soil Treatment	Groundwater Treatment	*In-Situ* Treatment	On-Site and/or Off-Site Treatment
Leichtweiss-Institut für Wasserbau c/o TU Braunschweig[a]	Braunschweig					
LFU, Labor für Umweltanalytik GmbH	Berlin	Full-Scale Demonstration	X		X	
Linde AG	Hollriegelskreuth	Research & Field Demonstration	X	X	X	X
Lurgi GmbH[a]	Frankfurt					
K. Massholder[a]	Heidelberg					
Messer Griesheim GmbH	Krefeld	Test Field Completed	X		X	
Philipp Holzmann AG	Düsseldorf	Pilot Study & Field (Shell Bioreg)	X	X	X	X
Preussag AG	Hannover	Laboratory & Pilot Studies	X		X	X
Probiotec GmbH[a]	Düren-Gúrzenich					
RAG, Ruhrkohle Umwelttechnik GmbH[a]	Essen					
Rethman Städtereinigung GmbH	Selm	Pilot Project	X			X
Richard Buchen GmbH[a]	Köln					
Fa. Sanexen Industrieentsorgung GmbH[a]	Duisburg					
Santec GmbH	Berlin	Pilot Test	X		X	
Senator Projekt Service GmbH	Düsseldorf	Demonstration Full-Scale Project (GDS Process)	X	X		

(a) These companies offer microbiological processes, but detailed information has not been collected.

TABLE 1. (Continued)

Company	Location	State of Development	Soil Treatment	Groundwater Treatment	*In-Situ* Treatment	On-Site and/or Off-Site Treatment
Sorbios, Verfahrenstechnische Geràte und Systeme GmbH	Berlin	Laboratory	X		X	
TGU Technologieberatung Grundwasser und Umwelt GmbH	Koblenz	Field Tests	X	X	X	
Thyssen Engineering GmbH	Essen	Development at Demonstration Scale	X			X
TÜV Stutgart e.V.[(a)]	Filderstadt					
Umweltschutz Nord GmbH	Ganderkesee	Full-Scale Demonstration Projects	X	X	X	X
Fa. L. Weiss GmbH u. Co.[(a)]	Crailsheim					
Xenex Gesellscheft	Iserlohn		X		X	

(a) These companies offer microbiological processes, but detailed information has not been collected.

TABLE 2. The Netherlands organizations active in bioremediation.

Company	Location	State of Development	Soil Treatment	Groundwater Treatment	*In-Situ* Treatment	On-Site and/or Off-Site Treatment
Delft Geotechnics	Delft	Experimental Field Project Completed	X	X	X	
Bodemsanering Nederland BV	Weert		X			X
Ecotechniek BV	Utrect	Research	X			X
Heidemij Uilvoering Mijeutechniek	Hertogenbosch	Developed Technology	X			X
Heijmans Milieutechniek	Rosmalen	Purification of Gases from Soil Treatment				X
HWZ-Bodemsanering	Gouda	Research	X		X	
Mourik Groot-Ammers	Groot-Ammers	Full-Scale Demonstration	X	X	X	X
De Ruiter Milieutechnologie	Halfweg & Zwanenburg	Full-Scale Demonstration	X		X	
DSM Research BV	Geleen	Laboratory Research	X		X	
Witteveen & Bos-Consulting Engineers	Deventer	Production Scale Trials	X			X
TAUW Infra Consult BV	Deventer	Full-Scale Demonstration	X		X	

TABLE 2. (Continued)

Company	Location	State of Development	Soil Treatment	Groundwater Treatment	*In-Situ* Treatment	On-Site and/or Off-Site Treatment
Rijksinstituut voor Volksgezondheid en Milieuhygiene (RIVM)	Bilthoven	Clean Up on Demonstration Scale	X	X	X	
TNO	Delft	Clean up on Demonstration Scale	X	X	X	
Ecolyse	Groningen	Small-Scale Demonstration	X	X	X	
Iwaco	Groningen	Clean Up on Demonstration Scale	X	X	X	

TABLE 3. United Kingdom and Wales organizations active in bioremediation.

Company	Location	State of Development	Soil Treatment	Groundwater Treatment	*In-Situ* Treatment	On-Site and/or Off-Site Treatment
Biotreatment Ltd. (subsidiary of Biotal)	U.K.		X	X		X
Groundwater Technology Int., Ltd.	Epsom, U.K.	Industrial-Scale Remediation	X	X	X	
Land Restoration Systems	Slough, U.K.	Experimental Installation in Operation	X	X	X	X
Biotal Ltd.	Cardiff, Wales	Development of Microbial Products				

TABLE 4. Other organizations active in bioremediation.

Company	Location	State of Development	Soil Treatment	Groundwater Treatment	*In-Situ* Treatment	On-Site and/or Off-Site Treatment
Castalia	Genova, Italy	Pilot Project	X			X
Biotecnologica	Broni, Italy	Research				
I.F.P.	Rueil-Malmaison, France	Research	X	X		
Cobadio	Vienne, France	Production of Bacteria	X		X	
Geoclean[(a)]	France					
BRGM + Elf	Orleans, France	Research Demonstration Project	X			
Water Resources Research Center	Budapest, Hungary		X			
A.S. Biol Jordens Kalundborg[(a)]	Denmark					
Biocentras	Vilnius, Lithuania	Full-Scale Demonstration	X		X	
NEO VAC	Crissier, Switzerland		X	X		X

(a) These companies offer microbiological processes, but detailed information has not been collected.

present a geographic overview of European organizations active in bioremediation and the dominant technologies.

The Netherlands and Denmark are leaders in establishing nationwide programs for decontaminating thousands of sites. In the Netherlands, for example, more than 6,000 sites have been cleaned up since 1982, and more than $1.5 billion has been spent over the past decade in cleaning up contaminated sites. Similarly, the Dutch are leading in bioventing work. For instance, work conducted by the Rijksinstituut voor Volksgezondheid en Milieuhygiene (RIVM) and J. van Eyk of Delft Geotechnics predates much of the U.S. work by Chevron, and the U.S. Air Force Engineering and Services Center-supported work by Rob Hinchee of Battelle Memorial Institute. The Dutch are also leaders in other areas including the development of bioreactors for treating organic contaminants in air streams.

Countries such as Germany and the United Kingdom have spent considerable time and money on identifying environmental problems and have recently begun decontamination projects. In some cases, their bioreclamation research efforts also outpace those in the United States. For example, work by Peter Werner and others at the University of Karlsruhe in Germany on the use of nitrate as an alternative electron acceptor dates to the early 1980s and was the stimulus for more recent work by the U.S. Environmental Protection Agency (EPA) at a demonstration site in Traverse City MI. And German efforts are not likely to diminish. Before reunification, West Germany was expected to triple its 1988 spending to $600 million by 1995 on ground-cleaning efforts.

The next tier of regions working to address contamination issues includes Italy, France, the Spanish province of Catalonia, and Switzerland. Italy, for example, has a strong environmental lobby that is concentrating its efforts on soil rehabilitation and reclamation. These regions have made efforts to identify contaminated sites, but have not yet defined nationwide contamination measures, selected technological approaches, or planned large decontamination projects.

Meanwhile, Spain, Portugal, Greece, and Ireland are only beginning to assess contamination problems and identify sites to be remediated.

As various pollution problems are identified and addressed throughout Europe, the scope and diversity of the remediation methods used will continue to grow—as will the number of companies involved. Numerous companies are establishing or expanding operations in Europe in anticipation of the 1992 European Common Market, which is expected to create broad economic opportunities. Groundwater Technology International Ltd., for example, which is headquartered in the United States, has expanded into Europe and developed the Enhanced Natural Degradation Process. This process, one of many that are emerging, treats extracted water by removing the dissolved contaminants, adding nutrients and oxygen, and reinjecting the water back into the contaminated zone.

Remediation Technologies

Among the remediation technologies that have been developed or enhanced by European efforts are extraction followed by thermal processing, solvent extraction or biological degradation, chemical oxidation, microbial treatment, and *in situ* treatment with air stripping. Microbial and *in situ* treatment methods, in particular, are gaining widespread recognition.

Microbial Treatments. Several companies have developed specific microbial systems for improved biodegradation. Umweltschutz Nord GmbH, for example, has developed the Terraferm Biosystem Soil, a biological soil regeneration method in which microorganisms degrade contaminants in a closed reaction room under controlled conditions. ESTE (Hamburg) has introduced the Bioreg (biological regeneration) system in which cultures of nonmanipulated pseudomonas supported on air-dried and comminuted pine bark are mixed with the contaminated soil. Another company active in microbial research is Biocentras, a Lithuanian company with offices in Czechoslovakia, which has developed the "Method Ecol" for treatment of contaminated soil and groundwater and "Putidoil," an oil-degrading bacterial preparation made up of pseudomonas putida cells dissolved in a fertilizer containing nitrogen, phosphorus, and potassium. Land Restorations Systems offers a package of remediation technologies including the Engineering Soil Bank system, the Leachate Recirculation System, and the Submerged Biological Fixed Film Reactor, among other technologies.

Still other companies are developing alternative microbial approaches. Rethmann Stadtereinigung GmbH, for instance, is proposing a system in which the contaminated soil is first treated with a microbe and nutrient solution in a pressure reactor at 15 to 20 bar. When microbial growth conditions are optimized, the material is pressed into bricks that are stored for one or two months to enable complete degradation of the pollutants. The bricks are then crushed and used as earth filler.

***In Situ* Treatments.** *In situ* biorestoration is the technology of choice when surface activities and structures prevent soil excavation or when the contamination is exceedingly deep or has seeped through to the groundwater. Once again, European companies are actively developing and applying *in situ* technologies. For example, De Ruiter Milieutechnologie, Halfweg, has conducted demonstration projects involving aliphatic or aromatic hydrocarbons, paying particular attention to the influence of parameters such as pH, nutrients (potassium, nitrate, and others), and the inoculation of adapted microorganisms. Germany's Argus Umweltbiotechnologie GmbH, another firm active in *in situ* remediation, uses infiltration of air and the addition of phosphates, ammonium, and manganese as nutrients for hydrocarbon degradation. Chemisches Laboratorium E. Wessling-Altenberge experimented successfully with the use of ozone in PAH degradation; ozone is blown through contaminated soil, and its oxidizing powder is used together with its nutrient value for microorganisms. Yet another example of European ingenuity is Delft Geotechnics, which has demonstrated the possibility of coupling venting-assisted evaporation and *in situ* biorestoration under field conditions. These and other companies have fostered significant developments in *in situ* bioreclamation technologies.

What Lies Ahead

As evidenced by numerous developments over the past few years, bioremediation has gained a new prominence in the European marketplace and abroad. Similarly, a wider recognition of environmental problems and of the need for mitigation is spreading throughout Eastern Europe and Third World countries.

An indication of this new global awareness of pollution is the recently established Regional Environmental Center for Central and Eastern Europe, the first of its kind, located in Budapest, Hungary. This U.S. EPA-funded, not-for-profit center will facilitate interaction among multiple governments, industries, and environmental groups in an effort to foster the establishment and growth of environmental institutions, as well as to help identify and address a myriad of pollution issues. With nearly 12.5 million people in Poland at risk from air and water pollution, with 1 in 10 Hungarians dying of pollution-related diseases, with the Danube, which traverses several counties, contaminated with untreated waste and metals—and these are only three examples of widespread pollution—the need for a regional, interdisciplinary approach to pollution in Europe is obvious and imperative.

Bioremediation will, for years to come, play an increasingly important role in the global solution to pollution.

Bioremediation of Fossil Fuel Contaminated Soils

Ronald M. Atlas[*]
University of Louisville

BIOREMEDIATION

Bioremediation involves the use of microorganisms and their biodegradative capacity to remove pollutants (Atlas & Pramer 1990). The byproducts of effective bioremediation, such as water and carbon dioxide, are nontoxic and can be accommodated without harm to the environment and living organisms. Using bioremediation to remove pollutants has many advantages. This method is cheap, whereas physical methods for decontaminating the environment are extraordinarily expensive. Over $1 million a day was spent in an attempt that was only partially successful to clean up the oiled rocks of Prince William Sound AK using water washing and other physical means after an Exxon tanker ran aground there. Neither government nor private industry can afford the cost to clean up physically the nation's known toxic waste sites. Therefore, a renewed interest in bioremediation has developed (Beardsley 1989). Whereas current technologies call for moving large quantities of toxic waste and its associated contaminated soil to incinerators, bioremediation can be done on site and requires simple equipment that is readily available. Bioremediation, though, is not the solution for all environmental pollution problems. Like other technologies, bioremediation has limitations.

Studies on the microbial degradation of hydrocarbons, including determination of the effects of environmental parameters on

* University of Louisville, 2301 South 3rd Street, Louisville KY 40292

biodegradation rates, elucidation of metabolic pathways and genetic bases for hydrocarbon assimilation by microorganisms, and examination of the effects of hydrocarbon contamination on microorganisms and microbial communities, have been areas of intense interest and the subjects of several reviews (Atlas 1981, 1984; Bartha 1986; Colwell & Walker 1977; Leahy & Colwell 1990; NAS 1985). Rates of biodegradation under optimal laboratory conditions have been reported to be as high as 2,500 to 100,000 $g/m^3/day$ (Bartha & Atlas 1987). Under *in situ* conditions petroleum biodegradation rates are orders of magnitude lower. *In situ* natural rates have been reported in the range of 0.001 to 60 $g/m^3/day$ (Bartha & Atlas 1987). The microbial degradation of petroleum in the environment is limited primarily by abiotic factors, including temperature, nutrients such as nitrogen and/or phosphorus, and oxygen (Atlas 1981, 1984; Leahy & Colwell 1990).

The biodegradation of petroleum and other hydrocarbons in the environment is a complex process, whose quantitative and qualitative aspects depend on the nature and amount of the oil or hydrocarbons present, the ambient and seasonal environmental conditions, and the composition of the indigenous microbial community (Atlas 1981; Leahy & Colwell 1990). Microbial degradation of oil has been shown to occur by attack on the aliphatic or light aromatic fractions of the oil. Although some studies have reported their removal at high rates under optimal conditions (Rontani *et al.* 1985; Shiaris 1989), high molecular weight aromatics, resins, and asphaltenes are generally considered to be recalcitrant or to exhibit only very low rates of biodegradation. In aquatic ecosystems, dispersion and emulsification of oil in slicks appear to be prerequisites for rapid biodegradation; large masses of mousse, tarballs, or high concentrations of oil in quiescent environments tend to persist because of the limited surface areas available for microbial activity. Petroleum spilled on or applied to soil is largely adsorbed to particulate matter, decreasing its toxicity, but possibly contributing to its persistence.

ASSESSING BIOREMEDIATION TECHNOLOGIES

Laboratory Efficacy Testing

To demonstrate that a bioremediation technology is potentially useful, it is important that the ability to enhance the rates of hydrocarbon biodegradation be demonstrated under controlled conditions.

This generally cannot be accomplished *in situ* and thus must be accomplished in laboratory experiments. Laboratory experiments demonstrate the potential a particular treatment may have to stimulate the removal of petroleum pollutants from a contaminated site (Bailey *et al.* 1973; Chianelli *et al.* 1991; Venosa *et al.* 1991). Laboratory experiments that closely model real environmental conditions are most likely to produce relevant results (Bertrand *et al.* 1983; Bragg *et al.* 1990; Buckley *et al.* 1980). In many cases this involves using samples collected in the field that contain the indigenous microbial populations. In such experiments it is important to include appropriate controls, such as sterile treatments, to separate the effects of the abiotic weathering of oil from actual biodegradation. Such experiments do not replace the need for field demonstrations, but are critical for establishing the scientific credibility of specific bioremediation strategies. They are also useful for screening potential bioremediation treatments.

The parameters typically measured in laboratory tests of bioremediation efficacy include enumeration of microbial populations, rates of microbial respiration (oxygen consumption and/or carbon dioxide production), and rates of hydrocarbon degradation (disappearance of individual hydrocarbons and/or total hydrocarbons). The methodologies employed in these measurements are critical. It is assumed, for example, that bioremediation of oil pollutants will result in elevated populations of hydrocarbon degraders. Many of the organisms that form colonies on agar-based hydrocarbon media grow on contaminants rather than on hydrocarbons. Therefore, confirmatory tests are needed. In some studies isolates have been tested in liquid culture with more rigorous criteria, such as measuring actual hydrocarbon disappearance, to establish that particular organisms are, in fact, hydrocarbon degraders. These tests often have shown that less than 30 percent of the organisms that form colonies on oil agar actually are capable of metabolizing hydrocarbons. To overcome these limitations, methods have been developed that utilize indicators of hydrocarbon metabolism. For example, dyes can be used to demonstrate the actual metabolism of aromatic hydrocarbons by specific organisms on agar plates or in liquid culture in microtiter plates (Shiaris & Cooney 1983). Colony hybridization procedures have also been employed to identify positively the colony forming units with the genetic capacity for degrading specific aromatic hydrocarbons, for example, by using gene probes for the naphthalene catabolic genes (Sayler *et al.* 1985). The production of radiolabeled carbon dioxide from radiolabeled hydrocarbon substrates also can be used to demonstrate hydrocarbon utilization (Caparello & LaRock 1975). Such production of $^{14}CO_2$ from radiolabeled hexadecane has been used with a most probable number format to enumerate hydrocarbon degraders with certainty that the number of organisms determined reflects the real number of hydrocarbon degraders in the sample (Atlas 1979).

Oxygen consumption rates have been used as a measure of microbial utilization of hydrocarbons (Gibbs *et al.* 1975; Venosa 1991). The measurement of such rates, however, may reflect the utilization of contaminants in the sample. This is especially critical since the addition of oil to the environment may result in the death of some organisms and the input of organic carbon from such organisms would be a ready source of carbon for aerobic microbial respiration. Since the consumption of oxygen can produce erroneous results, the production of carbon dioxide, particularly from radiolabeled hydrocarbon substrates, has been used as a more definitive measure of rates of microbial hydrocarbon metabolism (Atlas 1979; Caparello & LaRock 1975). If experiments are performed using simulated field conditions, the measurement of such respiration rates can be extrapolated to rates of hydrocarbon metabolism likely to appear at contaminated sites. Measurements of respiration, however, require that the system be closed. This, of necessity, creates an artificial condition that makes simulation of flow-through systems difficult. The actual measurement of the disappearance of hydrocarbons has been considered a definitive test of whether or not biodegradation has occurred. Interpretation of such data, however, depends upon the actual analytical procedures employed. Methods that measure gravimetrically the loss of hydrocarbons are difficult to interpret. This is possibly because of the inclusion of water in the extract, leading to underestimation of the result; the production of products, leading to a similar underestimation; or the failure to extract efficiently the sample so that an excessive loss of hydrocarbons is erroneously measured.

Most studies have turned to more definitive analytical procedures, such as gas chromatography and mass spectrometry (Fedorak & Westlake 1981; Schwall & Herbes 1979). Gas chromatographic analyses of the aliphatic fraction, often following column chromatographic separation of this fraction from other hydrocarbon fractions, are used most frequently. Such analyses permit the determination of specific hydrocarbon losses so that the degradation of individual hydrocarbons and individual classes of hydrocarbons can be determined. These detailed analyses, particularly when performed using capillary chromatography columns, allow for the inclusion of recovery standards so that the efficiency of hydrocarbon extraction and analysis can be determined. The aromatic hydrocarbon fraction can similarly be analyzed using capillary column gas chromatography. Most often these analyses are coupled with mass spectrometry using the selected ion monitoring mode to determine the fate of individual aromatic hydrocarbons and classes of aromatic hydrocarbons.

The problems with such analyses, however, are that not all compounds in an oil mixture can be resolved, so a significant unresolved hydrocarbon complex remains to be dealt with. Also, high molecular weight polynuclear aromatic hydrocarbons are very difficult to analyze. In some studies, high-pressure liquid chromatography has been used to assay the biodegradation of such compounds (Heitkamp & Cerniglia 1989).

Field Evaluations

The evaluation of hydrocarbon biodegradation *in situ* is far more difficult than in laboratory studies. Analyses that require enclosure, such as respiration measurements, typically are precluded from such field evaluations. Field evaluations, therefore, have relied upon the enumeration of hydrocarbon-degrading microorganisms and the recovery and analysis of residual hydrocarbons. This is especially complicated since the distribution of oil in the environment typically is patchy; therefore, a very high number of replicate samples must be obtained to yield statistically valid results. Even in partially enclosed containers, the patchiness of hydrocarbon distribution requires the analysis of multiple replicates (Haines & Atlas 1982). Movement of macroorganisms, such as polychaete worms, through sediments creates zones where oxygen incorporation favors biodegradation, while adding to the physical patchiness of the oil distribution. In open water situations, it is difficult to ascertain that appropriate sites are being resampled, especially when time-course determinations are being made to measure rates of hydrocarbon biodegradation.

Because of the problems with quantitation of hydrocarbon recovery from field sites, ratios of hydrocarbons within the complex hydrocarbon mixture have been used to assess the degree of biodegradation (Atlas *et al.* 1981). In particular, the fact that hydrocarbon-degrading microorganisms usually degrade pristane and phytane at much lower rates than straight-chain alkanes has permitted the use of pristane or phytane as internal recovery standards. These measurements assume that pristane and phytane remain undegraded; therefore, by determining the ratio of straight-chain alkanes to these highly branched alkanes, it is possible to estimate the extent to which microorganisms have attacked the hydrocarbons in the petroleum mixture. In situations, however, where pristane or phytane is degraded at rates similar to straight-chain alkanes, this assumption is invalid, and alternative internal standards, such as hopanes, are required (Atlas *et al.* 1981; Prince *et al.* 1990; Pritchard 1990).

Of equal importance to the problem of determining the appropriate measures to be used in assessing the effectiveness of a bioremediation treatment is the experimental design that includes appropriate controls. Often in a field bioremediation situation, the necessity for cleaning up the pollutants overshadows the need for leaving an untreated reference site that is comparable to the site being treated. Thus at the end of many bioremediation efforts, all areas have been treated, leaving no basis for comparison with what would have happened had no bioremediation treatment been employed. Given the natural degradation capacity of the indigenous microorganisms, this leaves in question the effectiveness of many bioremediation treatment strategies.

Ecological Effects Testing

In addition to demonstrating efficacy, it is essential to demonstrate that bioremediation treatments do not produce any untoward ecological effects (Colwell 1971; Doe & Wells 1978; O'Brien & Dixon 1976). The focus of ecological effects testing of bioremediation has been on the direct toxicity of chemical additives, such as fertilizers, to indigenous organisms. Standardized toxicological tests are used to determine the acute toxicities of chemicals. Chronic toxicities and sublethal effects may also be determined. Generally toxicity tests are run using a bivalve larvae, such as oyster larvae, and a fish, such as rainbow trout. Sometimes regionally important species, such as salmon or herring, are included. Additionally, tests are run to assess effects on algal growth rates to determine what levels of fertilizer application will stimulate oil biodegradation without causing algal blooms. No test protocols have been developed and implemented for testing the potential pathogenicity of seed cultures. Concerns have been voiced that seed cultures could cause disease among humans or plant and animal populations. These concerns have been put aside when indigenous microbes—to which these populations are naturally exposed—are employed.

APPROACHES TO ENHANCING MICROBIAL DEGRADATION

The microbial degradation of petroleum in aquatic and soil environments is limited primarily by nutrients, such as nitrogen and phosphorus, and oxygen availability. The initial steps in the biodegradation of hydrocarbons by bacteria and fungi involve the oxidation of the substrate by oxygenases for which molecular oxygen is required (Atlas 1984). Aerobic conditions are, therefore, necessary for this route of microbial oxidation of hydrocarbons in the environment. Conditions of oxygen limitation normally do not exist in the upper levels of the water column in marine and freshwater environments (Cooney 1984; Floodgate 1984). The availability of oxygen in soils, sediments, and aquifers is often limited depending on the type of soil and whether the soil is waterlogged (Bossert & Bartha 1984; Jamison *et al.* 1975; von Wedel *et al.* 1988). Anaerobic degradation of petroleum hydrocarbons by microorganisms also occurs (Grbić-Gallić & Vogel 1987; Vogel & Grbić-Gallić 1986; Ward & Brock 1978; Zeyer *et al.* 1986). The rates of anaerobic hydrocarbon biodegradation, however, are very low, and its ecological significance appears to be minor (Atlas 1981; Bailey *et al.* 1973; Bossert & Bartha 1984; Cooney 1984; Floodgate 1984; Jamison *et al.* 1975; Ward *et al.* 1980).

Several investigators have reported that concentrations of available nitrogen and phosphorus in seawater are severely limiting to microbial hydrocarbon degradation (Atlas & Bartha 1972; Bartha & Atlas 1973; Floodgate 1973, 1979; Gunkel 1967; LePetit & Barthelemy 1968; LePetit & N'Guyen 1976). Other investigators (Kinney *et al.* 1969), however, have reached the opposite conclusion: i.e., that nitrogen and phosphorus are not limiting in seawater. The difference in results is paradoxical and appears to be based on whether the studies are aimed at assessing the biodegradation of hydrocarbons within an oil slick or the biodegradation of soluble hydrocarbons. In an oil slick, a mass of carbon is available for microbial growth within a limited area. Since microorganisms require nitrogen and phosphorus for incorporation into biomass, the availability of these nutrients within the same area as the hydrocarbons is critical. When considering soluble hydrocarbons, nitrogen and phosphorus are probably not limiting since the solubility of the hydrocarbons is so low as to preclude establishment of an unfavorable C/N or C/P ratio. Investigators considering the fate of low-level discharges of hydrocarbons (soluble hydrocarbons), thus, have properly concluded that available nutrient concentrations are adequate to support hydrocarbon biodegradation.

Oxygenation to Enhance Oil Biodegradation

Biodegradation of petroleum hydrocarbons at rapid rates requires molecular oxygen. Under anaerobic conditions, hydrocarbons are not biodegraded in the environment at rates that can be used to remediate polluted sites. Therefore, for effective petroleum biodegradation, it is necessary to ensure an available supply of oxygen (Floodgate 1973; ZoBell 1973). The use of forced aeration and nutrient supplementation to stimulate the biodegradation of gasoline in groundwater has been reported by Jamison *et al.* (1975, 1976). This study was conducted on a groundwater supply that had been contaminated with gasoline following a pipeline break. The reservoir was found to contain microorganisms capable of growth on the hydrocarbons that are found in gasoline as the sole carbon source. It was the conditions for growth and metabolism of oil that were found to be limiting. Ammonium sulfate (58 metric tons), monosodium phosphate, and disodium phosphate (29 metric tons) were added to the groundwater. Air was also pumped into the groundwater with a small compressor. This treatment increased the number of hydrocarbon-utilizing microorganisms, and gasoline was degraded. It was estimated that 2×10^5 L of gasoline were removed by stimulated degradation and that the use of forced aeration and nitrogen and phosphorus addition significantly reduced the time necessary to remove the spilled gasoline from this groundwater reservoir. A patent was issued to Raymond (1974) for reclamation of hydrocarbon-contaminated groundwater using stimulated microbial degradation.

To overcome oxygen limitation, hydrogen peroxide may be added in appropriate and stabilized formulations (API 1987). The decomposition of hydrogen peroxide releases oxygen that can support aerobic microbial utilization of hydrocarbons. At concentrations that are too high, however, hydrogen peroxide is toxic to microorganisms and will actually lower rates of microbial hydrocarbon biodegradation. Also, hydrogen peroxide typically is not stable and decomposes rapidly upon addition to contaminated environments. Nevertheless, this treatment has been used effectively to stimulate microbial degradation of environmental hydrocarbon contaminants.

Berwanger and Barker (1988) investigated *in situ* biorestoration involving stimulating aerobic biodegradation in a contaminated anaerobic, methane-saturated groundwater situation using hydrogen peroxide as an oxygen source. Batch biodegradation experiments were conducted with groundwater and core samples obtained from a Canadian landfill. Hydrogen peroxide, added at a nontoxic level, provided oxygen that promoted the rapid biodegradation of benzene, toluene, ethyl benzene, and *o*-, *m*-, and *p*-xylene. In winter of 1983, Frankenberger *et al.* (1989) found a flow of approximately 4,000 L of diesel fuel along an asphalt parking lot of a commercial establishment toward a surface drain near an open creek. Investigations led to the discovery of a leaking underground diesel fuel storage tank. Hydrocarbon quantities ranged up to 1,500 mg/kg of soil. A laboratory study indicated fairly high numbers of hydrocarbon-oxidizing organisms relative to glucose-utilizing microorganisms. Bioreclamation was initiated in April 1984 by injecting nutrients (nitrogen and phosphorus) and hydrogen peroxide and terminated in October 1984 upon detection of no hydrocarbons (< 1 mg/kg). A verification boring within the vicinity of the contaminated plume confirmed that residual contamination had reached background levels.

Nutriation to Enhance Oil Biodegradation

Landfarming. Much interest has developed in the disposal of oily wastes by soil cultivation. Soil microorganisms have a high capacity for degrading petroleum hydrocarbons (Bossert & Bartha 1984). Several investigators have examined the feasibility of using landfarming for the removal of oily wastes (Bartha & Bossert 1984; Dibble & Bartha 1979a,b; Francke & Clark 1974; Gudin & Syratt 1975; Huddleston & Cresswell 1976; Jones & Greenfield 1991; Kincannon 1972; Lehtomake & Niemela 1975; Maunder & Waid 1973, 1975; Odu 1978; Raymond *et al.* 1976a,b). There have been some reports on mobilization of oil into the soil column (Verstraete *et al.* 1975), but in most cases little evidence has been found for significant downward leaching of oil (Dibble & Bartha 1979a,b; Raymond *et al.* 1976a,b).

Cook & Westlake (1974) reported a series of studies in which northern crude oils were spilled on soils in northern Alberta, Canada. The biodegradation of several different oils was examined. Oil biodegradation was found to be stimulated by application of nitrogen and phosphorus fertilizer. This conclusion was based on observed increases in microbial numbers and chromatographic analysis of the residual oil. For soil application many nitrogen- and phosphorus-containing fertilizers are available (Dotson et al. 1971) that can be tilled into the soil. The best fertilizers for soil application are forms of readily useable nitrogen and phosphorus and slow-release forms that provide a continuous supply of nutrients that are not leached from the oil-soil interface. Kincannon (1972) reported that biodegradation rates of heavy oily wastes in soils were as high as 21 L/m^2/month. The addition of nitrogen and phosphorus fertilizer resulted in a doubling of the oil biodegradation rate to 16 kg/m^3/month. Fertilizer was added before the oil and rototilled into the soil. It was recommended that monthly determinations of nitrogen and phosphorus levels in the soil and periodic fertilizer application when necessary would optimize the degradation process. Neither leaching of oil nor applied fertilizer into the soil column was observed. The cost of soil disposal of oily wastes was estimated at $0.02/per liter.

Wang and Bartha (1990) recently studied the effects of bioremediation on residues of fuel spills in soil (2.3 mL cm^3 of jet fuel, heating oil, and diesel oil). Persistence and toxicity of the fuel increased in the order of jet fuel < heating oil < diesel oil. Bioremediation treatment (fertilizer application plus tilling) strongly decreased fuel persistence and toxicity and increased microbial activity as compared to contaminated but untreated soil. Good correlations were found among fuel residue decline, microbial activity, and toxicity reduction. These findings indicate that bioremediation treatment can restore fuel spill contaminated soils in 4 to 6 weeks to a degree that can support plant cover. Recovery of the soil is complete in 20 weeks.

Wang *et al.* (1990) continued their studies on bioremediation treatment to remove the polycyclic aromatic hydrocarbon (PAH) components of diesel oil in soil. Bioremediation treatment, while increasing the rate of total hydrocarbon degradation, had an even greater effect on PAH persistence, almost completely eliminating these compounds in 12 weeks. Without bioremediation, 12.5 to 32.5 percent of the higher weight PAH was still present at 12 weeks. After substantial initial mutagenicity and toxicity, the contaminated soil approached the background level of uncontaminated soil after 12 weeks of bioremediation. Detoxification was complete in 20 weeks.

Marine Oil Spills. The spill of more than 4×10^7 L of crude oil from the oil tanker *Exxon Valdez* in Prince William Sound AK on March 24, 1989 (Hagar 1989), as well as smaller spills in Texas, Rhode Island, and the Delaware Bay (Anon. 1989), has focused attention on the problem of hydrocarbon contamination in marine and estuarine environments and the potential use of bioremediation to remove petroleum pollutants. The ability of environmental modification to stimulate microbial degradation of oil in marine ecosystems by indigenous microorganisms has been demonstrated in several cases. Atlas and Bartha (1973) developed an oleophilic nitrogen and phosphorus fertilizer that Atlas (1975), Atlas and Busdosh (1976), and Atlas and Schofield (1975) tested for its ability to stimulate petroleum degradation by indigenous microorganisms in several environments. This oleophilic fertilizer, paraffinized urea and octyl phosphate, has been tested in near-shore areas off the coast of New Jersey, in Prudhoe Bay, and in several ponds near Barrow AK, including *in situ* as well as *in vitro* experiments in each case. Each site included a naturally occurring microbial population that was capable of petroleum biodegradation when this oleophilic fertilizer was added to the oil, and in each case, the addition of oleophilic fertilizer stimulated biodegradative losses. The amount of stimulation varied for different crude oils tested, but oil degradation generally was 30 to 40 percent higher in oleophilic-fertilized oil slicks than in unfertilized slicks. Application of oleophilic fertilizer was not found to lead to undesirable algal blooms or to produce effects toxic to invertebrate bioassay organisms. Dibble and Bartha (1976) found additional stimulation with some crude oils when oleophilic iron and ferric octoate were added along with nitrogen and phosphorus. Greater stimulation was observed in nonpolluted than in polluted near-shore waters. Addition of oleophilic iron is likely to result in even greater stimulation of petroleum biodegradation in open ocean areas where iron concentrations are particularly low.

The *Exxon Valdez* spill formed the basis for a major study on bioremediation and the largest application of this emerging technology (Pritchard 1990; Pritchard & Costa 1991). The initial approach to the clean-up of the oil spilled from the *Exxon Valdez* was physical. Washing of oiled shorelines with high-pressure water was expensive, and cleaned shorelines became reoiled, forcing recleaning. Bioremediation, therefore, was considered as a method to augment other clean-up procedures. The U.S. EPA and Exxon entered into an agreement to explore the feasibility of using bioremediation. This historic effort considered a variety of approaches to optimizing microbial degradation of oil. The project focused on determining whether nutrient augmentation could stimulate rates of biodegradation. Three types of nutrient supplementation were considered: water soluble (23:2 N:P garden fertilizer fomulation); slow release (isobutylenediurea); and oleophilic (Inipol EAP 22 = oleic acid, urea, lauryl phosphate) (Chianelli *et al.* 1991; Safferman 1991; Tabak *et al.* 1991). Each was tested in laboratory simulations and in field demonstration plots to show the efficacy of nutrient supplementation. Consideration was also given to potential adverse ecological effects, particularly eutrophication from algal blooms and toxicity to fish and invertebrates. Application rates were adjusted to minimize undesirable ecological impact. Using a sprinkler system periodically to apply the water-soluble fertilizer at low tide stimulated rates of biodegradation without causing excessive algal growth. The application of the oleophilic fertilizer produced very dramatic results, stimulating biodegradation so that the surfaces of the oil-blackened rocks on the shoreline turned white and were essentially oil-free within 10 days after treatment. The use of Inipol and Customblen was approved for shoreline treatment and was used as a major part of the clean-up effort. A joint Exxon-U.S. EPA-Alaska monitoring effort followed the effectiveness of the bioremediation treatment, which was estimated to increase the rates of biodegradation at least threefold (Chianelli *et al.* 1991; Prince *et al.* 1990).

Microbial Seeding to Enhance Oil Biodegradation

Seeding involves the introduction of microorganisms into the natural environment for the purpose of increasing the rate or extent, or both, of biodegradation of pollutants. The rationale for this approach is that indigenous microbial populations may not be capable of degrading the wide range of potential substrates present in such complex mixtures as petroleum. However, the premises that the microorganisms naturally present in an environment subjected to contamination with oil would be incapable of extensively degrading petroleum and that added microorganisms would be able to do a superior job should be examined carefully. The criteria to be met by effective seed organisms include ability to degrade most petroleum components, genetic stability, viability during storage, rapid growth following storage, a high degree of enzymatic activity and growth in the environment, ability to compete with indigenous microorganisms, nonpathogenicity, and inability to produce toxic metabolites (Atlas 1977).

Some individuals have proposed that hydrocarbon-degrading microorganisms and their enzymatic capabilities may be critical limiting factors in the rates of hydrocarbon biodegradation (Atlas 1977). Clearly, there is an adaptive process following the introduction of oil into the environment, and if metabolically active hydrocarbon utilizers, capable of utilizing hydrocarbons in the petroleum pollutant, could be added quickly, the lag period before the indigenous population could respond would be reduced. Even if these organisms were subsequently replaced by competition with indigenous hydrocarbon utilizers, there might be some benefit to such seeding operations. However, if freeze-dried or otherwise metabolically inactive organisms were to be added or if delay were necessary to culture such organisms, the benefit of reducing the lag period before the onset of rapid hydrocarbon degradation might be negated. Additionally, there is the problem of finding microorganisms with the right metabolic capabilities to augment the activities of the indigenous populations, and the problem of adding microorganisms that could survive and favorably compete with the indigenous organisms.

Terrestrial ecosystems differ from aquatic ecosystems in that soils contain higher concentrations of organic and inorganic matter and, generally, larger numbers of microorganisms and have more variable physical and chemical conditions (Bossert & Bartha 1984). The microbial community of soils usually includes a significant hydrocarbon-utilizing component, which readily increases in response to hydrocarbon contamination (Atlas *et al.* 1980; Jensen 1975; Llanos & Kjoller 1976; Pinholt *et al.* 1979). The presence of indigenous microbial populations highly adapted to a particular soil environment would be expected to influence negatively the ability of seed microorganisms to complete successfully and survive; for this reason, soils are not widely considered to be amenable to improvements in rates of biodegradation through seeding alone (Atlas 1977; Bossert & Bartha 1984). Other potential problems associated with the inoculation of soils, as reviewed by Goldstein *et al.* (1985), include inadequate (i.e., extremely low) concentrations of the chemical of interest, the presence of inhibitory substances, predation, preferential metabolism of competing organic substrates, and insufficient movement of the seed organisms within the soil.

Many investigators have suggested that complex mixtures of hydrocarbon degraders would be necessary to degrade effectively all of the hydrocarbons in a complex petroleum mixture. Others have attempted to isolate organisms, which could be stockpiled for use in case of an oil spill or in the treatment of oily wastes, that are particularly effective at degrading hydrocarbons. In particular, some investigators have sought organisms capable of degrading specific components within an oil that usually are only slowly degraded. For example, some investigators have sought organisms that specifically degrade four-ring aromatic hydrocarbons, which are among the more resistant compounds found in petroleum (Heitkamp & Cerniglia 1989).

Microbial seeding of petroleum-contaminated aquatic environments has been attempted, with mixed results. Tagger *et al.* (1983) observed no increase in petroleum degradation in seawater inoculated with a mixed culture of hydrocarbon-degrading bacteria. Atlas and Busdosh (1976) reported increased degradation of oil in a saline Arctic pond after inoculation with an oil-degrading *Pseudomonas* sp., but no improvement in a freshwater pond. Horowitz and Atlas (1980) found that greater losses of oil in seawater in an open flow-through system occurred when octadecane-coated bacteria were applied 2 weeks after the addition of an oleophilic fertilizer to the system, than when the fertilizer alone was added. In the same study, no significant increases in the loss of gasoline from freshwater sediment were produced by seeding. Venosa *et al.* (1991) found several commercial cultures ineffective at degrading oil but that a few had some potential; field studies, however, could not demonstrate effectiveness of microbial seeding over the biodegradation capacities of indigenous marine microorganisms.

Mixed cultures have been most commonly used as inocula for seeding because of the relative ease with which microorganisms with different and complementary hydrocarbon-degrading capabilities can be isolated. A special culture collection was begun as a depository for hydrocarbon-utilizing microorganisms (Cobet 1974). Several commercial enterprises began to market microorganism preparations for removing petroleum pollutants. Commercial mixtures of microorganisms are being marketing for use in degrading oil in waste treatment lagoons. These commercial microbial seed mixtures are also intended for use in other situations for the removal of oil pollutants. The applicability of seeding-selected bacteria and fungi to oil spills has been patented by Azarowicz (1973). The literature supplied with seed bioremediation products is often the only information available about them. The full claims of their effectiveness remain to be proven.

Biotreatment of the *Mega Borg* spill off the Texas coast consisted of applying a seed culture with a secret catalyst produced by Alpha Corporation to the oil at sea (Mangan 1990). Claims were made that the treatment was successful at completely removing the oil (Mauro 1990a, b), but the effectiveness of the Alpha seeding to stimulate biodegradation has not been verified, nor has the effectiveness of the culture been confirmed by the U.S. EPA in laboratory tests (Fox 1991).

Another approach has been to engineer microorganisms genetically with the capacity to degrade a wide range of hydrocarbons. The potential for creating, through genetic manipulation, microbial strains able to degrade a variety of different types of hydrocarbons has been demonstrated by Friello *et al.* (1976). They successfully produced a multiplasmid-containing *Pseudomonas* strain capable of oxidizing aliphatic, aromatic, terpenic, and polyaromatic hydrocarbons. The genetic information for at least some enzymes involved in alkane and simple aromatic hydrocarbon transformation occurs on plasmids (Chakrabarty 1974; Chakrabarty *et al.* 1973; Dunn & Gunsalus 1973).

The use of such a strain as an inoculum during seeding would preclude the problems associated with competition between strains in a mixed culture. However, there is considerable controversy surrounding the release of such genetically engineered microorganisms into the environment, and field testing of these organisms must therefore be delayed until the issues of safety, containment, and potential for ecological damage are resolved (Sussman *et al.* 1988). A hydrocarbon-degrading pseudomonad was engineered by Chakrabarty and was the organism that the U.S. Supreme Court in a landmark decision ruled could be patented (Anon. 1975). The organism engineered by Chakrabarty is capable of degrading a number of low molecular weight aromatic hydrocarbons, but does not degrade the higher molecular weight persistent polynuclear aromatics, and thus has not been used in the bioremediation of oil spills.

Given the current regulatory framework for the deliberate release of genetically engineered microorganisms, it is unlikely that any such organism would gain the necessary regulatory approval in time to be of much use in treating an oil spill. Such organisms, however, could be useful in enclosed oily waste treatment systems that could be used to replace landfarming as an option for disposing of such residual oils.

SUMMARY

Microorganisms clearly have the potential for degrading a substantial portion of, but not all, the components of an oil that may pollute the environment. Demonstrating the effectiveness of bioremediation generally involves laboratory tests to show potential and field tests to confirm applicability. The rates of microbial hydrocarbon degradation can be enhanced several fold which forms the basis for bioremediation of oil polluted environments. Methods that overcome environmental limitations have proven most effective. These methods include oxygenation and nutriation. For the most part seeding with microbial cultures has produced ambiguous results.

REFERENCES

Anon. *Sci. News* **1975**, pp 108, 180.

Anon. *Oil Gas J.* **1989**, *87*, 22.

API (American Petroleum Institute). *Field Study of Enhanced Subsurface Biodegradation of Hydrocarbons using Hydrogen Peroxide as an Oxygen Source*; API Publ. 4448; American Petroleum Institute: Washington DC, 1987.

Atlas, R. M. In *Proceedings of the First Intersect. Congress*, International Association of Microbiological Studies; Science Council of Japan: Tokyo, 1975.

Atlas, R. M. *Crit. Rev. Microbiol.* **1977**, *5*, 371–386.

Atlas, R. M. *Measurement of Hydrocarbon Biodegradation Potentials and Enumeration of Hydrocarbon-Utilizing Microorganisms Using Carbon-14 Hydrocarbon-Spiked Crude Oil*; special technical report from American Society for Testing and Materials: Philadelphia, PA, 1979.

Atlas, R. M. *Microbial. Rev.* **1981**, *45*, 180–209.

Atlas, R. M., Ed.; *Petroleum Microbiology*; Macmillan: New York, 1984.

Atlas, R. M.; Bartha, R. *Biotechnol. Bioeng.* **1972**, *14*, 309–317.

Atlas, R. M.; Bartha, R. *Environ. Sci. Technol.* **1973**, *7*, 538–541.

Atlas, R. M.; Boehn, P. D.; Calder, J. A. *Est. Coastal and Shelf Science* **1981**, *12*, 598–608.

Atlas, R. M.; Busdosh, M. In *Proceedings of the Third International Biodegradation Symposium*; Sharpley, J. M.; Kaplan, A. M., Eds.; Applied Science: London, 1976; pp 79–86.

Atlas, R. M.; Schofield, E. A. In *Proceedings: Impact of the Use of Microorganisms on the Aquatic Environment*; Bourquin, A. W.; Ahern, D. G.; Meyers, S. P., Eds.; EPA 660-3-75-001; U.S. Environmental Protection Agency: Corvallis, OR, 1975; pp 183–198.

Atlas, R. M.; Pramer, D. *ASM News* **1990**, *56*, 7.

Atlas, R. M.; Sexstone, A.; Gustin, P.; Miller, O.; Linkins, P.; Everett, K. In *Proceedings of the Fourth International Biodeterioration Symposium*; Oxley, T. A.; Becker, G.; Allsop, D., Eds.; Pitman: London, 1980; pp 21–28.

Azarowicz, R. M. U.S. Patent 3 769 164, 1973.

Bailey, N.J.L.; Jobson, A. M.; Rogers, M. A. *Chem. Geol.* **1973**, *11*, 203–221.

Bartha, R. *Microb. Ecol.* **1986**, *12*, 155–172.

Bartha, R.; Atlas, R. M. In *The Microbial Degradation of Oil Pollutants*; Ahearn, D. G.; Meyers, S. P., Eds.; Publication No. LSU-SG-73-01; Center for Wetland Resources; Louisiana State University: Baton Rouge, 1973; pp 147–152.

Bartha, R.; Atlas, R. M. In *Long-Term Environmental Effects of Offshore Oil and Gas Development*; Boesch, D. F.; Rabalais, N. N., Eds.; Elsevier Applied Science: New York, 1987; pp 287–341.

Bartha, R.; Bossert, I. In *Petroleum Microbiology*; Atlas, R. M., Ed.; Macmillan: New York, 1984; pp 553–578.

Beardsley, T. *Sci. Amer.* **1989**, *261*(3), 43.

Bertrand, J. C.; Rambeloarisoa, E.; Rontani, J. F.; Guisti, G.; Mattei, G. *Biotechnol. Lett.* **1983**, *5*, 567–572.

Berwanger, D. J.; Barker, J. F. *Wat. Pollut. Res. J. Can.* **1988**, *23*(3), 460–475.

Bossert, I.; Bartha, R. In *Petroleum Microbiology*; Atlas, R. M., Ed.; Macmillan: New York, 1984; pp 434–476.

Bragg, J. R.; Roffall, J. C.; McMillen, S. *Column Flow Studies of Bioremediation in Prince William Sound*; Exxon Production Research Company: Houston, TX, 1990.

Buckley, E. N.; Pfaender, F. K.; Kylber, K. L.; Ferguson, R. L. In *Proceedings of a Symposium on Preliminary Results from the September 1979 RESEARCHER/PIERCE IXTOC-I Cruise*; National Oceanic and Atmospheric Administration: Boulder, CO, 1980.

Caparello, D. M.; LaRock, P. A. *Microb. Ecol.* **1975**, *2*, 28–42.

Chakrabarty, A. M. U.S. Patent 3 813 316, 1974; 922.

Chakrabarty, A. M.; Chou, G.; Gunsalus, I. C. *Proc. Natl. Acad. Sci. USA* **1973**, *70*, 1137.

Chianelli, R. R.; Aczel, T.; Bare, R. E.; George, G. N.; Genowitz, M. W.; Grossman, M. J.; Haith, C. E.; Kaiser, F. J.; Lessard, R. R.; Liotta, R.; Mastracchio, R. L.; Minak-Bernero, V.; Prince, R. C.; Robbins, W. K.; Stiefel, E. I.; Wilkinson, J. J.; Hinton, S. M. In *Proceedings of the 1991 International Oil Spill Conference*; San Diego, CA, 1991; pp 549–558.

Cobet, A. B. *Hydrocarbonoclastic Repository, in Progress Report Abstracts*; Microbiology Program, Office of Naval Research: Arlington, VA, 1974; p 131.

Colwell, E. B., Ed.; *The Ecological Effects of Oil Pollution on Littoral Communities*; Applied Science: London, 1971.

Colwell, R. R.; Walker, J. D. *Crit. Rev. Microbiol.* **1977**, *5*, 423–445.

Cook, F. D.; Westlake, D.W.S. *Information Canada* **1974**; Cat. No. R72-12774.

Cooney, J. J. In *Petroleum Microbiology*, Atlas, R. M., Ed.; Macmillan: New York, 1984.

Dibble, J. T.; Bartha, R. *Appl. Environ. Microbiol.* **1976**, *31*, 544–550.

Dibble, J. T.; Bartha, R. *Soil Sci.* **1979a**, *128*, 56–60.

Dibble, J. T.; Bartha, R. *Appl. Environ. Microbiol.* **1979b**, *37*, 729–739.

Doe, K. G.; Wells, P. G. In *Chemical Dispersants for the Control of Oil Spills*; McCarthy, L. T., Jr.; Lindblom, G. P.; Walter, H. F., Eds.; American Society for Testing and Materials: Philadelphia, PA, 1978.

Dotson, G. K.; Dean, R. B.; Cooke, W. B.; Kenner, B. A. In *Proceedings of the 5th International Water Pollution Research Conference*; Pergamon Press: New York, 1971.

Dunn, N. W.; Gunsalus, I. C. *J. Bacteriol.* **1973**, *114*, 974.

Fedorak, P. M.; Westlake, D.W.S. *Can. J. Microbiol.* **1981**, *27*, 432–443.

Floodgate, G. D. In *The Microbial Degradation of Oil Pollutants*; Ahearn, D. G.; Meyers, S. P., Eds.; Publ. No. LSU-SG-73-01; Center for Wetland Resources; Louisiana State University: Baton Rouge, 1973.

Floodgate, G. D. In *Proceedings of Workshop, Microbial Degradation of Pollutants in Marine Environments*; Bourquin, A. W.; Pritchard, P. H., Eds.; EPA-66019-79-012; Environmental Research Laboratory: Gulf Breeze, FL, 1979.

Floodgate, G. In *Petroleum Microbiology*; Atlas, R. M., Ed.; Macmillan: New York, 1984; pp 355–398.

Fox, J. E. *Bio/Technology* **1991**, *9*, 14.

Francke, H. C.; Clark, F. E. "Disposal of Oil Wastes by Microbial Assimilation"; Report Y-1934; U.S. Atomic Energy Commission: Washington, DC, 1974.

Frankenberger, W. T., Jr.; Emerson, K. D.; Turner, D. W. *Environ. Manage.* **1989**, *13*(3), 325–332.

Friello, D. A.; Mylroie, J. R.; Chakrabarty, A. M. In *Proceedings of the Third International Biodegradation Symposium*; Sharpley, J. M.; Kaplan, A. M., Eds.; Applied Science Publications: London, 1976.

Gibbs, C. F.; Pugh, K. B.; Andrews, A. R. *Proc. R. Soc. London Ser. B.* **1975**, *188*, 83–94.

Goldstein, R. M.; Mallory, L. M.; Alexander, M. *Appl. Environ. Microbiol.* **1985**, *50*, 977–983.

Grbić-Gallić, D.; Vogel, T. M. *Appl. Environ. Microbiol.* **1987**, *53*, 254–260.

Gudin, C.; Syratt, W. J. *Environ. Pollut.* **1975**, *8*, 107–112.

Gunkel, W. *Helgol. Wiss. Meeresunters.* **1967**, *15*, 210–224.

Hagar, R. *Oil Gas J.* **1989**, *87*, 26–27.

Haines, J. R.; Atlas, R. M. *Mar. Environ. Res.* **1982**, *7*, 91–102.

Heitkamp, M. A.; Cerniglia, C. E. *Appl. Environ. Microbiol.* **1989**, *55*, 1968–1973.

Horowitz, A.; Atlas, R. M. In *Proceedings of the Fourth International Biodeterioration Symposium*; Oxley, T. A.; Becker, G.; Allsopp, D., Eds.; Pitman: London, 1980; pp 15–20.

Huddleston, R. L.; Cresswell, L. W. In *Proceedings of the 1975 Engineering Foundation Conference: The Role of Microorganisms in the Recovery of Oil*; American Institute of Chemical Engineers; Washington, DC, 1976.

Jamison, V. M.; Raymond, R. L.; Hudson, J. O., Jr. *Dev. Ind. Microbiol.* **1975**, *16*, 305–312.

Jamison, V. M.; Raymond, R. L.; Hudson, J. O. In *Proceedings of the Third International Biodegradation Symposium*; Sharpley, J. M.; Kaplan, A. M., Eds.; Applied Science Publishers: London, 1976; pp 187–196.

Jensen, V. *Oikos* **1975**, *26*, 152–158.

Jones, M. A.; Greenfield, J. H. In *Proceedings of the 1991 International Oil Spill Conference*; sponsored by the United States Coast Guard, American Petroleum Institute, and U.S. Environmental Protection Agency: San Diego, CA, 1991; pp 533–540.

Kincannon, C. B. *Oily Waste Disposal by Soil Cultivation Process*; EPA R2-72-100; U.S. Environmental Protection Agency: Washington, DC, 1972.

Kinney, P. J.; Button, D. K.; Schell, D. M. In *Proceedings of 1969 Joint Conference on Prevention and Control of Oil Spills*; American Petroleum Institute: Washington, DC, 1969.

Leahy, J. G.; Colwell, R. R. *Microbiol. Rev.* **1990**, *54*(3), 305–315.

Lehtomake, M.; Niemela, S. *Ambio* **1975**, *4*, 126–129.

LePetit, J.; Barthelemy, M. H. *Ann. Inst. Pasteur Paris* **1968**, *114*, 149–158.

LePetit, J.; N'Guyen, M.-H. *Can. J. Microbiol.* **1976**, 22, 1364–1373.

Llanos, C.; Kjoller, A. *Oikos* **1976**, *27*, 377–382.

Mangan, K. S. *Chron. of Higher Education* **1990**, *37*(3), A5–A9.

Maunder, B. R.; Waid, J. S. In *Proceedings of the Pollution Research Conference*, 20–21 June 1973, Wairakei, New Zealand, Information series no. 97; New Zealand Department of Scientific and Industrial Research: Wellington, 1973.

Maunder, B. R.; Waid, J. S. In *Proceedings of the Third International Biodegradation Symposium*, 17–23 August; University of Rhode Island: Kingston, 1975; XXV-5.

Mauro, G. *Combating Oil Spills Along the Texas Coast: A Report on the Effects of Bioremediation*; paper presented at the Bioremediation Symposium, Oct. 9, Lamar University, Beaumont, TX, 1990a.

Mauro, G. *Mega Borg Spill off the Texas Coast: An Open Water Bioremediation Test*; paper presented at the Bioremediation Symposium, Oct. 9, Lamar University, Beaumont, TX, 1990b.

NAS (National Academy of Sciences). *Oil in the Sea-Inputs, Fates, and Effects*; National Academy Press: Washington, DC, 1985.

O'Brien, P. Y.; Dixon, P. S. *Br. Physol. J.* **1976**, *11*, 115–142.

Odu, C.T.I. *Environ. Pollut.* **1978**, *15*, 235–240.

Pinholt, Y.; Struwe, S.; Kjoller, A. *Holarct. Ecol.* **1979**, *2*, 195–200.

Prince, R.; Clark, J. R.; Lindstrom, J. E. *Bioremediation Monitoring Program*; joint report of Exxon, the U.S. EPA, and the Alaskan Dept. of Environmental Conservation, Anchorage, AK, 1990.

Pritchard, H. P. Presented at the 199th National Meeting of the American Chemical Society, Boston, MA, 22–27 April 1990; Abstract Environment 154.

Pritchard, H. P.; Costa, C. F. *Env. Sci. Technol.* 1991, in press.

Raymond, R. L. U.S. Patent 3 846 290, 1974.

Raymond, R. L.; Hudson, J. O.; Jamison, V. W. *Appl. Environ. Microbiol.* **1976a**, *31*, 522–535.

Raymond, R. L.; Jamison, V. W.; Hudson, J. O. In *Water—1976;* American Institute of Chemical Engineers: New York, 1976b.

Rontani, J. F.; Bosser-Joulak, F.; Rambeloarisoa, E.; Bertrand, J. C.; Giusti, G.; Faure, R. *Chemosphere* **1985,** *14,* 1413–1422.

Safferman, S. I. In *Proceedings of the 1991 International Oil Spill Conference;* sponsored by the United States Coast Guard, American Petroleum Institute, and U.S. Environmental Protection Agency: San Diego, CA, 1991; pp 571–576.

Sayler, G. S.; Shields, M. S.; Tedford, E. T.; Breen, A.; Hooper, S. W.; Sirotkin, K. M.; Davis, J. W. *Appl. Environ. Microbiol.* **1985,** *49,* 1295–1303.

Schwall, L. R.; Herbes, S. E. In *Methodology for Biomass Determinations and Microbial Activities in Sediments;* Litchfield, C. D.; Seyried, P. L., Eds.; American Society for Testing and Materials: Philadelphia, PA, 1979.

Shiaris, M. P. *Appl. Environ. Microbiol.* **1989,** *55,* 1391–1399.

Shiaris, M. P.; Cooney, J. J. *Appl. Environ. Microbiol.* **1983,** *45,* 706–710.

Sussman, M.; Collins, C. H.; Skinner, F. A.; Stewart-Tull, D. E., Eds. *Release of Genetically-Engineered Microorganisms;* Academic: London, 1988.

Tabak, H. H.; Haines, J. R.; Venosa, A. D.; Glaser, J. A.; Desai, S.; Nisamaneepong, W. In *Proceedings for the 1991 International Oil Spill Conference;* sponsored by the United States Coast Guard, American Petroleum Institute, and U.S. Environmental Protection Agency: San Diego, CA, 1991; pp 583–590.

Tagger, S.; Bianchi, A.; Julliard, M.; LePetit, J.; Roux, B. *Mar. Bio.* (Berlin) **1983,** *78,* 13–20.

Venosa, A. *Protocol for Testing Bioremediation Products Against Weathered Alaskan Crude Oil;* paper presented at the Bioremediation: Fundamentals and Effective Applications Symposium, 22 February, Lamar University, Beaumont, TX, 1991.

Venosa, A. D.; Haines, J. R.; Nisamaneepong, W.; Govind, R.; Pradhan, S.; Siddique, B. In *Proceedings of the 1991 International Oil Spill Conference;* sponsored by the United States Coast Guard, American Petroleum Institute, and U.S. Environmental Protection Agency: San Diego, CA, 1991; pp 563–570.

Verstraete, W.; Vanlooke, R.; deBorger, R.; Verlinde, A. In *Proceedings of the Third International Biodegradation Symposium;* Sharpley, J. M.; Kaplan, A. M., Eds.; Applied Science: London, 1975; pp 98–112.

Vogel, T. M; Grbić-Gallić, D. *Appl. Environ. Microbiol.* **1986,** *52,* 200–202.

von Wedel, R. J.; Mosquers, J. F.; Goldsmith, C. D.; Hater, G. R.; Wong, A.; Fox, T. A.; Hunt, W. T.; Paules, M. S.; Quiros, J. M.; Wiegan, J. W. *Water Sci. Technol.* **1988,** *20,* 501–503.

Wang, X.; Bartha, R. *Soil Biol. Biochem.* **1990,** *22*(4), 501–506.

Wang, X.; Yu, X.; Bartha, R. *Environ. Sci. Technol.* **1990,** *24*(7), 1086–1089.

Ward, D. M.; Brock, T. D. *Geomicrobiol. J.* **1978,** *1,* 1–9.

Ward, D.; Atlas, R. M.; Boehm, P. D.; Calder, J. A. *Ambio* **1980,** *9,* 277–283.

Zeyer, J.; Kuhn, E. P.; Schwarzenbach, R. O. *Appl. Environ. Microbiol.* **1986,** *52,* 944–947.

ZoBell, C. E. In *The Microbial Degradation of Oil Pollutants;* Ahearn, D. G.; Meyers, S. P., Eds.; Publ. No. LSU-SG-73-01; Center for Wetland Resources; Louisiana State University: Baton Rouge, 1973.

Hydrogeologic Considerations for *In Situ* Bioremediation

*L. Alföldi**
Water Resources Research Centre

INTRODUCTION

In situ bioreclamation technologies are those for which one cannot select either the location or the space in which the bioreclamation processes take place.

Biotechnologies, in normal operations, are usually applied in a closed system—in reactors, pipes, tanks, etc.—and can be controlled in a planned and predictable way. *In situ* biotechnologies, however, must be applied to a subsurface space with the following characteristic features:

- A structure that is never fully homogeneous
- A space that is open to atmosphere
- Spatial boundaries that are rather probabilistic or arbitrary in nature
- A space that is filled with substances of different physical states or phases
- Gaseous and liquiform substances that are mobile within the solid skeleton, forming a dynamic balance

* Water Resources Research Centre, Kvassay J. út 1., Budapest 1095, Hungary

- Nonsterile conditions, whether biological or microbiological
- Substances filling the space that are able to react with each other and with substances from external sources

It follows from the above considerations that the *in situ* application of any biotechnological procedure requires a thorough knowledge of the space. To be able to select the appropriate technology, the type and kind of polluting substance and the magnitude and the extent of pollution must also be known. The structure of the subsurface space in its original natural condition should be known, as well.

GEOLOGICAL FACTORS

The subsurface medium can possess a variety of conditions; it can be consolidated, solid, and fractured, or unconsolidated, loose, and porous.

Consolidated Formations

Consolidated, solid systems of pores can be of many types, but their main feature is gaps bordered by flat planes. The predetermined flow paths of liquids and gases are defined by the interconnection of these fissures, which can be broadened by karstic dissolution processes or by brecciated fracturing. From the point of view of applying biotechnological measures, the simultaneous occurrence of fissures of different orders of magnitude, causing differing flow conditions within the same space, is rather unfavourable. The system of gaps and fissures with differing detention times, recharged naturally or artificially, empty rather intermittently in a sequence defined by their order of magnitude. Complex and interwoven fissure networks of both saturated and unsaturated states occur (Alföldi 1986; Figure 1).

Unconsolidated Formations

Unconsolidated, loose, porous formations are generally of younger sediment. They occur in the shallow near-surface zones, covering consolidated solid formations in thick or thin layers. They commonly carry alluvial and sedimentary genetic marks. Knowing these typical features, the conditions that might substantially affect the design and

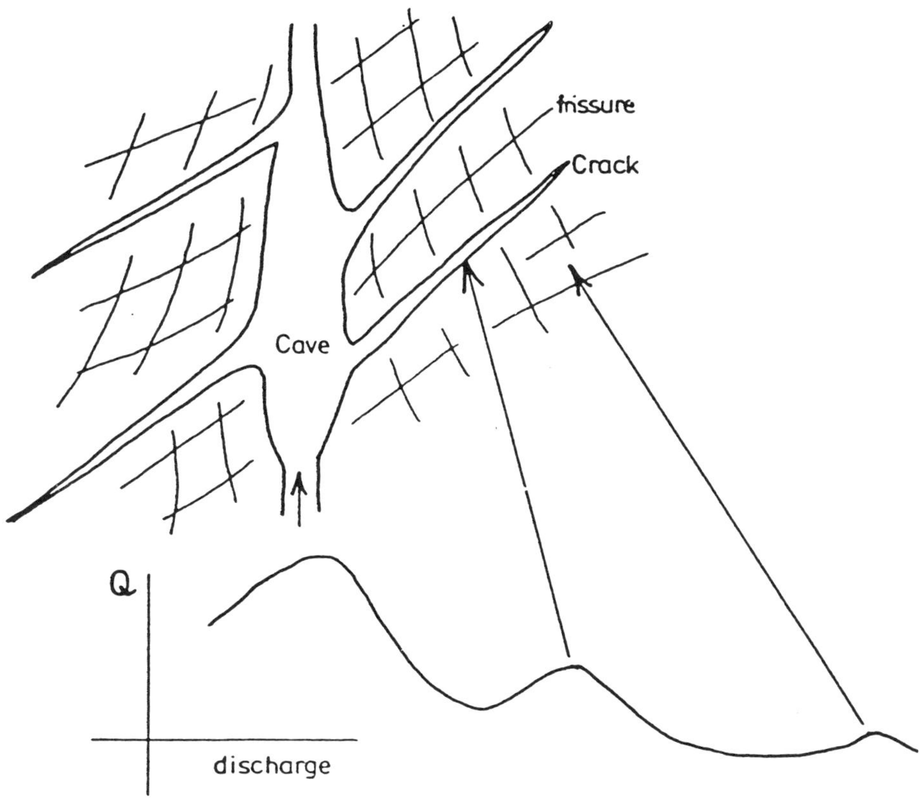

FIGURE 1. Horizontal section of the structure of typical karstic rock.

implementation of biotechnologies can be determined. The neglect of these typical features may, in extreme cases, result in unsuccessful or inefficient control measures.

Talus Slopes. Loose sedimentary formations can form only slight topographic variations. Solid consolidated rocks, on the other hand, form topographic variations on the order of mountains, and in continental conditions, their lower slope sides are covered by clastic formations that originate from the weathering or wearing of the rock (Figure 2). In applying biotechnological measures, these foot hill detrital cones may create more difficulties and hidden traps than the fissured or karstic regions. Rough rubble (several decimeters in size) is imbedded in clayey-silty fraction, in which irregular tubular eroded flow pathways may be formed; at other places they may be replaced by thin, tape-like, sloping permeable formations. Because of their sloping character, these

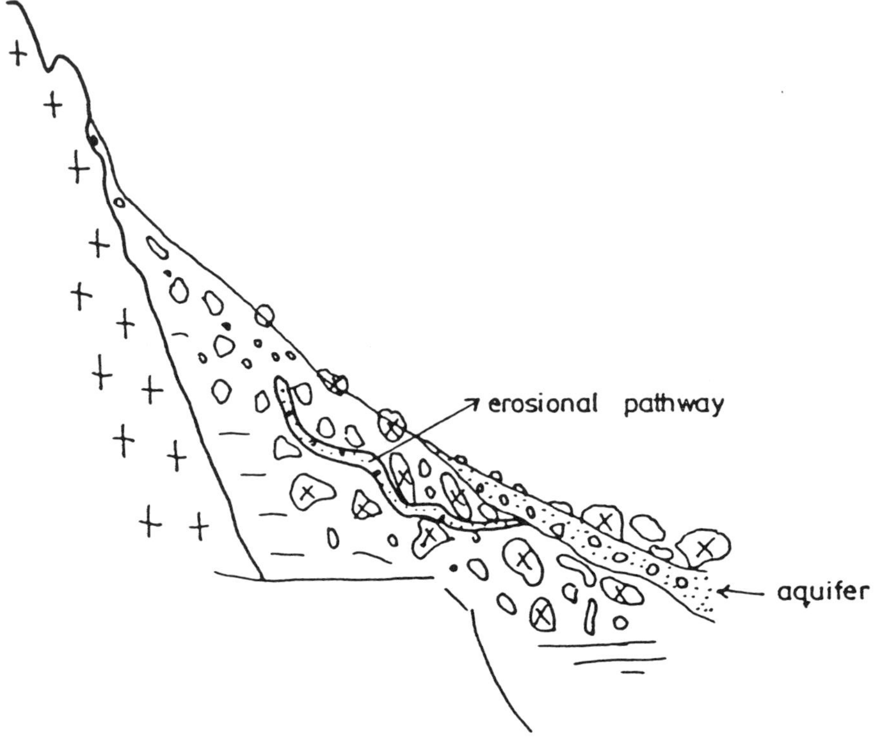

FIGURE 2. Mountain-foot debris cone.

aquifers may respond with extremely varying water levels and discharge rates to precipitation inputs, as a function of periodically varying magnitude and intensity. The exploration of flow pathways and tape-like permeable layers is rather difficult, so only biotechnologies of self-regulating character, based on a single seeding at the upper parts of the slope, can be expected to yield favourable results.

Simple Stratification. The structure of these deposits in the subsurface is, in most cases, inhomogeneous in both material and physical properties. In a simple case of stratification, the deposited material is arranged quasi-horizontally; that is, the strata are bordered by quasi-horizontal planes. Consequently, the horizontal permeability is an order, or several orders, of magnitude higher than the vertical one. These differences may result in flow patterns of neighbouring layers that are relatively independent from each other (Figure 3).

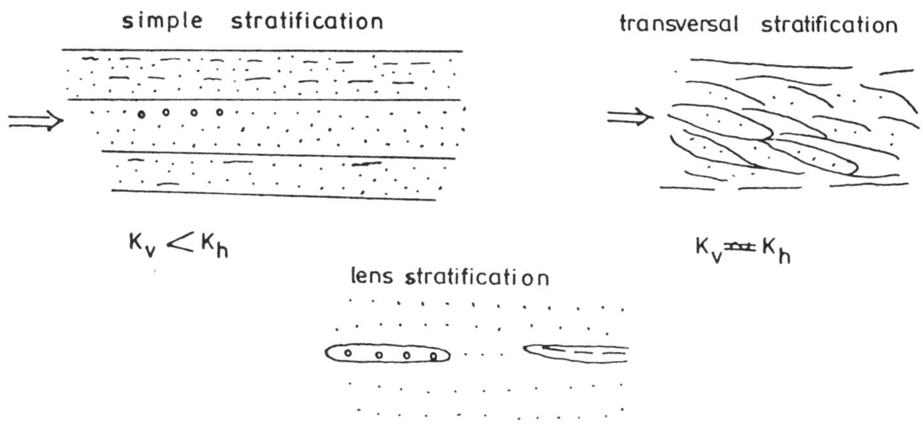

FIGURE 3. The basic types of stratification.

Transverse Stratification. Transverse stratification causes complex inhomogeneities with varying horizontal and vertical permeabilities that are not significantly different from each other. Although the individual groups of layers can be characterized by independent hydraulic parameters, they may seem homogeneous over larger regions, because of their spatially interwoven character.

Lens Stratification. The features of simple and transverse stratification occur in a mixed way. The characteristics of each are hidden and not easy to distinguish. In some cases over smaller regions, they can be misinterpreted to be those of simple stratification, whereas in other cases they exhibit typically lens-like or tape-like stratification characteristics.

Depending on the origin of these formations, the basic types described occur in combinations that can substantially affect the efficiency of bioreclamation.

Sedimentation

Layered Marine Sediments. A commonly known feature of the stratification of marine sediments is that the type, character, and thickness of superimposed layers change only slightly over several hundred meters of depth, thus making their characteristics identifiable with even a few drillings. In horizontal or nearly horizontal aquifers of this type, flow velocity is usually on an order of magnitude of several centimeters a year, and the water transfer between the neighbouring layers is very small. Waters of highly differing character can sometimes be found in aquifers directly neighbouring each other (Figure 4). In the upper part of layers that directly connect to the ground surface, the natural microflora is rich both in species and number; however, in pressured, confined layers, the flora rapidly diminishes.

Contaminants discharged into these layers propagate very slowly, even in the case of dissolved substances, and may stay in the same layer for years.

Even the construction of a single well (Figure 5) will, upon water extraction, accelerate flow velocities in the effective range of the well, resulting in increased exchange of water between different layers and rapid changes in the microbiological conditions. With water withdrawal, both known and unknown subsurface contaminants may be mobilized, and the flow directions thus induced may even be the reverse of the natural ones (Alföldi 1983). The pair of wells for extraction and recharge should be located in such a way that induces flow between the two wells in the natural direction. In the case of uneven and sloping surfaces, the siting of wells should be designed with special care, since the rhythmically varying stratification, frequently found in marine sediments, may cause the loss of efficiency of biotechnological measures (Figure 6).

In mountain foothills, a frequently encountered situation is that rhythmically layered marine sediments rest on hillsides with a mild slope. In such situations, two drillings made to identical depths from the surface will explore similar stratification and find different aquifers with different characteristics. This means that the aquifers explored by the two drillings are isolated from each other, although stratification indicate their linkage (Figure 7).

Fluvial Sediments. On a global scale, the most important water-bearing layers are probably found in the alluvial deposits of river valleys. These alluvial aquifers have good permeability and are a good source of water from recharge areas. However, they exhibit varying spatial and structural features. As sediments of the erosion base they can be easily contaminated, endangering significant water resources. Their typical features include lenses and tape-like imbedments with multifold inclined and transverse layer patterns. These features might present undetected barriers to the successful application of *in situ* water treatment technologies.

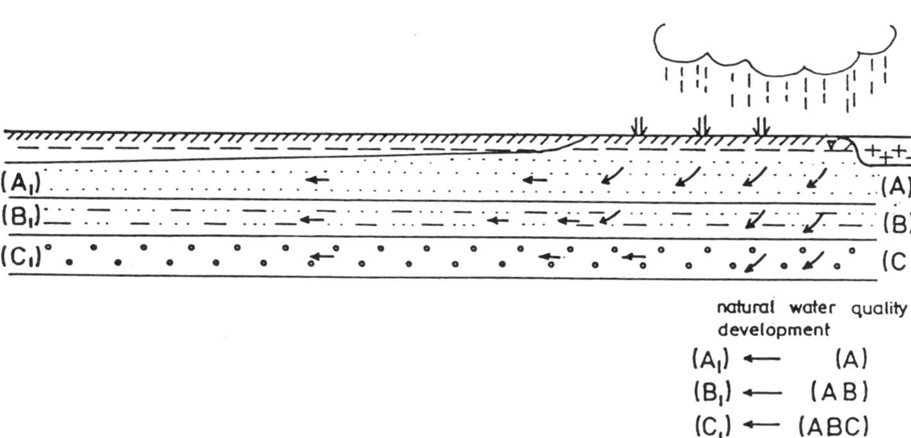

FIGURE 4. A common hydrogeological situation in a marine sediment.

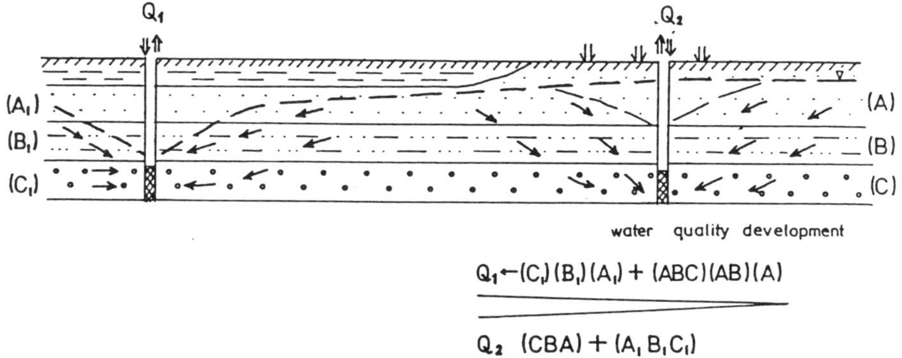

FIGURE 5. Underground water flow and quality transformation from double-well action in the same horizontal sedimentary units.

Clay lenses or other sediments of low permeability, frequently found in series, in alluvial sand and gravel strata, may not substantially alter the hydraulic communication of aquifers (Figure 8). This may even be the case when the profiles of exploration wells indicate a typically varying stratification.

Clayey lenses will affect both natural and induced flow patterns, thereby increasing the distance along the route of flow between two points. However, clay lenses may not result in isolation when the system is operated over longer periods of time.

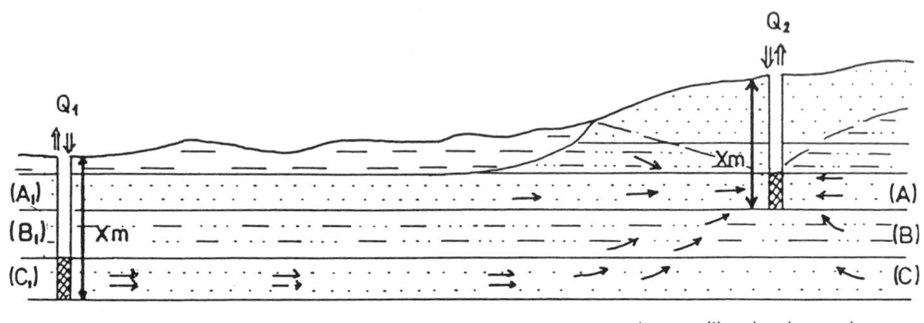

FIGURE 6. A typical error associated with double wells set in inclined sediments.

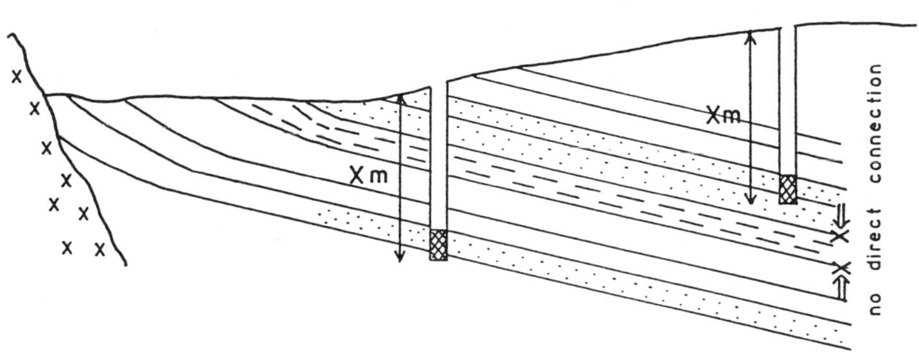

FIGURE 7. Underground water flow and quality transformation from double-well action in different horizontal sedimentary units.

An opposite case might occur, however, in the fine-structured and clayey sediments dominating in the fringes of river valleys. In this case, the imbedded tape-like aquifers, showing a lens-like shape in their vertical cross section, might be completely isolated from the surface and from each other (Figure 9). A frequent error is made when, on the basis

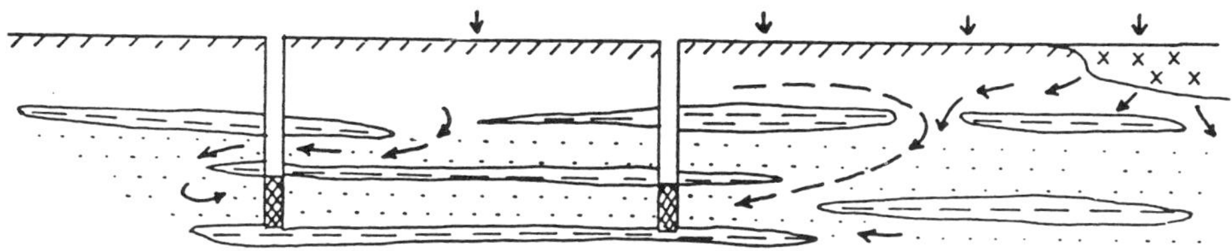

FIGURE 8. Hydrological connections between aquifers in fluvial sediments.

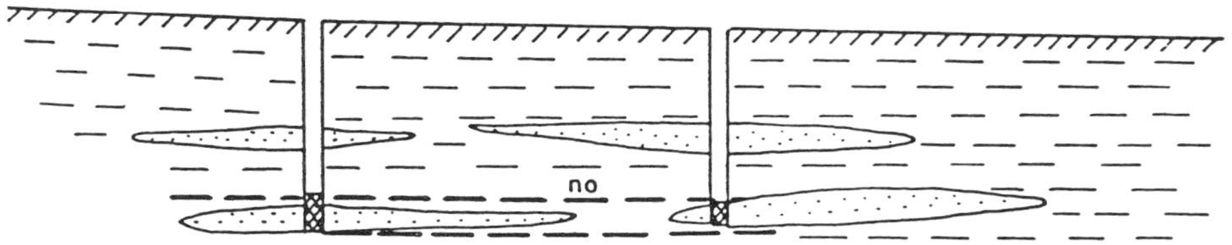

FIGURE 9. Stratigraphic features of lens-stratified aquifers.

of two drilling profiles of identical stratification, a typically layered, continuous structure is interpreted in the drawings, disregarding the alluvial character of the zone. Information tends to be insufficient even when geological profiles are drawn with due concern to the alluvial characters, since tape-like and lens-form imbedments cannot be distinguished from each other on the basis of their vertical cross section. Moreover, the vertical cross sections of tape-like sedimentary units will not indicate whether there is continuity between two such exploratory boreholes. Neither the possibility of such connection nor the complete separation of the two formations can be excluded (Figures 10 and 11).

In typical flat-lying alluvial formations, the depth of the alluvium is usually limited, seldom exceeding 20 m. Thus, they alone can be contaminated since the intrusion depth of artificial structures usually reaches to the bottom of such aquifers, or even penetrates beyond them into the underlying confining layers. Consequently, contamination from exploration may reach aquifers that previously had been isolated from the surface.

Sediment Cones and Clastic Formations. Streams originating in mountainous regions that flow down into flatland areas deposit coarser sediments, which form detrital cones and piles along the flatter section of the stream toward the flatlands.

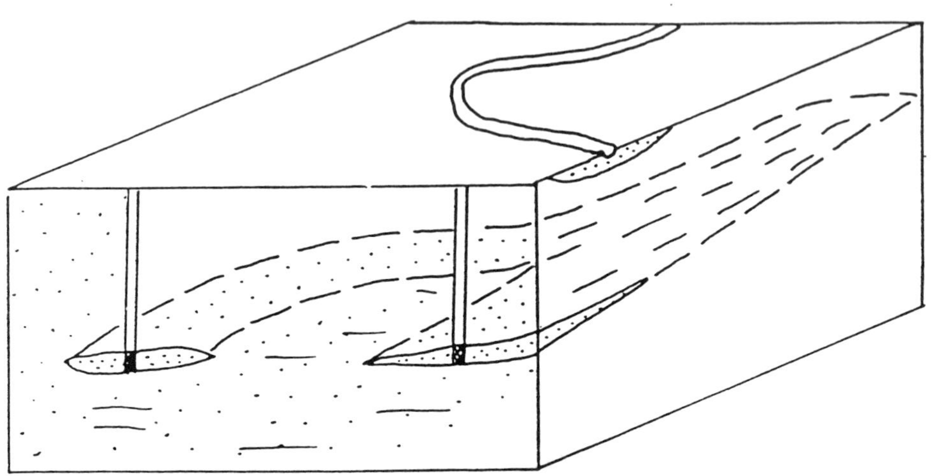

FIGURE 10. A typical alluvial aquifer.

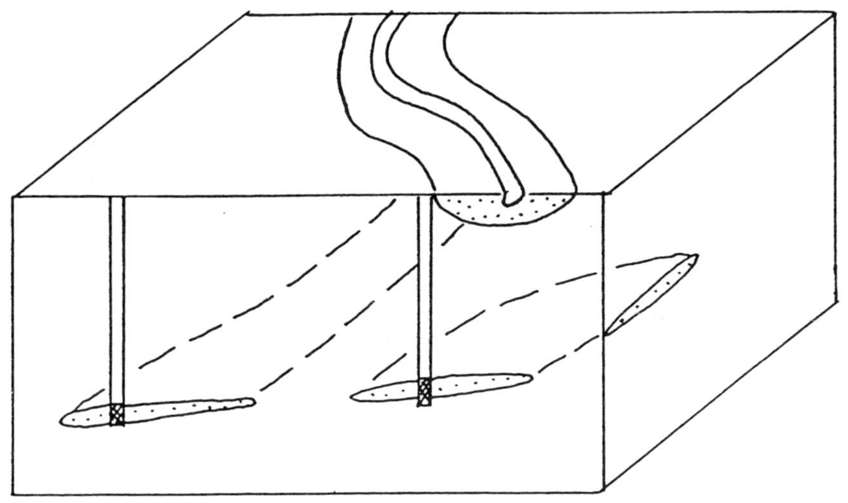

FIGURE 11. A special alluvial condition resulting in isolated aquifers.

Flowing on top of these detrital cones, the streams fork into branches and continue downslope in migrating channels. If the area subsides due to increased loads or other geological forces, the alluvial cones may fill the whole basin (Figure 12). These formations, frequently reaching several hundred meters depth, are of extremely varying structure and consist mainly of units of coarser particle size that have good permeability. Although the major laws of generation of varying

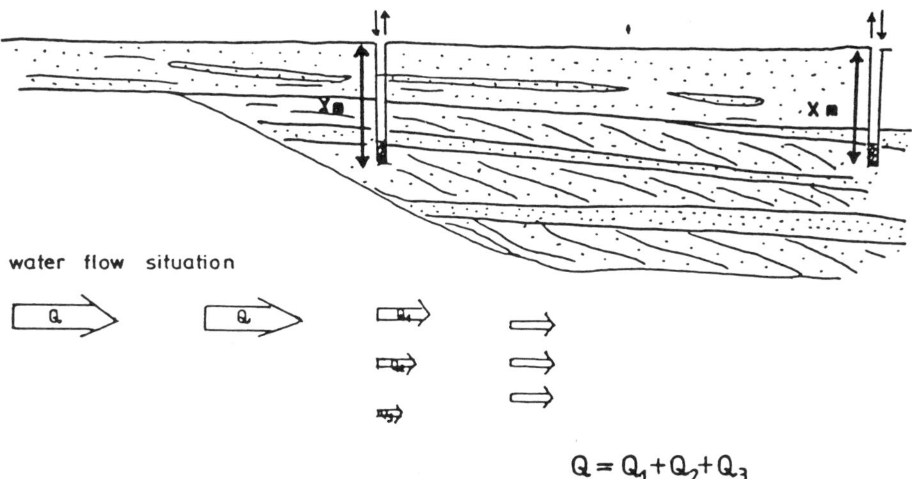

FIGURE 12. A typical hydrogeological situation in a basin-type sediment cone.

sedimentary formations upon the effects of geological and meteorological cycles can be described in rough approximation, they will not provide information of practical importance (Miall 1984). The variations of sloping, horizontal, lens-like, and tape-like layers cannot be identified on the basis of a few exploratory drillings. Extremely varying quality conditions may be created in the slowly or rapidly flowing or stagnant waters within layers of largely varying particle size and stratification; this may occur also as a function of the highly varying organic material content of the sediments and of their redox conditions. Similarly, subsurface ecological conditions are also highly variable in character. When water production and supply facilities are put in place, natural conditions will be altered and new flow patterns and water quality conditions will be induced. There are good possibilities for applying *in situ* bioreclamation techniques in such formations. For the design of such systems, however, typical geologic and hydrogeologic data will not provide sufficient information; on the contrary, this information may be the source of serious error. To avoid such error, the interaction of hydraulic and water quality conditions should be investigated by *in situ* measurements, and the data thus obtained used for models for the purpose of prediction and simulation.

Essentially similar is the case when land subsidence will not follow the deposition rate of sediments from the river, and thus the river will flow on the top of its own detrital talus cone, gradually building the cone (Figure 13). In this case the river is the main source

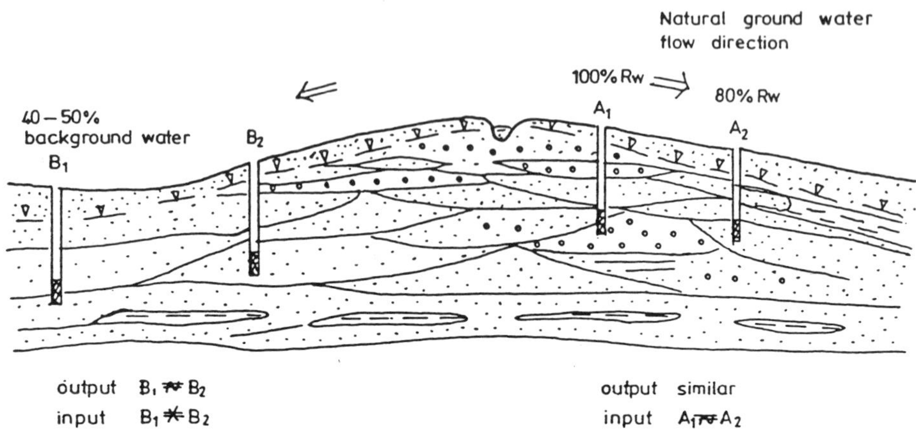

FIGURE 13. A typical hydrogeological situation in a fluvial sediment pile.

of recharge into underlying aquifers, and the flow pattern of groundwater will be determined by the slope and head conditions of the river and the basin. The improper consideration of natural flow conditions, sedimentary character, and the conditions of the terrain might result in a disadvantageous outcome when applying *in situ* bioreclamation techniques. The success of such technologies may be adversely affected by the usually increasing contamination levels of the river and by variation in the pollution loads.

Conditions in Bank-Filtered Aquifers. The interconnection between alluvial deposits and the stream channel provides wide possibilities for extracting river water, filtered through the sediments of the river bank, from wells in the form of groundwater. It is obvious that the nearer the well to the channel, the larger part of the extracted water comes from the river. The process termed *bank filtration* does not mean solely physical filtration, but the biologically active layer of a few centimeters thickness, formed on the channel bed, provides for biochemical filtration as well. One must not forget, however, that the hydraulic regime and natural water level fluctuation of the river will substantially affect the flow conditions of the bank-filtered zone. These variations affect the quality and temperature of water extracted, as well (Figure 14).

Production wells, operating in the vicinity of suspended channels and in river islands, extract filtered river water at all hydrological conditions—i.e., at low, medium, and high flow periods—reflecting all the consequences of the influence of the river. Under

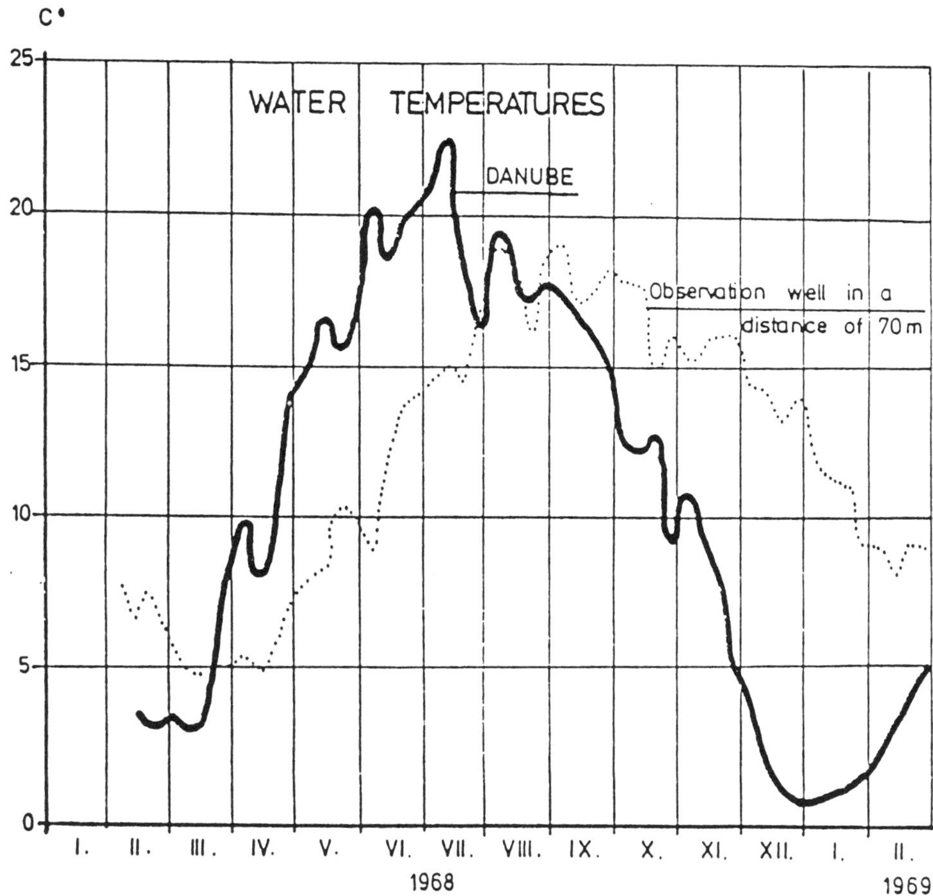

FIGURE 14. Variations in river water temperature compared to variations in temperatures in a river-connected well.

natural conditions, incised rivers recharge the near-river aquifers at high water conditions. The flow reverses at low water levels, discharging aquifer water into the river. Thus in the bank-filtered wells of such rivers, the ratio of filtered river water to the waters stemming from the background zone is continuously changing with the hydrological regime (Figure 15). This sometimes results in a seemingly paradoxical situation where the "background water" of the aquifer becomes more polluted than the water filtered through the channel bed. Consequently, the variation of the ratio of river water to background water causes the respective variation of quality of water extracted. At flood flow conditions, the water over the floodplain will cause a vertical recharge

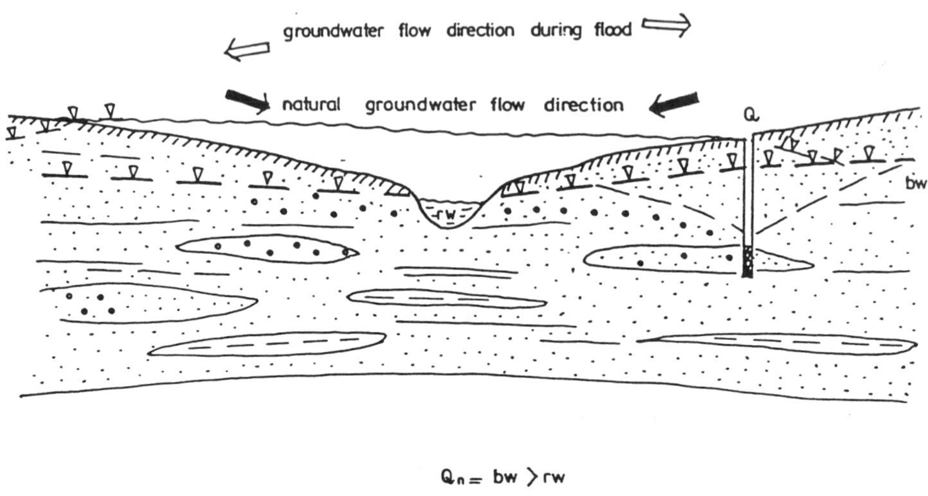

FIGURE 15. A typical hydrogeological situation in bank-filtered aquifers.

to the aquifer across the topsoil of the floodplain, in addition to the lateral inflow across the channel bed and the biologically active zone. It follows from the above considerations that the near-river aquifer and the one below the floodplain are closely related in ecological and microbiological terms as well.

HYDROGEOLOGICAL FACTORS

Infiltration

In hydrogeologic studies of vertical seepage, the soil zone is treated as a black box, and the processes occurring in it are disregarded; only its outputs as inputs to vertical seepage are considered (Figure 16). Only the movement of waters after they have crossed the soil zone is considered.

The unsaturated zone, located between the water table and the soil zone, can be characterized (as an annual average) as a zone not saturated with water. However, infiltration that follows precipitation events may periodically saturate or partially fill the pores within the particles, even several times a year. In the dry periods, without infiltration, the pores—with the exception of capillary pores—are filled with air, and interstitial pore water is attached to the particle surfaces.

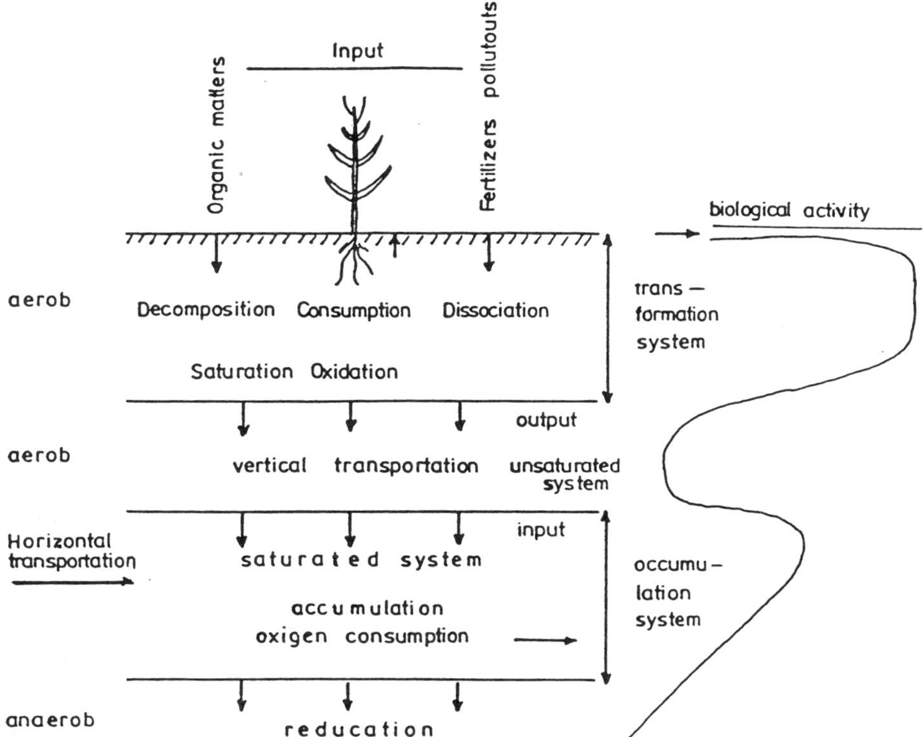

FIGURE 16. Hydrogeological conditions in near-surface soil.

Saturation and unsaturation vary with the periodicity of precipitation events. In fine sediments, downward flow may be of the piston type, pushing the not yet dissolved gaseous air in front of it in the saturated zone.

The conditions in recharge areas of near-surface aquifers differ basically from those of the aquifers having lateral recharge only. A special transitional zone is formed in the fine pores of the unsaturated zone by capillary water. This may induce reductive conditions inside the aerobic space. In deposits consisting of fine sediment particles (in eolithic loess, for example), the capillary rise can be several meters high, forming special hydraulic and biochemical conditions. In fine sediments, the capillary rise makes the determination of the boundaries of the saturated zone and the water table rather difficult. This capillary use can induce a reversed water motion, as compared to the direction of infiltration, toward the ground surface (Kovács 1981).

Fluctuation of Water Levels

Infiltration induced by precipitation cycles causes the rise of the water table, which in turn induces lateral and sometimes vertical flow. From the point of view of the technologies of bioreclamation, both processes may be of significance. The annual fluctuation of the water table may be in the range of several meters, causing repeated saturation and desaturation, and thus leaching, of a relatively thick zone. Natural horizontal flow velocities and their rate of change vary simultaneously.

GEOCHEMICAL AND MICROBIOLOGICAL FACTORS

The temperature of near-surface zones varies along with the fluctuation of ambient surface temperatures. The annual variation of temperature occurs to a depth of about 10 to 15 m in the aquifers. In addition to the annual variation of ambient air temperature, groundwater temperature is also influenced by recharge water. The density, viscosity, isothermal compressibility, and vapour pressure of the groundwater change as a function of temperature. These changes also affect the living conditions of microbes. Organisms transported into the aquifers with the recharge waters will find temporarily or permanently altered temperature conditions, which in turn will cause the reduction of their growth rates and other activities. This especially applies in temperate and cold climatic zones (Alföldi 1988).

Between the soil and the saturated zone is a transitional unsaturated zone. The rate of temperature change is the highest in this transitional zone. The organic material content of water and the nutrient supply decrease in comparison with those of the soil zone. There is a poor supply of nutrients for the heterotrophic bacteria, but the complete lack of organic matter occurs only in extreme situations (Thorn & Ventullo 1988). The concentration of inorganic substances, needed for securing an energy supply to chemo-autotrophic organisms, also decreases, and a shift in the ratio of major ions is a likely condition. In the unsaturated zone, in water attached to soil particles and in pore water, bacteria consisting of a few cells are found. Following nutrient gradients, bacterial colonies are formed whose growth rates and metabolic activity are diminished by the ongoing lack of nutrients or the intermittent, periodical nutrient supply (Marxsen 1988). During our experiments with the gravel deposits of the Danube River, we did not find significant difference between the species composition in the saturated and unsaturated zones, but the number of individuals and their activity were orders of magnitude higher in the saturated zone. In the shallow near-surface zone, we found bacterial populations of specific, independent character. The frequently and rapidly changing nutrient supply conditions of the vertical profiles, along with the varying pH and Eh conditions, play a selective rarifying role on the bacteria that are a potential food for animals living in the pores (Bärlocher & Mürdock 1989). Under flood conditions the river water recharging the aquifer, is likely to carry bacteria and detritus consumers into the saturated zone. In the Danube gravel, for example, the presence of two oligoheta species was detected in 12 to 15 m depth and at a distance of 200 m from the river channel. The two species were the larger *Limnodrillus hofmeisteri* Claparede and the smaller *Nais behningi* Michaelsen (Oláh *et al.* in press).

Substantial information on the bacterial communities of groundwaters has been available only since the latter 1980s. This field is likely to present some surprises in the future (Bone & Balkwill 1988; Danielopol 1989; Kölbe-Boelke *et al.* 1988.)

The presence of active and inactive bacteria in the various zones of bank-filtered systems is likely to create further possibilities for biotechnology, along with unexpected difficulties.

SUMMARY

For the successful application of known bioreclamation techniques and for the development of new ones, the complex investigation of water motion in subsurface space is required. This paper attempts to draw the attention of technologists and biologists to the complexity of the subsurface space and its hydraulic processes, emphasizing the indispensable need for professional exploratory work before the application of biotechnologies. The scope of this paper did not allow the detailed professional presentation and description of related hydrological and hydrogeological processes and thus could not support its statements with appropriate literature references. The single objective of this work was to draw the attention of all interested parties in this new field to the extreme hydrogeological complexities involved in this area.

REFERENCES

Alföldi, L. "Questions Related to the Regional Protection of Groundwater Resources." U.N. Seminar on Groundwater Protection Strategies and Practices, Athens, 1983; 8 p.

Alföldi, L. "Groundwater Protection in Porous, Fractured and Karstic Aquifer Systems;" general report 19th Congress of IAH, Karlovy Vary, Czechoslovakia, **1986**, 8–15, 9.

Alföldi, L. "Groundwater Microbiology Problems and Biological Treatment—State of the Art Report." *Wat. Sci. Tech.* **1988**, *20*(3), 1–31.

Bärlocher, F.; Murdock, J. H. "Hyperheic Biofilms—A Potential Food Source for Interstitial Animals." *Hydrobiologia* **1989**, *184*, 61–67.

Bone, T. L.; Balkwill, D. L. "Morphological and Cultural Comparison of Microorganismus in Surface Soil and Subsurface Sediments at a Pristina Study Site in Oklahoma." *Microb. Ecol.* **1988**, *16*(1), 49–64.

Danielopol, D. L. "Groudwater Fauna Associated with Riverine Aquifers." *I.N. Am. Benthol. Soc.* **1989**, *8*(1), 18–35.

Kölbel-Boelke, E.; Anders, M.; Nehrkorn, A. "Microbial Communities in the Saturated Groundwater Environment II. Diversity of Bacterial Communities in Pleistocene Sand Aquifer and Their *In Vitro* Activities." *Microb. Ecol.* **1988**, *16*(1), 31–48.

Kovács, G.; *et al. Subterranean Hydrogeology*; Water Resources Publications: Littleton CO, 1981; p 421.

Marxsen, J. "Investigation into the Number of Respiring Bacteria in Groundwater from Sandy and Gravelly Deposits." *Microb. Ecol.* **1988**, *16*(1), 65–72.

Miall, A. D. *Principles of Sedimentary Basin Analisis.* Springer Verlag: New York, Berlin, Heidelberg, Tokyo, 1984.

Oláh, J.; Alföldi, L.; Pekár, F.; Csizmarik, G.; Szabó, S.; Botos, M. "Nitrification in Danube Gravel Deposits." *Hidr. Közlöny*, in press.

Thorn, P. M.; Ventullo, R. M. "Measurement of Bacterial Growth Rates in Subsurface Sediments Using the Incorporation of Trithieted Thymidine into D.N.A." *Microb. Ecol.* **1988**, *16*(1), 3–16.

Formulation of Nutrient Solutions for *In Situ* Bioremediation

Pradeep K. Aggarwal, Jeffrey L. Means, Robert E. Hinchee*
Battelle Columbus Operations

INTRODUCTION

Adding nutrients and oxygen to enhance indigenous microbial activity for biodegrading hydrocarbons is a promising technique for *in situ* remediation of contaminated soils. When water is used as the oxygen carrier, hydrogen peroxide is commonly used as a source of oxygen, together with nitrogen and phosphorus sources, with or without micronutrients. Nitrogen sources include ammonium and nitrate salts, and phosphorus sources typically are orthophosphate salts of sodium, ammonium, or potassium. The amount of phosphate in the nutrient solutions generally is greater than may be required for enhancing microbial growth because phosphate is added in an attempt to decrease the rate of H_2O_2. Table 1 lists the composition of several nutrient formulations reportedly in use for *in situ* bioremediation.

Conflicting results have been reported in the literature on the success of H_2O_2-based nutrient applications at hydrocarbon contaminated sites (API 1987; Brown & Norris 1986; Hinchee *et al.* 1989, 1991; Huling *et al.* 1990). Two major factors have limited a broader application and success of the enhanced biodegradation technology: plugging of the subsurface formations caused by excessive precipitation and lack of sufficient oxygen caused by rapid decomposition of hydrogen peroxide. Aggarwal *et al.* (1991) have recently discussed aspects of H_2O_2

* Present address: Argonne National Laboratories, Environmental Research Division, Argonne IL 60439

TABLE 1. Composition of nutrient solutions used for field or laboratory bioreclamation.

	Nutrient Concentrations, mg/L											
Component	I	II	III	IV	V	VI	VII	VIII	IX	X	XI	XII
$NH4^+$	1,683	23	19,584	–	7,452	113	273	225	0.4	11	–	23
K^+	–	32	–	117	–	832	413	1,184	128	191	429	416
Ca^{2+}	–	0.4	16	100	–	7	–	7	12	–	3	17
Na^+	1,216	51	19,284	–	7,199	–	43	–	389	–	1,277	–
Cl^-	3,319	0.7	28	0.2	–	13	121	16	0.3	–	5	36
NO_3^-	–	78	–	465	–	388	–	775	0.1	39	2,920	78
SO_4^{2-}	–	9	105,158	96	39,744	78	728	–	160	78	40	480
PO_4^{3-}	2,904	93	52,155	48	19,665	1,153	646	19	1,114	233	1,444	577
CO_3^{2-}	–	57	253	–	–	–	–	–	0.2	–	–	–
BO_33-	–	–	–	2	–	–	–	–	0.09	–	–	–
VO_3^{3-}	–	–	–	–	–	–	–	–	–	–	–	–
Mg^{2+}	–	2	170	24	–	19	12	40	10	10	1	20
Mn^{2+}	0.6	13	0.06	–	–	0.6	–	–	0.3	–	–	–
Fe^{2+}	–	0.1	4.5	–	–	0.6	15	2	0.3	–	0.2	3
Zn^{2+}	–	–	–	0.1	–	–	–	–	0.1	–	–	–
Cu^{2+}	–	–	0.01	–	–	–	–	–	0.1	–	–	–
Co^{2+}	–	–	–	–	–	–	–	–	0.05	–	–	–
Mo^{2+}	–	–	–	0.02	–	–	–	–	0.12	–	–	–

Sources:

I - Restore™ 375 manufactured by FMC Chemicals.
II - Jhaveri and Mazzacca, 1983.
III - Raymond *et al*. 1978.
IV - Hoagland's Solution (Morholt *et al*. 1958).
V - Raymond *et al*. 1975.
VI - Flathman and Githens 1985.
VII - Liu and Wong 1975.
VIII - Olive *et al*. 1975.
IX - Zeyer and Kearney 1982.
X - Kim and Maier 1986.
XI - Swindoll *et al*. 1988.
XII - Goldstein *et al*. 1985.

decomposition and the performance of several additives in decreasing the rate of H_2O_2 decomposition in the subsurface. This paper presents an analysis of the plugging problem and proposes methods to select nutrient formulations that will minimize precipitation in the aquifer.

GEOCHEMISTRY OF THE SOIL-NUTRIENT SOLUTION INTERACTION

The geochemistry of natural groundwaters in shallow aquifers is governed by soil-water interactions under relatively aerobic conditions. Contamination of groundwaters by hydrocarbons provides an energy source and increases microbiological activity in the subsurface, resulting in anaerobic conditions with lower pH and higher alkalinity. Oxygen depletion and lower pH also result in increased solubility of redox-sensitive elements such as Fe or Mn. Elevated concentrations of these elements are commonly found in groundwaters downgradient of hydrocarbon-contaminated sites.

Adding oxygenated, phosphate-rich nutrient solutions to stimulate microbial activity increases the availability of oxygen and alters the chemical equilibrium between soil and groundwater in the aquifer.

These changes in geochemical conditions may result in the precipitation of hydroxides of Fe and phosphates of Fe and other cations (*e.g.*, Ca) commonly found in groundwaters. Excessive precipitation of these hydroxide and phosphate phases may plug the aquifer.

The extent of precipitation in an aquifer upon injecting phosphate-rich nutrient solutions may be estimated using computer codes for geochemical modeling (*e.g.*, EPA 1988; Kharak *et al.* 1988). These computer codes calculate chemical speciation and saturation states of aqueous solutions given pH, Eh, and analytical concentrations of the various elements. The saturation states of minerals are easily expressed in terms of the saturation index (SI):

$$SI = \log IAP - \log K \quad (1)$$

where K is the solubility product and IAP is the ion activity product for the dissolution reaction of the mineral. For example, the dissolution reaction of brushite (DCPD, $CaHPO_4 \bullet 2H_2O$) is

$$CaHPO_4 \bullet 2H_2O = Ca^{2+} + PO_4^{-3} + H^+ + 2H_2O \quad (2)$$

for which,

$$IAP = [Ca^{2+}] \bullet [PO_4^{3-}] \bullet [H^+] \quad (3)$$

and where square brackets denote thermodynamic activities of the aqueous species. Positive SI values indicate supersaturation (i.e., potential for precipitation), and negative SI values suggest under-saturation (i.e., potential for the solid phase to be solubilized). When SI equals zero, the solution is in chemical equilibrium with the given phase. The geochemical modeling codes can also be used to calculate the amount of precipitation or dissolution required to reach equilibrium with one or more minerals.

The extent of precipitation during an *in situ* bioremediation operation was estimated using site-specific data from a project conducted at Eglin Air Force Base (AFB), Florida (Hinchee *et al.* 1989) and the geochemical modeling program, MINTEQ (EPA 1988). Partial chemical composition of nutrient-spiked feed solutions and the modeling results are given in Table 2.

The above analysis of aquifer geochemistry upon nutrient injection agrees with field observations at Eglin where injection rates in wells decreased with time during several months of operation. The geochemical modeling results based on Eglin data are expected to be

TABLE 2. Geochemical modeling for precipitation at the Eglin site.

1. Concentration of Major Components in Reinjected Groundwater.

Component	Concentration (mg/L)
Fe	2.5
Ca	19.0
PO_4	160.0
Cl	400.0
Na	160.0
NH_4	168.0
pH	7.0
Alkalinity (as HCO_3^-)	28.0

2. Nutrient Solution Volume

Nutrient delivery rate	= 19 L/min
Period of application	= 12 h/week
Nutrient delivery per week	= ~ 13,600 L

3. Modeling Results

Expected minerals in the precipitate: Ca and Fe phosphates

Precipitate volume = 0.02 cm^3/L
(assuming an average molar volume of 160 cm^3/mole for solid phases)

Weekly precipitate volume = 273 cm^3

If precipitation occurs in 50 percent of the pore spaces, the initial 20 percent porosity of a 30 × 30 cm patch of an aquifer will be reduced by half to a depth of ~3 cm in each weekly cycle of nutrient delivery.

applicable also to sites where the Ca and Fe contents of the groundwater are similar. For example, injection rates during enhanced bioreclamation operations near Granger, Indiana, decreased from ~150 L/min to ~38 L/min within 1 week (API 1987). The Granger site is located in a silty sand and gravel aquifer. Although water chemistry data are unavailable, precipitation reactions at Granger were probably in sites with limestone lithologies and would be expected to be even greater than in sandy lithologies because of the greater amounts of calcium that would be available for phosphate mineral precipitation.

APPLYING MODELING RESULTS TO FIELD OPERATIONS

While geochemical modeling results correctly predict the observed precipitation in the field, a commercial laboratory test, recommended by the manufacturer of the nutrient formulation to determine the *in situ* chemical stability of the formulation, was an inaccurate indicator of aquifer plugging at Eglin. The commercial laboratory test consisted of a column filled with a composite of the soil samples from the site. Deionized water was passed through the column bed under constant head to establish a baseline flow rate. Restore™ 375 nutrient formulation was dissolved in deionized water at a concentration of 500 mg/L and passed through the soil at a flow rate of 3.0 to 5.0 mL/min until the concentration of nutrients in the recovered water approached that of the feed. Phosphate, which was not detected in the effluent until 4 pore volumes of solution had been collected, approach steady state after ~11 pore volumes. These results were interpreted to indicate that the transport of phosphate could be achieved without plugging the aquifer at the Eglin site.

Experimental studies of the kinetics of calcium phosphate precipitation (*e.g.*, van Kemenade & de Bruyn 1987) suggest that the precipitation of a stable phase, hydroxyapatite (HAP) or chlorapatite, is preceded by the formation of a more soluble, metastable phase, octacalcium phosphate (OCP) or brushite (DCPD); the precursor phase is transformed upon aging to the stable phase by solution-mediated reactions. The time lag for the initiation of precipitation depends on the degree of supersaturation of the precursor phase and decreases with increasing supersaturation. Because of the higher solubility of the precursor phase, significant precipitation may not be observed in the commercial tests described above if the experiment is not conducted over a long period of time. However, field nutrient injection operations are usually carried out over a period of several weeks or months and would most likely result in the transformation of the precursor phase to a stable phase.

The validity of existing commercial tests for predicting the extent of precipitation in the field was examined using long-term column experiments with the Eglin soil. Approximately 200 mL of nutrient solution was recirculated through the column from bottom to top, with a residence time of ~90 min in the column bed. The nutrient solutions were prepared by mixing site-specific groundwater and Restore 375 or a similar formulation. Three experiments used a formulation similar to

Restore 375, but with no polyphosphate, and the other two used Restore 375 (Table 3).

TABLE 3. Nutrient formulations used in experiments to validate modeling results.

	Concentrations (mg/L)				
	Simulated RESTORE			RESTORE™ 375	
Component	RE-1 Uncont.	RE-2 Uncont.	RE-7 Cont.	RE-8 Cont.	RE-9 Cont.
Ca^{2+}	9	9	9	13	29
PO_4^{3-}	146	155	150	172	172
$P_3O_5^-$ (as PO_4)	–	–	–	92	92
NH_4^+	168	168	168	168	168
CO_3^{2-}	28	28	28	28	40
Na^+	160	160	160	160	200
K^+	18	18	18	18	18
Fe^{3+}	2.5	2.5	2.5	1	1

Measured concentrations of orthophosphate during the column experiments are given in Table 4. The saturation indices of various phosphate phases in the experimental solutions were calculated using the MINTEQ code. These indices are listed in Table 4 and are plotted on a total-phosphate versus total calcium concentration diagram along with the stability fields of the phosphate phases (Figure 1).

An inspection of Table 4 and Figure 1 indicates that phosphate concentrations did not decrease significantly in experiments where the solutions remained undersaturated with the metastable phases (DCPD and OCP). Minor precipitation occurred in solutions that were undersaturated with DCPD, the most soluble phase, but oversaturated with OCP. Maximum precipitation occurred in experiment RE-9 which was undersaturated with DCPD and OCP, but oversaturated with HAP. This apparent oversaturation with HAP is probably because of the fine-grained size of the precipitate, which is likely to have a higher solubility compared with a well-crystallized phase assumed for calculations. Given sufficient time for crystal growth, the chemical composition of

TABLE 4. Ca and P concentrations and saturation states[a] of phosphate minerals in solutions from experiments to validate modeling results.[b]

	RE-1					RE-7					RE-8						RE-9					
			SI					SI						SI						SI		
Time (h)	OP	Ca	DCPD	OCP	HAP	OP	Ca	DCPD	OCP	HAP	OP	TP	Ca	DCPD	OCP	HAP	OP	TP	Ca	DCPD	OCP	HAP
0	147	9	-0.8	-2.0	2.8	150	9	-0.8	-2.0	2.8	172	264	13	-0.6	0.7	2.9	175	265	29	-0.3	1.0	2.3
2	123					131					174						143	168	5	-0.3	1.7	3.2
4	134					133					173											
24	142	15[c]	-0.9	-1.2	2.9	108					214	214	14	0.5	0.2	2.6						
68																	70	84	12	-1.0	-1.5	3.1
96											190	199	25	-0.3	0.6	2.7	62	77	18			
117																	53	65	13			
189						111	47	-0.3	1.4	2.9							41	43	14			
360																	22	20	12			
405																	19	20	12			
453																	19	20	13	-1.8	-4.0	2.72

OP = orthophosphate; TP = total phosphate.
(a) The MINTEQ code was used to calculate saturation indices (SI).
(b) Concentrations of OP, TP, and Ca are in mg/L; Run RE-1 used uncontaminated Eglin soil and all other runs used contaminated soil.
(c) Assumed for modeling purposes.

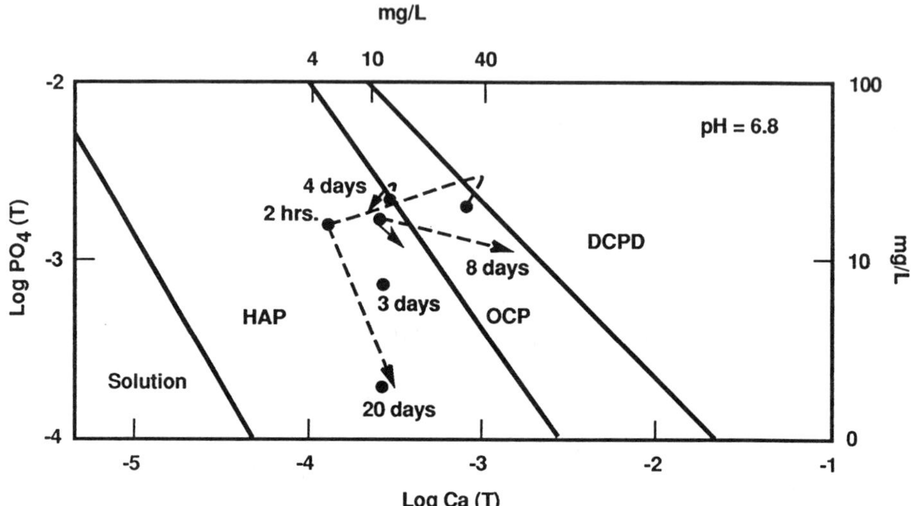

FIGURE 1. Stability fields of several Ca-phosphate phases and the disposal of experimental solutions from column studies. Note that significant phosphate precipitation appears to have occurred only in Run RE-9, which was likely supersaturated with DCPD during the early part of the experiment.

the solution would probably approach equilibrium with a well-crystallized HAP phase.

Results of the long-term precipitation experiments discussed above suggest that in most cases the short-term commercial tests will underestimate the extent of precipitation from nutrient applications in the field. A more realistic estimate of subsurface precipitation may be obtained using equilibrium geochemical modeling calculations.

ALLEVIATING THE PLUGGING PROBLEM

Existing nutrient formulations frequently contain excessive orthophosphate to decrease the rate of peroxide decomposition. This role of phosphate in the nutrient solutions may be rather limited and may only inhibit the relatively minor, inorganic catalysts or peroxide decomposition (Aggarwal *et al.* 1991; Hinchee *et al.* 1991). Thus, the phosphate concentration may be lowered to meet only the microbial requirements, and stabilization of peroxide, if at all possible, may be achieved by adding other inhibitors of peroxide decomposition (Aggarwal *et al.* 1991).

The concentration of orthophosphate required to optimize microbial growth under oxygen-limited conditions, such as those expected in bioremediation operations, is not well known. Most laboratory studies are conducted under aerobic conditions in which oxygen is not the limiting nutrient. In oxygen-limited systems, less biomass will be produced and the nutrient requirements may be lower than those in laboratory studies. Miller and Hinchee (1990) found at a bioventing site in North Florida the addition of nutrients had no significant effect, it is possible that at many sites nutrients may not be required. The maximum orthophosphate concentration that may avoid significant precipitation in most geochemical environments is ~10 mg/L. This concentration of orthophosphate may be sufficient in many cases to provide excess phosphorous for microbial growth under oxygen-limiting conditions (Aggarwal *et al.* 1990). However, to achieve an orthophosphate concentration of ~10 mg/L at some distance away from the point of injection, much higher concentrations of phosphate in the feed solutions may be required because orthophosphate is easily adsorbed onto soil components. This higher orthophosphate concentration may also result if excessive precipitation, and therefore, the plugging problem may not be alleviated in orthophosphate-based nutrient formulations.

Alternative Phosphate Source

Polyphosphates [*e.g.*, pyrophosphate [$P_2O_7^{-4}$], tripolyphosphate [$P_3O_{10}^{-5}$], trimetaphosphate [$P_3O_9^{-3}$]) are an alternative source of phosphate in nutrient formulations. Sodium tripolyphosphate is used in the Restore 375 formulation, but in addition to orthophosphates. To the authors' knowledge, use of polyphosphates as the sole source of phosphorous nutrient formulations has not been documented in the biodegradation literature, although it has apparently been used in commercial practice (Brown, R. A. Groundwater Technology, Inc., Project Meeting, Chadds Ford PA, personal communication to R. Hinchee of Battelle, April 14, 1989).

The kinetics of polyphosphate hydrolysis reactions are influenced by several factors: those most important to biodegradation are pH, the ionic composition of the solution, microbial activity, concentration of the enzyme polyphosphatase, and chain length (Blanchar & Hossner 1969; Blanchar & Riego 1976; Busman & Tabatabai 1985; Dick & Tabatabai 1986; Gilliam & Sample 1968; Hons *et al.* 1986). The half-lives of polyphosphate hydrolysis in natural soils are found to vary from approximately 1 day to tens of days. Thus, polyphosphates can in effect be used as *in situ*, slow-release sources of orthophosphate, and the rate of orthophosphate production may be geochemically controlled to avoid plugging problems.

Experimental Evaluation of Alternative Phosphate Sources

The performance of three polyphosphate species, pyrophosphate (PP), tripolyphosphate (TPP), and trimetaphosphate (TMP), as nutrient sources was evaluated with batch and flow-through experiments. In the batch experiments, ~25 g Eglin soil and ~250 mL aqueous polyphosphate solutions (prepared with site-specific groundwater) were reacted in glass vessels on a mechanical shaker. Samples were withdrawn periodically with a syringe and analyzed for dissolved orthophosphate and total phosphate concentrations.

Data from batch experiments show significant sorption (adsorption and precipitation) of orthophosphate on Eglin soils (Aggarwal *et al.* 1990). Sorption of polyphosphate species (PP and TPP) was insignificant; however, the rapid hydrolysis of these species resulted in the production of orthophosphate that was sorbed onto the soils. The performance of TMP was much better than that of PP and TPP; TMP contains PO_4 groups linked in a ring structure, adsorbs on the soils to a lesser extent, and has a higher solubility (Blanchar & Hossner 1969; Dick & Tabatabai 1986).

The potential performance of TMP in the field application of nutrient solutions was evaluated with flow-through experiments in an aquifer simulator (Figure 2).

Nutrient solutions in the aquifer simulator experiments consisted of various proportions of TMP and citric acid along with other additives for decreasing the rate of hydrogen peroxide decomposition (Table 5). The nutrient solutions were injected for ~450 h, with a ~50-h break after 218 h. To evaluate the performance of TMP and the extent of precipitation in the aquifer, selected samples were analyzed for ortho- and total P, Ca, Mg, Na, K, Fe, and Mn concentrations using colorimetric methods or an inductively coupled plasma (ICP) spectrophotometer. Results of H_2O_2 stability experiments are presented by Aggarwal *et al.* (1991).

As the solutions moved through the soil zone, the orthophosphate content first decreased slightly because of sorption (adsorption

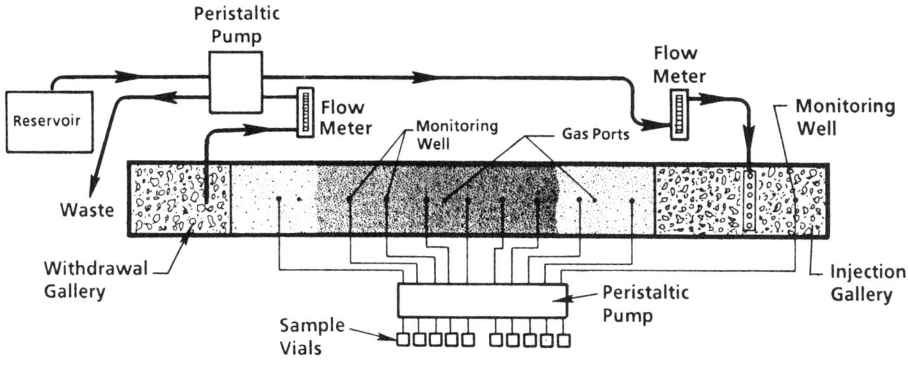

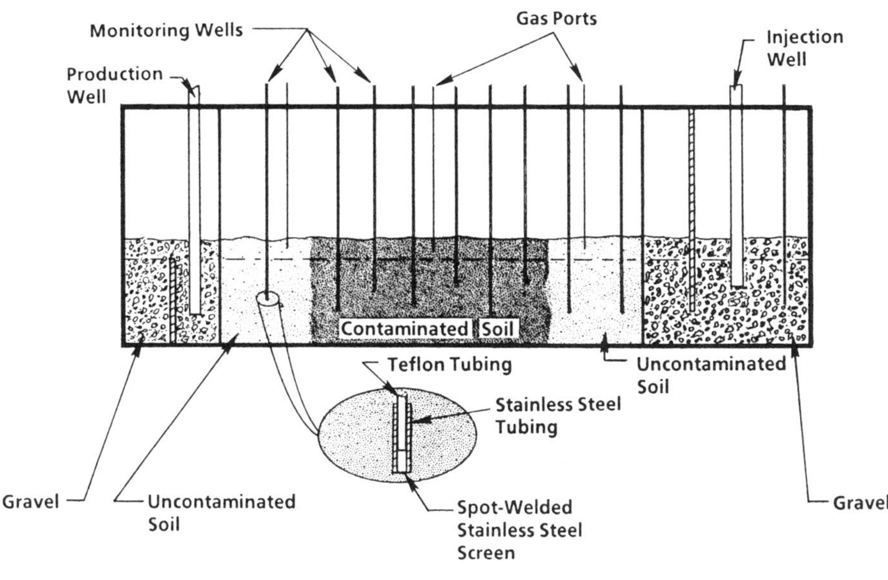

FIGURE 2. Two views of the aquifer simulator. The distances between the injection gallery and the wells are as follows: 9 = 5 cm, 8 = 20 cm, 7 = 30 cm, 6 = 40 cm, 5 = 52 cm, 4 = 60 cm, 3 = 72 cm, 2 = 82 cm, and 1 = 102 cm. Hydraulic conductivity in the soil zone = 2.2×10^{-2} cm/s.

TABLE 5. Input compositions of nutrient solutions used in the aquifer simulator experiments.

	Solution Composition[a]			
Time (h)	TMP (mg/L)	Citric Acid (mg/L)	Na_3PO_4 (mg/L)	Peroxide (%)
0	306	206		
52	300	206		0.09%
59	97	67	28	
123	97	67	28	0.065
145	97	206	28	0.065
197[b]	97	–	28	0.065
218	Injection of nutrients stopped for 50 hours			
291	306	206	28	0.12
323	0.08% RESTORE 375 Formulation			0.065

(a) Other additives (mg/L): KCl = 20; NH_4Cl = 50.
(b) Instead of Citric Acid, Ascorbic Acid (352 mg/L) and $CuSO_4 \bullet 5H_2O$ (2.5 mg/L) were added as peroxide stabilizers.

and precipitation), then hydrolysis of TMP in the soil zone increased orthophosphate relative to polyphosphate. Sorption and microbial growth processes, however, decreased the total phosphate content of the solutions. An increase in orthophosphate concentration occurred nearly 30 cm into the soil zone. Input solutions for the first 59 h contained <3 mg/L of orthophosphate. To increase the availability of orthophosphate for microbial growth before significant hydrolysis of TMP occurred, ~20 mg/L orthophosphate were added to the input solutions used after 59 h.

Retardation of phosphates in the aquifer simulator did not appear significant because the hydraulic conductivity did not decrease during the experiment. Data for Ca concentration also are consistent with minimal precipitation. The Ca concentration decreased from input values of ~30 to ~15 mg/L toward the output gallery. Geochemical modeling of solution compositions in the aquifer simulator suggests that the solutions near the production gallery (Wells 1, 2, 3, and 4) are saturated with DCPD, a metastable phosphate phase. Measured Ca values near the production gallery were lower in the beginning (50 to 158 h) compared with values in the latter part of the experiment. The lower Ca concentrations reflect greater precipitation of Ca phosphates early in the experiment because of a higher concentration of TMP (total P = ~280 mg/L as PO_4) in the input solutions from 0 to 54 h. Hydrolysis of TMP in these solutions will produce a higher concentration of orthophosphate, resulting in a higher degree of supersaturation and, therefore, a lower kinetic barrier to phosphate mineral precipitation.

In summary, the bench-scale experiments suggest that using trimetaphosphate in nutrient formulations may alleviate the excessive precipitation of phosphates that is likely when orthophosphates are used. Hydrolysis of trimetaphosphate appears to be sufficiently fast to provide an adequate concentration of orthophosphate for microbial growth and sufficiently slow to avoid plugging.

SUMMARY AND CONCLUSIONS

The preceding discussion on the geochemistry of nutrient solution and groundwater interactions demonstrations that the presence of orthophosphate appears to be largely responsible for the excessive precipitation and consequent aquifer plugging during bioremediation operations. Based on the results of this study, the following criteria may be used for choosing a phosphate concentration for field applications of phosphate-based formulations:

- **Available Phosphorus Content.** According to the laboratory tests performed using contaminated soils from the Eglin AFB (Aggarwal *et al.* 1990), an available (or exchangeable) soil phosphate content of ~20 mg/L may be sufficient to provide excess phosphorus for microbial growth under oxygen-limiting conditions.

- **Type of Soil.** If phosphate is to be added, either because of P content of the soil is low or based on a conservative estimate of microbial requirements, soil characteristics should be considered. Calcareous soils have a high Ca content and a high sorption capacity. In such soils, phosphate should not be added; nearly all added phosphate in calcareous soils will probably be adsorbed or precipitated. In sandy (quartz) soils, such as at the Eglin AFB, phosphate may be added in nutrient formulations. A combination of orthophosphate and TMP may be used to minimize precipitation. Phosphate levels up to ~20 mg/L may be achieved by adding phosphate in the form of orthophosphate. For higher concentrations, phosphate may be added as TMP or as a mixture of TMP and orthophosphate.

When phosphates are added in the nutrient formulations, the impact of phosphate mineral precipitation at a site may be evaluated on the basis of long-term, bench-scale tests of *at least 30 days* duration. Short-term tests are likely to be deceptive and may suggest a much lower precipitation potential than what may occur *in situ*. Alternatively, a conservative estimate of the volume (V) of precipitated phosphate minerals per unit volume of nutrient solution may be calculated by the following equations:

$$^{V}\text{total (cm)}^3 = {}^{V}\text{Ca-phos.} + {}^{V}\text{Fe-phos.} \tag{4}$$

$$^{V}\text{Ca-phos.} = \frac{M_{Ca}}{8} \times 0.16 \text{ when } M_{Ca} < \frac{3M_{PO_4}}{5} \tag{5}$$

or

$$= \frac{M_{PO_4}}{31.7} \times 0.16 \text{ when } M_{Ca} > \frac{3M_{PO_4}}{5} \tag{6}$$

$$^{V}\text{Fe-phos.} = \frac{M_{Fe}}{57} \times 0.065 \text{ when } M_{Fe} < M_{PO_4} \tag{7}$$

where M_i is the total concentration (mg/L) of the subscripted species expected in the aquifer. For Ca and PO_4, this concentration should be nearly the same as that in the nutrient formulation. For Fe, however, the groundwater Fe concentration in contaminated areas should be considered. The total volume of phosphates precipitated during the entire operation is then ^{V}total × L where L is the total quantity of nutrient solutions (in liters) that are to be injected during the operation. A more precise statement of the volume of expected precipitation in the aquifer may be obtained by using geochemical modeling codes.

REFERENCES

Aggarwal, P. K.; Means, J. L.; Hinchee, R. E.; Headington, G. L.; Gavaskar, A. R.; Scowden, C. M.; Arthur, M. F.; Evers, D. P.; Bigelow, T. L. "Methods to Select Chemicals for *In-Situ* Biodegradation of Fuel Hydrocarbons"; final report to the

Air Force Engineering and Services Center on Contract F08635-85-C-0122, subtask 3.05, 1990.

Aggarwal, P. K.; Means, J. L.; Downey, D. C.; Hinchee, R. E. "Use of Hydrogen Peroxide as an Oxygen Source of *In-Situ* Biodegradation: Part II. Laboratory Studies." *Journal of Hazardous Materials* **1991**, *27*(3).

API (American Petroleum Institute). *Field Study of Enhanced Subsurface Biodegradation of Hydrocarbons Using Hydrogen Peroxide as an Oxygen Source*; API Publication #4448, **1987**.

Blanchar, R. W.; Hossner, L. R. "Hydrolysis and Sorption of Ortho-, Pyro-, Tripoly-, and Trimetaphosphate in 32 Midwestern Soils." *Soil Sci. Society Amer. Proc.* **1969**, *33*, 622–625.

Blanchar, R. W.; Riego, D. C. "Tripolyphosphate and Pyrophosphate Hydrolysis in Sediments." *Soil Sci. Society Amer. J.* **1976**, 225–229.

Brown, R. A.; Norris, R. D. "Field Demonstration of Enhanced Bioreclamation"; *Sixth Natl. Symp. and Exp. on Aquifer Restoration and Ground Water Monitoring*, Columbus, OH, May 19–22, 1986.

Busman, L. M.; Tabatabai, M. A. "Hydrolysis of Trimetaphosphate in Soils." *Soil Sci. Society Amer. J.* **1985**, *45*, 630–636.

Dick, R. P.; Tabatabai, M. A. "Hydrolysis of Polyphosphte in Soils." *Soil Science* **1986**, *142*, 132–140.

EPA (U.S. Environmental Protection Agency). *MINTEQA2, An Equilibrium Metal Speciation Model, User's Manual*; EPA600/3-87/012; NTIS PB88-144-167, 1988.

Flathman, P. E.; Githens, G. D. "*In Situ* Biological Treatment of Isopropanol, Acetone, and Tetrahydrofuran in the Soil/Groundwater Environment." In *Groundwater Treatment Technology*; Nyer, E. K., Ed.; Van Nostrand Reinhold: New York, 1985; pp 173–185.

Gilliam, J. W.; Sample, E. C. "Hydrolysis of Pyrophosphate in Soils: pH and Biological Effects." *Soil Science* **1968**, *106*, 352–357.

Goldstein, R. M.; Mallory, L. M.; Alexander, M. "Reasons for Possible Failure of Inoculation to Enhance Biodegradation." *Appl. Env. Microbiol.* **1985**, *50*(4), 977–983.

Hinchee, R. E.; Downey, D. C.; Slaughter, J. K.; Westray, M. "Enhanced Biodegradation of Jet Fuels, A Full Scale Test at Eglin Air Force Base, Florida"; Air Force Engineering and Services Center Report ESL/TR/88-78, 1989.

Hinchee, R. E.; Downey, D. C.; Aggarwal, P. K. "Use of Hydrogen Peroxide as an Oxygen Source of *In-Situ* Biodegradation: Part I. Field Studies." *Journal of Hazardous Materials* **1991**, *27*(3).

Hons, F. M.; Stewart, W. M.; Hossner, L. R. "Factor Interactions and Their Influence on Hydrolysis of Condensed Phosphates in Soils." *Soil Science* **1986**, *141*, 408–416.

Huling, S. G.; Bledsoe, B. E.; White, M. V. "Enhanced Bioremediation Utilizing Hydrogen Peroxide as a Supplemental Source of Oxygen: A Laboratory and Field Study"; unpublished report, U.S. EPA, Ada, OK, 1990.

Jhaveri, V.; Mazzacca, A. "Bioreclamation of Ground and Groundwater Case History"; *Fourth Nat. Conf. Management of Uncontrolled Hazardous Waste Sites*, Washington, DC, October 31–November 2, 1983.

Kharak, Y. K.; Gunter, W. D.; Aggarwal, P. K.; Perkins, E. H.; DeBraal, J. D. "SOLMINEQ.88: A Computer Program for Geochemical Modeling of Water-Rock Interactions"; U.S. Geological Survey, Water Resources Investigations Report No. 88-4227, 1988.

Kim, C. J.; Maier, W. J. "Biodegradation of Pentachlorophenol in Soil Environments." *Proceedings of the 41st Industrial Waste Conference* **1986**, 303–312, May 13–15.

Liu, D.; Wong, P.T.S. "Biodegradation of Bunker GC Fuel Oil." *Proc. Third Int. Biodeg. Symp.*, Appl. Sci. Pub.: Kingston, RI, August 17–23, 1975.

Miller, R. N.; Hinchee, R. E. "Enhanced Biodegradation Through Soil Venting." U.S. Air Force AFESC, Final Report Contract F08635-85-L-0122, Tyndall AFB, FL, 1990.

Morholt, E.; Bradwine, P. F.; Josef, A. *Source Book for Biological Sciences*; Hercourt Brace: New York, 1958; pp 113.

Olive, W. E. Jr.; Cobb, H. D.; Atherton, R. M. "Biological Treatment of Cresylic Acid Laden Waste Water." *Proc. Third Int. Biodeg. Symp.*, Appl. Sci. Pub. Ltd.:74 London, Kingston, RI, August 17–23, 1975.

Raymond, R. L.; Jamison, V. W.; Hudson, J. O.; Mitchell, R. E.; Farmer, V. E. "Beneficial Stimulation of Bacterial Activity in Ground Waters Containing Petroleum Products." *AIChE Symp. Se.* **1975**, *73*, 390–404.

Raymond, R. L.; Jamison, V. W.; Hudson, J. O.; Mitchell, R. E.; Farmer, V. E. "Field Application of Subsurface Biodegradation of Gasoline in a Sand Formation"; final report to the American Petroleum Institute on Project No. 307-77, 1978.

Swindoll, C. M.; Aelion, C. M.; Pfaender, F. K. "Influence of Inorganic Nutrients on Aerobic Biodegradation and on the Adaptation Response of Subsurface Communities." *Appl. Env. Microbiol.* **1988**, *54*(1), 212–217.

van Kemenade, M.J.J.M.; de Bruyn, P. L. "A Kinetic Study of Precipitation from Supersaturated Calcium Phosphate Solution." *J. Colloid and Interface Science* **1987**, *118*, 564–585.

Zeyer, Jl; Kearney, P. C. "Microbial Degradation of para-Chloroaniline as Sole Carbon and Nitrogen Source." *Biochem. Physiol.* **1982**, *17*, 215–223.

Hydraulic Fracturing to Improve Nutrient and Oxygen Delivery for *In Situ* Bioreclamation

Wendy J. Davis-Hoover[*]
U.S. Environmental Protection Agency
Lawrence C. Murdoch, Stephen J. Vesper
University of Cincinnati
Herbert R. Pahren
National Urban League
Omar L. Sprockel, Ching L. Chang, Ajax Hussain, W. A. Ritschel
University of Cincinnati Medical Center

INTRODUCTION AND BACKGROUND

The *in situ* delivery of nutrients and oxygen in soil is a serious problem in implementing *in situ* biodegradation. Current technology requires ideal site conditions to provide the remediating organisms with the nutrients and oxygen required for their metabolism, but the shortage of oxygen in many subsurface sites is the factor that most frequently limits biological activity (Brown *et al.* 1984; Hinchee *et al.* 1984; Wilson *et al.* 1986).

Several strategies have been utilized to solve this problem of subsurface delivery. In some cases oxygen, as a gas or dissolved in

[*] U.S. Environmental Protection Agency, Risk Reduction Engineering Laboratory, Waste Minimization, Destruction Research Division, Municipal Solid Waste and Residuals Management Branch, Soils and Residuals Section, Center Hill Laboratory, Cincinnati OH 45268

Mention of trade names or commercial products does not constitute endorsement or recommendation for use.

water or hydrogen peroxide, has been injected into the subsurface. Delivery of oxygen or peroxide by injection into common groundwater wells requires the subsurface to be fairly permeable, with 10^{-4} cm/sec as the lower limit of hydraulic conductivity. Moreover, preferred flowpaths, such as fractures or macropores, will channel fluids and leave large blocks unexposed to injected oxygen.

These procedures have other problems. They generally require continuous pumping, which means continuous expense and extended time at a site. In the case of hydrogen peroxide, the microbial toxicity limits the exposure concentration. Also, populations of bacteria in the vicinity of the injection site limit the usefulness of the peroxide by destroying it through the catalase reaction.

Vertical wells are typically used to access subsurface soils. At the Center Hill Laboratory, hydraulic fracturing techniques have been developed to create flat-lying lenses of solid, granular material at shallow depths in unlithified glacial drift. When filled with sand, hydraulic fractures act as permeable channels that increase the rate and area of delivery of fluids to the subsurface (or recovery therefrom), a quality that will benefit a variety of remedial technologies including bioremediation. When filled with granules of slow-dissolving nutrients or oxygen-releasing chemicals, hydraulic fractures could produce a reservoir of these compounds that would enhance bioremediation. Investigating these possibilities is the purpose of the research described here.

The Process of Hydraulic Fracturing

The process of hydraulic fracturing at a well begins with the injection of fluid into the well, typically using a constant rate pump. The pressure of the fluid increases until it exceeds a critical value and a fracture is nucleated. A granular material, termed a *proppant*, is simultaneously pumped into the fracture as it grows away from the well. Sand is commonly used as a proppant, but granules of nutrients or oxygen-releasing compounds could be used as well. Transport of proppant is facilitated by using a viscous fluid, usually a gel formed from guar gum and water, to carry the proppant grains into the fracture. After pumping, the proppant holds the fracture open while the viscous gel breaks down into a thin fluid. The thinned gel is then pumped out of the fracture, leaving a layer of proppant grains in the subsurface.

Since 1987, the Center Hill Laboratory has conducted a coordinated program of laboratory, theoretical, and field investigations into the use of hydraulic fractures to improve remediation. Field testing of the program was conducted during June 1988 (Murdoch 1990) and during June and July of 1989 (Murdoch *et al.* 1990a, b) at an uncontaminated site 10 km north of Cincinnati, OH. A method of creating sand-filled hydraulic fractures at shallow depths in unlithified material was evaluated during the 1989 field tests. Details of the hydraulic fracturing technique were presented by Murdoch (Murdoch *et al.* 1990b).

Recently, we have initiated a program to develop solid materials that could facilitate *in situ* bioremediation with the intention of delivering them by injection into hydraulic fractures. Nutrient compounds in granular form that release gradually over several weeks are commercially available. Efforts to create solids that slowly release oxygen have focused on encapsulating granules of peroxides in ethylcellulose, a process adapted from pharmaceutical applications.

Hydraulic Fractures Created in the Field

The vicinity of 19 hydraulic fractures created during the 1989 field tests was excavated and the fractures mapped in detail. The site consists of a bench 10 to 15 m wide bounded to the southwest by a 5-m tall, steep ascent and to the northeast by a 10-m deep descent. Fractures were created in unsaturated, overconsolidated glacial drift at depths ranging from 0.9 to 1.9 m.

Hydraulic fractures exposed by excavation at the site were remarkably similar in form. Three fractures created at one borehole, EL6, are a typical example (Figure 1). Each of the three fractures was created during a separate application of the fracturing technique. The fractures are essentially horizontal and equant to slightly elongate in view (Figure 2). They are highly asymmetric with respect to the borehole, however, with a preferred direction of propagation roughly parallel to the slope of the overlying ground surface. The fractures are stacked one on top of another at a spacing of 30 cm, which is maintained from borehole to leading edge (Figure 2). The major axes of the fractures range from 5.5 to 8.5 m, and the maximum thickness of sand is 1.3 to 1.4 cm. The sand proppant is thickest near the centers of the fractures and thins toward the edges (Figure 3).

Each fracture at the borehole EL6 contained 270 kg (600 lb) of sand. The middle level fracture is bigger than the others because it was intersected by another fracture created from EL7, a neighboring borehole.

Other fractures created at a depth of 2 m were filled with 540 kg of sand, and at slightly greater depth (3.8 m) we created a fracture filled

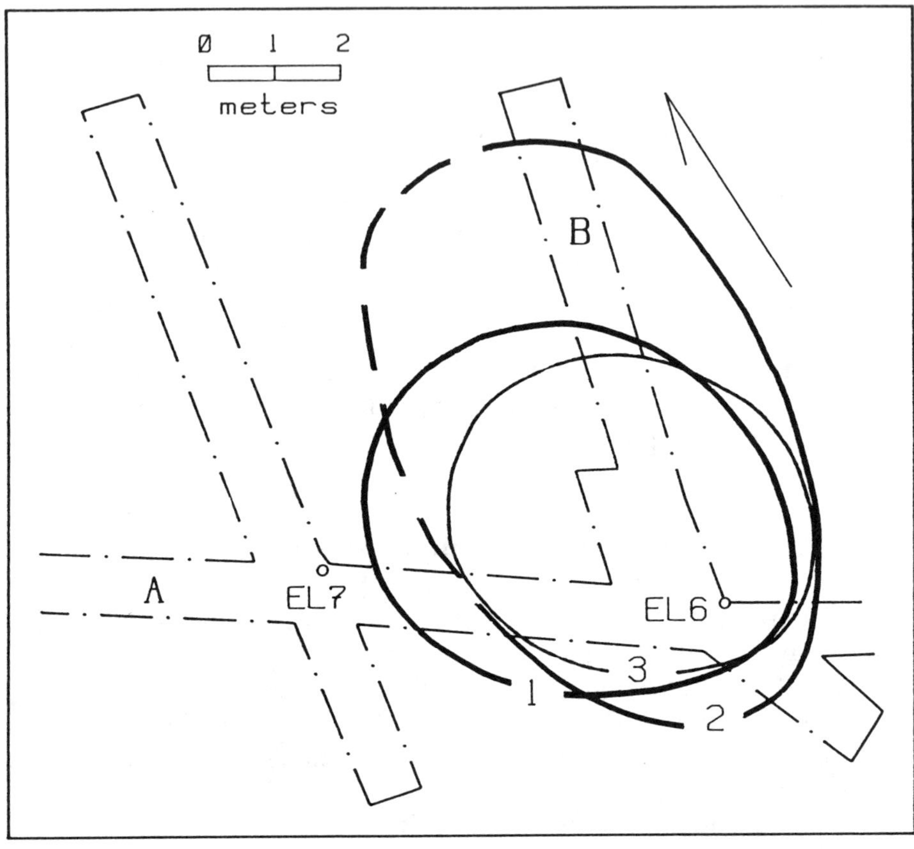

FIGURE 1. Map of three fractures created at borehole EL6. Dash-dot line is the wall of a trench. Dashed line is where fracture 2 intersects a neighboring fracture (not shown) created from borehole EL7.

with 1,600 kg of sand. As many as four fractures were stacked together, and more fractures could have been added at greater depths. Vertical spacings of stacked fractures were as close as 15 cm.

Two possible strategies for using hydraulically created fractures in bioremediation are now in development. These are the facilitated injection of fluid and the delivery of solids.

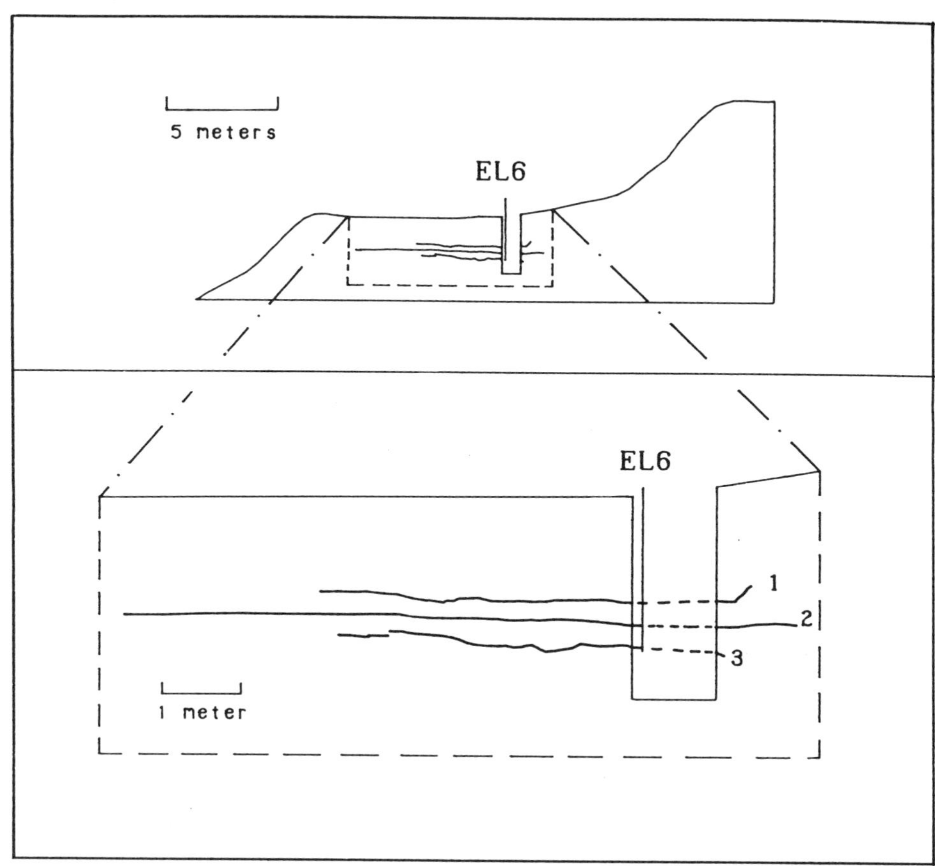

FIGURE 2. Cross-sections of three fractures created at borehole EL6, showing topographic profile (upper) and details of fracture traces (lower). Section from eastern wall of trench B.

Facilitated Injection of Fluid

A simple application is to use hydraulic fractures to facilitate the injection of nutrient- or oxygen-bearing fluid through wells. Hydraulic fractures filled with sand will act as permeable channels to increase both the rate of delivery and the area affected by each well.

Inflow tests were conducted at the site to obtain a preliminary estimate of the magnitude of the increase in injection rate. The tests were conducted using a Guelph permeameter (Elrich *et al.* 1988), a device that yields the flow rate required to hold a constant water level in a borehole. Water levels were held at 1.0 m above the bottom of

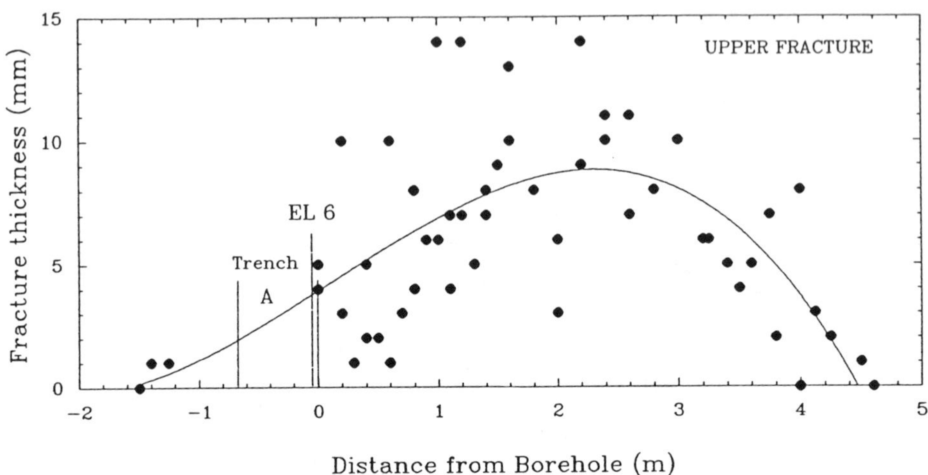

FIGURE 3. Distribution of sand in the upper fracture, as exposed on the walls of trench B. Borehole EL6 and trench A are shown for reference.

open boreholes during all tests. The average steady-state rate of flow into three boreholes in unfractured ground (silty to clayey glacial drift) was 0.055 L/min. The rate of inflow into boreholes intersecting hydraulic fractures was initially 0.25 to 2.5 L/min, but decreased to between 0.175 and 0.5 L/min at steady state (data from Murdoch *et al.* 1990b). The initial rate of inflow increased by more than an order of magnitude, and the steady-state rate of inflow increased by a factor between 3.2 and 9.1 as a result of the creation of the fractures. Similar increases in flow rate are seen following hydraulic fracturing operations at oil wells, gas wells, and water wells (references in Murdoch *et al.* 1990b). Based on that information, we expect that hydraulic fractures could improve the effectiveness of existing methods of delivering hydrogen peroxide, oxygenated water, or air into wells for bioreclamation.

Delivery of Solids

Hydraulic fracturing is one of the few techniques of placing substantial masses of solid compounds in the subsurface. As mentioned above, fractures containing hundreds of kilos of material have been created within a few meters of the ground surface, and much bigger fractures are certainly possible. Stacking flat-lying fractures, as in Figure 2, offers the possibility of dissecting a contaminated site with closely spaced reservoirs of nutrients and oxygen, released slowly from solid phases. Since oxygen is the most important limiting factor found at subsurface sites, a program was begun to encapsulate solid peroxides to create a slow-release oxygen source. The desire was to release oxygen over an extended period of time (3 months). This would limit microbial exposure to peroxide to subtoxic levels and extend the time available for bioremediation.

MATERIALS AND METHODS

Solid Peroxide Encapsulation. An emulsion/solvent evaporation process was used for microencapsulation of sodium percarbonate. Ethyl cellulose (10 g) was dissolved in acetone (80 mL), and the percarbonate powder (10 g) (Fluka Chemical, Ronkonkoma NY) was dispersed in this solution. This dispersion was emulsified in a manufacturing vehicle comprised of liquid paraffin (150) and polyoxyethylene sorbitan monooleate (1% v/v). The solvent was removed over a 30-min period and the microcapsules collected and washed free of the oil phase with hexane.

Release of O_2 from Solid Peroxide in Model Subsurface Soil Studies. A soil column was used in these studies to simulate a hydraulic fracture in the subsurface soil (Figure 4). It was a standard, 10 × 15 cm (diameter x height) soil permeameter slightly modified by placing a stainless steel access port halfway up the side. The soil used was a 50:50 (wt:wt) mixture of Center Hill Soil and Este Fine Sand totaling 1800 g of air-dried material into which was thoroughly mixed 300 mL of deionized water. The soil was packed into the column. A sand-filled hydraulic fracture or "pseudofracture" was simulated by adding 250 g of coarse sand at the location of the side access plug during column filling.

Oxygen release studies of sodium percarbonate were completed in the setup shown in Figure 5, which was held at 12 C (subsurface soil temperature) in a controlled environment room (Labconco, Kansas City MO). Deionized water was vigorously bubbled with oxygen-free nitrogen (Linde, Cleveland OH) to created low dissolved oxygen (DO) conditions in the water reservoir (b). This water was pumped through the soil column (d) at a rate of 10 to 12 mL/min. The oxygen level (0–20 ppm) was continuously monitored with a DO probe (Nester Inst., North Wales PA). After the soil column had reached a low DO steady state, the sodium percarbonate was introduced into the pseudofracture

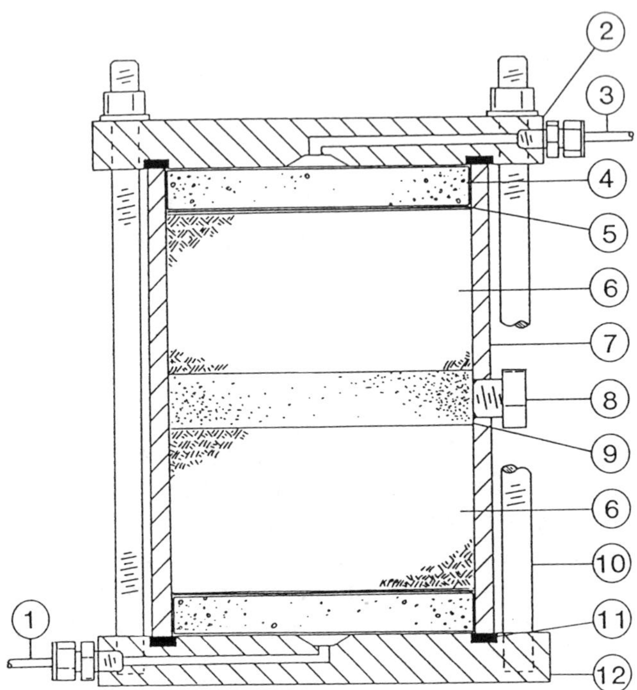

FIGURE 4. Model soil column. The following items make up the soil column: 1 inlet, 2 top plate, 3 outlet, 4 dispersion stone, 5 filter paper (Whatman #1), 6 top soil layer, 7 stainless steel side, 8 plug to pseudofracture, 9 pseudofracture (sand layer), 10 screw clamp, 11 Teflon O-ring, 12 base plate.

by pumping (with a peristaltic pump) the peroxide confined in the guar gum matrix (about 10 mL) (described above). For these studies, 1 g of encapsulated peroxide was used (i.e., 0.5 g sodium percarbonate) and 0.5 g of nonencapsulated sodium percarbonate. The negative control was guar gum matrix alone.

Peroxide Release and Toxicity. The release of peroxide from encapsulated and nonencapsulated sodium percarbonate was studied by monitoring the change in absorbance at 240 nm (Cohen *et al.* 1970). Fifty milliliters of deionized water or 50 mL of a salts solution (Dang *et al.* 1989) were placed in 125-mL dilution bottles and held at 12 C. To these were added 14.5 mg of powdered sodium percarbonate or 29 mg of encapsulated sodium percarbonate. The dilution bottles were placed on

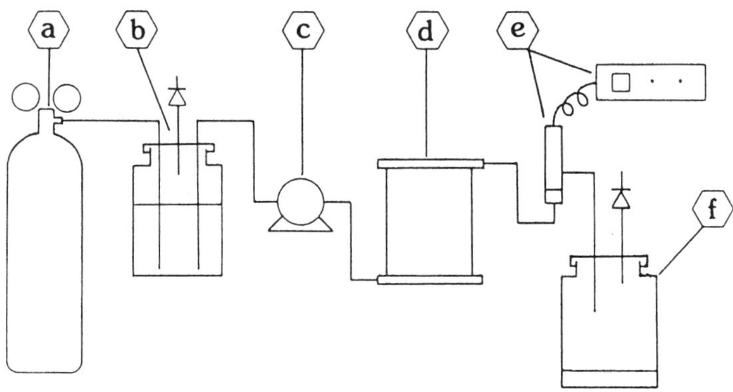

FIGURE 5. Soil column setup for oxygen studies. The following items make up the system: (a) oxygen-free nitrogen tank, (b) water reservoir, (c) peristaltic pump, (d) soil column, (e) dissolved oxygen probe and meter, and (f) receiver.

a reciprocating shaker and the absorbance at 240 nm measured hourly on a Lamda 2 Spectrophotometer (Perkin-Elmer, Norwalk CT).

The toxic effects of exposure to peroxide were monitored using selected hydrocarbon-degrading bacteria. Hydrocarbon-degrading microbes were isolated from enrichments of cultures exposed to gasoline. Isolations for fluorescent pseudomonads were made using the selective S2 medium (Gould *et al.* 1985). The isolate's hydrocarbon-degrading ability was determined by inoculating it on a salts medium (Dang *et al.* 1989) solidified with 2 percent agar. These plates were placed in a closed desiccator vessel and exposed to gasoline vapors. Two cultures (A2 and B1) were selected for further study.

To test the toxicity of the peroxides and produce kill curves, cultures of A2 and B1 were grown in the salts medium containing 200 µL of gasoline in 100 mL of medium. After 3 weeks' growth, the cultures were diluted to approximately 10^5 cfu/mL. Ten milliliters of each diluted culture were placed in sterile 16 x 125 mm test tubes (Becktin/Dickinson, Lincoln Park NJ), to which were added 2.9 mg of powdered sodium percarbonate or 5.8 mg of encapsulated sodium percarbonate. The tubes were incubated at 12 C on their sides on the reciprocating shaker. Hourly samples (200 µL) were plated on R2A medium (Difco, Detroit MI) using the Spiral Plater (Spiral Systems, Cincinnati OH).

RESULTS

Evaluation of Solid Peroxide Encapsulation. Figure 6 indicates that encapsulating sodium percarbonate in ethyl cellulose resulted in a decrease in the oxygen release rate in distilled and deionized water as compared with the unencapsulated powder. This effect appears to be limited to the first 24 h. These same peroxides were next tested in a model soil column (Figure 7). The negative control (no peroxide) indicates that the column can be run for an extended period of time with only a small amount of dissolved oxygen drift. If the nonencapsulated sodium percarbonate is introduced into the pseudofracture, oxygen release is measured in the first few hours and quickly exceeds saturation conditions, but within 48 h is depleted. Oxygen release from the encapsulated percarbonate was not measured for about 6 h; release then proceeds quickly to saturation, followed by a gradual decline in release.

Peroxide Release and Toxicity. Figure 8 compares the release of peroxide from the encapsulated and nonencapsulated sodium percarbonate in both distilled water and the salts solution. The nonencapsulated powder quickly releases the peroxide, whereas the encapsulated sodium percarbonate releases it much more gradually. However, in about 4 h the levels of free peroxide are about the same in each treatment.

The toxic effects of the released peroxide were then measured by looking at the kill curves for two hydrocarbon degrading fluorescent pseudomonads (Figure 9). In both cases, the nonencapsulated sodium percarbonate had a more rapid killing effect than the encapsulated form. The B1 organism is somewhat more resistant, but in all cases, the protection from the peroxide is only delayed by a matter of hours by encapsulation.

DISCUSSION

Solutions to subsurface soil contamination problems require a multidisciplinary approach. Geologists and engineers have demonstrated the creation of hydraulic fractures is possible under near-surface conditions. When filled with sand, hydraulic fractures should facilitate the distribution of nutrients by providing permeable conduits, which would increase the radius of influence and decrease the number of wells required to cover a site. It could be possible, for example, to dissect a contaminated region with an array of fractures stacked at vertical spacings of 30 cm, as in Figure 2. This would place every point in the

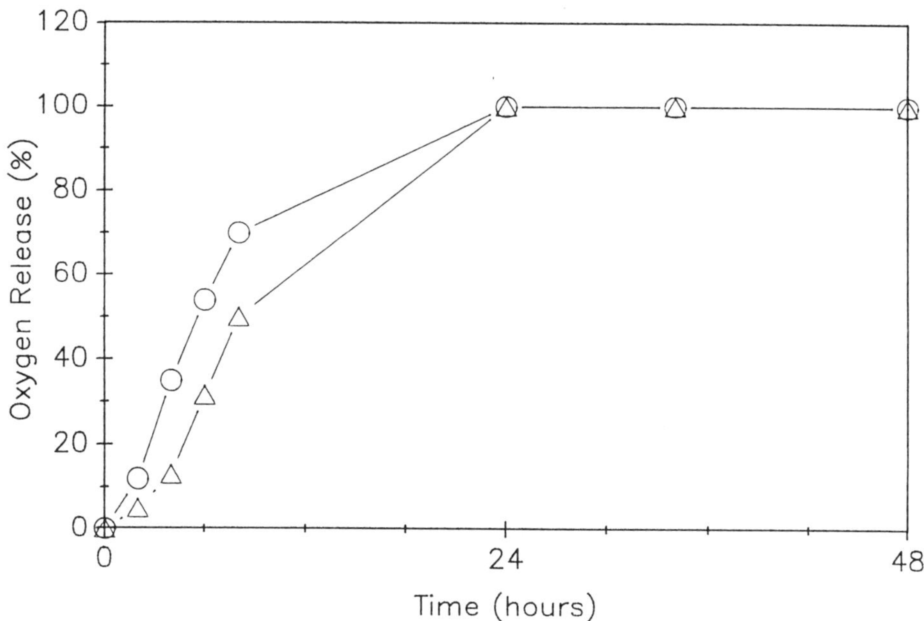

FIGURE 6. Oxygen release from sodium percarbonate in deionized water. To 50 mL of distilled deionized water was added 200 mg of encapsulated (open triangle or nonencapsulated (open circle) sodium percarbonate. The flask was mixed with rapid stirring. The percent of released oxygen was calculated from volume displacement.

contaminated region within 15 cm of a potential source of nutrients and oxygen. That scenario would clearly offer advantages over injection using wells alone, although it should be pointed out that it is an augmentation of current techniques.

Experts in microencapsulation are creating slow-release forms of nutrients and oxygen. The initial attempts here demonstrate the possibilities, but a longer lasting formula that is nontoxic to the microbes must be devised. Creating subsurface reservoirs of solid compounds designed to leach essential constituents at required rates offers an exciting departure from current techniques. It could be possible, for example, to fill hydraulic fractures with the ingredients necessary to sustain bioreclamation so that constant or periodic replenishment through wells is unnecessary. In that case, surface facilities, maintenance, and labor costs are markedly reduced.

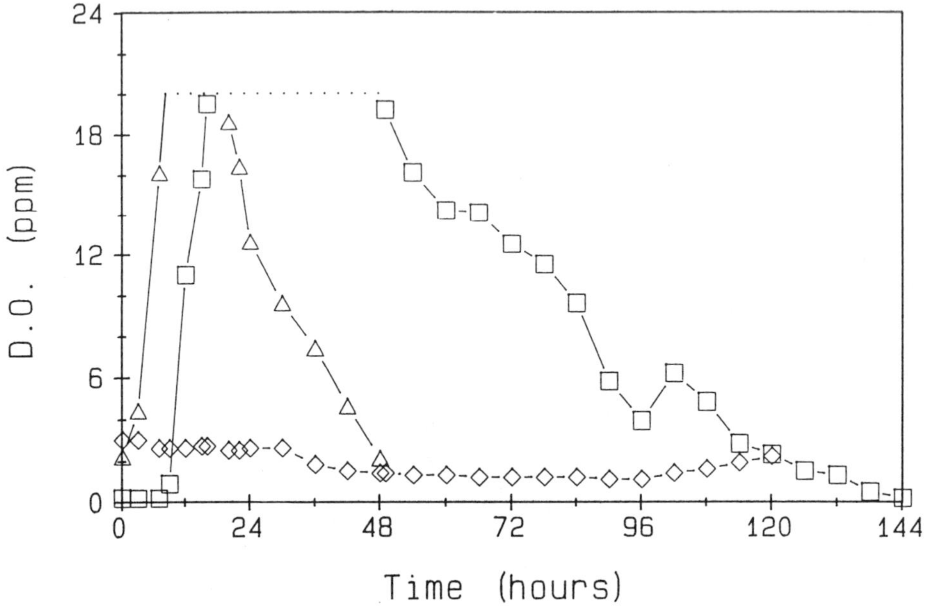

FIGURE 7. Oxygen release from sodium percarbonate in subsurface soil column. Graphs represent sample dissolved oxygen data from control (open diamond), unencapsulated sodium percarbonate (open triangle), and ensapsulated sodium percarbonate (open square). Experiments were carried out at 12 C with water pumping rates of 10 to 12 mL/h.

Microbiologists are devising subsurface remediation strategies for enhancing and evaluating microbiological systems that degrade contaminants. The concentration of oxygen present in subsurface soils is critical to bioremediation. As Tseng and Montville (1990) and Teraguchi *et al.* (1987) have demonstrated, changes in oxygen concentrations have a profound effect on the enzyme activities of microorganisms. This implies that changes in oxygen concentrations can affect rates of biodegradation. In fact, Mille *et al.* (1988) demonstrated this, although only three levels of oxygen concentrations and hydrocarbon degradation in aquatic environments were studied. However, getting adequate concentrations of oxygen and nutrients to the subsurface is very complicated.

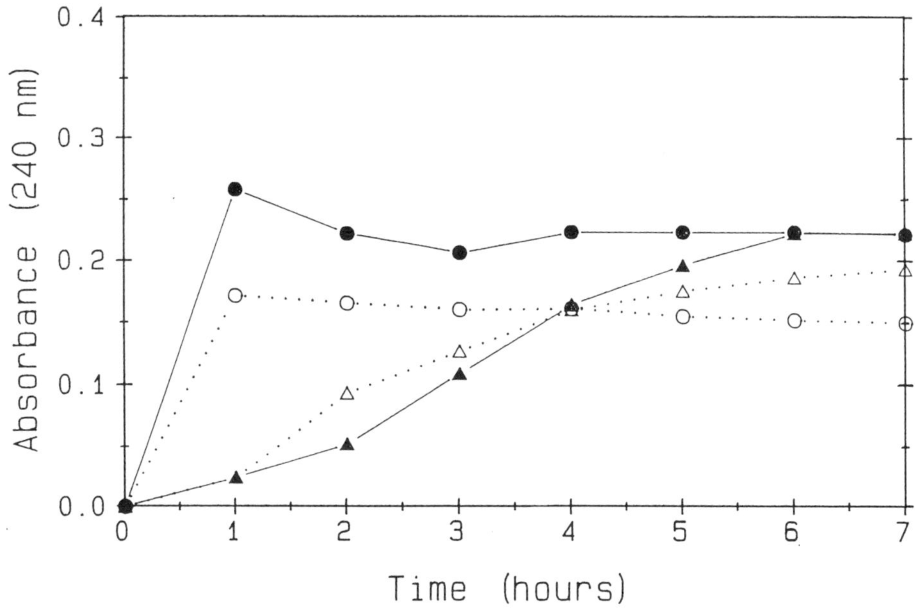

FIGURE 8. Peroxide release studies from sodium percarbonate. Results of peroxide release from unencapsulated sodium percarbonate (open and closed circle) or encapsulated sodium percarbonate (open and closed triangle) into deionized water (open figures) or salts solution (solid figures). Experiments were carried out at 12 C with reciprocated mixing. The release was monitored at 240 nm.

Two problems have arisen when trying to enhance bioremediation with oxygen: what oxygen bearing compound to use, and how to deliver it to the subsurface.

Most investigators attempt to use air sparging or hydrogen peroxide. Air sparging is inefficient and expensive. The air cannot be delivered very far into the soil, and the technique must be delivered continuously. Hydrogen peroxide can be toxic to the bacteria needed for biodegradation, as Wang and Latchaw (1990) demonstrated. We believe that by controlling the release of oxygen from a solid peroxide or similar compound, we can create aerobic conditions without the intensive continuous labor of air sparging or the potentially toxic effects of H_2O_2.

Physically delivering the oxygen to the contaminated soil and bacteria can be solved with the hydraulic fracturing technique. The

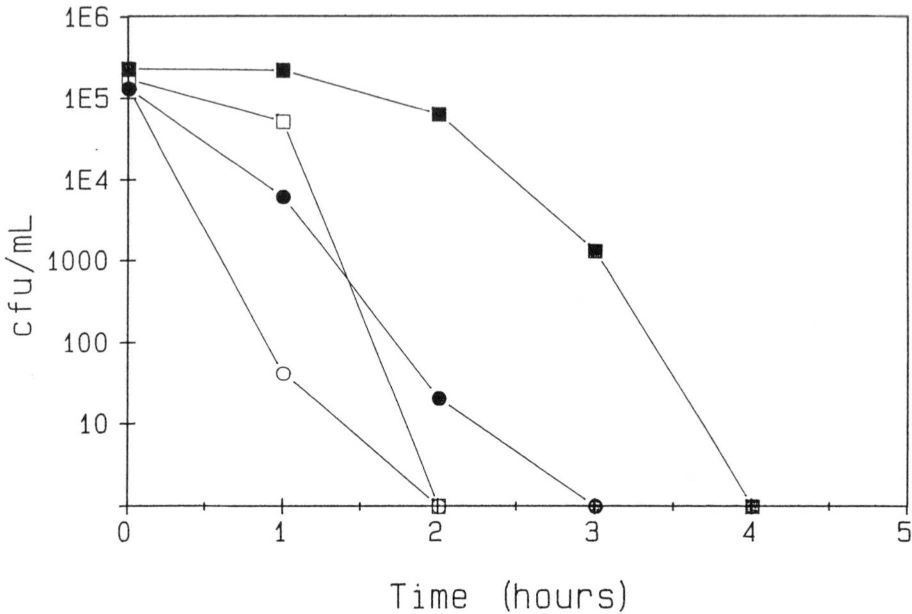

FIGURE 9. Toxicity of peroxide to two hydrocarbon-degrading bacteria. Two fluorescent pseudomonads (A2 - circles and B1 - squares) were tested for their survival in the presence of nonencapsulated sodium percarbonate (open circle and open square) and encapsulated sodium percarbonate (closed circle and closed square). Equal amounts of sodium percarbonate were added to the 10 mL of diluted cultures of each organism. The number of colony forming units (CFU) was determined every hour by plating with a Spiral System Plater onto R2A medium.

feasibility of implementing this technique at a site itself depends on two issues, one technical, one economic. The technical issue centers on whether hydraulic fractures such as the ones shown here can be created at the site. In particular, to be well suited to remediation the fractures should be flat-lying. We know that the orientation of hydraulic fractures depends on the *in situ* state of stress: where lateral exceeds vertical stress flat-lying fractures result, and where vertical exceeds lateral stress steeply-dipping fractures are formed that climb upward and vent. The state of stress depends on various details of the geologic history of the site and cannot be predicted without testing. As a result, at some sites it will be possible to create flat-lying hydraulic fractures; at others it will be possible (with available techniques) to create vertical fractures. Currently, *in situ* testing is required to determine site feasibility.

The cost of creating hydraulic fractures is difficult to assess because we have used the technique as a research tool. In general, costs will involve mobilization of equipment (which resembles equipment used for grouting), labor (a crew of several people), materials (sand is the major expense), and design considerations. Filling the fractures with sophisticated slow-release oxygen and nutrient sources is another cost that will eventually need to be factored in. Compared with the potential benefits and expensive alternatives, we expect that costs of hydraulic fracturing could be attractive.

We are currently investigating bacterial movement in the subsurface soil to determine fracture spacing requirements, and we will continue to pursue slow-release oxygen and nutrient sources. In the near future we plan to demonstrate these techniques at a contaminated site.

Acknowledgments

The field study was facilitated by John Stark and the management of the ELDA landfill. Contributions from Pete Paris, Elizabeth Spencer, and our colleagues at the Center Hill Research Facility have been invaluable.

REFERENCES

Brown, R. A.; Norris, R. A.; Raymond, R. L. "Oxygen Transport in Contaminated Aquifers with Hydrogen Peroxide." *Proceedings*, Petroleum Hydrocarbons and Organic Chemicals in Groundwater—Prevention, Detection and Restoration Conference, Houston, TX, and National Water Well Association, Worthington, OH, 1984.

Cohen G.; Dembiec, D.; Marcus, J. "Catalase Activity in Tissue Extracts". *Analytical Biochemistry* **1970**, *34*, 30–38.

Dang, J. S.; Harvey, D. M.; Jobbagy, A.; Grady, J. "C.P.L." *Research Journal WPCF* **1989**, *61*, 1711–1721.

Elrich, D. E.; Reynolds, W. D.; Tan, K. A. "A New Analysis of the Constant Head Well Permeameter Technique." *Proceedings*, Conference on Validation of Flow and Transport Models for the Unsaturated Zone; Albuquerque, NM, May 23–26, 1988.

Gould, W. D.; Hagedorn, C.; Bardinelli, T. R.; Zablotowicz, R. M. "New Selective Media for Enumeration and Recovery of Fluorescent Pseudomonads from Various Habitats." *Applied and Environmental Microbiology* **1985**, *49*, 28–32.

Hinchee, R. E.; Downey, D. C.; Coleman, E. J. "Enhanced Bioreclamation Soil Venting and Groundwater Extraction: A Cost-Effectiveness and Feasibility

Comparison." *Proceedings,* Petroleum Hydrocarbons and Organic Chemicals in Groundwater—Prevention, Detection and Restoration Conference, Houston, TX, and National Water Well Association, Worthington, OH, 1984.

Mille, G.; Mulyono, M.; el-Jammal, T.; Bertrand, J. C. "Effects of Oxygen on Hydrocarbon Degradation Studies *in vitro* in Surficial Sediments." *Estuarine, Coastal, and Shelf Science* **1988,** *27,* 283–295.

Murdoch, L. C. "A Field Test of Hydraulic Fracturing in Glacial Till." *Proceedings,* USEPA 15th Annual Research Symposium, Cincinnati, OH, 1990, EPA/600/9-90/006.

Murdoch, L. C.; Losonsky, G.; Klich, I.; Cluxton, P. "Hydraulic Fracturing to Increase Fluid Flow." *Proceedings,* Third Annual KFK/TNO Conference on Contaminated Soils, Karlsruhe, FRG, 1990a.

Murdoch, L. C.; Losonsky, G.; Cluxton, P.; Patterson, B.; Klich, I.; Braswell, B. "The Feasibility of Hydraulically Fracturing Soil to Improve Remedial Actions." Final Report, 1990b, U.S. EPA contract #68-03-3379-08 (in press).

Teraguchi, S.; Ono, J.; Kiyoswana, I.; Okonog, S. "Oxygen Uptake Activity and Metabolism of *Streptococcus thermophilus* STH450." *Journal of Dairy Science* **1987,** *70,* 514–523.

Tseng, C-P.; Montville, T. J. "Enzyme Activities Affecting End Products Distribution by *Lactobacillus plantarum* in Response to Changes in pH and O_2." *Applied and Environmental Microbiology* **1990,** *56,* 2761–2763.

Wang, Y-T.; Latchaw, J. L. "Anaerobic Biodegradability and Toxicity of Hydrogen Peroxide Oxidation Products of Phenols." *Research Journal Water Pollution Control Federation* **1990,** *62,* 234–238.

Wilson, J. T.; Leach, L. E.; Henson, M.; Jones, J. N. "*In Situ* Biorestoration as a Groundwater Remediation Technique." *Ground Water Monitoring Review* **1986,** *6,* 56–64.

The Feasibility of Utilizing Hydrogen Peroxide as a Source of Oxygen in Bioremediation

*Scott G. Huling**, *Bert E. Bledsoe*
U.S. Environmental Protection Agency
Mark V. White
Mantech Environmental Technology Inc.

INTRODUCTION

In aerobic respiration, free molecular oxygen accepts electrons from an electron donor, usually carbon, and is reduced to a lower oxidation state. An important aspect of these biochemical redox reactions is their irreversibility; dissolved oxygen is always consumed and never produced as a result of bacterial metabolism (Rose & Long 1988). Oxygen, if not present in adequate concentration, will limit the ability of aerobic microorganisms to degrade contaminants.

Hydrogen peroxide injection into the groundwater has recently become a popular method of introducing oxygen to contaminated, low dissolved oxygen zones. Two moles of decomposed hydrogen peroxide, ideally, yield two moles of water and one mole of oxygen (Equation [1]), thereby introducing oxygen into groundwater. Hydrogen

* U.S. Environmental Protection Agency, Robert S. Kerr Environmental Research Laboratory, P.O. Box 1198, Ada OK 74820

Disclaimer: The research described in this article has been funded by the U.S. Environmental Protection Agency. It has not bee subjected to the Agency's peer and policy review. Therefore, it may not necessarily reflect the view of the Agency and no official endoresment should be inferred.

peroxide decomposition can be characterized by the net result reaction (Schumb *et al.* 1955):

$$2H_2O_2 \longrightarrow 2H_2O + O \tag{1}$$

The stoichiometry of Equation (1) indicates that 47.1 percent by weight of decomposed hydrogen peroxide will be oxygen. The two main mechanisms for hydrogen peroxide decomposition are enzymatic and nonenzymatic reactions. Enzymatic decomposition reactions are catalyzed by hydroperoxidases, catalase and peroxidase (Britton 1985). Catalase, found in almost all aerobic bacteria, is primarily responsible for catalytically decomposing cell-synthesized hydrogen peroxide, thus preventing the accumulation of hydrogen peroxide to a toxic level. Catalase is outstandingly effective in this process, being active at low hydrogen peroxide concentrations and at a rate far exceeding that of most other catalysts (Schumb *et al.* 1955). Significant hydrogen peroxide decomposition observed in an infiltration gallery at an enhanced bioremediation pilot study was largely attributed to the catalase driven reaction (Spain *et al.* 1989).

One reviewer of nonenzymatic iron-decomposition reactions (Schumb *et al.* 1955) indicated that the most notable are those in the presence of iron salts and that the generally accepted mechanism is a series of complex chemical reactions involving hydroxyl and perhydroxyl radical intermediates along with both ferric and ferrous iron. The overall stoichiometry for both catalase and iron decomposition of hydrogen peroxide is equivalent to that described in Equation (1).

Nonenzymatic decomposition of hydrogen peroxide was investigated in laboratory studies to determine the effects of a potassium phosphate "stabilizer" and pH (Britton 1985). The presence of a 0.01-M solution of monobasic potassium phosphate was observed to significantly inhibit hydrogen peroxide decomposition. In the same laboratory study, hydrogen peroxide was used as the main source of oxygen for hydrocarbon-degrading bacteria and plate counts were used as an indicator of microbiological activity. It was reported that the maximum concentration tolerated by a mixed culture of gasoline degraders was 0.05 percent hydrogen peroxide. The tolerance was increased to 0.2 percent by incrementally raising the hydrogen peroxide concentration. Tolerance was determined to occur when the number of colony forming units in the test column were essentially the same for the control column. One important observation made during this study was that nonviable cell material catalyzed the decomposition of hydrogen peroxide as well as viable cell material.

The objectives of this laboratory study were to confirm that hydrogen peroxide can be used to supply oxygen in the bioremediation process, assess the tolerance of the system to hydrogen peroxide, and estimate the overall oxygen demand based on stoichiometric degradation of hydrocarbon. An *in situ* bioremediation pilot study in which hydrogen peroxide and nutrients were injected into contaminated aquifer material provided the opportunity to study laboratory observations in the field.

LABORATORY STUDY

Materials and Methods

A sample of contaminated aquifer material, characterized as a fine to medium-grained sand, was retrieved from a thick, glacial-deposit aquifer in Traverse City, Michigan (Leach *et al.* 1988; Twenter *et al.* 1985). An aseptic, undisturbed sample was retrieved from the heart of an aviation gasoline plume using a modified hollow stem auger drilling tool (Leach *et al.* 1988). Aviation gasoline is the targeted bioremediation substrate in the study. Approximately 476 g of wet, contaminated soil was placed in each column. An abiotic control column was used to investigate the mechanism of hydrogen peroxide decomposition, but not to differentiate between the biotic and abiotic oxygen demand. A schematic of the laboratory apparatus is shown in Figure 1. The glass columns were approximately 18 cm long (4 cm i.d.). Soil columns were kept in a constant temperature chamber at 12 C.

A mixture of feed water (pH 6.5 to 7.0), nutrients, and hydrogen peroxide was pumped using a peristaltic pump through the columns. The columns were operated in a continuous up-flow mode. Feed solutions were mixed in an in-line mixing coil prior to introduction to the columns. Effluent samples from the column were retrieved in-line using a syringe pump. This enabled the retrieval of an aqueous sample in a closed system without losing volatiles and without aerating/deaerating the column effluent. Inverted centrifuge tubes were installed in-line to capture and quantify the gas produced from the column.

Chemical analyses were performed in accordance with EPA methods (U.S. EPA 1979). Fuel carbon analyses were performed by gas chromatography (Vandegrift & Kampbell 1988). Available oxygen (Equation [2]) from both the hydrogen peroxide and the dissolved oxygen (DO) was measured using the Winkler azide modification method (EPA Method No. 360.2).

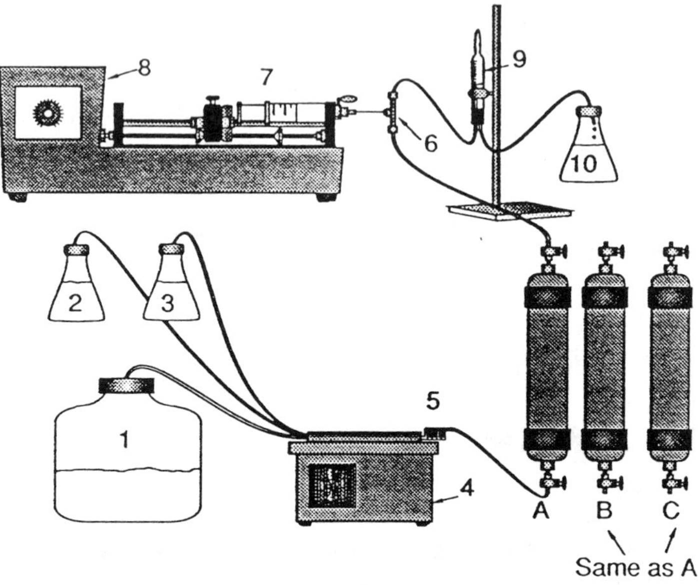

FIGURE 1. Schematic of soil columns. (1) feed water, (2) nutrient solution, (3) hydrogen peroxide solution, (4) peristaltic pump, (5) mixing coil, (6) septa sampling port, (7) glass syringe, (8) syringe pumpl, (9) gas trap, and (10) waste collection.

$$\text{Available oxygen} = [(\text{DO}) + 0.471\ (H_2O_2)] \qquad (2)$$

where: DO = dissolved oxygen, mg/L
H_2O_2 = hydrogen peroxide concentration, mg/L

Hydrogen peroxide analysis was determined using a peroxytitanic acid colorimetric procedure. The columns were pretreated with a phosphate-rich nutrient solution (128.0 mg/L orthophosphate as P) prior to the introduction of hydrogen peroxide. This pretreatment step was a precaution taken to prevent iron decomposition of hydrogen peroxide. The influent phosphate concentration was decreased (89.2 mg/L O-P as P) after 20 h of pretreatment when 98 percent breakthrough of O-P (as P) had occurred. The influent nutrient concentrations to the columns were 400.0 mg/L ammonium chloride, 200.0 mg/L monobasic potassium phosphate, 200.0 mg/L sodium phosphate, and 100.0 mg/L magnesium sulfate. Hydrogen peroxide was introduced at 15 mg/L.

Results

Because of the low oxygen demand observed during the first 8 days of operation, the flow rate was reduced from 80.0 mL/h to 45.0 mL/h. The increased hydraulic residence time in the column resulted in greater oxygen consumption and increased the accuracy in determining the oxygen demand. The oxygen demand exerted on the influent was calculated as follows:

$$\text{Oxygen demand} = [\text{Influent DO} + 0.471(H_2O_2)] - \text{Effluent (Available Oxygen)} \quad (3)$$

where the effluent available oxygen concentration, as determined by the Winkler method, detects both DO and oxygen from hydrogen peroxide. Approximately 2 weeks was required before significant oxygen consumption was observed in all three columns (Figures 2 through 4). This response was interpreted as characteristic of microbial acclimation to a new chemical or physical environment. The concentrations of nitrates and nitrites in the column effluent were consistently equivalent to the background concentration (<1.0 mg/L) in the feed water.

Hydrogen peroxide was increased from 15 mg/L to 30 mg/L to 100 mg/L after the oxygen demand exceeded approximately 80 percent of the available oxygen. Gas was captured in the in-line gas traps soon after hydrogen peroxide was introduced at 100 mg/L, indicating that a loss of oxygen from the system was occurring.

During a period of 15 days following the hydrogen peroxide injection of 100 mg/L, effluent available oxygen (AO) remained constant (AO avg. = 24.6 mg/L, n = 29, st. dev. = 1.25 mg/L) in all three columns. In-line gas traps were used to capture and quantify the gas produced from the columns (Figure 1). The average rate of gas generation during this period from columns B and C was 1.17 mL/h. Gas chromatography indicated that the captured gas was composed of approximately 65 to 70 percent oxygen and 30 to 35 percent nitrogen.

A mass balance was performed on the influent and effluent available oxygen in the system. Mass balance results are included in Table 1. The oxygen mass balance indicates that roughly 44 and 45 percent of the influent available oxygen was recovered in the aqueous and gaseous phases, respectively, for a total recovery of 89 percent. The unrecovered oxygen was consumed by both biotic and abiotic mechanisms. Since an abiotic column was not used, it was not possible to determine the exact mechanism. It is expected that the oxygen demand is primarily biotic. The cumulative total and cumulative adjusted

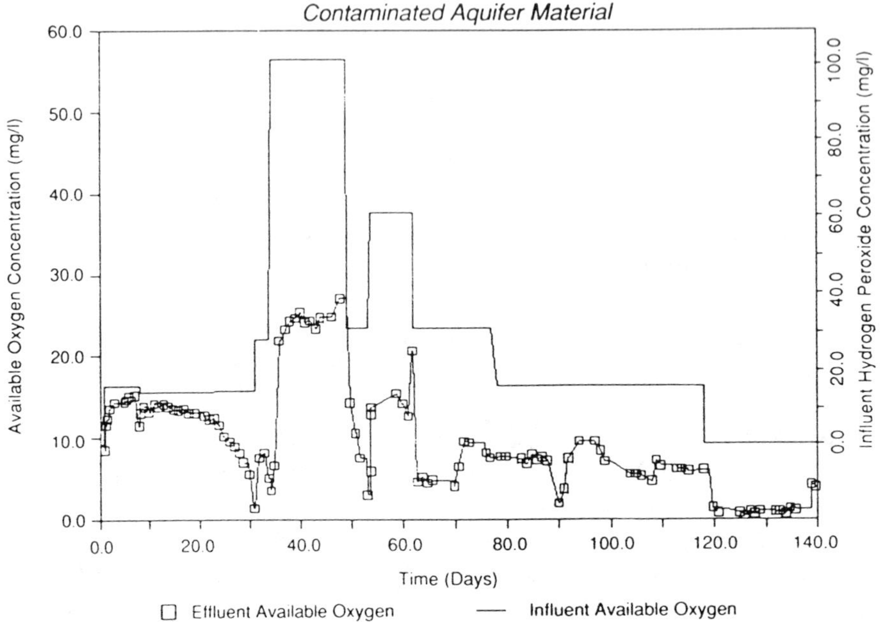

FIGURE 2. Oxygen response curve, column A.

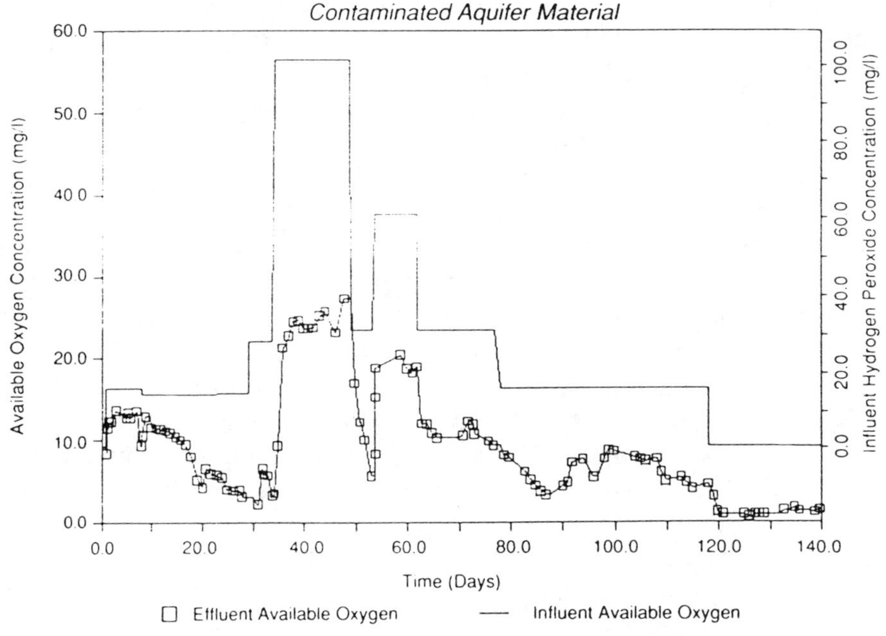

FIGURE 3. Oxygen response curve, column B.

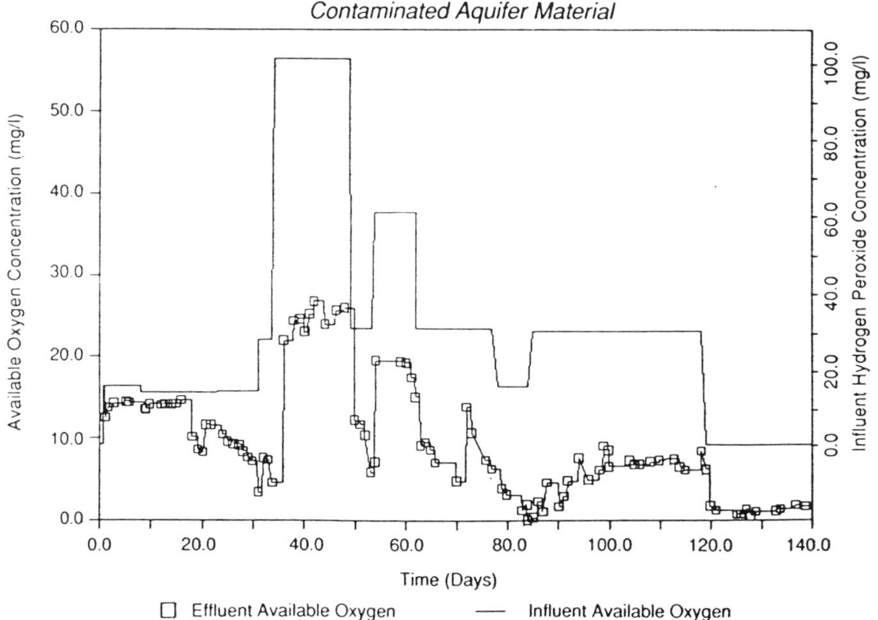

FIGURE 4. Oxygen response curve, column C.

TABLE 1. Column influent and effluent flux of available oxygen[(a),(b)] in both aqueous and gaseous phases.

	Columns		
	A	B	C
Influent			
aqueous	7.78E-5	7.78E-5	7.78E-5
Effluent			
aqueous	3.41E-5 (43.8%)	3.40E-5 (43.7%)	3.42E-5 (44.0%)
gaseous	3.50E-5[(c)] (45.0%)	3.71E-5 (47.7%)	3.29E-5 (42.3%)

(a) Flux rate, moles oxygen/h.
(b) Values in parentheses indicate percent effluent of total influent flux.
(c) Average of columns B and C.

oxygen demand curves for each column are presented in Figures 5 through 7. The cumulative adjusted oxygen demand curve is the difference between the cumulative total oxygen demand and the oxygen lost from the system due to degassing. These curves demonstrate the potential of hydrogen peroxide, at 100 mg/L, to rapidly decompose, resulting in the production of oxygen gas. The slopes of the cumulative adjusted oxygen demand curves (Figures 5 through 7) during the 100 mg/L hydrogen peroxide period are less than the slope at lower hydrogen peroxide concentration periods. Three times less oxygen was consumed during the 100 mg/L period than during the 30 mg/L period. A decrease in the oxygen consumption rate indicated that inhibition of bacterial respiration during this period had occurred.

The average cumulative total oxygen demand from the columns was 1940 ± 127 mg oxygen, and the average cumulative total oxygen demand, adjusted for the oxygen degassing, was 1360 ± 67 mg oxygen. During the various operating scenarios (i.e., varying influent hydrogen peroxide concentration) of this study, degassing accounted for approximately 30 percent of the total oxygen demand.

The slopes of the cumulative adjusted oxygen demand curves in Figures 5 through 7 represent the rate of oxygen consumption. Linear regression analysis ($r = 0.99$) of the cumulative adjusted oxygen demand versus time, after the 100 mg/L hydrogen peroxide injection, yielded values of 11.4, 10.4, and 13.5 mg oxygen/day for columns A,B, and C, respectively. Gas chromatography of the initial and final aquifer material (Vandegrift & Kampbell 1988) and the column effluent was performed to calculate an approximate mass balance of hydrocarbons in the system (Table 2). Based on the mass of aquifer material in each column, the initial and final average fuel carbon (FC) concentrations, and an estimate of the effluent FC, the amount of FC degraded was estimated using Equation (4). Based on the average of all three columns, 36 percent of the initial mass of fuel carbon leached from the aquifer material, 10 percent remained on the aquifer material, and 54 percent degraded.

$$M_D = M_I - M_F - M_E \tag{4}$$

where: M_D = fuel carbon degraded
M_I = fuel carbon initial
M_F = fuel carbon final
ME = fuel carbon in column effluent

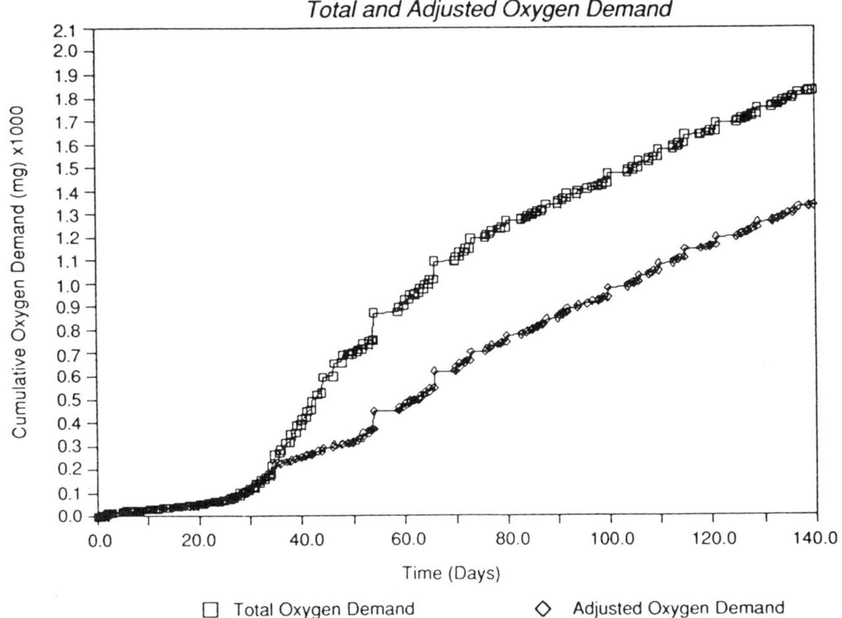

FIGURE 5. Cumulative oxygen demand, column A.

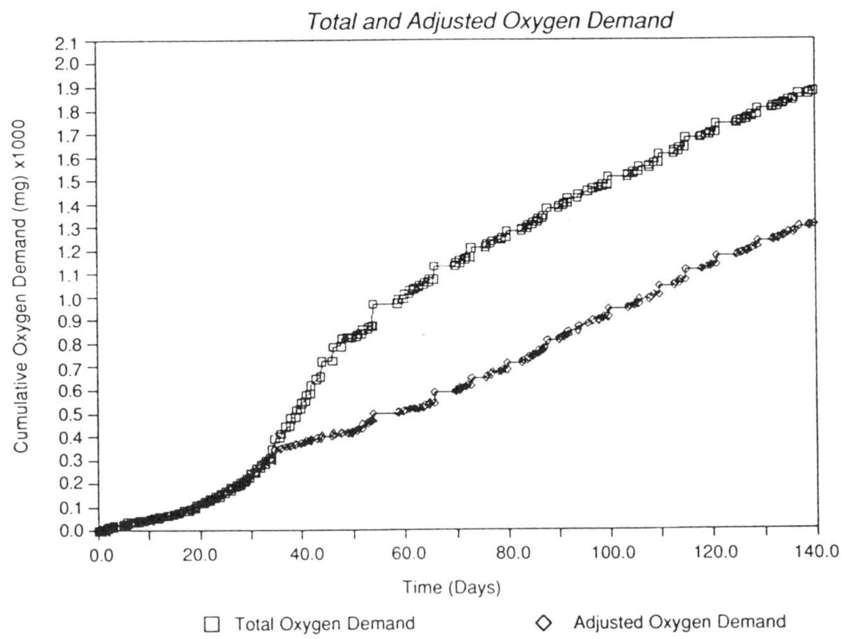

FIGURE 6. Cumulative oxygen demand, column B.

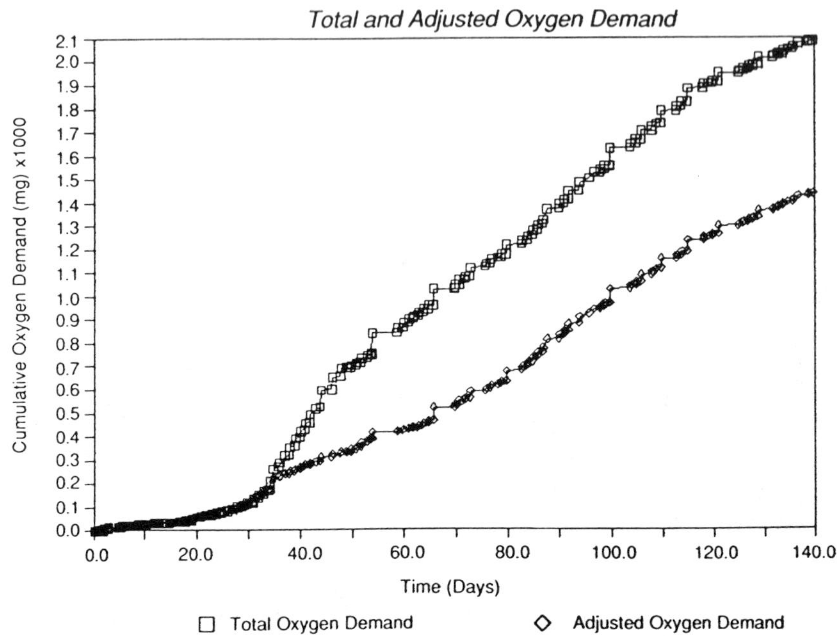

FIGURE 7. Cumulative oxygen demand, column C.

The empirical carbon and hydrogen content of the aviation gasoline was determined to be approximately 2.16 parts hydrogen per part carbon, or roughly 85 percent carbon and 15 percent hydrogen (Powell *et al.* 1988). The ideal stoichiometric biological conversion of hydrocarbon to carbon dioxide and water is approximately 3.48 parts oxygen per part hydrocarbon as described in the following equation:

$$CH_{2.16} + 1.54\ O_2 \longrightarrow CO_2 + 1.08\ H_2O \qquad (5)$$

The ratio of the estimated oxygen consumed to estimated aviation gasoline degraded in this study was greater than the stoichiometric conversion ratio (Table 3). Measured conversion ratios greater than the stoichiometric prediction were likely to occur due to the endogenous respiration, the abiotic oxygen demand, and due to errors in the mass balance analysis. Again, it was not possible to differentiate between the biotic and abiotic oxygen demand in this study.

Colorimetric hydrogen peroxide analysis of the column effluent was performed to determine the persistence of hydrogen peroxide in contact with the contaminated aquifer material. Prior to terminating the experiment, hydrogen peroxide was injected at 50, 100, and 200 mg/L. These concentrations were injected for 7, 10, and 13 h, respectively,

TABLE 2. Hydrocarbon mass balance.

	Columns		
	A	B	C
Aquifer Material Analyses			
Mass (kg)	0.479	0.475	0.474
Initial [FC][a] (mg/kg)	906	906	906
Initial mass FC[b] (mg) (M_I)	434	430	429
Final [FC] soil[c] (mg/kg)	136	60	76
Mass FC final[d] (mg) (M_F)	65	29	36
Column Effluent Analyses			161
Effluent FC aqueous[e] (mg) (M_E)	154	150	
Estimated FC Degraded			
Mass FC degraded[f] (mg) (M_D)	215	251	232

(a) Average of triplicate analyses of contaminated aquifer material from which the column material was derived (790, 1045, and 884 mg/kg), initial fuel carbon concentration in column aquifer material = 906 mg/L.
(b) Mass FC initial = 906 (mg/kg) × soil mass (kg).
(c) Average of replicate analyses for each column of fuel carbon in the final aquifer material.
(d) Mass FC final = Final [FC] mg/kg × soil mass (kg).
(e) Estimated FC in column effluent.
(f) Mass FC biodegraded, $M_I - M_F - M_E$.

prior to collecting effluent samples. Breakthrough of hydrogen peroxide, for all three concentrations, was less than 11 percent (Table 4). The introduction of phosphate as a nutrient into the columns was expected to minimize the nonenzymatic catalysis of hydrogen peroxide. Therefore, the rapid decomposition of hydrogen peroxide in this system appeared to be due to enzymatic catalysis.

An abiotic column experiment was initiated to investigate the mechanism of hydrogen peroxide decomposition. Contaminated aquifer material was selected from the same stock sample used in the biotic column study and was subjected to 500 C for 1.0 h. Since viable and

TABLE 3. Conversion ratios (O_2 (mg)/aviation gasoline (mg)).

	Columns		
	A	B	C
Mass O_2 consumed[(a)] (mg)	1322	1307	1434
Mass aviation gasoline[(b)] degraded (mg)	253	295	274
Conversion ratio (mg O_2/mg av. gasoline)	5.23	4.43	5.23

(a) Adjusted cumulative oxygen demand, Figures 5 through 7.
(b) Mass aviation gasoline degraded = mass FC degraded/0.85.

TABLE 4. Hydrogen peroxide breakthrough.

	Columns								
	A	B	C	A	B	C	A	B	C
$[H_2O_2]i$ (mg/L)		50			100			200	
$[H_2O_2]e$ (mg/L)	ND	ND	ND	9.6	10.4	2.5	16.5	17.0	9.3
H_2O_2 Break-through (%)	<10	<10	<10	9.6	10.4	2.5	8.3	8.5	4.7

$[H_2O_2]i;[H_2O_2]e$ – influent and effluent hydrogen peroxide concentration, respectively

Detection limits:

$[H_2O_2]_{50}$ = 5.0 mg/L, $[H_2O_2]$100,200 = 2.5 mg/L

ND = not detected.

nonviable bacterial cells contain catalase that is capable of catalyzing the decomposition of hydrogen peroxide, thermal treatment was necessary to remove the biological material from the sample. The aquifer material was placed in a column similar to the biotic column study, and nutrients were introduced at the same concentration and flow rate as in the biotic column study. Hydrogen peroxide was introduced at 100 mg/L. Roughly 3.1 mL of gas were captured in the in-line gas trap after 5 days of operation. The volume of gas produced during this test was too small to analyze and was significantly less than the gas produced during a comparable time frame with biological material present, i.e., 140 mL. While the thermal treatment is expected to have changed the oxidative state of inorganic ions in the aquifer material, this data conditionally indicates that the hydrogen peroxide decomposition is dominated by biological (enzymatic) catalysis.

FIELD STUDY

Background

Twenty years ago, aviation gasoline spilled into a shallow, sandy, water table aquifer at the U.S. Coast Guard Station in Traverse City, Michigan. Extensive coring at this site indicated that the majority of the contamination was distributed within a narrow interval between 4.6 and 5.2 m (15 and 17 ft) below the land surface (Wilson *et al.* 1989), corresponding to the seasonal water table fluctuation.

In 1988, the U.S. Coast Guard and the U.S. EPA began the operation of a pilot-scale *in situ* bioremediation project in the area of the original spill (Solar 1989; TGI 1988, 1989). A series of five deep wells (I1–I5) were used to inject clean water beneath the plume area in an effort to raise the water table and subsequently saturate the contaminated "smear zone" (Figure 8). Raising the water table allowed delivery of soluble nutrients to the targeted zones of contamination. Five chemical feed wells (CF1-CF5) were used to inject nutrients and hydrogen peroxide in the shallow, contaminated layer (Figures 8 and 9). A series of downgradient monitoring wells and subsurface sampling lines were installed to monitor groundwater quality. The sampling lines consist of several vertical sampling ports (0.3 m [1-ft] stainless steel screens) at 0.3 m (1-ft) intervals.

Nutrient and oxygen enriched water was injected beginning in March 1988 at a total flow rate of approximately 11 gpm in the five chemical feed wells. The nutrients in the injected water contained approximately 380 mg/L ammonium chloride, 190 mg/L disodium phosphate, and 190 mg/L potassium phosphate. Oxygen and hydrogen peroxide were injected according to the schedule given in Table 5. Pretreatment with phosphate was also initiated to complex the iron found in the aquifer material in an effort to minimize the iron-catalyzed hydrogen peroxide decomposition reaction.

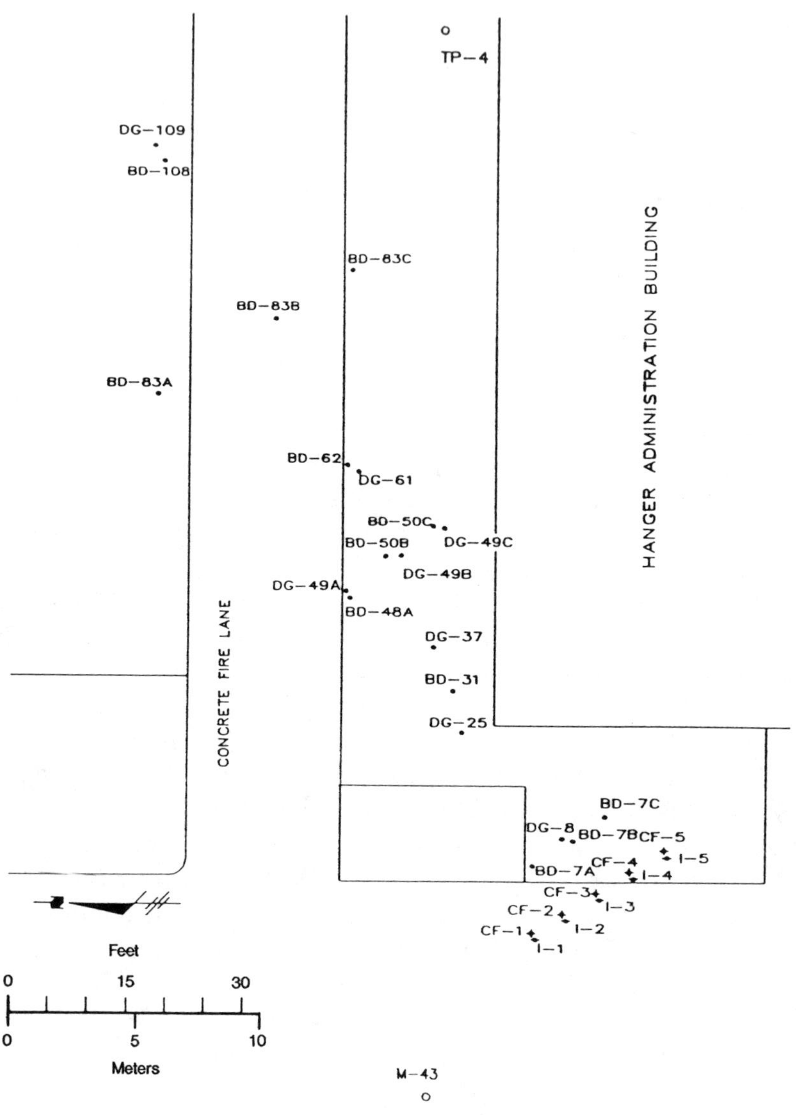

FIGURE 8. U.S. Coast Guard Station, Traverse City MI. Pilot-scale aerobic biodegradation project, monitoring and injection wells.

Both soil gas and groundwater analyses were used to investigate the fate of hydrogen peroxide in the *in situ* bioremediation field-scale pilot study. The laboratory study results indicated that hydrogen

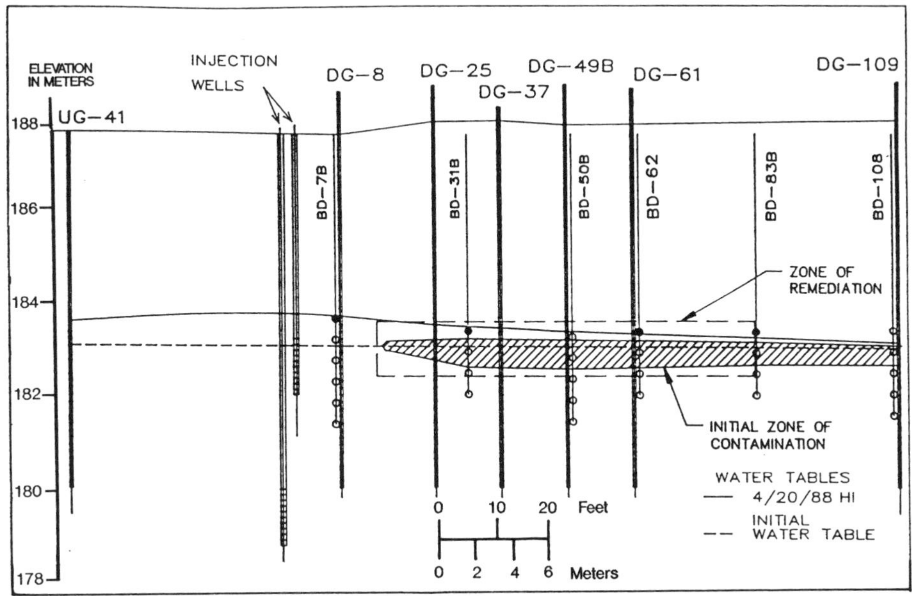

FIGURE 9. Cross section of pilot study wells.

TABLE 5. Oxygen/hydrogen peroxide injection schedule.

Date	Time (days)	Cumulative Time (days)	Oxygen Source	Concentration
3/2/88	90	90	Liquid oxygen	40[(a)]
6/2/88	5	95	Hydrogen peroxide	50
6/7/88	7	102	Hydrogen peroxide	110
6/14/88	64	166	Hydrogen peroxide	250
8/17/88	106	272	Hydrogen peroxide	500
12/3/88	179[(b)]	451	Hydrogen peroxide	750

(a) Concentration as dissolved oxygen, mg/L.
(b) 179 days as of 5/31/89.

peroxide injected into the subsurface will decompose rapidly, resulting in the liberation of oxygen gas into the unsaturated zone. The oxygen content of the soil gas in the injection area was measured to evaluate this hypothesis.

Materials and Methods

Soil gas samples were obtained using a series of 1-cm i.d. stainless steel tubes (3/8 in.) could be coupled together and driven into the subsurface to various depths. Soil gas was pumped with a hand-held positive displacement pump into a sample vessel containing an oxygen detector. The oxygen detector is a GasTech Model LO2 OxyTechTor galvanic cell that measures the concentration of oxygen from 0 to 100 percent ± 5 percent of the reading. Oxygen was also measured in the headspace in groundwater monitoring wells by lowering the detector to various levels within the well.

Groundwater analyses for available oxygen performed throughout the pilot study (Solar 1989; TGI 1988, 1989) offered the opportunity to evaluate the performance of the bioremediation system. Available oxygen was measured using the Winkler azide modification method (EPA Method No. 360.2),

Results

During the field investigation (August 1989), hydrogen peroxide had been injected (11 gpm) into groundwater at 750 mg/L for 179 days. The water table in the injection area was approximately 4.1 to 4.4 m (13.5 to 14.5 ft) below grade. Numerous soil gas samples were retrieved and analyzed for the concentration of oxygen. These data were used to map the horizontal distribution of oxygen as a function of depth (0.9, 1.8, 2.7–3.0 m [3, 6, 9–10 ft]), as shown in Figures 10 through 12. The concentration of oxygen in the unsaturated zone, as indicated in these figures, increased with depth and was clearly greater than both atmospheric and background soil oxygen concentrations, 20.9 percent and 20.7 percent, respectively.

The concentration of oxygen in the headspace of nearby monitoring wells was obtained by lowering the oxygen detector into the monitoring wells. Although the concentration of oxygen in the monitoring wells may not be representative of the concentration of oxygen in the unsaturated zone, elevated oxygen concentrations, as high as 78 percent oxygen, demonstrate the rapid rate of hydrogen peroxide decomposition. The same general trend occurred in the wells as in the soil gas; the oxygen concentration increased with depth in the injection area.

The concentration of available oxygen in groundwater downgradient from the injection zone clearly indicated that oxygen was being

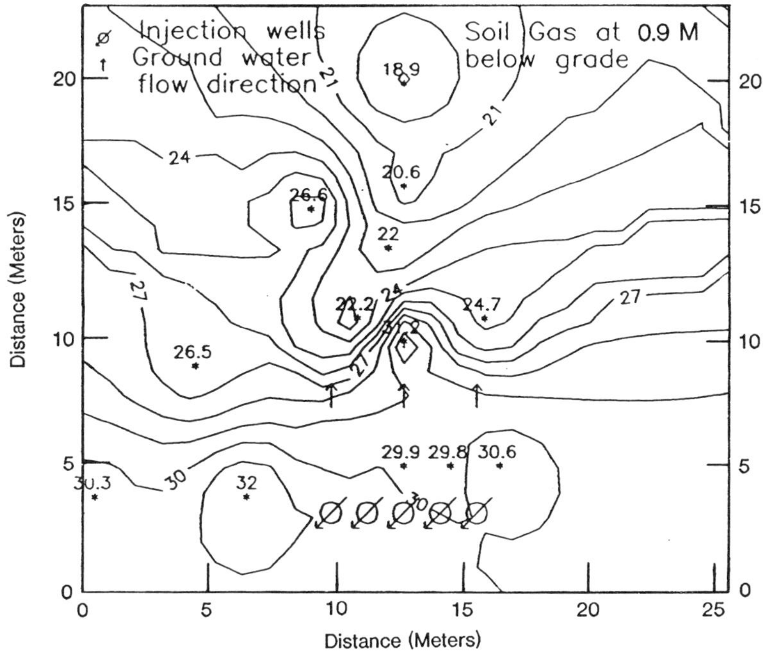

FIGURE 10. Percent oxygen in the unsaturated zone, *in situ* bioremediation pilot study, aviation gasoline fuel spill (0.9 m [3 ft]) Traverse City MI (8/89).

delivered to the saturated zone. Available oxygen was detected in sampling ports located between 2.1 to 33 m (7 to 108 ft) from the injection point. The concentration value ranged from greater than 250 mg/L at 2.1 m (7 ft) to values greater than 35 mg/L at 25 and 33 m (83 and 108 ft). It was difficult to calculate the rate of hydrogen peroxide decomposition or the fraction of available oxygen that was liberated into the unsaturated zone. This was largely due to the variability of the influent hydrogen peroxide concentration as a function of time.

CONCLUSIONS

Hydrogen peroxide was shown to decompose rapidly and produce oxygen. Due to precautions taken to minimize nonenzymatic decomposition in this study, the data indicate that hydrogen peroxide decomposition to oxygen and water was primarily the result of enzymatic catalysts. Oxygen provided by the hydrogen peroxide decomposition

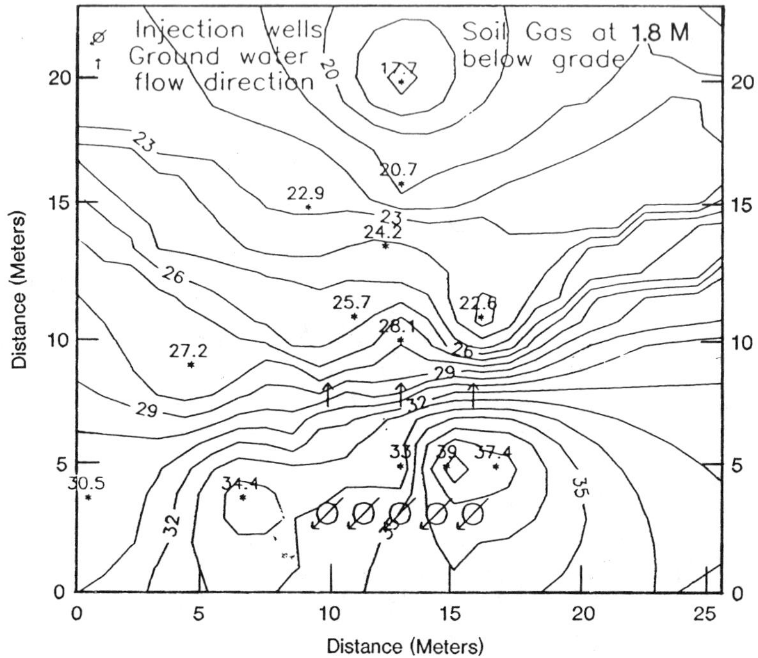

FIGURE 11. Percent oxygen in the unsaturated zone, *in situ* bioremediation pilot study, aviation gasoline fuel spill (1.8 m [6 ft]), Traverse City MI (8/89).

reaction was consumed in the columns. It was not possible to distinguish between abiotic and biotic oxygen demand in this study.

The injection of hydrogen peroxide at 100 mg/L had two notable effects. Firstly, the rapid rate of hydrogen peroxide decomposition and the subsequent liberation of oxygen gas occurred at a rate faster than could be utilized biologically, although the solubility of dissolved oxygen had not been exceeded. Approximately 45 percent of the available oxygen injected into the columns was transferred to the gaseous phase. Secondly, the rate of oxygen consumption decreased, indicating that bacterial inhibition had occurred.

Mass balances of oxygen and hydrocarbons were calculated to quantify the mass of oxygen consumed and the mass of hydrocarbon degraded. The ratio of estimated oxygen consumed to aviation gasoline degraded was found to be 25 to 50 percent greater than the stoichiometric prediction.

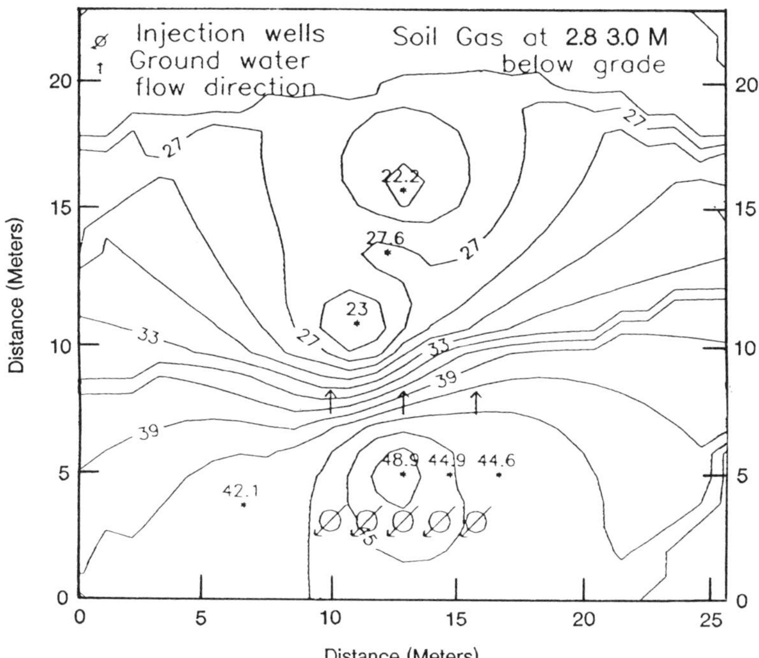

FIGURE 12. Percent oxygen in the unsaturated zone, *in situ* bioremediation pilot study, aviation gasoline fuel spill (2.8 to 3.0 m [9 to 10 ft]), Traverse City MI (8/89).

Hydrogen peroxide, introduced into the biologically active columns at 50, 100, and 200 mg/L, at approximately the same groundwater flow velocity as in the pilot study, did not exceed 11 percent breakthrough, although the columns were 18 cm in length.

Injecting hydrogen peroxide into the aquifer at the pilot study area resulted in (1) increasing the concentration of available oxygen in downgradient wells, (2) rapid decomposition of hydrogen peroxide, and (3) the liberation of oxygen gas into the unsaturated zone. The concentration of oxygen in soil gas in the injection area was found to be much greater than background. The rate of hydrogen peroxide decomposition at the site was unknown, but was expected to be rapid because of the elevated concentration of oxygen gas measured in the pilot study area.

REFERENCES

Britton, L. N.; Texas Research Institute. *Feasibility Studies on the Use of Hydrogen Peroxide to Enhance Microbial Degradation of Gasoline*; Publication No. 4389, API: Washington, DC, 1985.

Leach, L. E.; Beck, F. P.; Wilson, J. T.; Kampbell, D. H. "Aseptic Subsurface Sampling Techniques for Hollow-Stem Auger Drilling." *Proceedings*, 2nd National Outdoor Action Conference on Aquifer Restoration, May 23–26, 1988, NWWA, Dublin, OH; Vol. 1, pp 31–51.

Powell, R. M.; Kampbell, D. H.; Bledsoe, B. E.; Callaway, R. W.; Michalowski, J. T.; Vandegrift, S. A.; White, M. V.; Wilson, J. T. "Comparison of Methods to Determine Bioremediation Oxygen Demand of a Fuel Contaminated Aquifer." *International J. of Environmental and Analytical Chemistry* **1988**, *34*, 253–266.

Rose, S.; Long A. "Monitoring Dissolved Oxygen in Ground Water: Some Basic Considerations." *Ground Water Monitoring Review* **1988**, *8*, 93–97.

Schumb, W. C.; Satterfield, C. N.; Wentworth, R. L. *Hydrogen Peroxide*; Reinhold Publishing Corp.: New York, 1955.

Solar Universal Technologies, Inc. Presented at Coordination Meeting, Traverse City Bioremediation Projects Hydrogen Peroxide Biodegradation Project (BIO I), Oklahoma City, OK, Feb. 3, 1989.

Spain, J. C.; Milligan, J. D.; Downey, D. C.; Slaughter, J. K. "Excessive Bacterial Decomposition Of H_2O_2 during Enhanced Biodegradation." *Ground Water* **1989**, 27(2).

TGI (Traverse Group, Inc.). Quarterly Report, USCG/EPA Robert S. Kerr Environmental Research Laboratory, Pilot Scale Biodegradation Project, USCG Air Station, Traverse City, MI, March, June, Sept., Dec. 1988, March, June 1989.

Twenter, F. R.; Cummings, T. R.; Grannemann, N. G. *Ground-Water Contamination in East Bay Township, Michigan*; USGS Water Resources Investigation Report 85-4064; U.S. Geological Survey, 1985.

U.S. EPA (Environmental Protection Agency). *Methods for Chemical Analysis of Water and Waste*; 1979.

Vandegrift, S. A.; Kampbell, D. H. "Gas Chromatographic Determination of Aviation Gasoline and JP-4 Jet Fuel in Subsurface Core Samples." *J. of Chromatographic Sci.* **1988**, *26*, 566–569.

Wilson, J. T.; Leach, L. E.; Michalowski, J.; Vandergrift, S.; Calloway, R. *In-Situ Bioremediation of Spills from Underground Storage Tanks: New Approaches For Site Characterization, Project Design, and Evaluation of Performance*; U.S. Environmental Protection Agency, July 1989; EPA/600/2-89/042.

Effectiveness and Kinetics of Hydrogen Peroxide and Nitrate-Enhanced Biodegradation of Hydrocarbons

E. R. Barenschee[*]*, P. Bochem, O. Helmling, P. Weppen*
Degussa Corporation

INTRODUCTION

Techniques are rapidly developing for *in situ* biodegradation processes of contaminations in lower water bearing formations. It is well known that underground biodegradation is limited by the lack of electron acceptors and nutrients.

Hydrogen peroxide and nitrate are the most commonly applied electron acceptors to enhance *in situ* bioremediation of contaminated soils (Barenschee & Bochem 1990; Battermann & Werner 1984; Downey *et al.* 1988; Huling *et al.* 1990; Hutchins *et al.* 1991; Raymond *et al.* 1986; Riss & Schweisfurth 1985).

This work had two purposes. The first was to compare the effects and rates of degradation processes with hydrogen peroxide and nitrates as electron acceptors despite of differences in type of metabolism. The second was to develop a bioreactor system that allows consideration of the effects of mass transportation and evaluation of the overall kinetics for biodegradation processes in soils.

[*] Degussa AG, Department IC-ATAO, P.O. Box 1345, D-6450 Hanau 11, Federal Republic of Germany

MATERIALS AND METHODS

Experiment 1: Comparison of Hydrogen Peroxide and Nitrate as Electron Acceptors

Laboratory trials were conducted with columns filled with artificially contaminated sand.

Fifty kilograms of washed sand with a mean diameter of 0.6 mm was mixed in a mixing machine with 1 kg of a fine coarse soil sample from a contaminated site and 250 g diesel fuel for 1 h. Samples were taken from different parts of the mixture and analyzed to make sure that the contamination was homogenous. Due to losses of high volatile compounds, the remaining concentration of hydrocarbons was 3,800 mg/kg. The addition of the soil sample from a contaminated site was done to achieve proper starting conditions for the bacterial activity.

Colony counts of water and soil samples of heterotrophic bacteria were determined on TGE media agar plates (Merck) and the denitrifying bacteria with the MPN technique according to Riss (1988) (Experiment 1).

In Experiment 2 heterotrophic bacteria were determined with the spiral plater technique on standard I media (Merck). The hydrocarbon oxidizers were counted on agar plates with a mineral medium and vacuum gas oil as the only C-source.

Esterase activity in soil and water samples were analyzed according to Obst and Holzapfel-Pschorn (1988) with fluorescinedi-acetate (FDA).

Nine milliliters of buffer solution (Sörensen) at pH 7 was added to 1 g soil and 40 µl FDA solution (20 mg FDA in 10 mL acetone) and mixed. After 20 h incubation at 20 C, the samples were centrifuged (10 min at 10,000 rpm) and measured at 490 nm against a blank (sterilized soil with FDA in buffer solution). Soil samples of approximately 10 g were taken every week at the end of each column.

The analyses of nitrate, nitrite, phosphate, and ammonia were made according to *German Standard Methods* (1989).

Hydrogen peroxide was analyzed by titration with Cer(IV) sulfate. Dissolved CO_2 was measured by an CO_2-sensitive electrode (Ingold) and titration with NaOH from pH 8.4 to 4.5.

Two soil columns were continously operated with a bottom-to-top flow pattern in parallel. This simulated water-saturated conditions and allowed assessment of mass balances of nutrient consumption, electron acceptors, and the formation of degradation products. Table 1 shows the monitoring program. The parameters were recorded at the inlet and outlet of both columns. Table 2 shows the initial conditions.

TABLE 1. Monitoring program.

	Liquid Phase	Solid Phase (Soil)
Daily	O_2	
	H_2O_2	
	pH	
2 - 3 Days	CO_2	
	NO_3^-	
	NO_2^-	
	NH_4^+	
	PO_4^{3-}	
	TIC	
	TOC	
Weekly	CFUs (HC Oxidizer)	CFUs (HC Oxidizer)
	HC	Esterase-Activity
	Esterase-Activity	

Electron acceptors and nutrients were added to tap water with a natural nitrate concentration of 19 ppm. One column was operated with increasing concentrations of hydrogen peroxide in the range of 50 to 250 mg/L. Table 3 shows the time course of hydrogen peroxide dosing. The other was operated with nitrate at a concentration of 70 mg/L (50 mg/L added to the tap water). The water was purged with nitrogen to lower concentrations of oxygen below 0.9 mg/L.

The columns were operated for a period of 345 days. At the end of this time, the columns were opened and the contents separated into five layers and analyzed.

TABLE 2. Initial conditions.

Reactor	Experiment 1 H_2O_2	Experiment 1 Nitrate	Experiment 2 1	Experiment 2 2	Experiment 2 3	Experiment 2 4
Reactor Size						
Diameter [cm]	8			8		
Packing Height [cm]	50.0		30.0	22.5	15.0	7.5
Soil Weight [kg]	4.00		3.00	2.25	1.50	0.75
Contamination						
Diesel Fuel [mg/kg]	3800 each			3410 each		
Flow Rate [mL/h]	50			50		
Nutrients						
NH_4^+ [mg/L]	2 - 5			5		
PO_4^{3-} [mg/L]	2			5		
Nitrate [mg/L]		70				
H_2O_2 [mg/L]	50 - 250			50 - 250		

TABLE 3. Time course H_2O_2 dosing.

Experiment 1 Day	H_2O_2 [mg/L]	Flow [L]	O_2 [mg]	Experiment 2 Day	H_2O_2 [mg/L]	Flow [L]	O_2 [mg]
0 - 20	-	35	0	0 - 15	-	18	0
21 - 27	50	7	164	13 - 21	20	6	56
28 - 34	100	7	329	22 - 29	70	8.4	276
35 - 111	150	83	5851	30 - 37	150	8.4	592
112 - 194	210	125	12337	37 - 314	250	332.4	39060
195 - 345	250	210	24675				
Σ from H_2O_2			43360	Σ from H_2O_2			39980
+ O_2 from Diss. O_2 (Inlet)			- 4140	+ O_2 from Diss. O_2 (Inlet)			- 3770
- O_2 from Diss. O_2 (Outlet)			- 2000	- O_2 from Diss. O_2 (Outlet)			- 50
	ΣΣ		45500			ΣΣ	43700

The total concentrations of the hydrocarbons were analyzed with IR spectrometry according to German Standard Methods H 18 as a summarized parameter. Fifty grams of wet soil sample and 20 g Na_2SO_4

were given into closed bottles and extracted three times with freon (1,1,2-trichlorotriflouroethane) at room temperature for 2 h. One half of the combined organic phases were filtered over Al_2O_3 to remove polar components. The hydrocarbon concentrations were determined in both samples with the IR spectrophotometer. The difference between Al_2O_3 and untreated samples gave information about the amount of polar degradation products. The concentrations of hydrocarbons were calculated as mg/kg dry matter.

Analyses of remaining metabolites in soil were made with alkaline and acid extraction of soil samples, derivating with diazomethane and analysis with GC-FID. The analyses were made by Battelle Europe, Frankfurt, Germany.

Experiment 2: Determination of Kinetic Data

To evaluate kinetic data, a system of soil columns with different heights was constructed. The system consisted of two installations of four different column heights, operated in parallel. These installations were operated and monitored similar to the first experiment. Figure 1 shows the experimental setup.

The diesel fuel concentration was set on 3,410 mg/L. Hydrogen peroxide was used as an electron acceptor with the concentration ranging from 50 to 250 mg/L. The timetable of hydrogen peroxide dosing is given in Table 3.

By operating columns of four different heights under the same conditions, it was possible to get information about time-dependent changes of the monitored parameters at different sections of the reactor. This allowed for development of kinetic data.

Running two identical columns of each height allowed for analysis of the remaining hydrocarbons at different times in the same reactor section. This data yielded information not only about the degradation rates, but also about the degree of mineralization that can be achieved at the point of oxygen breakthrough.

The data on dissolved carbon dioxide in the effluent compared with the consumption of oxygen and the disappearance of hydrocarbons was used for determination of the extent of mineralization. Operating the system in the described manner also allowed us to assess the velocity of the moving degradation zone along the column length.

Sterile controls were operated parallel to the microbial active columns to determine the fraction of hydrocarbons that disappeared by

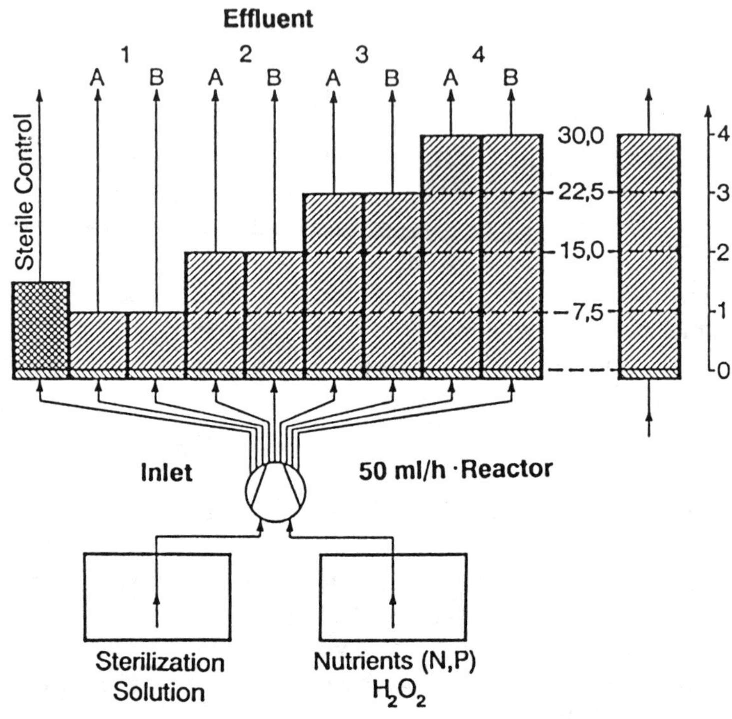

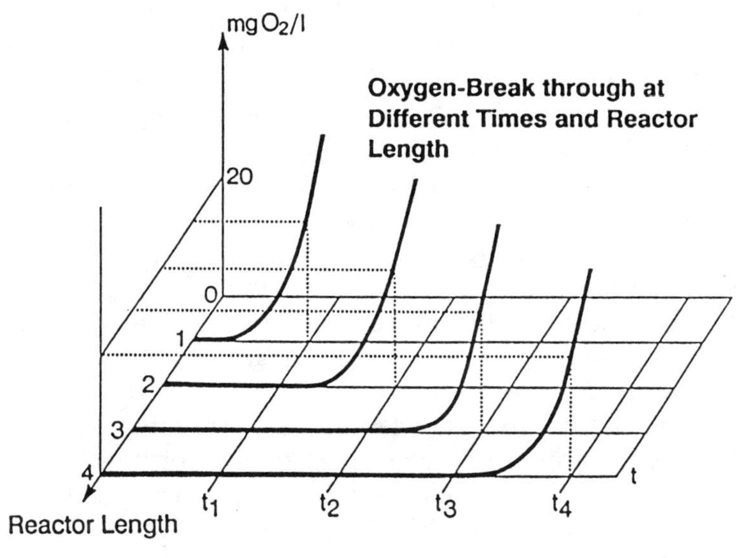

Experimental Set-up **Model**

FIGURE 1. Experimental setup.

nonbiological processes (i.e., leaching). The sterile control was poisoned with 0.1N $AgNO_3$ solution.

RESULTS

Experiment 1: Comparison H_2O_2 and Nitrate

As the experiment progressed, differences between the parameters of the H_2O_2- and nitrate-treated columns were observed.

Mass Balances of Electron Acceptors and Products

H_2O_2-Treated Column. The H_2O_2 concentration was stepwise increased from 50 to 250 mg/L (25 to 125 mg/L O_2), but over a period of 230 days no dissolved oxygen or H_2O_2 could be observed in the effluent. The oxygen breakthrough on the 230th day indicated that oxygen was no longer the limiting factor for the degradation process. The majority of the contaminants should have been degraded at that time—mass balances are given in Table 4. Figure 2 shows the oxygen concentration in the effluent and the change of the oxygen consumption of that column.

At the same time as the oxygen breakthrough, the CO_2 formation rate—which during the first 230 days was very high and at a constant level of 60 to 70 ppm—slowed down over the next 100 days.

The pH value, as well, decreased during the first 230 days from 7.8 to 6.0 from the formation of acidic products like CO_2 and other metabolites. When oxygen breakthrough occurred, the pH value slowly increased to 6.5 (data not shown).

Nitrate-Treated Column. The nitrate-treated column exhibited totally different mass balances, as shown in Table 4 and Figure 3.

Nitrate consumption was very low, and the difference between inlet and outlet was in the range of 5 to 10 mg/L. Mean value of consumption was 9 mg/L. In addition, nitrite production was very low except during a short period around the 60th day. At that time, the nitrate consumption increased to 30 ppm, corresponding with a sharp peak of nitrite formation.

Nearly the same results were found for CO_2 formation. The CO_2 concentration was on a low and constant level in the range of 5 to 10 mg/L.

TABLE 4. Mass balances for Experiment 1 (comparison H_2O_2 and nitrate).

Reactor	H_2O_2		Nitrate	
Soil Weight (Start) [kg]	4.00		4.00	
Soil Weight (End) [kg]	3.00		3.05	
Run Time [d]	345		345	
Hydrocarbons [mg]		%		%
Start (4.0 kg)	15540	100.0	15540	100.0
End (3.0 kg)	2450	15.7	9590	61.8
Leaching	1190	7.7	1190	7.7
Sampling (1.0 kg)	820	5.3	3150	20.3
Biodegradation	11080	71.3	1610	10.4
Without Consideration of Soil Sampling				
Start (3.0 kg)	11670	100.0	11690	100.0
End (3.0 kg)	2450	21.0	9590	82.0
Leaching	1190	10.2	1190	10.2
Biodegradation	8030	68.8	910	7.8
Electronacceptors [mg]				
Oxygen Consumption	45500			
Nitrate Consumption (as NO_3^-)			4300	
Max. Electronacceptor Consumption [mg/h]	6.4		0.52	
$\left(\frac{\text{Electronacceptor}}{\text{Hydrocarbon}}\right)$	4.1 - 5.7		2.7 - 4.7	

Colony Counts. After a few weeks, the colony forming units (CFU) of the heterotrophic bacteria of the H_2O_2-treated column increased and by the end of the experiment were tenfold to 100-fold higher than those of the nitrate-treated column.

In that column, the CFUs in the effluent increased rapidly to 10^7 CFUs during the first 70 days and then gradually decreased to 5×10^5 over the next 100 days. The graph of the CFUs in the nitrate column effluent looked very similar, but on a lower level (data not shown).

A noticeable difference between the two columns was observed with regard to changes in CFUs in soil over time. In the H_2O_2-treated

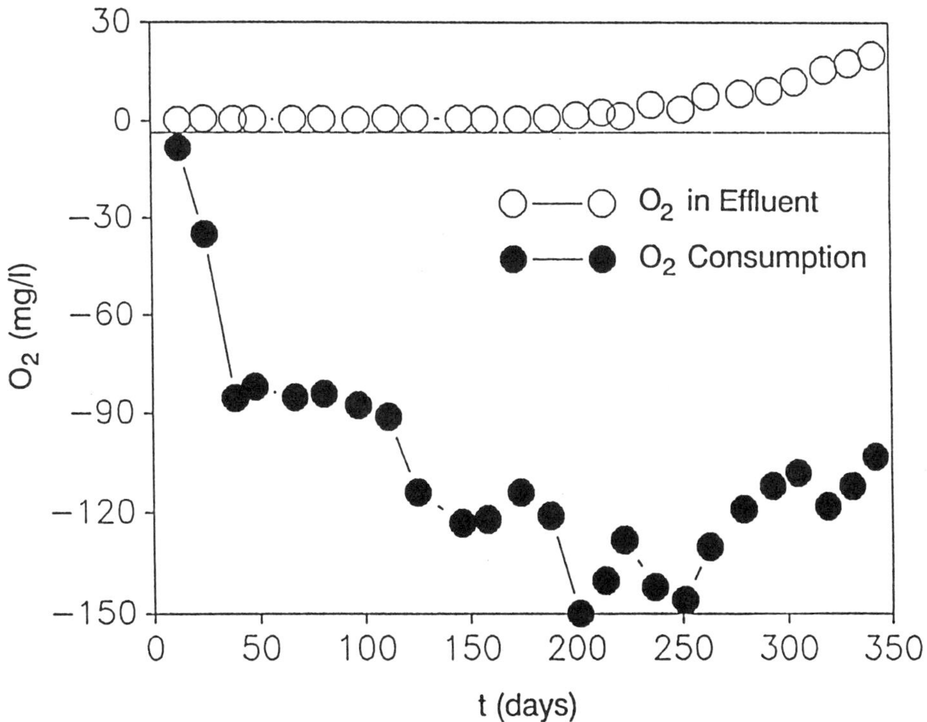

FIGURE 2. Oxygen consumption in H_2O_2-treated column.

column the CFUs improved strongly with time up to 10^8/g, whereas in the nitrate-treated column the CFUs remained at 5 × 105 to 10^6/g until the 250th day. From that peak the CFUs decreased. Figure 4 shows the CFUs in soil.

Esterase Activity. The esterase activity, used as an indicator for generally heterotrophic activity, also showed noticeable differences when comparing H_2O_2 and nitrate.

The esterase activity of the effluent of the H_2O_2-treated column advanced during the first half of the experimental time and reached a peak around the 200th day. In the second period, the esterase activity slowly decreased. However, the activity in soil improved strongly during the entire period.

Compared with the activity of the effluent and the soil, the nitrate-treated columns were totally different. No activity increase was observed in the effluent; rather, activity remained near the detection level. The soil exhibited only poor activity growth over the entire

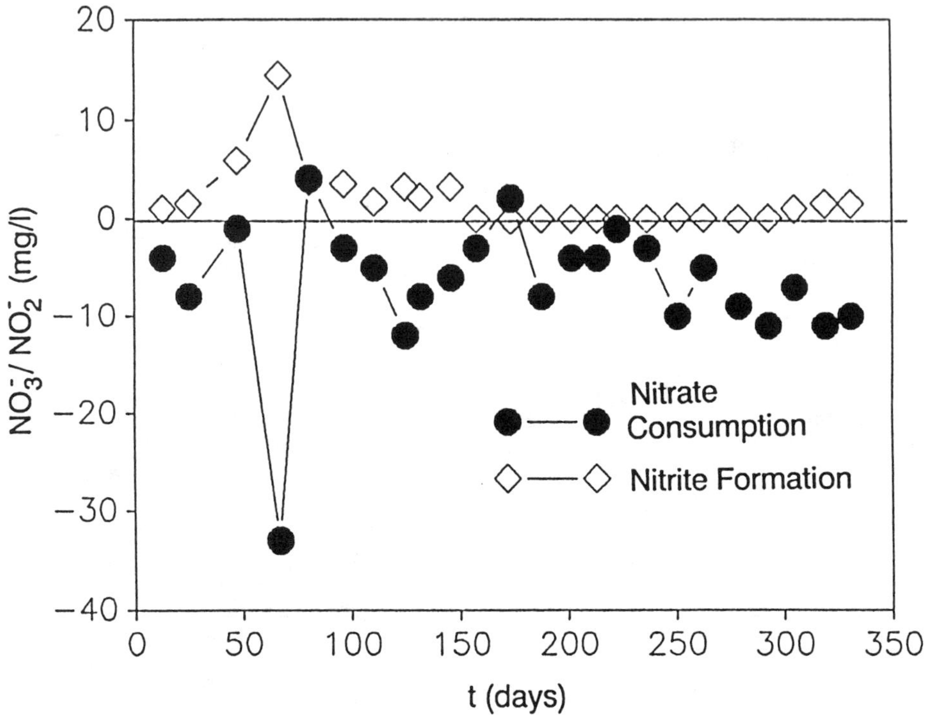

FIGURE 3. Nitrate consumption and nitrite formation in nitrate-treated column.

period. Figure 5 shows the esterase activity in soil. The activity growth of the H_2O_2-treated column was four times higher.

Compared to the H_2O_2-treated column, all monitored parameters of the nitrate column indicated a lower level of microbial degradation of diesel fuel. Not only were colony counts and esterase activity in the effluent and in the soil lower in the nitrate-treated column, but also the conversion rates of the electron acceptors and the formation rates of CO_2 remained lower.

Residual Hydrocarbons. The same tendencies were shown by the analysis data of the remaining concentrations of hydrocarbons.

At the end of the experiment, both columns were opened and the contents separated into five layers. Figure 6 compares the remaining concentrations of hydrocarbons of both columns. (See also Table 4.)

The H_2O_2-treated column exhibited no observable concentration gradient between inlet and outlet. The degree of hydrocarbon removal

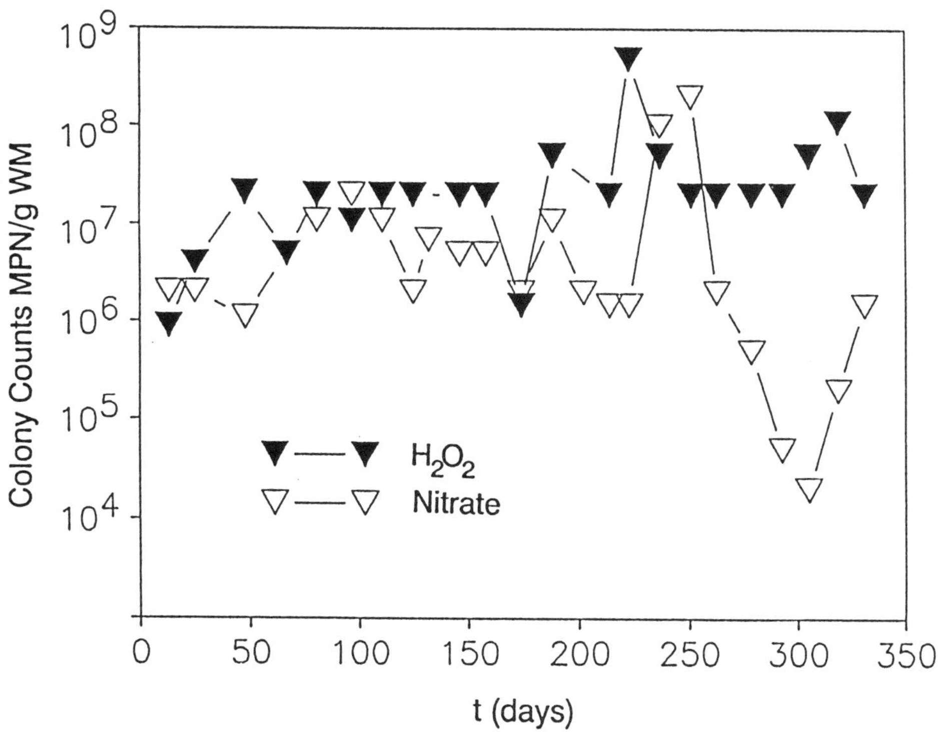

FIGURE 4. Heterotrophic bacterial counts in soil.

was constant and in the range of 80 to 85 percent. However, the esterase activity showed a low gradient: the activity behind the inlet was slightly lower than in the middle or at the end of the column length.

In contrast, the column treated with nitrate showed a strong gradient from inlet to outlet, not only for the remaining hydrocarbons but also for esterase activity. The degree of hydrocarbon removal was in the range of 35 percent at the inlet and decreased to 10 to 15 percent along the column length.

The gradient of esterase activity mirrored that of hydrocarbon activity: the highest activity was at the column inlet. These findings are in correlation with the other time-dependent results described earlier.

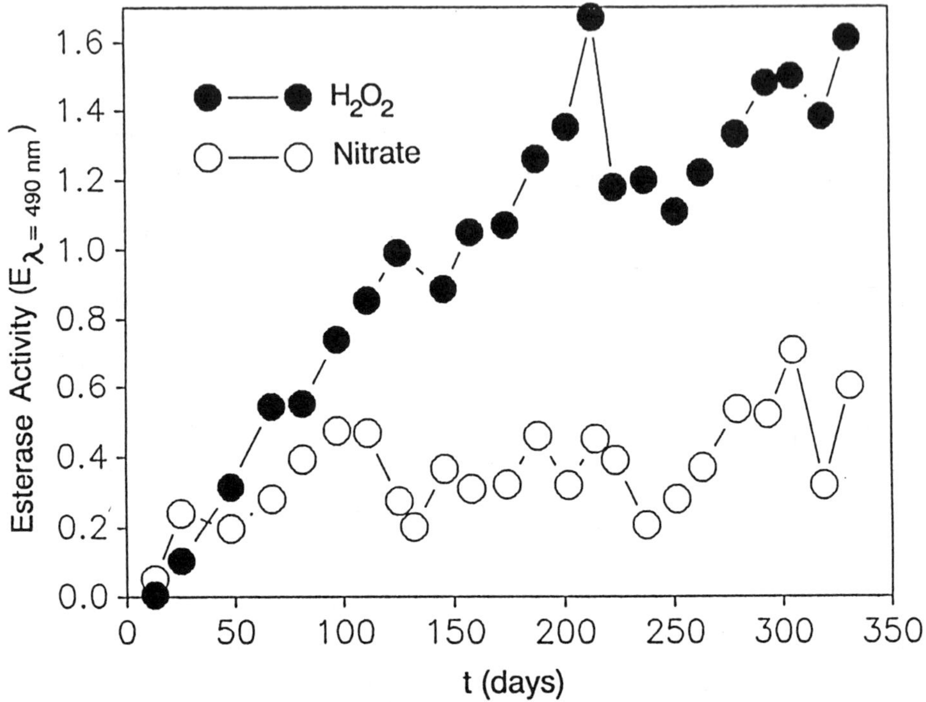

FIGURE 5. Esterase activity in soil.

Experiment 2: Determination of Kinetic Data

Independent methods were utilized to correlate and assess changes in the monitored parameters with respect to time and the conversion rates of all reactants. Taking into consideration stoichiometric aspects, determination of molar- and mass-related rates of consumption and/or production of oxygen, hydrocarbons, and carbon dioxide was possible.

The simplified stoichiometry for the conversion of saturated hydrocarbons with hydrogen peroxide without consideration of the buildup of biomass is

$$(CH_2)_x + \frac{3x}{2}\ O_2 \longrightarrow x\ CO_2 + x\ H_2O$$

$$2\ H_2O_2 \longrightarrow 2\ H_2O + O_2$$

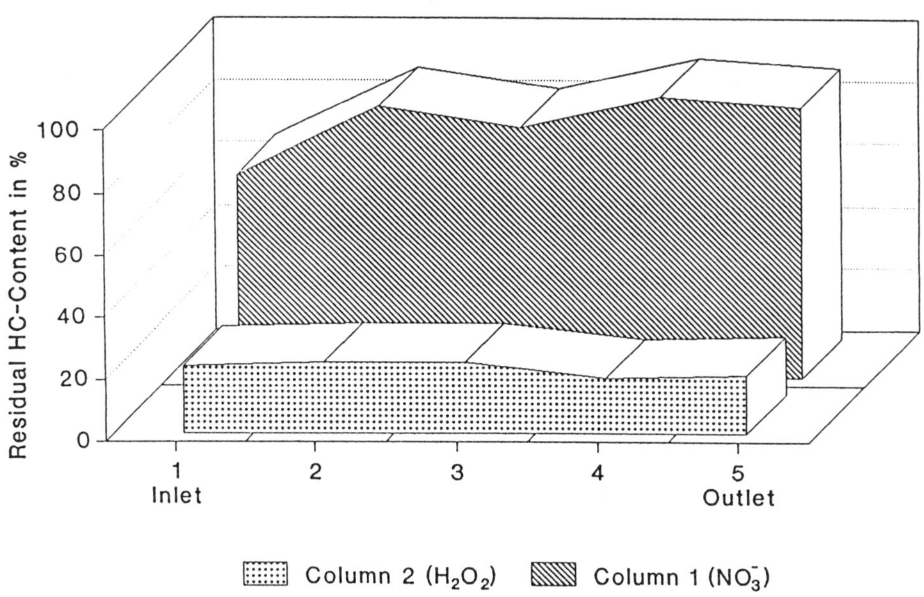

FIGURE 6. Residual hydrocarbon content.

This equation showed the maximum oxygen demand for hydrocarbon oxidation.

The buildup of biomass was neglected because the design of these experiments allowed no determination of these data. With consideration of buildup of biomass, the specific oxygen demand is lower.

Table 5 lists the results of the mass balances and conversion rates; Table 6 lists the specific ratios of consumption and formation of the reaction products.

The degradation of hydrocarbons in a fixed-bed reactor showed three distinct kinetic relevant stages.

The first stage was under oxygen-limited conditions and was characterized by an excess of hydrocarbons. With the constant oxygen feeding rate, a linear dependency between the conversion of hydrocarbons and time was observed.

The second was limited in both oxygen and carbon source. This stage was characterized by a complex kinetic and a short time of duration. The kinetic in this stage was not further investigated.

The final stage was limited in carbon source or hydrocarbons and was characterized by an excess of oxygen. Again, a linear dependency between the conversion of hydrocarbons and time was observed, but with a much lower slope than in stage 1.

TABLE 5. Mass balances for Experiment 2 (kinetic studies with H_2O_2).

Reactor	1		2		3		4	
Packing Height [cm]	30.0		22.5		15.0		7.5	
Soil Weight [kg] (Start)	3.00		2.25		1.50		0.75	
Soil Weight [kg] (End)	2.57		1.78		1.28		0.58	
Run Time Until Oxygen Break Through [d]	314		259		174		97	
Hydrocarbons [mg]		%		%		%		%
Start (3.41 g/kg)	10230	100.0	7670	100.0	5120	100.0	2560	100.0
End	2730	26.7	2410	31.4	1400	27.3	610	23.8
Losses Through Leaching	1080	10.6	880	11.5	550	10.7	280	10.9
Losses Through Soil Sampling	1010	9.9	970	12.6	510	10.0	410	16.0
Biodegraded	5410	52.9	3410	44.5	2660	52.0	1260	49.2
CO_2 Ø [mg/d]	53.4		50.8		44.2		35.6	
CO_2 [mg]	16770		13160		7700		3450	
Hydrocarbons Calculated from CO_2	5410		4250		2480		1110	
Oxygen Consumption [mg]	43700		35300		22300		10500	

TABLE 6. Ratios from mass balances, Experiment 2.

Ratios	Method	Reactor 1	Reactor 2	Reactor 3	Reactor 4
$\left(\frac{O_2}{HC}\right)$	a) HC-Analysis	8.1	10.3	8.4	8.3
	b) O_2 -Zone Migration	7.7			
$\left(\frac{O_2}{CO_2}\right)$	$\frac{O_2 \text{ Consumption}}{CO_2 \text{ Formation}}$	2.61	2.68	2.90	3.04
$\left(\frac{CO_2}{HC}\right)$	HC-Analysis (with Consideration of Leaching)	3.10	3.86	2.89	2.74
$\left(\frac{HC}{HC}\right)$	$\frac{CO_2 \text{ Formation}}{\text{HC-Analysis}}$ (with Consideration of Leaching)	1.00	1.25	0.93	0.88
$\left(\frac{\Delta O_2}{\Delta CO_2}\right)$	$\frac{\text{Decrease in } O_2 \text{ Consumption}}{\text{Decrease in } CO_2 \text{ Formation}}$	2.6			

(In the printed table, the values 7.7 and 2.6 are centred across all four reactor columns.)

Oxygen-Limited Stage. In Figure 7, the dependency of the oxygen content at different reactor lengths was plotted against time. The velocity of the moving degradation front was calculated from oxygen breakthrough at the end of each reactor section. Up to that time, all available oxygen was consumed in the reactor and the degradation was oxygen limited. Table 7 shows the relationship between the oxygen concentration in the effluent of both reactor installations and time. Results showed a linear relationship between the time of the oxygen breakthrough and reactor length, indicating a constant conversion rate in the moving degradation front and therefore a linear relationship between the hydrocarbon concentration on soil with time. (See also Table 5.)

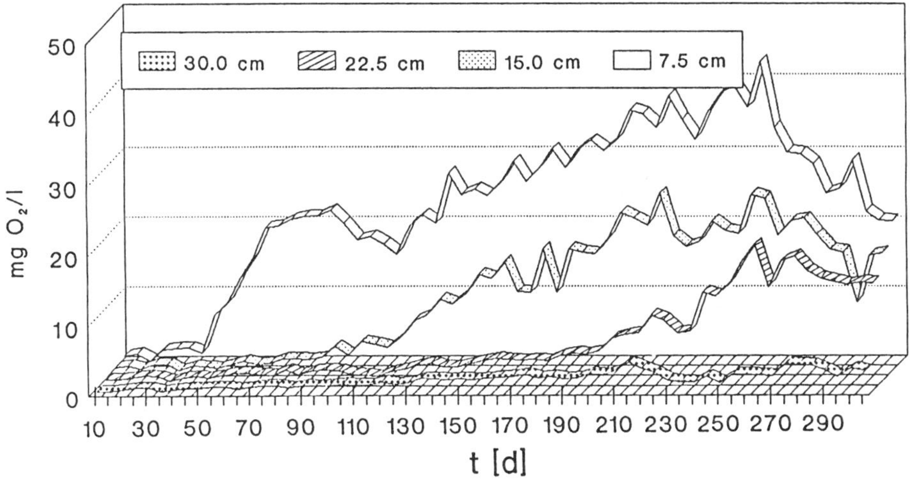

FIGURE 7. Oxygen content in effluent.

In this case, the hydrocarbon concentration was in high excess. Oxygen was the limiting factor and was available at a constant rate, the feeding rate. The degradation rate was characterized by a zero-order reaction for the hydrocarbon concentration.

$$-\frac{dc_{HC}}{dt} = k' \qquad k' = k'' \bullet \dot{m}_{O_2}$$

The disappearance of hydrocarbons was linear with time and was proportional to the oxygen feeding rate.

TABLE 7. Oxygen breakthrough as f (time).

Reactor		1		2		3		4	
Installation		a	b	a	b	a	b	a	b
Packing Height [cm]		29		20		12		6	
Oxygen Break Through [d]									
at	O_2 > 3 mg/L	205	185	182	(160)	100	97	41	51
	O_2 > 5 mg/L	280	196	200	205	126	137	50	55
	O_2 > 10 mg/L	314	210	208	217	135	157	61	63

The mass related proportional factor k″ was found to be in the range of 0.09 to 0.13 and, from the stoichiometry, a factor of 0.29 could have been expected.

The ratio of oxygen consumption to hydrocarbon degradation rate O_2/HC was found to be 2.3 to 3.0 times higher than stoichiometric calculations.

The same results were obtained by comparing oxygen consumption and CO_2 formation rates in this stage and by the decrease in O_2 consumption and CO_2 formation ratio.

In every case, oxygen consumption was approximately two to three times higher than the disappearance of hydrocarbons or the formation of CO_2 (see also Table 6).

The difference between experimental and theoretical results cannot with certainty be attributed to oxygen losses from H_2O_2 decomposition and effects of biodegradation because the design of the experimental setup did not focus on measurement of oxygen losses through degassing. Works by Hinchee *et al.* (1988 and 1991) indicated that oxygen losses through degassing appear when the oxygen consumption rate is lower than the decomposition rate of hydrogen peroxide caused by microbial catalase.

A similar result was obtained in work by Cole *et al.* (1974) when H_2O_2 was used as the sole oxygen source in biological wastewater treatment plants.

The ratio of CO_2 formation and hydrocarbon disappearance was an indicator for the degree of mineralization. Taking into consideration a 10 to 15 percent decrease by leaching, the ratio of CO_2 formation and hydrocarbon disappearance was approximately 3:1, the theoretical value. That indicated that the degree of mineralization was approximately 100 percent. No polar degradation products were observed on soil and in the effluent with the exception of CO_2. This was determined by GC (soil samples) and IR spectrometry (water samples).

One of the two identically operated columns was thoroughly analyzed to determine the residual hydrocarbon content at the moment of oxygen breakthrough. All sections of the reactor indicated levels near 30 percent. This further showed that only 70 percent of the contamination removal was oxygen limited. The degradation of the residual 30 percent was hydrocarbon limited and resulted primarily from low availability for the microorganisms or low degradability.

Transitional Stage. The transitional stage was short in duration. This was the period of time when not all of the available oxygen was consumed in the reactor, and the oxygen concentration in the effluent increased.

The changes in oxygen consumption rate could be calculated from the slope of the oxygen increase or the decrease of the CO_2 formation rate. The changes of degradation rates also indicated the thickness of the moving degradation front.

Further experiments are necessary to evaluate the dependency between the oxygen feeding rate, an one hand, and the velocity and thickness of the moving degradation front, on the other.

Figure 8 shows the colony counts of hydrocarbon-oxidizing microorganisms on the soil surface for each reactor section.

The initial colony count was 10^7/g dry matter and increased to >10^8/g dry matter. A maximum of colony forming units was noted, which collapsed at the time of oxygen breakthrough. After having passed through the oxygen-enriched zone, the microbial counts decreased by 0.5 to 1.0 order of magnitude. Obviously, the highest microbial activity occurred during the transition phase from oxygen to hydrocarbon limitation.

Hydrocarbon-Limited Stage. The hydrocarbon-limited stage was reached when the majority of the contaminants were degraded. Operating one of the two parallel installations for the entire period of the experiment allowed determination of the hydrocarbon conversion rate under conditions of excess oxygen. During this period, oxygen concentration was no longer the rate-limiting factor; rather, the conversion rate depended on the concentration or availability of hydrocarbons.

In Figure 9 the conversion of hydrocarbons is plotted against reactor length. One graph (rectangles) shows the conversion of

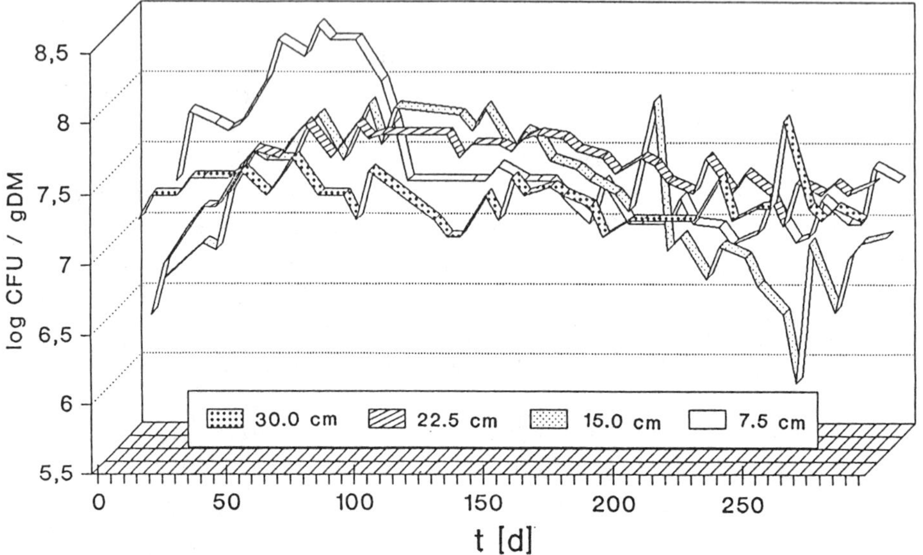

FIGURE 8. Hydrocarbon-oxidizing bacteria in soil.

hydrocarbons at different sections of the reactor at the time of oxygen breakthrough. In this case, the conversion was constant over the whole reactor length.

The other graph (crosses) shows the conversion of hydrocarbons at the same sections, but after 314 days (the end of the experiment). It illustrates the fact that the increase in conversion is slowed under hydrocarbon-limited conditions. For example, the conversion at 60 mm was approximately 70 percent at the time of oxygen breakthrough after 60 days, but approximately 250 more days were needed to increase the conversion from 70 to 80 percent under conditions of oxygen in excess.

Again a linear dependency between disappearance of hydrocarbons and time was found, but with a lower slope than in the first stage. In this case with an excess of oxygen, the degradation rate no longer depended on the oxygen feed rate. The reaction was characterized by a zero-order for oxygen.

GC analyses have shown a shift in the composition of the hydrocarbon mixture during the biodegradation process. The residual hydrocarbons were mainly long chains with very low water solubility.

During the biodegradation process, hydrocarbons were available at a low and constant rate probably because of their low water

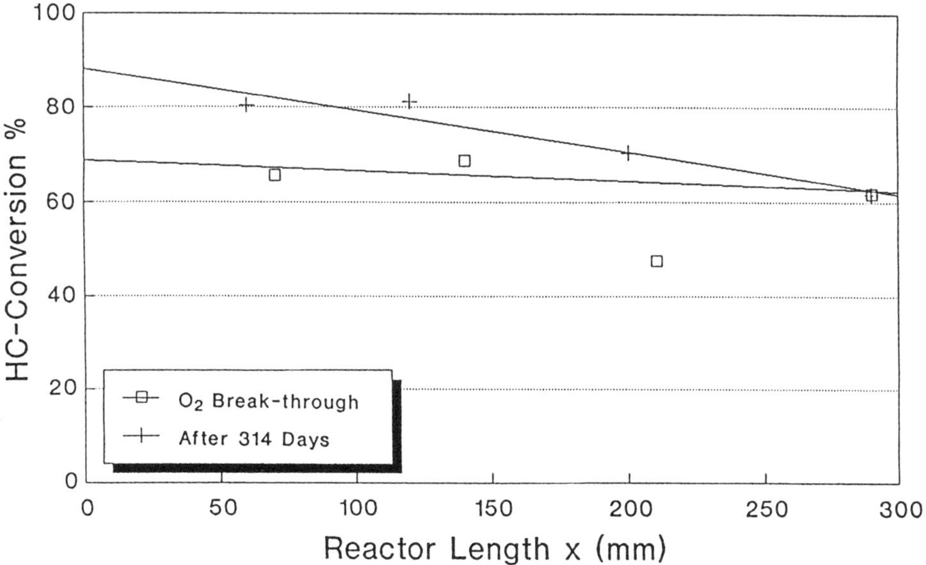

FIGURE 9. Hydrocarbon conversion as f(x).

solubility. In this stage, the rate of solution seemed to be rate limiting. The degradation rate was found to be

$$-\frac{dc_{HC}}{dt} = -\frac{1}{\mu} \bullet \frac{dc_{O_2}}{dt} = k$$

$$k = 1.21 \bullet 10^{-3} \text{ (gHC/kg DM} \bullet \text{ d)}$$

Figure 10 shows the time-dependent conversion of hydrocarbons in a qualitative plot for the three stages described above.

Zero-order kinetics or linear biodegradation has been observed frequently. According to Alexander and Scow (1989), linear transformations may occur if the nutrients that limit the growth of the active population become available at a constant rate. This is the case when oxygen limitation occurs because of high substrate concentrations in soil and the oxygen feeding rate is low and constant (stage 1). On the other hand, linear degradation occurs when the population is growing on certain C-compounds that have low water solubilities and the amount

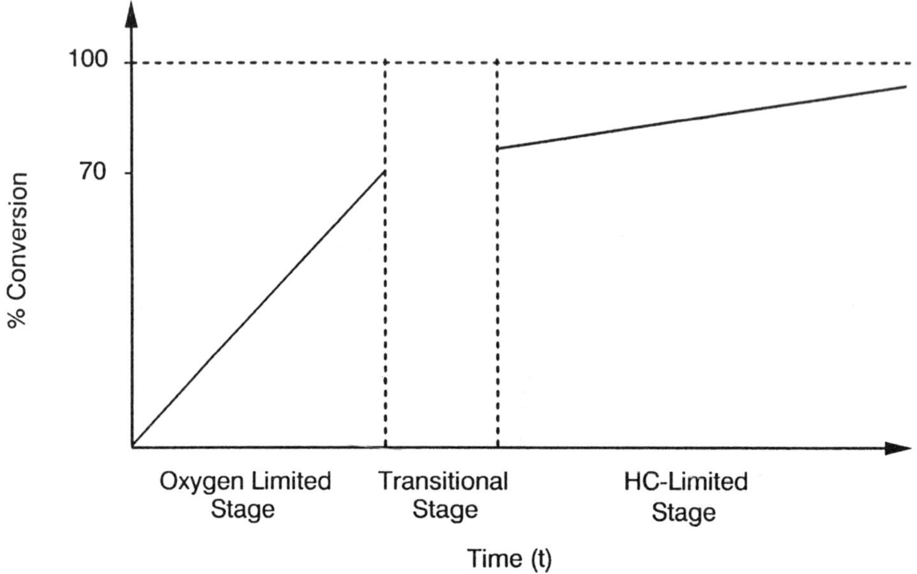

FIGURE 10. Conversion of hydrocarbons as f(t).

in aqueous solution has been totally consumed. This is the case in the C-source limited stage (stage 3).

Further investigations are necessary to answer several outstanding questions. What kind of hydrocarbons are in the residue of the third stage? Why is the degradation of these chemicals difficult? Is the problem really related to low solubility, or is it an energy problem as Zehnder and Schraa (1988) assume?

SUMMARY

Compared with nitrate as an electron acceptor, with hydrogen peroxide

- Diesel fuel degradation rates were four to seven times higher than with nitrate; within the same time frame, degradation with H_2O_2 was 70 to 80 percent, whereas that with nitrate showed only 10 to 20 percent.

- Heterotrophic bacterial activity (measured as esterase activity) was increased from 0.2 to 1.4 (relative units); with nitrate, esterase activity leveled off at 0.2.

- Colony counts of hydrocarbon oxidizing bacteria in the soil were found to be 100 times higher.

- CO_2 formation rate was four to five times higher.

- Total mineralization of the eliminated hydrocarbons was obtained with respect to leaching, as determined by a sterile control.

- The rate of oxygen consumption was up to 3.0 times higher than the stoichiometrically calculated ratio. The sole causes were biomass buildup and degassing effects.

With H_2O_2, three kinetic relevant stages were determined for the biodegradation stages:

1. The state of limited oxygen. Characteristics of this stage were a zero-order reaction rate for the hydrocarbons and a linear conversion of hydrocarbons over time at a constant oxygen feeding rate. The rate of hydrocarbon conversion was found to be proportional to the oxygen input rate at a factor of 0.09 to 0.13. From the stoichiometry, a factor of 0.29 would be expected.

2. A transitional stage between the oxygen- and hydrocarbon-limited stages. This short stage was characterized by incomplete oxygen consumption and maximum colony counts.

3. The state of limited carbon. This stage also exhibited a zero-order reaction, but in the oxygen. The dependence of hydrocarbon conversion was linear with respect to time and was dependent on the solubility and availabilty of hydrocarbons. In comparison to the first stage, the reaction velocity was greatly reduced.

Acknowledgments

Some of the research reported in this paper was supported by the Government of Germany: BMFT Bundesminister für Forschung und

Technologie, Projektnr. 146 0583, Federal Secretary for Research and Technology.

REFERENCES

Alexander, M.; Scow, K. M. *SSSA Spec. Publ.* **1989**, *22* 243–249.

Barenschee, E. R.; Bochem, P. *WLB Wasser, Luft und Boden* **1990** , 11-12, 86–89.

Battermann, G.; Werner, P. *gwf-Wasser/Abwasser* **1984**, *125*, 366–373.

Cole, C. A.; Ochs, D. L.; Funnell, F. C. *Journal WPCF* **1974**, *46*(II), 2579–2592.

Downey, D. C.; Hinchee, R. E.; Westray, M. S.; Slaughter, J. K. 1988 NWWA/API Conference Proceedings, pp 627–645.

German Standard Methods; Deutsche Einheitsverfahren zur Wasser-, Abwasser- und Schlammuntersuchung, *1–22* Lieferung; VCH Weinheim; 1989.

Hinchee, R. E.; Downey, D. C. 1988 NWWA/API Conference Proceedings, pp 715–722.

Hinchee, R. E.; Downey, D. C.; Aggarwal, P. K. *Journal of Hazardous Material*, in press.

Huling, S. G.; Bledsoe, B. E.; White, M. V. Report EPA/600/2-90/DO6, Feb. 1990.

Hutchins, S. R.; Sewell, G. W.; Kovacs, D. A.; Smith, G. A. *Environ. Sci. Technol.* **1991**, *25*, 68–76.

Obst, U.; Holzapfel-Pschorn, A. *Enzymatische Tests für die Wasseranalytik*; R. Oldenbourg; München und Wien, 1988.

Raymond, R. L.; Brown, R. A.; Norris, R. D.; O'Neill, E. T. U.S. Patent 4 588 506, 1986.

Riss, A. Ph.D. Thesis, University of Homburg, Germany, 1988.

Riss, A.; Schweisfurth, R. *Water Supply* **1985**, 2, 27–34.

Zehnder, A. J. B.; Schraa, G. *gwf-Wasser/Abwasser* **1988**, *129*, 369–373.

Laboratory Evaluation of the Utilization of Hydrogen Peroxide for Enhanced Biological Treatment of Petroleum Hydrocarbon Contaminants in Soil

Paul E. Flathman, John H. Carson, Jr., S. Jeanne Whitehead*
OHM Remediation Services Corporation
Khalique A. Khan
AeroVironment Inc.
Denise M. Barnes
Naval Civil Engineering Laboratory
John S. Evans
Science Applications International Corporation

INTRODUCTION

The objective of the laboratory study was to evaluate the benefit of hydrogen peroxide (H_2O_2) addition to soil as a source of molecular oxygen for enhanced biological treatment of petroleum hydrocarbon (PHC) contaminants (JP-5, diesel fuel, and lubricating oil) by the indigenous microflora. Hydrogen peroxide is a strong oxidant and a potential source of molecular oxygen for enhanced biological treatment. Following mineral nutrient addition and pH adjustment (if required), the availability of molecular oxygen is thought to be the factor limiting aerobic biodegradation of available organics in the environment (Atlas 1981).

* OHM Remediation Services Corp., 16406 U.S. Route 224 East, Findlay OH 45839-0551

The analytical results from a laboratory study performed by Britton and the Texas Research Institute (1985) on enhanced biological treatment of gasoline-contaminated soil suggested the benefit of H_2O_2 addition. In their study, comparable volumes of gasoline were removed from soil in test and control columns receiving Dworkin-Foster medium. Dworkin-Foster medium is a basal salts mixture used to support microbial growth on available organics. Bacterial population density in test columns, however, was two orders of magnitude greater than in the control. With the conversion of gasoline to biomass, this observation suggested the benefit of H_2O_2 addition for enhanced biological treatment.

In this laboratory study, the benefit of H_2O_2 addition to soil was evaluated using synthetic and field-contaminated samples. Upflow soil columns containing PHC-spiked soil (sand and humus) and previously contaminated sandy soils were used. Changes in bacterial population density and concentration-independent indicators of PHC biodegradation (n-C_{17}/pristane, n-C_{18}/phytane, and resolved alkane/unresolved alkane ratios) between test and control columns were used as the test parameters (Atlas 1981; Blumer & Sass 1972; Halmo 1985; Jordan & Payne 1980; Senn & Johnson 1985). The study was designed to evaluate PHC biodegradation potential by the indigenous microflora in clean sand and humus that had been spiked with JP-5, diesel fuel, or lubricating oil (i.e., synthetic soil samples) and in PHC-contaminated soil that had been collected from three contaminated soil staging areas that were used for the temporary staging of soils containing JP-5, diesel fuel, or waste lubricating oil.

MATERIALS AND METHODS

Synthetic PHC-Contaminated Soil

Soil Characterization. The clean sand used for performing the laboratory study was obtained on November 7, 1988, from an area in the vicinity of the PHC-contaminated soil staging areas. The humus required for the study was purchased from a local garden supply store. Prior to spiking the soils with PHCs and initiating a determination of PHC biodegradation in the spiked soils, the soils were analyzed to quantify the following parameters: organic matter, texture (i.e., percent sand, silt, and clay), porosity, field capacity, moisture, cation exchange capacity (CEC), pH, total kjeldahl nitrogen (TKN), available mineral nutrients (NH_3-N, NO_3-NO_2-N, and PO_4-P), and aerobic heterotrophic bacterial population density (Bordner *et al.* 1978; Dahnke 1980; Klute 1986; Page 1982; Richards 1954).

Biodegradation Potential of PHCs. The experimental design and laboratory apparatus were similar to that utilized by the Texas Research Institute (Britton & TRI 1985). Twelve, two-piece upflow, 40 mm × 600 mm columns (Corning 38460-40) were utilized (Figure 1).

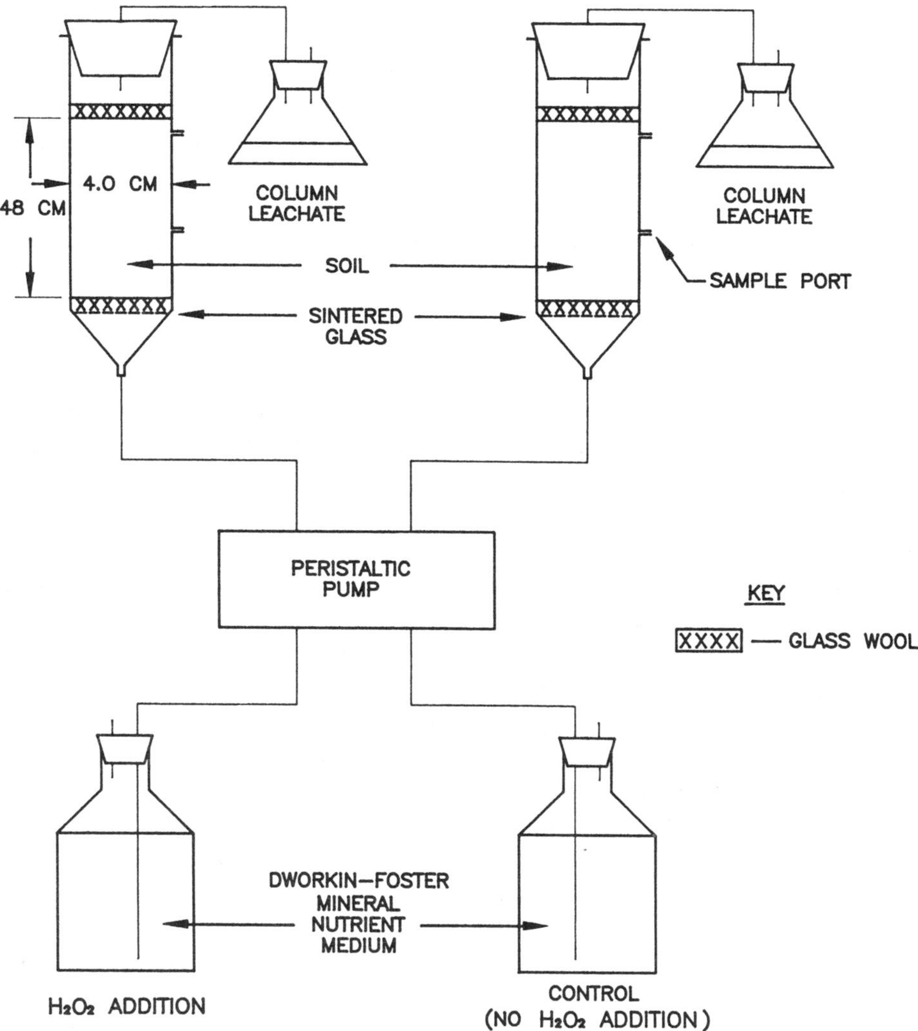

FIGURE 1. Laboratory apparatus for test and control upflow soil columns containing PHC-contaminated soil.

Six columns received sand that had been spiked with JP-5, diesel fuel, or lubricating oil. The remaining six columns received humus that had been similarly spiked with PHCs. Soil depth in each column was 48 cm. PHCs were spiked into sand and humus until uniform mixtures of the PHCs in the soils were obtained at concentrations that could be experienced in the field. JP-5, diesel, and lubricating oil were spiked into sand to final concentrations of 4.023, 4.146, and 5.193 percent (by weight), respectively. The final concentration of these PHCs in humus was 43.71 percent (by weight).

For each PHC product (i.e., JP-5, diesel fuel, or lubricating oil), two soil columns containing PHC-spiked sand and two soil columns containing PHC-spiked humus were prepared. With each of the two soil types, one of the two columns received H_2O_2, the second column did not and was utilized as a control to evaluate the benefit of H_2O_2 addition to the PHC-contaminated soil (Figure 1). Dworkin-Foster medium (Dworkin & Foster 1958) was used to supply the mineral nutrients required for enhanced microbial growth on the PHCs. The medium was prepared using procedures adapted by the Texas Research Institute (Harding, G. L. Texas Research Institute, Austin, TX, personal communication, January 14, 1988).

In addition to supporting growth on PHCs, Dworkin-Foster medium was selected because it had been shown to reduce appreciably the nonenzymatic breakdown of H_2O_2, particularly in the presence of ferric ions (Britton & TRI 1985). The mineral nutrients and H_2O_2 (J. T. Baker 2186, Phillipsburg NJ) were supplied to the upflow columns from 4-L reservoirs. The H_2O_2 was used as received and was standardized by iodometric titration.

An Ismatec multichannel peristaltic pump (Cole-Parmer Instrument Company, Chicago IL) was used to move the reagents through the soil columns. Flow rate to each column was 0.12 mL/min. Tygon tubing of 1.30 mm i.d. was used through the pump. Tubing through the pump was changed a minimum of once every 2 weeks; tubing outside the pump was changed at the same frequency. With 1.30-mm-i.d. tubing through the pump, flow rates ranging from 0.02 to 1.67 mL/min could be provided. Flow through these columns was initiated on May 23, 1989.

Each column had two sampling ports equally spaced along the column. After a high bacterial population density was observed in the leachate from the soil columns, H_2O_2 was added to the appropriate 4-L reservoirs containing Dworkin-Foster medium. H_2O_2 concentration in those reservoirs was increased incrementally until an H_2O_2 concentration approximately 500 mg/L was observed in leachate collected from the lower sampling port. H_2O_2 concentration in those samples was quantified using a Vacuette Kit (Chemetrics, Calverton VA) having a working range from 10 to 1,000 mg/L H_2O_2. The accuracy of the Vacuette Kit was periodically confirmed by iodometric titration.

On July 31, 1989, day 69, flow through the upflow columns was terminated. Soil in each column was sectioned by depth from the top (i.e., 0 to 12 cm, 12 to 24 cm, 24 to 36 cm, and 36 to 48 cm from the top) and replicate VOA vial samples were collected from each section. Aerobic heterotrophic bacterial population density was quantified immediately in one of the replicate pairs. The other replicate was preserved by refrigeration at 1 to 4 C prior to PHC analysis. One sample from each column was analyzed for PHCs, and the sample having the highest bacterial population density was the sample selected for PHC analysis.

A gas chromatograph (GC) equipped with a capillary column and a flame ionization detector (FID) was used to quantify PHCs in each sample following extraction, clean-up, and separation of the PHCs into aliphatic, aromatic, and polar fractions. The analytical procedures used for the analysis of the PHCs were developed by James R. Payne and his staff at Science Applications International Corporation (San Diego CA), based on U.S. EPA Method SW-846. For each sample analyzed, the following parameters were quantified:

- Aliphatics: Resolved—n-C_{17}/pristane, n-C_{18}/phytane, pristane/phytane; Unresolved
- Aromatics: Resolved; Unresolved
- Polars: Resolved; Unresolved

Total PHC (TPHC) content in each sample was the sum of the resolved and the unresolved portions for the three fractions analyzed.

Field-Contaminated Soil

Soil Characterization. The soils were analyzed for the same parameters as previously described for the clean sand and humus prior to their utilization in the upflow soil columns.

Biodegradation Potential of PHCs. The experimental design for this phase of the study was similar to that utilized to evaluate PHC biodegradation in the synthetic PHC-contaminated soils. Twelve soil columns were prepared using the equipment and procedures previously described. Each soil evaluated required the utilization of two soil columns (Figure 1). The soil in one of the two columns received H_2O_2; the second column did not receive H_2O_2 and was utilized as a control to evaluate the benefit of H_2O_2 addition to the PHC-contaminated soil. Flow through the columns began on March 23, 1989.

On July 31, 1989, day 130, flow through the columns was terminated. Soil in each column was sectioned using procedures previously described, and replicate VOA vial samples were collected from each section. Aerobic heterotrophic bacterial population density was quantified immediately in one of the replicate pairs. The other replicate was preserved for PHC analysis. One sample from each column was analyzed for PHCs. The sample having the highest bacterial population density was the sample selected for PHC analysis. For each sample analyzed, the parameters previously described for PHC analysis of the synthetic PHC-contaminated soils were quantified.

RESULTS AND DISCUSSION

With the exception of the humus, the soils used for the laboratory study were classified as sand according to the textural triangle for soil textural analysis using the USDA classification scheme (Klute 1986). To determine if H_2O_2 treatment had an effect on bacterial population density in the columns containing the different soils and petroleum hydrocarbon products at the four depths selected for analysis, a four-way analysis of variance (ANOVA) was utilized to analyze the plate count data.

Count data are typically well described by a Poisson distribution. The counts represented by the plate count data were forced to be between 30 and 300 by sample dilution. The dilution error was a positive multiplicative error believed to be approximately normally distributed and small in magnitude relative to the count error. The resulting error distribution for the data, i.e., the product of a truncated Poisson random variable and a truncated positive normal random variable (with small variance compared to that of the truncated Poisson), is approximately Poisson. In Poisson distributions the variance is equal to the mean; mean and variance, therefore, are not independent. To analyze the data using ANOVA, which requires independence of mean and variance, a fourth root transformation of the data was performed.

The fourth root transformation was determined empirically by fitting a model to the data and inspecting the residuals for heteroscedasticity, which would indicate dependence of mean and variance. The residuals were what remained after estimates from the fitted model were subtracted from the data (i.e., residual = data – fit). Discrepancies of many different kinds between a tentative model and the data can be detected by studying residuals. When assumptions concerning the adequacy of the model are true, the residuals are expected to vary randomly and sum to zero within treatments (Box *et al.* 1978; Dowdy & Wearden 1983; Hoaglin *et al.* 1983). If the distribution of plate count data was Poisson, the data would have met the assumptions of analysis of variance through a square root transformation. The fourth root transformation was required because the data were only approximately Poisson.

Initially, the model included all main effects, i.e., soil type, petroleum hydrocarbon product, depth, and treatment, as well as all two-factor and three-factor interactions between the main effects. The F-tests from the ANOVA table resulted in the two-factor interactions of soil type with treatment, petroleum hydrocarbon product with treatment, and depth with treatment as statistically significant with p-values of 0.002, 0.00013, and 0.011, respectively. A significance level of 0.05 was used in those tests. Interaction is defined as the dependence of the effect of one factor on the level of a second factor (Sokal & Rohlf 1981). Biologically, a significant interaction would be the observed result of synergism or interference of a phenomenon.

The model containing all main effects and significant two-factor interactions was fit to the data, and the residuals were examined for validation of model assumptions (i.e., independence, normality, and homogeneity of variance). The scatter plot of the residuals versus the estimated values from the model exhibited random scatter about the zero line, which supported the assumptions of independence and homogeneity of variance. No gross departures from normality were evident from the probability plot or histogram of the residuals. Both plots exhibited symmetry, but with slightly heavier tails than the normal distribution.

The interactions of soil type with treatment and petroleum hydrocarbon product with treatment were highly significant. The interaction of depth with treatment was also significant. For each of the interactions, Tukey's HSD test for paired comparisons was performed to determine which combinations of factor levels were responsible for the significant interactions observed (Wilkinson 1988). Mean bacterial population density in test columns containing PHC-spiked humus was significantly higher (p-value = 0.00022) than in the controls (Figure 2). Although mean bacterial population density was appreciably higher in

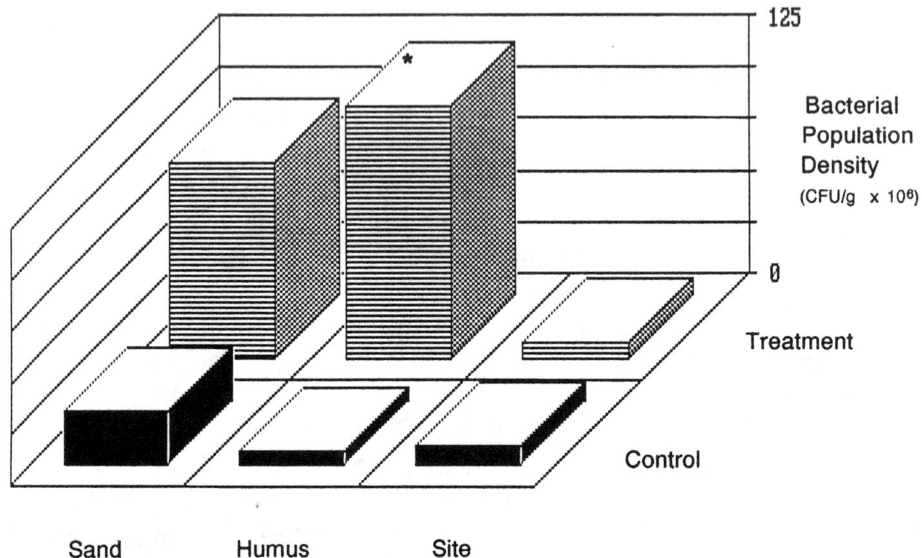

FIGURE 2. Geometric mean of bacterial population density in test and control columns as a function of soil type.

the test columns containing the PHC-spiked sand, the difference between test and control columns was not statistically significant. H_2O_2 addition to test columns containing the site matrix soils did not have an appreciable effect on bacterial population growth.

H_2O_2 addition to test columns containing diesel fuel had a significant effect (p-value = 0.00013) on bacterial population growth (Figure 3). Although mean bacterial growth in test columns containing JP-5 was appreciably higher than in the controls, the difference was not statistically significant. Mean bacterial growth in test columns containing lubricating oil was slightly less than in the control; the difference in the group means, however, was not statistically significant.

A relationship was observed between bacterial population density and depth in test columns (Figure 4). At the base of the test columns, i.e., at a depth of 36 to 48 cm, the difference in mean bacterial population density between test and control columns was appreciable. The biocidal property of hydrogen peroxide at the higher concentrations introduced into the columns was the assumed cause of the lower mean bacterial population density observed in test columns compared to the controls. At the 24- to 36-cm depth, mean bacterial growth in test columns was appreciably greater than in the controls. This increase was assumed to result from the enzymatic breakdown of H_2O_2 and the

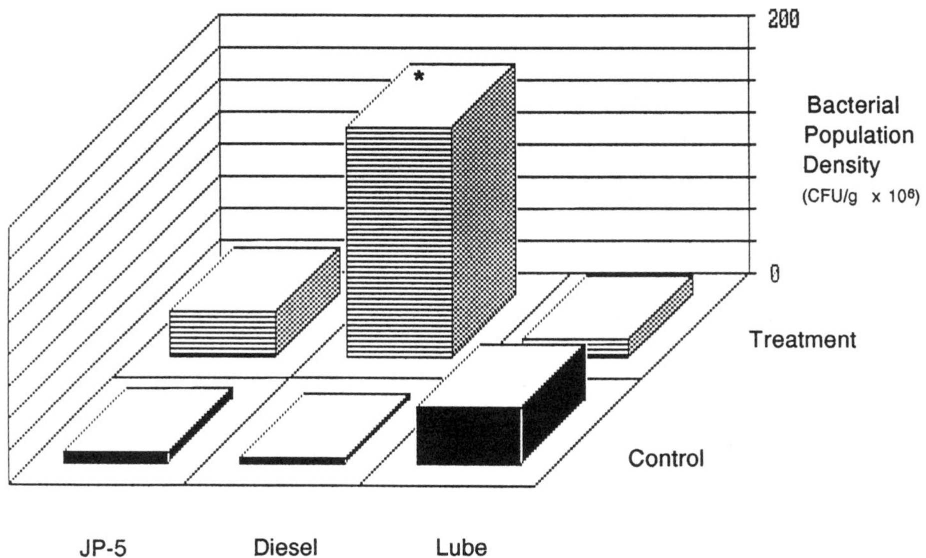

FIGURE 3. Geometric mean of bacterial population density in test and control columns as a function of petroleum hydrocarbon product.

release of molecular oxygen to support the aerobic biodegradation of available organics. The difference in mean bacterial population density between test and control columns became statistically significant (p-value = 0.032) at 12- to 24-cm depth from the top of the column. At 0- to 12-cm depth, the difference between test and control columns was appreciable, but not statistically significant. Mean bacterial population density in control columns remained essentially constant as a function of depth.

Changes in concentration-independent indicators of PHC biodegradation (i.e., n-C_{17}/pristane, n-C_{18}/phytane, and resolved alkane/unresolved alkane ratios) between test and control columns were used to determine the effect of H_2O_2 treatment (Figure 5). The n-C_{17} and n-C_{18} are normal alkanes that are amenable to biodegradation; pristane and phytane are isoprenoids that are resistant to biodegradation. The n-C_{17} and pristane peaks appear very close together on a gas chromatogram (Senn & Johnson 1985), and both compounds have similar volatility; likewise for n-C_{18} and phytane. With biodegradation of the PHCs, the n-C_{17}/pristane and the n-C_{18}/phytane ratios will decrease over time (Atlas 1981; Blumer & Sass 1972; Halmo 1985; Senn & Johnson 1985). For reductions in the values of these ratios to be used as valid measures

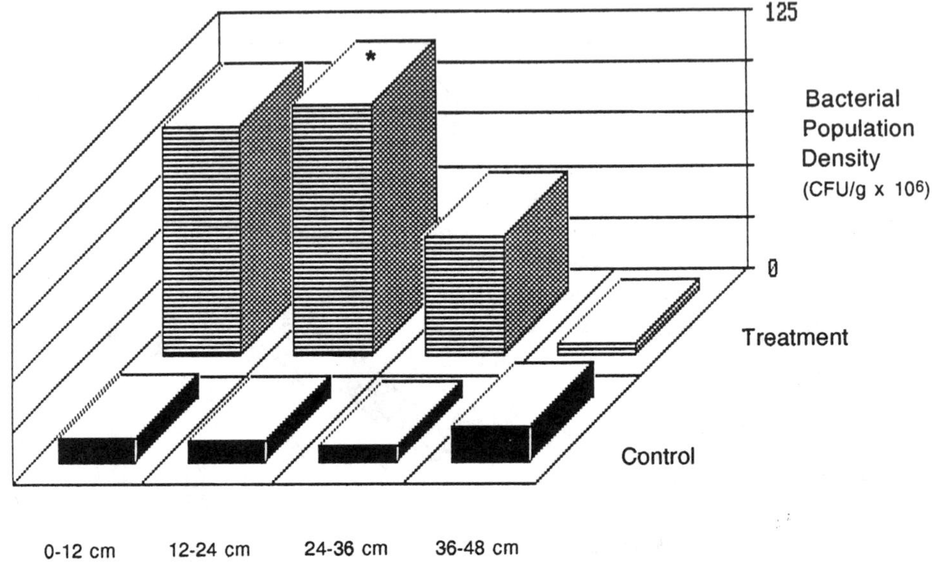

FIGURE 4. Geometric mean of bacterial population density in test and control columns as a function of soil depth.

of PHC biodegradation, the pristane/phytane ratio must remain constant over time. Reduction of the resolved PHCs/unresolved PHCs ratio as a function of time is also used as a concentration-independent indicator of biodegradation (Atlas 1981). The *n*-alkanes are readily resolvable and considered the most readily degraded component in a PHC mixture.

The initial PHC concentrations for the synthetic PHC-contaminated soils were determined by calculation following analysis of each PHC product that was mixed with the sand and humus. For the field-contaminated soils, initial PHC concentrations were determined by analyses of composites of the soils used in the upflow columns. The final concentrations were obtained by PHC analysis of a representative subsample from the 30-cm section of each column with the greatest bacterial population density.

The *n*-C_{17}/pristane, *n*-C_{18}/phytane, and resolved alkane/unresolved alkane ratios were initially analyzed by a three-way ANOVA using soil type, PHC product, and treatment as factors. The difference in the log of those ratios between initial and final PHC analyses was used as the response variable. For the *n*-C_{17}/pristane

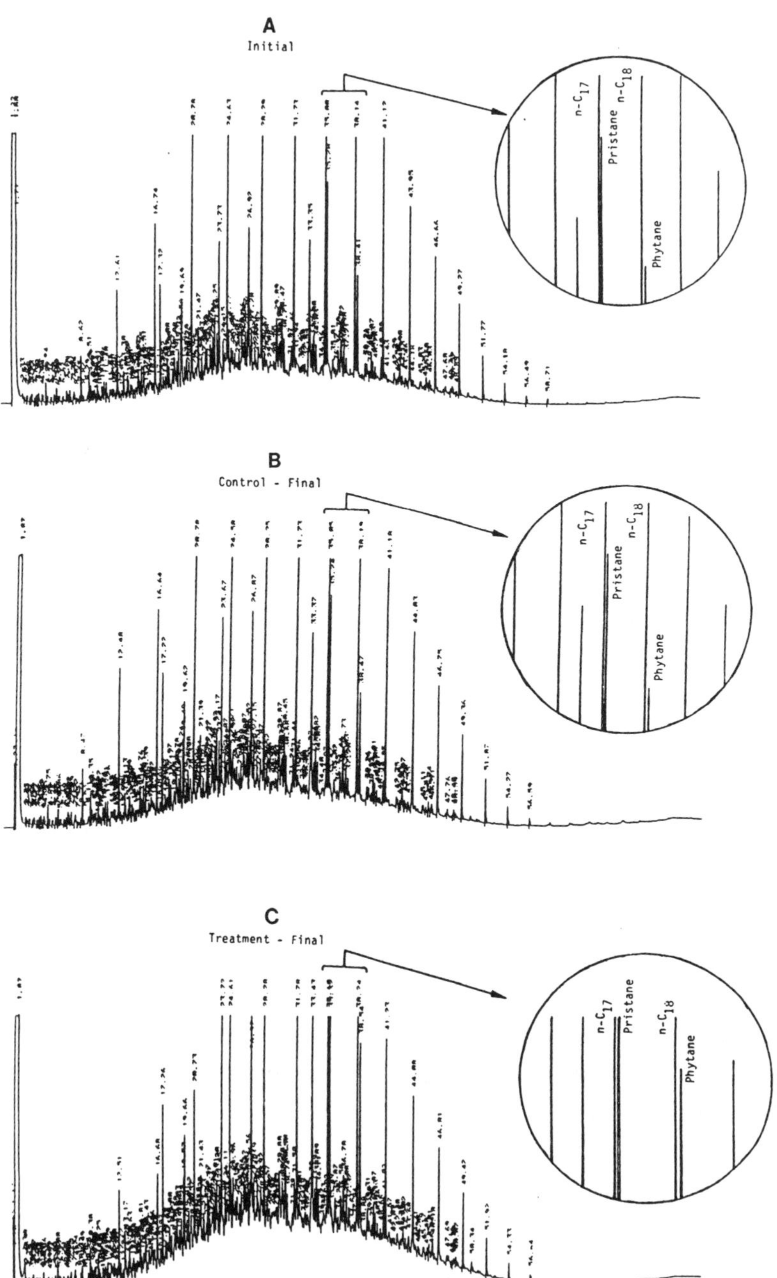

FIGURE 5. GC-FID chromatograms of PHCs from the diesel fuel area (A) and from control (B, no H_2O_2 addition) and test (C, H_2O_2 addition) columns following biological treatment.

ratio, for example, the response variable was calculated using the following relationship:

$$LR\text{-}17 = \log\,(n\text{-}C_{17}/\text{pristane})_{final} - \log\,(n\text{-}C_{17}/\text{pristane})_{initial}$$

For the n-C_{18}/phytane and resolved alkane/unresolved alkane ratios, the response variables were similarly determined. For brevity these response variables will be referred as LR-17 (log-ratio for n-C_{17}/pristane), LR-18 and LR-AL (log ratio for alkanes). Since the response variables were represented by ratios, their error distributions were determined to assess the appropriateness for analysis. The statistical distribution was derived based on the assumptions that the measurements would be positive and that measurement errors would be normally distributed and independent. The statistical distribution of the log ratios (*e.g.*, log [n-C_{17}/pristane]) was not symmetric; however, the distribution of the difference of log ratios was approximately symmetric.

Even prior to including the censoring imposed by analytical detection limits, the resulting distribution was too difficult to work with analytically, but was further studied by Monte Carlo simulation. Monte Carlo simulation is a method for studying the behavior of random variables and solving a variety of mathematical and physical problems approximately by the simulation of random quantities (Sobol 1974). Several different combinations of mean and variance in the numerator and denominator of the ratio of normal distributions were included in the simulation. The simulations indicated that the error distributions of the dependent variables (i.e., response variables) resembled a t-distribution with small to moderate (i.e., 3 to 15) degrees of freedom and that a three-way ANOVA using soil type, PHC product, and treatment as factors might be a reasonable method of analysis. The variability of the response variable and heteroscedasticity (i.e., unequal error variances), however, did not permit the data to be adequately analyzed by ANOVA.

When ANOVA was used, problems with heteroscedasticity (i.e., unequal error variances) and with the degree of variability of the response variables did not permit firm conclusions based on the analysis. Although not conclusive, the results of ANOVA were suggestive. The treatment effect and the interaction between treatment and PHC product were significant for both LR-17 and LR-18. For LR-AL, treatment was the only significant effect. There were not sufficient data to use a multiple paired comparisons test to determine which combinations of treatment and PHC product were responsible for the significance of the interaction. However, the residual from the ANOVA associated with the JP-5 field-contaminated soil was a statistical outlier. It suggested the likelihood of a significantly greater treatment effect for that soil. The greatest decrease in LR-AL was also observed for that sample.

As a result of the problems encountered in the ANOVA, a new response variable was obtained from the three ratios, LR-17, LR-18, and LR-AL, using the following relationship:

R-17 = log (LR-17)treatment – log (LR-17)control

R-18 = log (LR-18)treatment – log (LR-18)control

R-AL = log (LR-AL)treatment – log (LR-AL)control

Since the response for the treatment group was much more variable than that for the control, the new response variable was more homogeneous with respect to error variance and was amenable to testing with a one sample t-test (Mood *et al.* 1974). If the mean of the new response variable is significantly less than zero, the effect of treatment is significant. Only R-AL was significantly lower following H_2O_2 treatment; i.e., the 95 percent upper confidence limit for the mean was –0.126. The means for the new response variables were very similar. The standard error of the means for the responses associated with the R-17 and R-18 were larger than that associated with R-AL. Since each measurement of resolved and unresolved alkanes represented the sum of many individual measurements, the lower variability for the response variable associated with R-AL was not unexpected. It is speculated that the treatment effect was similar for R-17, R-18, and R-AL and that the latter effect was significant because of differences in variability of the data.

The boxplot (Hoaglin *et al.* 1983) provides a visual impression of the distribution of the data. Boxplots of the new response variables are presented in Figure 6. The horizontal lines on the box portion of the plots represent the third quartile, median, and first quartile from top to bottom. The vertical lines extending from the ends provide an indication of moderately extreme observations. Observations indicated by asterisks outside the vertical lines were extreme values; very extreme values were indicated by circles. The boxplots for R-17 and R-18 exhibited greater variation than that for R-AL. All boxes were located at approximately the same height on the graph. The entire box for R-AL, including the vertical lines, was below zero, which supported the significant treatment effect determined in the one-sample t-test (Table 1).

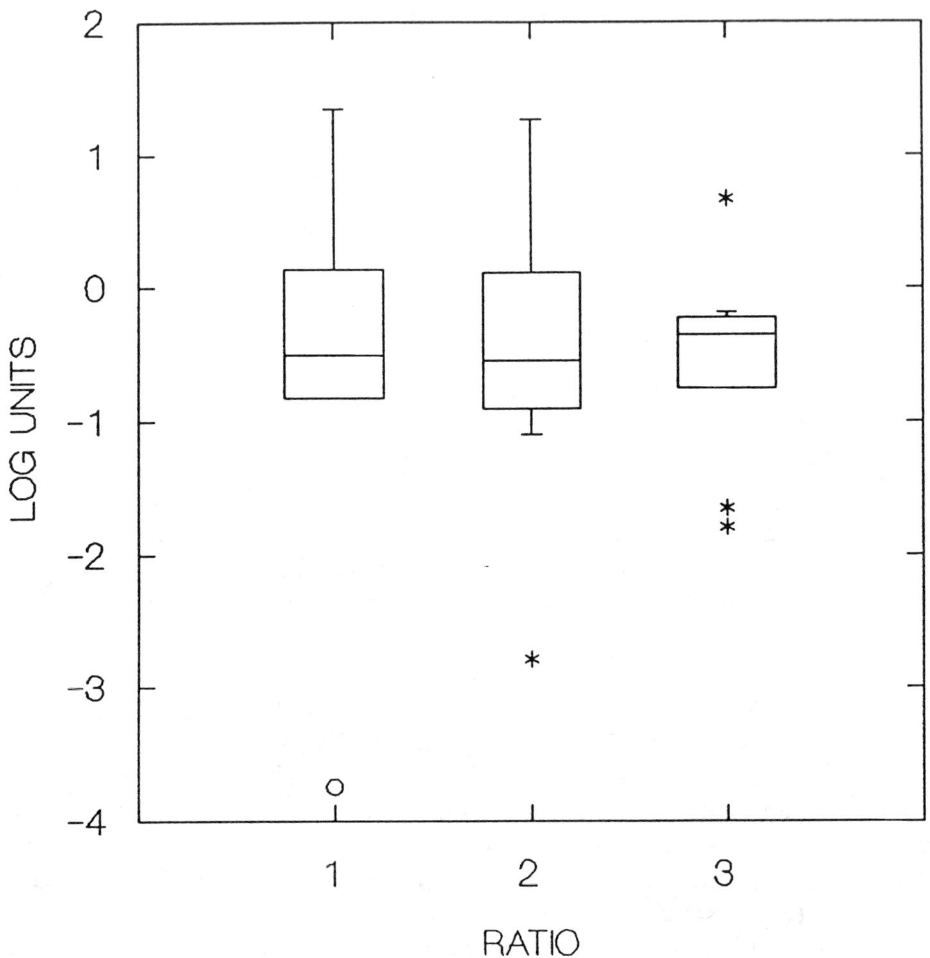

FIGURE 6. Boxplots of the difference in log ratios between treatment and control for the *n*-C_{17}/pristane (1), *n*-C_{18}/phytane (2), and resolved alkane/unresolved alkane PHC (3) ratios.

CONCLUSIONS

Based on the bacterial population densities observed in test and control columns at the end of the laboratory study, the addition of H_2O_2 to mineral nutrient supplemented, previously uncontaminated sand and

TABLE 1. Analytical results for characterization of the soil used in the upflow soil columns.

	Organic Matter (%)	Texture Sand (%)	Texture Silt (%)	Texture Clay (%)	Porosity (%)	Field Capacity (%)	Moisture (%)	Saturation (%)	Cation Exchange Capacity (meg/100g)	pH (SU)	TKN[a] (mg/kg)	Available Mineral Nutrients[a] NH_3-N (mg/kg)	NO_3-NO_2-N (mg/kg)	PO_4-P (mg/kg)	Aerobic Heterotrophic Bacterial Population Density[a] (CFU/g)
Sand	0.40	97.3	0.82	1.92	40.0	14.4	0.37	2.56	10.4	7.4	1,800	<100	1,300	25.0	1.0×10^7 2.2×10^6 1.4×10^6 3.1×10^6 GM[b] = 3.1×10^6
Humus	59.8	10.0	87.1	2.92	40.0	63.5	60.8	95.7	39.1	4.5	27,000	<100	2,040	10.2	8.9×10^6 2.1×10^7 GM = 1.5×10^7
JP-5 Area	0.90	86.7	6.37	6.92	ND[c]	14.8	5.29	35.7	45.5	7.9	1,200	<100	630	19.0	2.1×10^7 2.7×10^7 3.0×10^7 5.7×10^7 GM = 3.1×10^7
Diesel Fuel Area	1.90	88.1	5.97	5.92	ND	8.0	3.98	49.8	28.9	7.4	940	<100	<100	29.0	5.2×10^6 3.7×10^6 5.3×10^6 6.2×10^6 GM = 5.0×10^6
Waste Lube Oil Area	3.60	87.3	7.74	4.92	ND	9.80	9.23	94.2	18.0	8.4	5,400	<100	4,300	450	1.2×10^7 1.2×10^7 1.6×10^7 1.3×10^7 GM = 1.3×10^7

(a) Dry-weight basis.
(b) GM = geometric mean.
(c) ND = not determined.

humus were beneficial for enhancing the biological removal of PHCs. The difference between test and control columns was statistically significant for the PHC-spiked humus. For the field-contaminated site soils, however, H_2O_2 addition did not appear to have an appreciable effect on the enhanced removal of PHCs.

The benefit of H_2O_2 addition appeared greater in soils contaminated with JP-5 and diesel fuel than in soils contaminated with lubricating oil. The difference in bacterial population density between test and control columns was statistically significant in the diesel fuel-contaminated columns.

Changes in the n-C_{17}/pristane, n-C_{18}/phytane, and resolved alkane/unresolved alkane ratios suggested that the addition of hydrogen peroxide was beneficial for enhanced biological removal of PHCs. The difference in the change of the resolved alkane/unresolved alkane ratio between test and control columns supported a statistically significant treatment effect resulting from the addition of hydrogen peroxide.

In summary, the benefit of H_2O_2 addition to soil for enhanced biological treatment of PHCs has been supported by the results of this laboratory study. The utilization of the n-C_{17}/pristane, n-C_{18}/phytane, and resolved alkane/unresolved alkane ratios as concentration-independent indicators of PHC biodegradation has been demonstrated. The recommendation is made to evaluate further the benefit of H_2O_2 addition to soil for enhanced biological treatment of PHCs by utilizing the same test parameters in addition to total PHC concentration and by analyzing an appropriate number of samples for PHCs with each treatment to quantify statistically any real differences resulting from the effect of treatment.

Acknowledgments

We are indebted to the following individuals at OHM Remediation Services Corp., Findlay OH, for their support of this project: Maria Baker, Debra S. Gray, Brian P. Greenwald, Anne L. Hermiller, Douglas E. Jerger, and Patrick M. Woodhull. The funding for this project to AeroVironment was provided by the Naval Civil Engineering Laboratory, Port Hueneme CA, through Contract No. N62474-87-C-3062. OHM conducted this portion of the project as a subcontractor to AV.

REFERENCES

Atlas, R. M. "Microbial Degradation of Petroleum Hydrocarbons: An Environmental Perspective." *Microbiological Reviews* **1981**, *45*(1), 180–290.

Blumer, M.; Sass, J. "Oil Pollution: Persistence and Degradation of Spilled Fuel Oil." *Science* **1972**, *176*, 1120–1122.

Bordner, R.; Winter, J.; Scarpino, P., Eds. *Microbiological Methods for Monitoring the Environment—Water and Wastes*; Environmental Monitoring and Support Laboratory, U.S. Environmental Protection Agency: Cincinnati, OH, 1978; EPA 600/8-78-017.

Box, G.E.P.; Hunter, W. G.; Hunter, J. S. *Statistics for Experimenters: An Introduction to Design, Data Analysis, and Model Building*; John Wiley & Sons: New York, 1978; 653 pp.

Britton, L. N.; TRI (Texas Research Institute). *Feasibility Studies on the Use of Hydrogen Peroxide to Enhance Microbial Degradation of Gasoline*; American Petroleum Institute: Washington, DC, 1985; Publication No. 4389.

Dahnke, W. C., Ed. *Recommended Chemical Soil Test Procedures for the North Central Region*; North Dakota Agricultural Experiment Station, North Dakota State University: Fargo, ND, 1980; 33 pp.

Dowdy, S.; Wearden, S. *Statistics for Research*; John Wiley & Sons: New York, 1983; 537 pp.

Dworkin, M.; Foster, J. W. "Experiments with Some Microorganisms which Utilize Ethane and Hydrogen." *Journal of Bacteriology* **1958**, *75*, 592–603.

Halmo, G. "Enhanced Biodegradation of Oil." In *Proceedings*, 1985 Oil Spill Conference (Prevention, Behavior, Control, Cleanup), Los Angeles, CA, February 25–28, 1985; American Petroleum Institute (API); United States Environmental Protection Agency, United States Coast Guard: Washington, DC, 1985; pp 531–537; API Publication No. 4385.

Hoaglin, D. C.; Mosteller, F.; Tukey, J. W. *Understanding Robust and Exploratory Data Analysis*; John Wiley & Sons: New York, 1983; 447 pp.

Jordan, R. E.; Payne, J. R. *Fate and Weathering of Petroleum Spills in the Marine Environment: A Literature Review and Synopsis*; Ann Arbor Science Publishers: Ann Arbor, MI, 1980; 174 pp.

Klute, A. Ed. *Methods of Soil Analysis, Part 1, Physical and Mineralogical Methods*, 2nd Ed.; American Society of Agronomy and Soil Science Society of America: Madison, WI, 1986; 1188 pp.

Mood, A. M.; Graybill, F. A.; Boes, D. C. *Introduction to the Theory of Statistics*, 3rd Ed.; McGraw-Hill Book Company: New York, 1974; 564 pp.

Page, A. L., Ed. *Methods of Soil Analysis, Part 2, Chemical and Microbiological Properties*, 2nd Ed.; American Society of Agronomy and Soil Science Society of America: Madison, WI, 1982; 1159 pp.

Richards, L. A., Ed. *Agricultural Handbook Number 60, Diagnosis and Improvement of Saline and Alkali Soils*; U.S. Department of Agriculture: Beltsville, MD, 1954; 160 pp.

Senn, R. B.; Johnson, M. S. "Interpretation of Gas Chromatography Data as a Tool in Subsurface Hydrocarbon Investigations"; In *Proceedings*, Conference and Exposition, Petroleum Hydrocarbons and Organic Chemicals in Ground Water—Prevention, Detection, and Restoration, Houston, TX, November 13–15,

1985; American Petroleum Association: Washington, DC; National Water Well Association: Dublin, OH, 1985; pp 331–357.

Sobol, I. M. *The Monte Carlo Method*; The University of Chicago Press: Chicago, IL, 1974; 63 pp.

Sokal, R. R.; Rohlf, F. J. *Biometry: The Principles and Practice of Statistics in Biological Research*, 2nd Ed.; W. H. Freeman and Company: San Francisco, CA, 1981; 859 pp.

U.S. Environmental Protection Agency. *Test Methods for Evaluating Solid Waste, Physical/Chemical Methods*, 3rd Ed., SW-846; Office of Solid Waste and Emergency Response, U.S. Environmental Protection Agency: Washington, DC, 1986.

Wilkinson, L. *SYSTAT: The System for Statistics*; SYSTAT, Inc.: Evanston, IL, 1988; 822 pp.

Soil-Induced Decomposition of Hydrogen Peroxide

Bernard C. Lawes[*]
Du Pont Chemicals

INTRODUCTION

Based on low toxicity and the conversion

$$2H_2O_2 \rightarrow 2H_2O + O_2,$$

hydrogen peroxide has been used in place of air for biodegrading organic material in wastewater (Cole *et al.* 1979), on sand columns (Britton 1985), and in contaminated aquifers by *in situ* bioreclamation (ISB) (API 1987), including the removal of chlorinated hydrocarbons (Lawes & Litchfield 1988; Thomas & Ward 1989). Because H_2O_2 decomposition is not instantaneous, a more sustained release, and thus deeper penetration, of oxygen into a soil formation is possible (Yaniga & Smith 1984). The extent of penetration depends on hydraulic transport and the rate of H_2O_2 decomposition. Since hydraulic transport in ISB may be typically only a few meters per day, fast decomposition may be of more concern than slow decomposition. Both heavy metal ions (Britton 1985) and catalase buildup (Spain *et al.* 1989) have been blamed for too rapid H_2O_2 decomposition and consequently wasteful off-gassing of oxygen (Spain *et al.* 1989).

A preliminary study (Lawes 1990) using batch tests found no reliable indicators of soil activity for decomposing H_2O_2, save possibly manganese content, and suggested that soils efficiently converted H_2O_2

[*] Du Pont Chemicals, Chestnut Run Plaza, P.O. Box 80709, Wilmington, DE 19880-0709

to oxygen by biotic and abiotic catalysis, based on rate decreasing by autoclaving and phosphate amending, with some soils strongly favoring biotic decomposition (Lawes 1990; Spain *et al.* 1989). Ferric iron is known to catalytically decompose H_2O_2 (Eary 1985). It was shown that orthophosphate could remove most, if not all, iron-induced decomposition (Britton 1985). Information on soil activity and biotic versus abiotic decomposition could help in early ISB planning decisions on whether to use phosphates for H_2O_2 stabilization and what type of H_2O_2 injection strategy to use (*e.g.*, more injection wells), or other broader distribution means for especially active soils.

Static batch tests using a high (7.5:1) or low (1:3 or 1:4) ratio of 0.1 percent H_2O_2 solution to soil were developed to measure decomposition rates (Lawes 1990); the former are more useful for quickly scouting many samples, but the latter more closely simulate a liquid-saturated soil formation. The more realistic, and faster, decomposition rates from a low-ratio test, unfortunately, required a rather cumbersome gas collection procedure because supernatant quantity was small to nil and because liquid transport between any supernatant and a thick soil layer would be impeded. Time effects were not assessed; only a single peroxide amendation per test was used.

This study had three objectives. First was to compare high- and low-ratio tests with alternatives that might yield low-ratio rates using titration monitoring and, in so doing, to explore reaction conditions and define "high activity" and "low activity" soils. Second was to confirm high conversion efficiency of peroxide to oxygen, and to see how rate is affected by repeat peroxide amendations. Third was to better quantify biotic versus abiotic decomposition from sterilizing and phosphate treatments.

METHODS AND MATERIALS

Soil samples were unconsolidated and stored in closed containers at 4 C (39 F) until use. TYSUL WW50 hydrogen peroxide (Du Pont) was either diluted to make 0.1 or 1.1 percent H_2O_2 solutions, or added by precision pipette to a stirred mixture of water and soil to achieve a starting H_2O_2 concentration. Unless otherwise specified, tests were without agitation, at ambient pH, typically 6 to 7, at ambient temperature, typically 21 to 24 C (70 to 75 F); a ratio such as 1:4 or 7.5:1 referred to the weight ratio of 0.1 percent H_2O_2 solution to undried soil weight; autoclaving meant steam autoclaving for 1 h at 121 C (250 F). For a low-ratio test using phosphates, soil samples were pretreated as follows before placing in flask: a 400-g soil sample in a glass column was gravity percolated with a once-through charge of 750 g of water (control) or phosphate compound (0.2% as PO_4) and then completely air-dried and sieved (to break up lumps) before charging into a 500-mL flask containing 0.1 percent H_2O_2.

In high-ratio tests, peroxide disappearance was followed by standard iodometric titration of supernatant samples, but could also be followed (somewhat less accurately, perhaps) by collecting oxygen gas in a closed system into an inverted burette. Theoretical oxygen was determined using 30 mg of catalase instead of soil, with complete decomposition occurring within 10 min under mild agitation. Low-ratio tests were run in the smallest possible flask to minimize gas collection errors from freeboard volume changes caused by changes in ambient temperature. More experimental detail is given in the preliminary study (Lawes 1990). Relative rates were taken from rate constants from plots of the natural log of $(H_2O_2)/(H_2O_2)_o$ versus time.

In high-ratio complete mix tests, beaker contents were paddle stirred at 200 rpm, which produced the same rate as 150 or 500 rpm. At least five samples were withdrawn for iodometric titration. In a shallow layer (19-cm petri dish) test, 100 to 200 g of 0.1 percent H_2O_2 was added as evenly as possible within 5 sec over a uniformly spread layer of 100 g of soil. At least ten random titration samples of 10/1 to 2 g from the entire supernatant surface were withdrawn to minimize error from nonuniform supernatant thickness. Sample lots were homogenized by mixing and screening to remove stones and twigs.

In scouting tests with glass columns, gravity flow of dilute H_2O_2 solution through soil stopped within minutes, even with mild suction applied to outflow (Lawes 1990). Peroxide-induced loss of permeability in soils has been reported in studies on septic tank drainfield rejuvenation (Hargett *et al.* 1983) and *in situ* uranium mining (Lawes 1978), with sodium silicate (Lawes & Watts 1982) and potassium salts (DeVries & Lawes 1982; Lawes 1981) being used to minimize loss, which was said to be caused by changes in clay cell structure (Hargett *et al.* 1983).

RESULTS AND DISCUSSION

Test Method Comparisons

Before comparing test methods, it may help to broadly define "high activity" and "low activity" soils, using 1:3 to 1:4 low-ratio tests, with the latter simulating a saturated sandy soil. At the very low end of the activity spectrum were sandy soils from sites in South Carolina, southern New Jersey, the Florida Gulf Coast, and from a Long Island site that was remediated by H_2O_2-assisted ISB (Lee & Raymond 1991). These soils decomposed only up to about half the peroxide after 6 h in a low-ratio test. High activity soils, on the other hand, decomposed half to three-quarters (or more) of the peroxide in the 1:4 test within about 15 min, the rest decomposing much more slowly over 2 to 6 h. The top curves in Figures 1 and 2 exemplify rates from high activity soils. High-ratio tests also showed a wide range of activity, although, as shown below, at much slower rates. Of some 25 soils evaluated (Lawes 1990; Lawes, submitted), no particular activity range (high, low, or intermediate) predominated.

Table 1 compares key features of the high- and low-ratio tests with two modified titration tests. Rates from the four tests are compared in Figure 1 for a subsurface soil from a Stockton CA gas station and in Figure 2 for an atypically coarse sand from a southern California

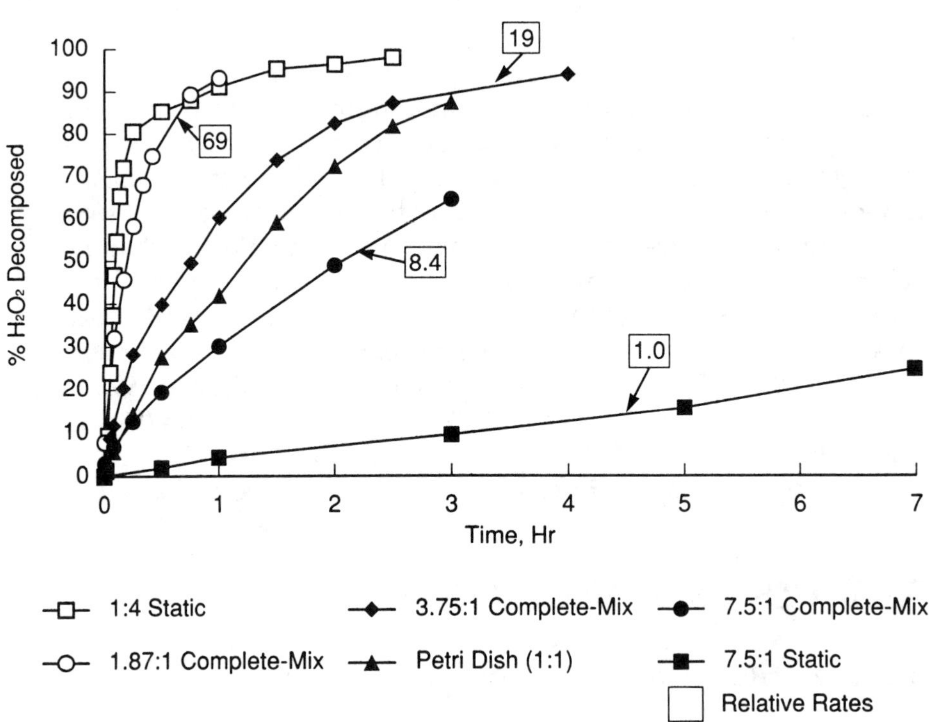

FIGURE 1. Decomposition tests on 0.1 percent H_2O_2 soil, Stockton, CA.

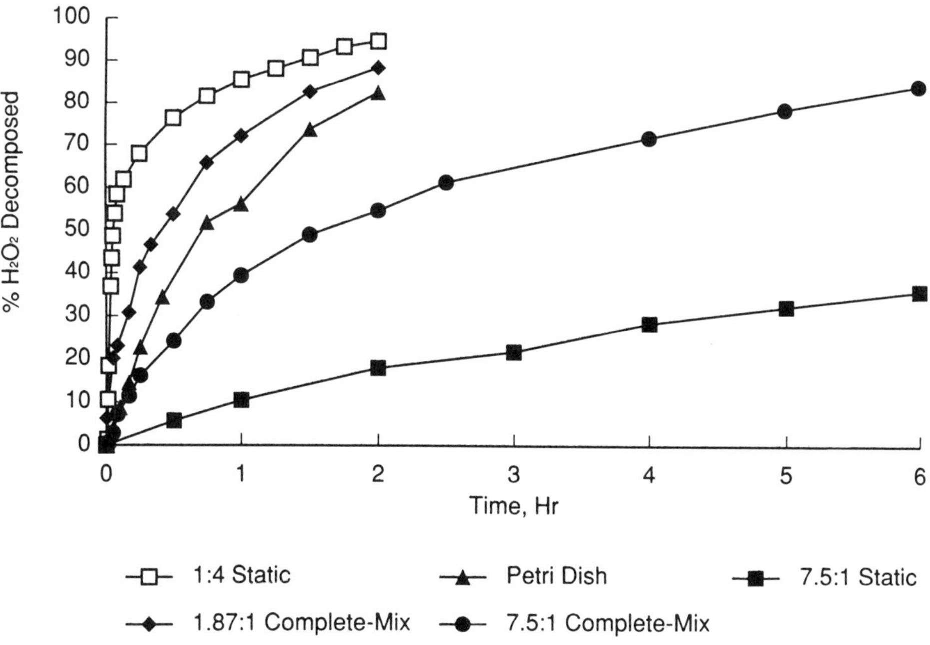

FIGURE 2. Decomposition tests on 0.1 percent H_2O_2 southern California coarse sand.

TABLE 1. H_2O_2 decomposition tests.

Test Name	Typical Ratio	Quantities, g 0.1% H_2O_2	Soil	Agitation	Analysis by
High-Ratio	7.5:1	150	20	None	Titration
Low-Ratio	1:4	100	400	None	Gas Coll.
Petri Dish	1:1	100	100	None	Titration
Complete-Mix	1.87:1	150	80	200 rpm	Titration

site that was evaluated for H_2O_2-assisted ISB (Flathman *et al.* 1991). Rates from the shallow layer (petri dish) tests were closer to those from the 1:4 than from the 7.5:1 static tests, although much less so for the much smaller grained Stockton soil. More promising as a titration-based surrogate for the 1:4 test was a complete mix test on a 1.87:1

liquid-to-soil ratio, in that curves for both soils came fairly close to approximating the rate curve from the 1:4 test. Data from these figures will be discussed further in the next two sections.

Site and sample homogeneity could influence results from decomposition tests. For example, the most and least active of several samples from a Texas Gulf plant site gave high-ratio rates that differed more than fivefold. For soils from three sites, more than twofold rate differences were seen from samples from the same sample bag or even from opposite ends of a sampling canister.

Test Factors and Modifications

Liquid to Soil Ratio. The large rate increase in going from static high- to low-ratio tests in Figures 1 and 2 is not surprising, since in low-ratio tests most of the peroxide solution is in direct contact with the soil rather than being in an out-of-contact supernatant. As already noted, a 1:4 ratio providing little or no supernatant should simulate a saturated sandy soil formation. A large rate difference was also seen for sandy soil from the U.S. EPA test site at Traverse City MI ISB (Lawes 1990). Decreasing the ratio from 1:3 to 1:4 for soil from the Texas Gulf plant site increased the amount of peroxide destroyed in the initial faster stage, *e.g.*, from half to three-quarters.

As noted above for Figures 1 and 2, the petri dish decomposition rate was much closer to the 1:4 rate for the California sand (Figure 2). A possible explanation is better liquid transport between the supernatant and the relatively very coarse California sand. Liquid transport should not be a rate limiter under the nondiffusion control conditions in complete mix stirring.

Agitation. Removing diffusion control by running high-ratio tests under complete mix agitation should increase decomposition rate; how much depended on the soil. On one extreme, less than a twofold increase was seen for very low activity soils such as the Long Island sandy soil, discussed above. On the other, about a sixtyfold rate increase was seen for the most active of 10 topsoils tested. Less extreme, but still manyfold, increases were seen for the high activity soils in Figures 1 and 2, particularly as the ratio was decreased from 7.5:1 to 1.87:1. For a sufficiently active soil, closer approximation of a complete mix curve to a 1:4 curve than that shown in Figure 1 may not be possible since the complete mix curves showed generally decent first order kinetics (R^2 = 0.99; F = 611 to 2,120, relative rates in Figure 1), whereas the low-ratio curve did not (R^2 = 0.88; F = 107). The 1:4 rate in Figure 1 actually was fairly well represented by two first order segments, the first achieving about 80 percent decomposition and having a slope approximately 7 times that of the segment exceeding 80 percent decomposition, i.e., k = 7.3/h (R^2 = 0.97) vs. 1.0/h (R^2 = 0.99).

Temperature. Using the Stockton CA soil, 1:2 ratio tests showed that adjustment will be needed to correct ambient temperature test results to a lower underground temperature. Comparing an ambient 21 C (75 F) with 10 C (50 F), 75 and 56 percent peroxide, respectively, was decomposed after 180 min. Thus, the lower temperature run gave an average decomposition 73 percent of that from the higher temperature run, i.e., 0.42 percent/min versus 0.31 percent/min.

pH. Though most tests were run at ambient, generally neutral pH, data from repeat peroxide amending (Figure 3) suggested that alkaline pH increases decomposition rate. More direct evidence came from a subsurface soil from a northwest Vermont gas station that had been treated by H_2O_2-assisted ISB (Kelly, M., Groundwater Technology, Inc., personal communication). Peroxide half-lives from high-ratio tests decreased from 250 to 55 min as pH was varied from 6.5 to 9.6. See also the negating effect of alkaline pH on phosphate stabilizing (Figure 6).

MECHANISM OF DECOMPOSITION

Efficiency of Oxygen Gas Release

Earlier work (Lawes 1990) suggested that soils were quite efficient in catalytically converting 0.1 percent H_2O_2 to O_2 according to the equation above. Even stronger evidence came from a 2:1 complete mix test using 10 times more H_2O_2 (1.1%) and a topsoil having a particularly high percent of organic matter (6.1% versus 1.5 to 5.5% for the nine other topsoils analyzed). The amount of oxygen release in two runs was 97.5 and 97.6 percent of that released by catalase (640 mL O_2). When the completed reaction mixture was reamended to the original H_2O_2 concentration, the figures were 99 and 98.6 percent. Thus little, if any, hydrogen peroxide was "wasted" by nonoxygen-releasing oxidations of natural organic (humic) debris in soil. Subsurface soils should raise even less concern; six random subsurface soils were analyzed and found to contain only 0.1 to 0.2 percent organic matter.

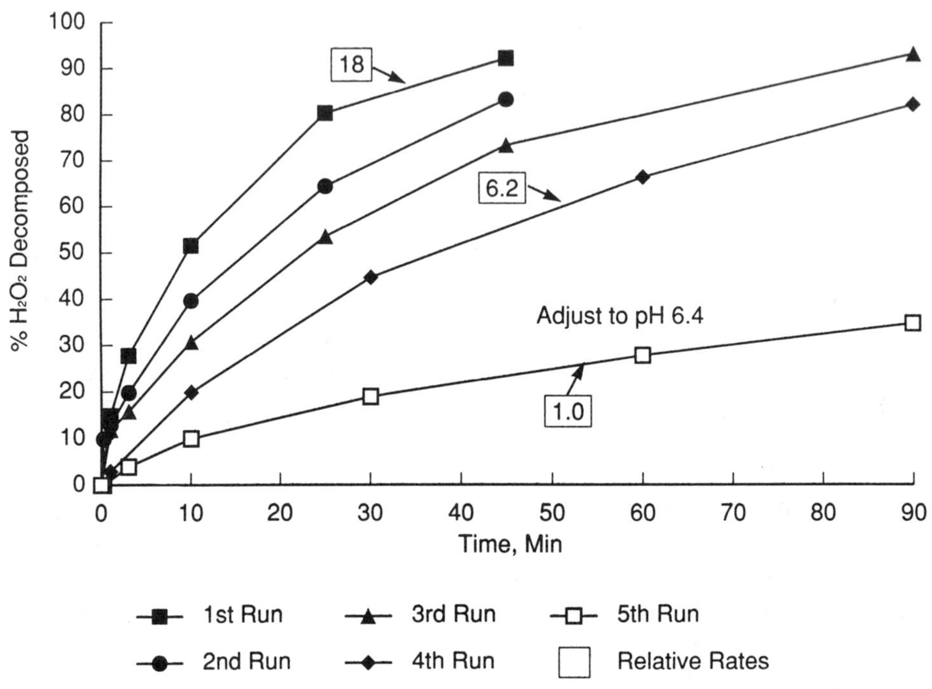

[a] Complete-mix 7.5:1 ratio pH 9.6
[b] At end of each run, add 50% H_2O_2 to reach starting concentration of 0.1%.

FIGURE 3. H_2O_2 decomposition[a] after successive amending.[b]

This does not rule out other peroxide consumables, such as the ferrous iron in some soils. It has been shown that in dilute H_2O_2 solution both ferric and ferrous iron completely decomposed peroxide, with the former causing almost stoichiometric release of oxygen and the latter, little to none (Britton 1985).

It was not surprising that the same amount of oxygen was obtained when the spent reaction mixture was amended back to the starting peroxide concentration. However, decomposition was slower the second time (H_2O_2 half-life 6 vs. 3 min). Figure 3 shows similar successive rate decreases for a subsurface soil from a Lodi CA gas station using five successive peroxide amendations to 0.1 percent H_2O_2. The rate of the fourth successive decomposition was only about one-third that of the first based on rate constants. This was a particularly alkaline soil (pH 9.6). When the pH was dropped to 6.5 (sulfuric acid) for the final run, the rate dropped more precipitously to only about one-sixth that of the preceding run.

Effect of Sterilizing and Phosphate Treatment

The use of sterilizing in conjunction with phosphate treatment on a high activity loamy topsoil from Farmington MN (Figure 4) provided evidence that soils can catalytically decompose H_2O_2 biotically and abiotically, that steam autoclaving may delineate the two modes of decomposition, but that phosphate treatment could not be used to delineate biotic from abiotic decomposition because it provided only partial protection against abiotic decomposition. Phosphates are reported to deactivate catalytic ions such as Fe^{+++} (Britton 1985; Schumb *et al.* 1955).

Evidence for biotic decomposition was the substantial decrease in decomposition rate from autoclaving. This single autoclaving treatment apparently removed all enzymatic activity, including that in dead cells, because no further decrease was found under more exhaustive autoclaving (on 3 successive days), and because some soils (Lawes 1990; Spain *et al.* 1989) lost virtually all decomposition activity from a single

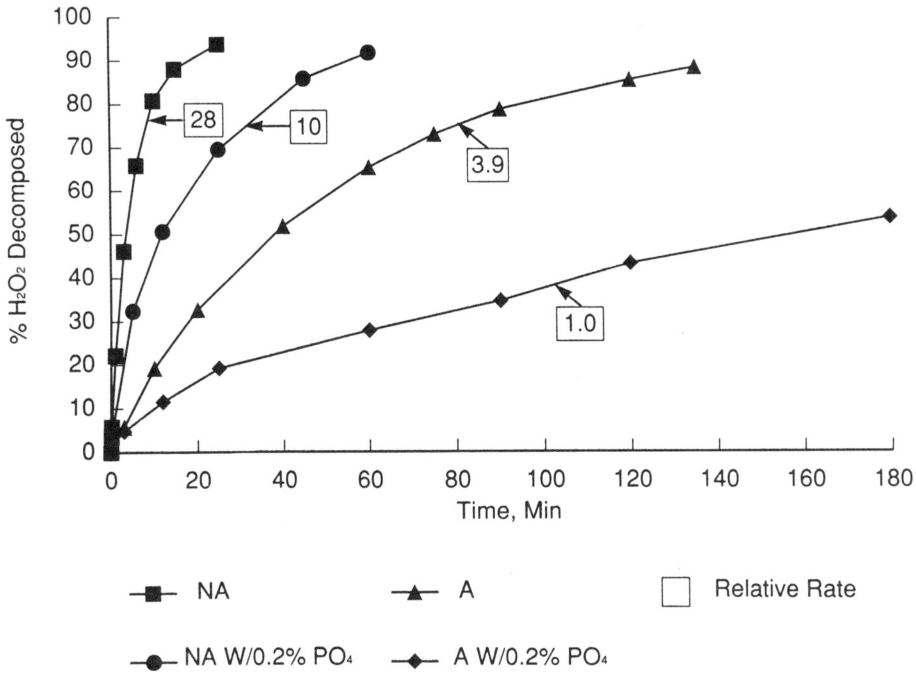

[a] Using complete-mix, 7.5:1 ratio on Farmington MN topsoils at pH 7.0.

FIGURE 4. Effect of PO_4 on decomposition of 0.1 percent H_2O_2 by autoclaved (A) and nonautoclaved (NA) soil.[a]

autoclaving treatment. That autoclaving might thus be a useful tool to distinguish biotic from abiotic activity was corroborated for a Canadian glacial topsoil even more strongly affected by autoclaving than the Farmington topsoil. It was found that an entirely different sterilizing method, namely mercuric chloride at pH 3.8, provided a rate curve very similar to that from autoclaving (Figure 5). Thus, autoclaving and acidic mercuric chloride were either both clearly delineating biotic and abiotic decomposition, or they were remarkably producing the same effect coincidentally. Ethylene oxide or mercuric chloride at near neutral pH was presumably much less effective because catalase in dead cells unaffected by these treatments continued to decompose H_2O_2.

Evidence for abiotic decomposition (Figure 4) was the substantial and similar decrease in decomposition rate caused by orthophosphate for both autoclaved and nonautoclaved soil. Actually, the rate decrease was a little more for the autoclaved soil. Alhough data in an earlier study suggested that under some conditions H_2O_2 decomposition by

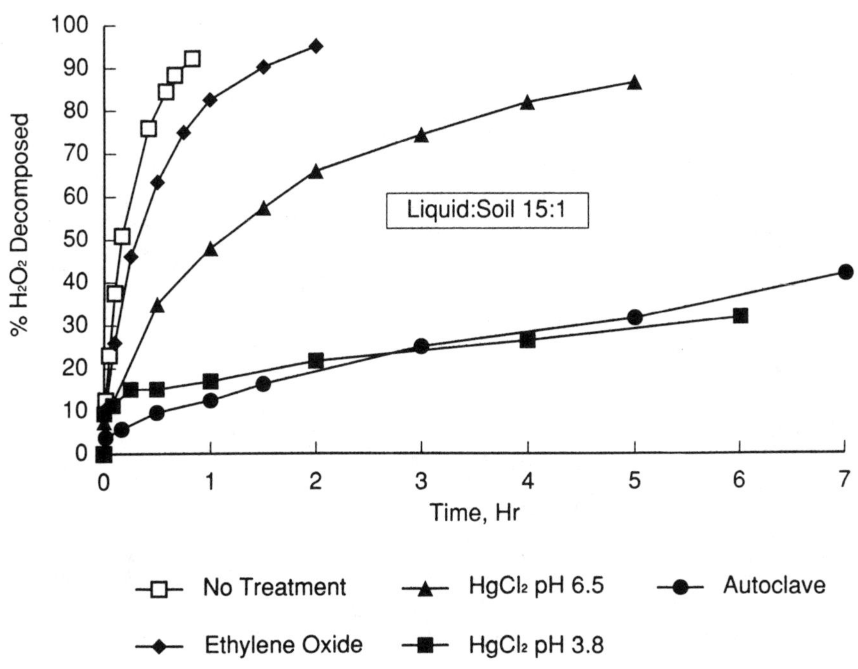

[a] Using complete-mix, 15:1 ratio and topsoil from Saskatchewan, Canada

FIGURE 5. Effect of sterilizing on decomposition of 0.1 percent H_2O_2.[a]

iron- and salt-amended sand could be completely stopped by phosphate (Britton 1985), this clearly was not the case for the topsoil in Figure 4. Phosphate removed only part of the decomposition activity. One reason may be less effectiveness against other heavy metals in this soil known to decompose H_2O_2 (including over 0.03%manganese) (Lawes 1990). Pyrophosphate, for example, was said to be able to hold in check 10 mg/L of Fe^{+++}, but not trace levels of Cu^{++} (Schumb *et al.* 1955). Another reason may be the probable inability of phosphates to stabilize against heterogeneous catalytic decomposition (Schumb *et al.* 1955).

High-ratio tests on the Vermont subsurface soil showed that complex phosphates, potassium tripolyphosphate (KTPP) or tetrasodium pyrophopshate (TSPP), could not bridge the stabilization gap left by orthophosphate; all three produced similar rate decreases in high-ratio tests. Results are illustrated for orthophosphate and KTPP in Figure 6, and as noted above, alkaline (but not acidic pH) diminished phosphate stabilization. Complex phosphates were of interest because they reportedly could inactivate metal ions by complexing as well as by precipitation (Schumb *et al.* 1955).

The foregoing test used high concentrations of phosphate (0.1 to 0.3% solution, as PO_4) and, therefore, do not suggest practical levels for ISB or the long-term effect of phosphate nutrient addition. Limited high-ratio tests on an active Concord NB topsoil showed a relatively small, but still noticeable, stabilization compared to that from 0.2 percent phosphate. In a single low-ratio (1:3) test using soil pretreated with orthophosphate or KTPP, both phosphates (especially orthophosphate) showed little if any stabilizing. A small amount of stabilizing by KTPP was not removed by alkaline pH (as it was in the complete mix high ratio test).

CONCLUSIONS

1. Soils can vary greatly in peroxide decomposition activity. Rates from batch tests can be especially increased by decreasing the ratio of dilute H_2O_2 solution to soil (down to 1:4, simulating a saturated soil formation) and by operating under complete mix conditions. Alkaline pH also accelerated decomposition.

2. In a 1:4 ratio test, high activity soils decompose most of the H_2O_2 in a 0.1 percent solution within about 15 min, whereas

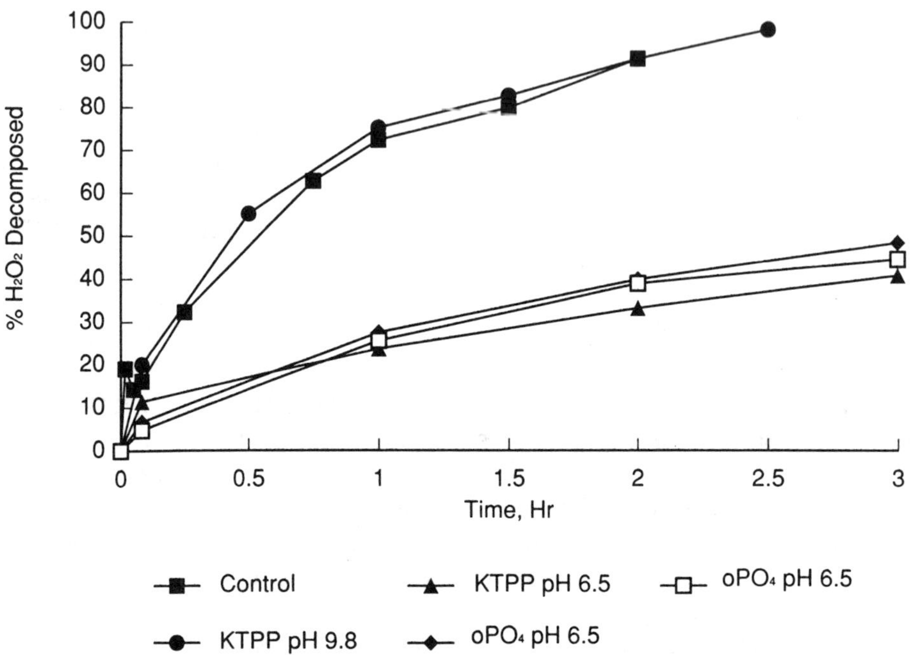

[a] Complete-mix at 7.5:1 ratio on subsurface soil from northwest Vermont.

FIGURE 6. Effect of phosphates on decomposition of 0.1 percent H_2O_2.[a]

low activity soils can take several hours to decompose half the H_2O_2.

3. As titration-monitoring surrogates for 1:4 tests, complete mix tests at about a 1.9:1 ratio (or lower) hold more promise than the 1:1 ratio petri dish test. Further work is needed to show general utility and to determine an optimum ratio.

4. An adequate number of samples must be taken to characterize properly the soil activity of a site, and care must be taken to homogenize lots for studying different conditions.

5. Hydrogen peroxide needed for ISB will not be "wasted" by reacting with (nonpollutant) soil organics.

6. Decomposition rates can change with time, but even repeat dose tests give only "snapshots" that do not take into

account microbial population dynamics or the long-term effects of nutrient and oxidant addition. Extended column or field tests are needed to assess such factors.

7. Soil decomposes H_2O_2 to the oxygen needed for ISB by both biotic and abiotic catalysis. Autoclaving or treatment with acidic mercuric chloride should be useful in estimating how much decomposition occurs by either mode.

8. Phosphates only partially protect against abiotic decomposition by soils and cannot be used to estimate the extent of abiotic decomposition.

9. How phosphates are applied and tested are important factors in evaluating phosphates for stabilizing in ISB. Other complexing compounds, *e.g.*, those used for stabilizing peroxide in *in situ* uranium mining (Lawes & Watts 1981), may be worth evaluating to improve stabilization by phosphate.

Acknowledgments

This contribution would not have been possible without the skillful laboratory efforts of our technician, Herb Doughty, and the valuable consultations with others in Du Pont having an active interest in bioreclamation. These included Drs. Maimu Yllo (analytical chemistry), Calvin Chien (hydrology), Henn Kilkson (kinetics), and Dwight A. Holtzen (mathematics). Many thanks also to Groundwater Technology, Inc., Chadds Ford PA; OHM Remediation Services, Findlay OH; and Du Pont Environmental Services, Aston PA for providing subsurface samples. Also, special thanks to Dr. C. D. Litchfield for helpful discussions in microbiology.

REFERENCES

API. "Field Study of Enhanced Subsurface Biodegradation of Hydrocarbons Using Hydrogen Peroxide as an Oxygen Source," API Publication No. 4448, 1987.

Britton, L. N. "Feasibility Studies on the Use of Hydrogen Peroxide to Enhance Microbial Degradation of Gasoline," API Publication 4389, May 1985.

Cole, C. A.; Ochs, D.; Funnell, F. C. *J. Water Pollut. Control Fed.* **1979**, *46*, 2579.

DeVries, F. W.; Lawes, B. C. U.S. Patent 311 341, 1982.

Eary, L. E. *Metallurgical Transactions B* **1985**, *16B*, 181.

Flathman, P.; Khan, K.; Barnes, D. Presented at *In Situ* and On-Site Bioreclamation International Symposium, Battelle, San Diego, CA, March 19–21, 1991.

Hargett, D. L.; Tyler, E. J. Converse, J. C.; "Chemical Rehabilitation of Soil Wastewater Absorption Systems Using Hydrogen Peroxide: Effects on Soil Permeability"; a report from the Department of Soil Science, Department of Agricultural Engineering, Wisconsin Geological and Natural History Survey, University of Wisconsin-Madison, 1983.

Lawes, B. C. *In Situ* **1978**, 2, 75.

Lawes, B. C. Canadian Patent 1 100 035, 1981.

Lawes, B. C. In *Petroleum Contaminated Soils*; Kostecki, P. T.; Calabrese, E. J., Eds.; Lewis Publishers: Chelsea, MI, 1990; Vol. 3, Chapter 19.

Lawes, B. C. Continuing work in paper submitted for publication in primary scientific journal.

Lawes, B. C.; Litchfield, C. D. U.S. Patent 4 749 491, 1988.

Lawes, B. C.; Watts, J. C. U.S. Patent 4 302 429, 1981.

Lawes, B. C.; Watts, J. C. U.S. Patent 4 320 923, 1982.

Lee, M.; Raymond, R. Presented at *In Situ* and On-Site Bioreclamation International Symposium, Battelle, San Diego, CA, March 19–21, 1991.

Schumb, W. C.; Satterfield, C. N.; Wentworth, R. L. *Hydrogen Peroxide*; ACS Monograph; Reinhold: New York, 1955, pp 540, 541.

Spain, J. C.; Milligan, J. D.; Downey, D. C.; Slaughter, J. K. *Groundwater* **1989**, 27, 163.

Thomas, J. M.; Ward, C. H. *Environ. Sci. Technol.* **1989**, 23, 760.

Yaniga, P. M.; Smith, W. *Proceedings*, NWWA/API Conference Petroleum Hydrocarbons and Organic Chemicals in Groundwater, National Water Well Association, Houston, TX, November 1984; pp 451–472.

Laboratory and Field Studies on BTEX Biodegradation in a Fuel-Contaminated Aquifer under Denitrifying Conditions

*S. R. Hutchins**, *J. T. Wilson*
U.S. Environmental Protection Agency

INTRODUCTION

Leaking underground storage tanks are a major source of groundwater contamination by petroleum hydrocarbons. Of the approximately 1.4 million underground tanks storing gasoline in the United States, some petroleum experts estimate that 75,000 to 100,000 are leaking (Feliciano 1984). Gasoline and other fuels contain benzene, toluene, ethylbenzene, and xylenes (collectively known as BTEX), which are hazardous compounds regulated by the U.S. Environmental Protection Agency (EPA 1977). Although these aromatic hydrocarbons are relatively water soluble, they are contained in the immiscible bulk fuel phase that serves as a slow-release mechanism for sustained groundwater contamination. Pump-and-treat technology alone is economically impractical for renovating aquifers contaminated with bulk fuel because the dynamics of immiscible fluid flow result in prohibitively long time periods for complete removal of the organic phase (Bouchard *et al.* 1989; Wilson & Conrad 1984).

Aerobic biorestoration, in conjunction with free product recovery, has been shown to be effective for many fuel spills (Lee *et al.* 1988; Thomas *et al.* 1987). However, success is often limited by the inability to provide sufficient oxygen to the contaminated intervals due to the

* U.S. Environmental Protection Agency, Robert S. Kerr Environmental Research Laboratory, P.O. Box 1198, Ada OK 74820

low solubility of oxygen (Barker *et al.* 1987; Wilson *et al.* 1986). Nitrate can also serve as an electron acceptor; this results in anaerobic biodegradation of organic compounds associated with the processes of nitrate reduction and denitrification (Tiedje 1988). Because nitrate is much more soluble than oxygen, it may require less time and hence be more economical to restore fuel-contaminated aquifers under denitrifying rather than aerobic conditions. Several investigators have observed biodegradation of aromatic fuel hydrocarbons under denitrifying conditions (Kuhn *et al.* 1988; Major *et al.* 1988; Mihelcic & Luthy 1988). However, the process is not well understood at field scale where several other processes, including aerobic biodegradation, can proceed concomitantly.

The use of nitrate to promote biological removal of fuel aromatic hydrocarbons was investigated for a JP-4 jet fuel spill at Traverse City, Michigan, through a field demonstration project in cooperation with the U.S. Coast Guard. This report compares the biodegradation rates observed in laboratory microcosms with those observed in the field demonstration project.

LABORATORY STUDIES

Methods and Materials

Core samples were obtained at several locations beneath the water table and contaminated zone using aseptic sampling techniques under nonoxidizing conditions to prevent oxygen intrusion and thus better maintain the microbial community structure and chemical integrity of the cores (Leach *et al.* 1989). Microcosms were prepared aseptically in an anaerobic glovebox to preclude intrusion of oxygen. All preparations were made when the atmospheric oxygen concentration in the glovebox was less than 10 ppm (v/v) as measured by an oxygen monitor. Test chemicals were reagent grade, and all glassware and preparation supplies were sterilized. Dilution water, used to prepare stock solutions and to transfer core material, consisted of distilled water mixed with groundwater from a spring near Ada to simulate the groundwater at the Traverse City site. The dilution water was then sterilized and aseptically purged with nitrogen gas prior to transfer into the glovebox.

Microcosms were prepared by adding 10.0 g core material to 12-mL serum bottles. Core material was rinsed into the serum bottles using a small quantity of water, and each sample was amended with nutrients to provide solution concentrations of 10 mg/L ammonia-nitrogen and 10 mg/L phosphate-phosphorus. The denitrifying microcosms were further amended with potassium nitrate to yield 100 mg/L nitrate-nitrogen. Positive controls were not amended with nitrate; poisoned controls contained nitrate with 250 mg/L mercuric chloride and 500 mg/L sodium azide as biocides to inhibit microbial growth. Each microcosm was then spiked with an aqueous stock containing benzene, toluene, ethylbenzene, *m*-xylene, *o*-xylene, and 1,2,4-trimethylbenzene to yield final solution concentrations of 1 to 4 mg/L for each compound and a headspace volume of 3.5 mL. Immediately after spiking, the microcosms were sealed using Teflon®-lined butyl rubber septa, mixed, and removed from the glovebox. The headspace gas for each microcosm was replaced by inserting a needle alongside the septum, without puncturing the material, and purging the headspace for 30 sec at 100 mL/min with the appropriate gas. Compressed air was used for aerobic microcosms, and helium was used for the denitrifying microcosms and the positive controls. Microcosms were inverted and incubated in an anaerobic glovebox in the dark at 12 C, the temperature of the groundwater at Traverse City.

Sampling and Analysis

Three replicates from each set were sacrificed at designated time intervals. Each microcosm was mixed and centrifuged at 1,500 rpm for 30 min to clarify the supernatant. The volatile aromatic hydrocarbons were analyzed by purge-and-trap gas chromatography using a Tekmar LSC-2000 liquid sample concentrator and an HP5890 GC with a flame ionization detector. Hydrocarbons were purged onto a Tenax trap for 6 min at 34 C followed by a 2-min dry purge and desorbed for 4 min at 180 C. Samples were chromatographed using a 30 m × 0.32 mm megabore DB-5 capillary column with a 1.0-µm film thickness. The injector temperature was 120 C, and the oven temperature was programmed from 32 C (4 min) to 110 C (1 min) at 8 C/min with a flow rate of 5 mL/min. The quantitation limit for these compounds was 0.001 mg/L. Samples were also analyzed for aqueous nitrate, nitrite, and ammonia concentrations using standard EPA methods (Kopp & McKee 1979).

Results

Biodegradation was observed for all of the test compounds under both aerobic and denitrifying conditions, with the exception of benzene under denitrifying conditions (Figure 1). No biodegradation was observed in the positive controls within the incubation period, indicating that either nitrate or oxygen were required for biodegradation of the compounds (data not shown). Under aerobic conditions, all of the compounds were degraded within 7 days, although detectable concentrations of *o*-xylene remained until day 14. In general, biodegradation under denitrifying conditions resulted in either decreased rates or increased lag periods (Figure 1). For purposes of comparison, reaction rates were assumed to be zero order and the reaction was considered to commence after the observed lag period. Kinetic data for this and the field demonstration project were derived for the labile compounds (Table 1). Zero-order rate constants ranged from 0.073 to 0.20 mg/L/day under denitrifying conditions. It was not possible to obtain specific rate constants under aerobic conditions because in most cases the reaction was complete before the first time step; in general, rate constants were greater than 0.3 mg/L/day.

FIELD DEMONSTRATION PROJECT

The field demonstration project is described in detail elsewhere (Hutchins *et al.* 1990). In brief, a shallow water table aquifer at a U.S. Coast Guard facility in Traverse City, Michigan, was contaminated with JP-4 jet fuel. The Coast Guard is evaluating nitrate-based biorestoration for cleanup of the aquifer through a field demonstration project on a 10 × 10 m section of the spill. An infiltration gallery was constructed to perfuse the study area with groundwater supplemented with nitrate and nutrients (Figure 2a). The study area contains five cluster wells to monitor groundwater quality. Each cluster well is constructed of 6.4-mm OD stainless steel tubing and contains six discrete sampling points with 40 × 40 mesh stainless steel screens. The sample points for each cluster well are spaced 0.6 m apart with interspersed bentonite seals to prevent channeling of groundwater recharge. The sample points for each cluster well are located both within and beneath the contaminated zone to allow monitoring of biodegradation and contaminant transport within the different regions.

Computer modeling of the hydraulics was conducted to design the infiltration system (Downs *et al.* 1989). Results of the model

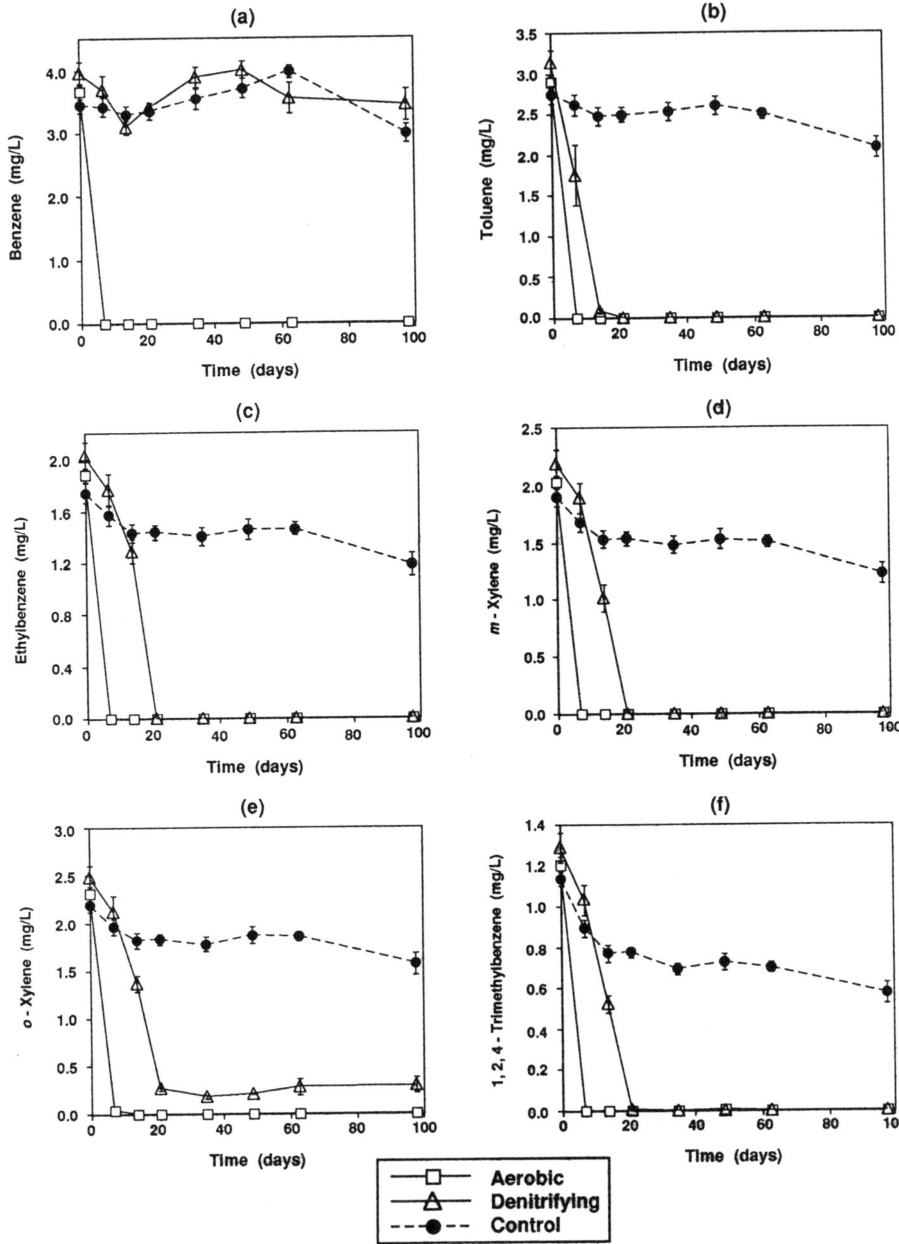

FIGURE 1. **Biodegradation of (a) benzene, (b) toluene, (c) ethylbenzene, (d) *m*-xylene, (e) *o*-xylene, and (f) 1,2,4-trimethylbenzene in laboratory microcosms under aerobic and denitrifying conditions.**

TABLE 1. Rates of BTEX removal under aerobic and denitrifying conditions in laboratory microcosms and in field demonstration project.

Compound	Rate Constant (mg/L/day)		
	Laboratory, aerobic	Laboratory, denitrifying	Field, denitrifying mean (range)
Benzene	>0.5[a]	–[b]	NQ[c]
Toluene	>0.4	0.20	0.32 (0.09–0.54)
Ethylbenzene	>0.3	0.13	0.15 (0.05–0.31)
m-xylene	>0.3	0.14	0.47[d] (0.10–0.95)
o-Xylene	0.3	0.13	0.11 (0.00–0.62)
1,2,4-trimethylbenzene	>0.2	0.073	NQ

(a) Degraded before first sampling period; boundary limit only.
(b) Not degraded.
(c) Not quantitated.
(d) Rate constant determined for *m*- and *p*-isomer.

indicated an infiltration rate of 1,090 m^3/day over the study area was required to create a water table mound encompassing the contaminated zone (Figure 2b). This enabled the contaminated unsaturated zone within the study area to be completely saturated and thus allowed even distribution of nitrate and nutrients. A preexisting line of four interdiction wells were pumped at 82 m^3/day each with the effluents routed to a carbon treatment system. These wells are screened throughout the aquifer and include the contaminated zone. In addition, five pumping wells were installed to recirculate the recharge water back to the infiltration gallery. These wells are screened below the contaminated zone to avoid influx of free product. The pumping wells recirculate the water at 1,090 m^3/day, and the interdiction wells provide a net discharge of 330 m^3/day to retain nitrate and contaminants within the facility boundaries.

Infiltration, using recirculated groundwater without amendments, commenced 20 April 1989 and continued until the design recirculation rate was attained; this rate was maintained for 2 weeks to achieve hydraulic equilibrium. A tracer study indicated that vertical

(a)

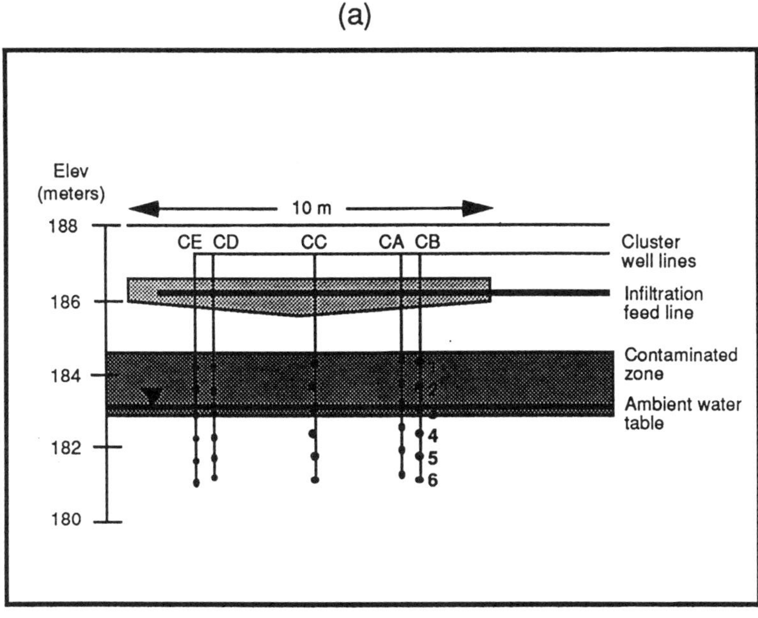

(b)

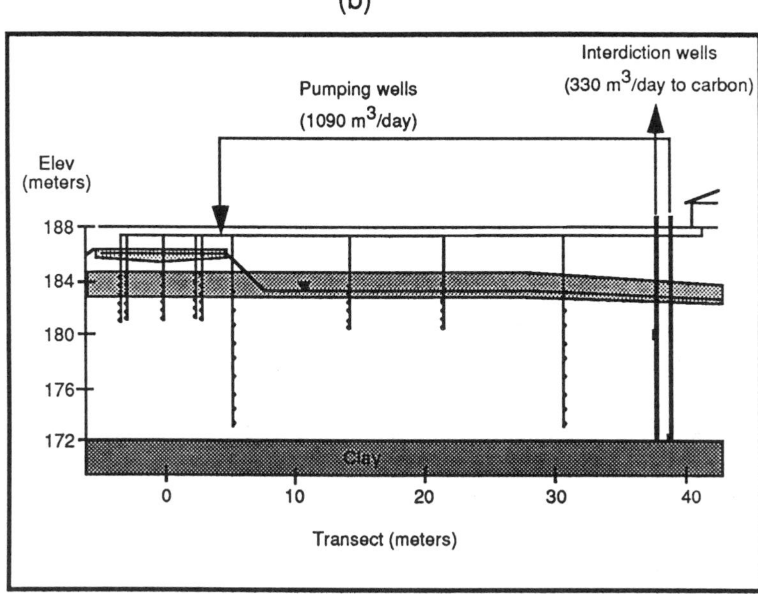

FIGURE 2. Cross-section of (a) infiltration gallery and (b) field demonstration project showing locations of monitoring wells, contaminated interval, and down-gradient wells.

flow was uniform beneath the infiltration gallery with a hydraulic residence time of 8 h within the contaminated zone. Nitrate and nutrients were then applied at design concentrations of 62 mg/L sodium nitrate, 10 mg/L monobasic potassium phosphate, 10 mg/L disodium phosphate, and 20 mg/L ammonium chloride, and full operation of the system commenced 31 May 1989 and continued for 2 months. Performance of the pilot project was assessed in part by monitoring aqueous BTX and inorganic chemical concentrations as recharge water passed through the contaminated zone beneath the infiltration gallery. Core samples were also taken periodically and are discussed in detail elsewhere (Hutchins *et al.* 1990).

Results

Water quality data were collected weekly throughout the system during the project. These analyses included dissolved oxygen, nitrate, nitrite, and BTX from selected cluster monitoring well points. As shown in Figure 2a, there are six well points for each cluster monitoring well, with the shallowest well point being designated Level 1 and the deepest being Level 6. The following discussion will focus on water quality at Level 2 (within the contaminated zone, above the original water table), Level 4 (just beneath the contaminated zone), and Level 6 (farther beneath the contaminated zone). Data were averaged for each of the five cluster monitoring well points across a given level. This was done to allow a concise description of changes in water quality during infiltration across the contaminated zone. A more complete data set has been published (Hutchins *et al.* 1990).

The dissolved oxygen and nitrate-nitrogen profiles for the injection water and the water at different levels are shown in Figure 3. Infiltration began on day 3, and nitrate and nutrient addition began on day 44. Samples were collected from the well points at day 0 to provide background information. Oxygen concentrations were initially high beneath the contaminated zone and in fact were even substantial at the bottom of the contaminated zone with values approximating 3 mg/L. However, oxygen concentrations dropped rapidly as water was recirculated through the contaminated zone (Figure 3a). Oxygen concentrations stabilized at 0.5 to 1.0 mg/L within the monitoring wells by day 7. It appears that limited oxygen consumption occurred from day 20 to day 64 with little oxygen removal afterward, but this may not be significant given the resolution of the data.

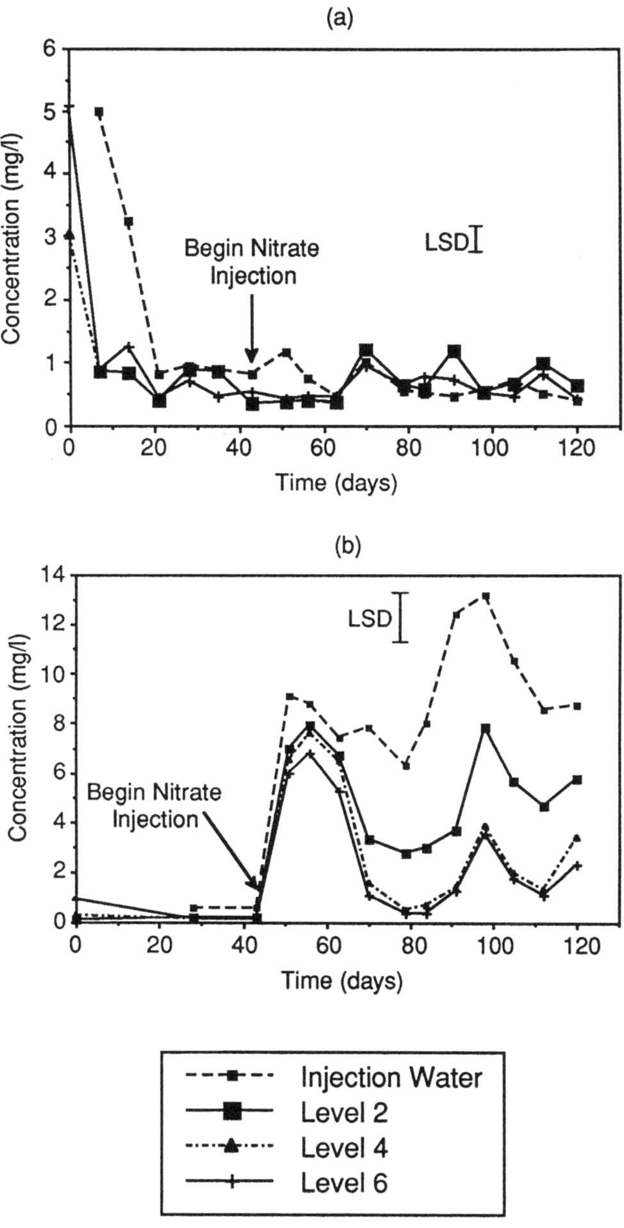

FIGURE 3. Profiles of (a) dissolved oxygen and (b) nitrate-nitrogen beneath infiltration gallery in field demonstration project. LSD is the least significant difference between means at the 90 percent confidence level.

When nitrate and nutrients were added to the infiltration water, there was a lag period of about 20 days before significant nitrate removal occurred (Figure 3b). This is consistent with the observation that denitrifying bacteria generally require a finite time for enzyme induction. However, nitrate removal was rapid and consistent after this time, with a removal rate of about 1 mg nitrate-nitrogen/L/h. This served to keep nitrate-nitrogen concentrations well below 10 mg/L once the water reached the interdiction field, in compliance with the permit granted by the Michigan Department of Natural Resources. There was essentially no nitrate removal between Levels 4 and 6, that is, after the infiltrating water had already passed through the contaminated zone. This implies that most of the nitrate was consumed during oxidation of organic matter associated with the contaminated interval. Nitrite production was transitory, and concentrations stabilized to 0.1 to 0.5 mg/L nitrite-nitrogen by day 70 (data not shown).

Based on the measured infiltration rates, the mass of oxygen added can be calculated using the following equation:

$$M = \int CF\,dt$$

where M is the total mass in mg, C is the concentration in mg/L, F is the infiltration rate in L/day, and t is the time in days. The integral is evaluated numerically using a trapezoidal algorithm. Mass estimates can be generated for each level throughout the vertical profile, and the difference between oxygen mass supplied in the infiltration water and that found at Level 6 corresponds to oxygen uptake within the contaminated zone, assuming that oxygen removal is due primarily to aerobic biodegradation within the contaminated zone. This equation was also used to calculate mass fluxes for the other parameters, and the mass estimates are tabulated in Table 2 before and after nitrate addition. Prior to nitrate addition, 92 kg oxygen were supplied in the recharge water and 55 kg oxygen were removed by the time the infiltrating water reached Level 6 (Table 2).

There were significant nitrate concentrations in the feed prior to nitrate addition; a net total of 25 kg nitrate-nitrogen were consumed as compared to 55 kg oxygen (Table 2). Thus, there may have been some denitrifying activity prior to exogenous nitrate addition, but it is not possible to distinguish this from nitrate uptake for cellular synthesis. After nitrate addition, however, approximately 500 kg nitrate-nitrogen were consumed. Given the limited oxygen availability at this time, it is probable that most of this uptake was in conjunction with denitrification. Supporting evidence is provided by analysis of selected core

TABLE 2. Calculated mass of nutrients in infiltration water before and after nitrate addition, and changes in mass after passage through contaminated zone. Data are in kg.

Parameter	Before Nitrate Addition (Day = 3 to 44)		After Nitrate Addition (Day = 44 to 112)	
	Total mass supplied in recharge	Mass change from feed to level 6	Total mass supplied in recharge	Mass change from feed to level 6
Dissolved Oxygen	92	–55	48	–4
Nitrate-Nitrogen	33	–25	690	–499
Nitrite-Nitrogen	<2	–	66	–25
Ammonia-Nitrogen	<2	–	301	–18
Phosphate-Phosphorus	<2	–	97	–20

samples before and after nitrate addition; although total microbial numbers did not increase appreciably, there was a significant increase in the denitrifier population, especially in those intervals that had been contaminated with fuel (Hutchins *et al.* 1990). Approximately 20 kg of ammonia-nitrogen and phosphate-phosphorus were also consumed throughout the study, presumably for cellular synthesis (Table 2).

In the initial part of the test, then, both nitrate and oxygen are available for respiration, whereas aerobic metabolism is probably minimal after exogenous nitrate addition. The stoichiometry of BTEX biodegradation can be described as follows:

$$C_{61}H_{67} + 77.75\ O_2 \rightarrow 61\ CO_2 + 33.5\ H_2O$$

$$C_{61}H_{67} + 62.2\ NO_3^- + 62.2\ H^+ \rightarrow 61\ CO_2 + 31.1\ N_2 + 64.6\ H_2O$$

Based on these equations, about 3.1 kg oxygen are required for each kg of BTEX mineralized under aerobic conditions, and 1.1 kg nitrate-nitrogen are required for each kg BTEX mineralized under denitrifying conditions. Therefore, oxygen consumption prior to nitrate addition was sufficient to account for the removal of 18 kg BTEX, whereas nitrate consumption after nitrate addition was sufficient to account for the removal of 450 kg BTX. These observations indicate that metabolic activity within the contaminated zone switched from primarily aerobic to denitrifying after the addition of exogenous nitrate.

BTEX profiles are shown in Figure 4 for the infiltration feed water and the Level 2 monitoring wells. Benzene removal was rapid and complete prior to nitrate addition (Figure 4a). By day 14, benzene levels were typically less than 0.001 mg/L throughout the contaminated zone. Significant benzene concentrations were present in the recirculated infiltration water up to day 35; these concentrations are the result of continued contact of the groundwater with the fuel down-gradient of the study site (Figure 2b). Removal of benzene occurred before the water reached the Level 2 well points. Based on core analyses, approximately 0.4 kg benzene was removed from the contaminated interval prior to nitrate addition (data not shown). There was clearly enough oxygen consumption in the system to account for aerobic biodegradation of benzene at this time.

Toluene removal was more complex (Figure 4b). By day 21, the system had equilibrated hydraulically and toluene concentrations were similar throughout. Between day 21 and day 50, some toluene removal became apparent as the infiltration water reached the Level 2 well points. After this time, toluene concentrations dropped rapidly and were below 0.001 mg/L at all levels even though concentrations persisted in the recirculated infiltration water. Unlike toluene, however, there was no significant removal of *m,p*-xylene prior to nitrate addition (Figure 4c). After nitrate addition, *m,p*-xylene concentrations rapidly dropped by the time the infiltrating water reached the Level 2 well points. A similar pattern was seen with ethylbenzene (data not shown). The profiles for *o*-xylene were unique, indicating little loss of the compound during infiltration (Figure 4d). There was a gradual decline in total mass due to dilution effects during recirculation of the infiltration feed, but little loss that could be attributed to biological activity.

Based on the hydraulic residence time and the removal of BTEX observed at the Level 2 well points, zero-order rate constants were calculated for each compound after day 56; this corresponds to approximately 20 days after nitrate addition to avoid the effects of the lag period. These data are summarized in Table 1. Average rate constants ranged from 0.11 to 0.47 mg/L/day and were in good agreement with those derived under denitrifying conditions, with the exception of *m*-xylene. This may be due to co-elution of *p*-xylene in the field project, leading to an inappropriate comparison.

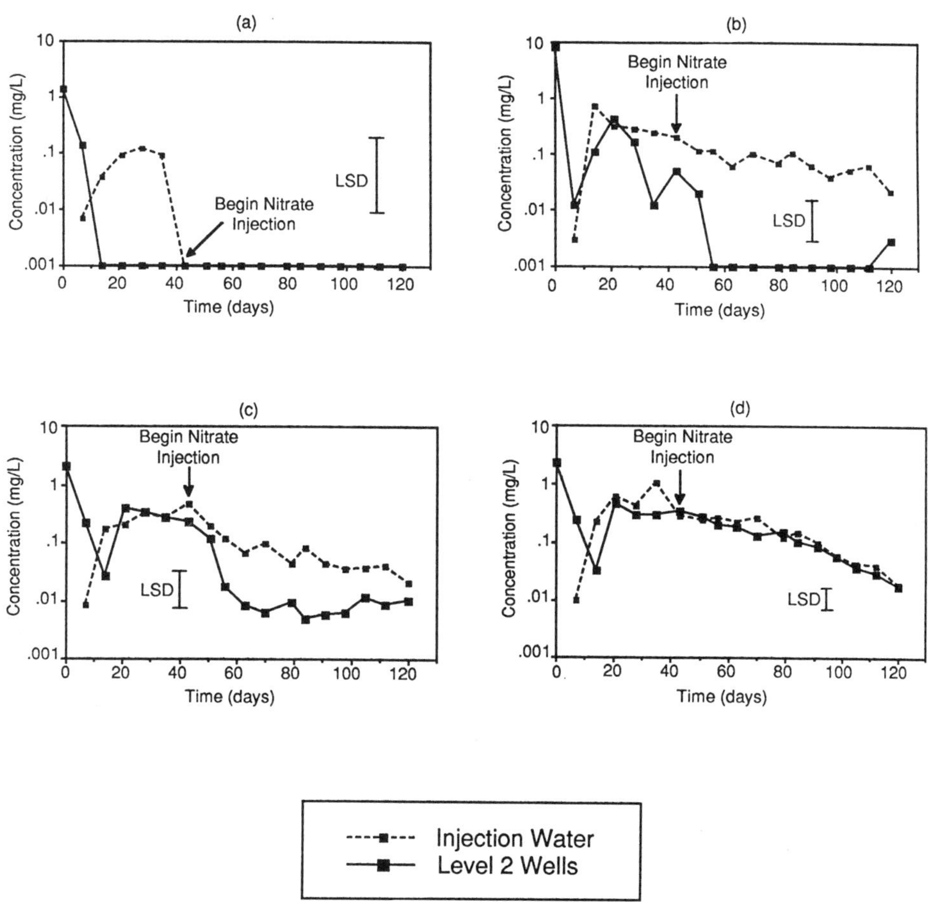

FIGURE 4. Profiles of (a) benzene, (b) toluene, (c) *m,p*-xylene, and (d) *o*-xylene beneath infiltration gallery in field demonstration project. LSD is the least significant difference between means at the 90 percent confidence level.

DISCUSSION

The microcosm data demonstrate that certain alkylbenzenes can be degraded under denitrifying conditions, although benzene is recalcitrant. Benzene biodegradation under denitrifying conditions is enigmatic; some studies report benzene to be recalcitrant (Bouwer &

McCarty 1983; Kuhn *et al.* 1988; Zeyer *et al.* 1986), whereas other studies indicate that benzene is rapidly degraded (Batterman 1986; Berry-Spark *et al.* 1986; Major *et al.* 1988). In some of these latter systems, the possibility of oxygen intrusion into microcosms cannot be discounted based on available information concerning the experimental design, but it is also possible that subsurface microbial populations and previous exposure differ sufficiently to account for the discrepancy. Even for the compounds that are degraded, however, rates of removal are less under denitrifying conditions than would occur under aerobic conditions. For some of these compounds, at least, these differences may not be significant. Further comparisons are required with different aquifer material to determine if this is consistent.

One of the problems in extrapolating these results to the field is that similar degrees of control are rarely possible in field situations, and therefore interpretation of results becomes more difficult. In the field demonstration project, for example, sorption, volatilization, abiotic degradation, and other microbial processes can be occurring concomitantly with nitrate-based respiration. However, hydraulic and chemical equilibrium appeared to have been attained by day 20 as evidenced by the monitoring well data. For ethylbenzene and the xylene isomers, at least, the concentrations were fairly uniform across the contaminated interval until day 42, just before nitrate addition commenced. Denitrifying activity was most likely responsible for the increased removal rates observed afterward; this is also supported by the observed decreases in nitrate concentrations across the contaminated zone, transient nitrite production, and increases in denitrifier populations. Futhermore, *o*-xylene is similar to the other isomers on a physical chemical basis and yet is more recalcitrant under denitrifying conditions in the field as well as under strict laboratory conditions. Mass balances also indicate that nitrate consumption was sufficient to account for alkylbenzene biodegradation after nitrate addition, whereas oxygen consumption was not (Hutchins *et al.* 1990).

Therefore, although it is not known to what extent BTEX was degraded solely by the denitrifying bacteria, these data are consistent with results from laboratory microcosms prepared with the same material under strictly denitrifying conditions. Given the number of processes that can be operating simultaneously in the field demonstration project, it is surprising that the zero-order rate constants were similar between the laboratory and field. Further work is needed in this area to determine whether this is consistent among other field sites and to better assess the potential for nitrate-based biorestoration of fuel-contaminated aquifers.

Acknowledgments

The field demonstration project was a multidisciplinary effort involving many more people than those listed as authors. In particular, the authors wish to thank Wayne Downs, Garmon Smith, Don Kampbell, Lowell Leach, Mike Cook, and Montie Frasier of the Robert S. Kerr Environmental Research Lab; Dave Kovacs, Dennis Fine, Mark White, Lynda Pennington, and Alton Tweedy of ManTech Environmental Technology, Inc.; Rob Douglass of The Traverse Group, Inc.; and Dan Hendrix and Shanna Shea of Solar Universal Technologies for their combined technical assistance.

Disclaimer

Although the research described in this paper has been funded wholly or in part by the U.S. Environmental Protection Agency through the Biosystems Technology Development Program (IAG DW69933299 from RSKERL to the 9th District, U.S. Coast Guard) and through the United States Air Force (MIPR N-89-44 from HQ AFESC/RDXP, Tyndall AFB, FL, to RSKERL), it has not been subjected to Agency review and therefore does not necessarily reflect the views of the Agency or the Air Force, and no official endorsement should be inferred.

REFERENCES

Barker, J. F.; Patrick, G. C.; Major, D. *Ground Water Monitoring Review* **1987**, *7*, 64–71.

Battermann, G. In *1985 International TNO Conference on Contaminated Soil;* Assink, J. W.; van den Brink, W. J., Eds.; Nijhoff: Dordrecht, 1986; pp 711–722.

Berry-Spark, K. L.; Barker, J. F.; Major, D.; Mayfield, C. I. In *Proceedings, Petroleum Hydrocarbons and Organic Chemicals in Ground Water: Prevention, Dectection, and Restoration;* NWWA/API; Water Well Journal Publishing: Dublin, OH, 1986; pp 613–623.

Bouchard, D. C.; Enfield, C. G.; Piwoni, M. D. In *Reactions and Movement of Organic Chemicals in Soil*; Sawhney, B. L., Brown, K., Eds.; Soil Science Society of America and American Society of Agronomy, SSSA Special Publication No. 22, 1989; pp 349–371.

Bouwer, E. J.; McCarty, P. L. *Applied and Environmental Microbiology* **1983**, *45*, 1295–1299.

Downs, W. C.; Hutchins, S. R.; Wilson, J. T.; Douglass, R. H.; Hendrix, D. J. In *Proceedings, Petroleum Hydrocarbons and Organic Chemicals in Ground Water: Prevention, Detection, and Restoration*; NWWA/API; Water Well Journal Publishing: Dublin, OH, 1989; pp 219–233.

EPA. *Serial No. 95-12;* U.S. Government Printing Office: Washington, DC, 1977.

Feliciano, D. *Congressional Research Service Report;* U.S. Library of Congress: Washington, DC, January, 1984.

Hutchins, S. R.; Downs, W. C.; Smith, G. B.; Wilson, J. T.; Hendrix, D. J.; Fine, D. D.; Kovacs, D. A.; Douglass, R. H.; Blaha, F. A. *RSKERL Research Report;* U.S. Environmental Protection Agency: Ada, OK, 1990.

Kopp, J. F.; McKee, G. D. *Manual—Methods for Chemical Analysis of Water and Wastes; U.S. Environmental Protection Agency, 1979;* EPA-600/4-79-020.

Kuhn, E. P.; Zeyer, J.; Eicher, P.; Schwarzenbach, R. P. *Applied and Environmental Microbiology* **1988**, *54*, 490–496.

Leach, L. E.; Beck, F. P.; Wilson, J. T.; Kampbell, D. H. In *Proceedings, Second National Outdoor Action Conference on Aquifer Restoration, Ground Water Monitoring and Geophysical Methods, Vol. 1;* **1989**; pp 31–51.

Lee, M. D.; Thomas, J. M.; Borden, R. C.; Bedient, P. B.; Wilson, J. T.; Ward, C. H. *Critical Reviews in Environmental Control* **1988**, *18*, 29–89.

Major, D. W.; Mayfield, C. I.; Barker, J. F. *Ground Water* **1988**, *26*, 8–14.

Mihelcic, J. R.; Luthy, R. G. *Applied and Environmental Microbiology* **1988**, *54*, 1188–1198.

Thomas, J. M.; Lee, M. D.; Bedient, P. B.; Borden, R. C.; Canter, L. W.; Ward, C. H. *RSKERL Publication;* U.S. Environmental Protection Agency, 1987; EPA 600/2-87/008.

Tiedje, J. M. In *Bioloqy of Anaerobic Microorganisms;* Zehnder, A.J.B., Ed.; John Wiley and Sons: NY, 1988; pp 179–244.

Wilson, J. L.; Conrad, S. H. In *Proceedings, Petroleum Hydrocarbons and Organic Chemicals in Ground Water: Prevention, Detection, and Restoration*; NWWA/API; Water Well Journal Publishing: Dublin, OH, 1984; pp 274–298.

Wilson, J. T.; Leach, L. E.; Henson, M.; Jones, J. N. *Ground Water Monitoring Review* **1986**, *6*, 56–64.

Zeyer, J.; Kuhn, E. P.; Schwarzenbach, R. P. *Applied and Environmental Microbiology* **1986**, *52*, 944–947.

Hydraulic Circulation System for *In Situ* Bioreclamation and/or *In Situ* Remediation of Strippable Contamination

B. Herrling, J. Stamm, W. Buermann*
University of Karlsruhe

INTRODUCTION

Researchers of different fields endeavour to remediate contaminated groundwater zones from organic compounds that have found their way into the subsurface in every industrial country. A physical method to clean the groundwater is to use stripping techniques for volatile substances. Biological degradation by microorganisms is another method to remove organic contaminants. From the early work of Raymond *et al.* (1976) to a practical application by Battermann and Werner (1987), both concerning petroleum products, to very complex biodegradation of chlorinated ethenes by Semprini *et al.* (1990), excellent research has been carried out. Review papers of biorestoration in aquifers have recently been published by Lee *et al.* (1988), McCarty (1988), and Wilson *et al.* (1986).

To reduce the investment and operating costs, *in situ* remediation methods are favored. The authors have investigated an *in situ* method that can remove strippable substances, *e.g.*, volatile chlorinated hydrocarbons, and BTEX, from the subsurface (groundwater zone, capillary fringe, and unsaturated zone); it is currently being used at

* Institute of Hydromechanics, University of Karlsruhe, Kaiserstrasse 12, D-7500 Karlsruhe 1, Federal Republic of Germany

numerous locations in Germany (Herrling *et al.* 1990, in press). This technology is an alternative to conventional hydraulic remediation measures (pumping, off-site cleaning, and reinfiltration of groundwater). The contaminated groundwater is stripped *in situ* by air in a below atmospheric pressure field within a so-called "vacuum vaporizer well" (German: *U*nterdruck-*V*erdampfer-*B*runnen, UVB). The used air, charged with volatile contaminants, is cleaned using activated carbon.

The UVB technique produces a vertical circulation flow in the area surrounding the well, which catches the total aquifer. The vertical velocity component yields a desired flow through the horizontal structure of a native aquifer. Numerical results demonstrate the size of the sphere of influence and the capture zone of a well or well field; extended field measurements have been and continue to be taken.

The advantages of the UVB technique concerning the vertical circulation system around the wells instigated thought about other applications, even without stripping the groundwater. The realization of *in situ* biodegradation is such an example and seems to be an appropriate alternative to other existing hydraulic systems (*e.g.*, McCarty *et al.* 1989). The different nutrients and/or electron acceptors needed for biological activity can be added when the groundwater passes the well casing. The added quantity depends directly on the quality of the groundwater entering the well through the inflow screen. For sites with natural groundwater flow, a part of the treated groundwater, together with contaminated groundwater coming from upstream, will pass the well casing again. Another part of the treated groundwater will flow directly downstream after having been remediated by *in situ* biodegradation. If biofouling clogs the aquifer behind the outflow screen of the well, the flow direction within the well can be reversed.

This paper presents the UVB technique for *in situ* removal of strippable contaminants, then explains the UVB method using only the hydraulic circulation system for biorestoration. The circulation system, sphere of influence, and capture zone of a UVB or UVB field as essential components of the hydraulic flow system are discussed in detail.

OPERATION OF THE WELL FOR *IN SITU* STRIPPING

The UVB helps to remove volatile substances from the groundwater, the unsaturated zone, and the capillary fringe. When using the UVB method, a special well with two screen sections is employed, one at the aquifer bottom and one at the groundwater surface (Figure 1). The borehole reach between the two screen sections should be made impermeable. One well should be used to remediate only one aquifer (phreatic or confined) and should not connect different aquifers.

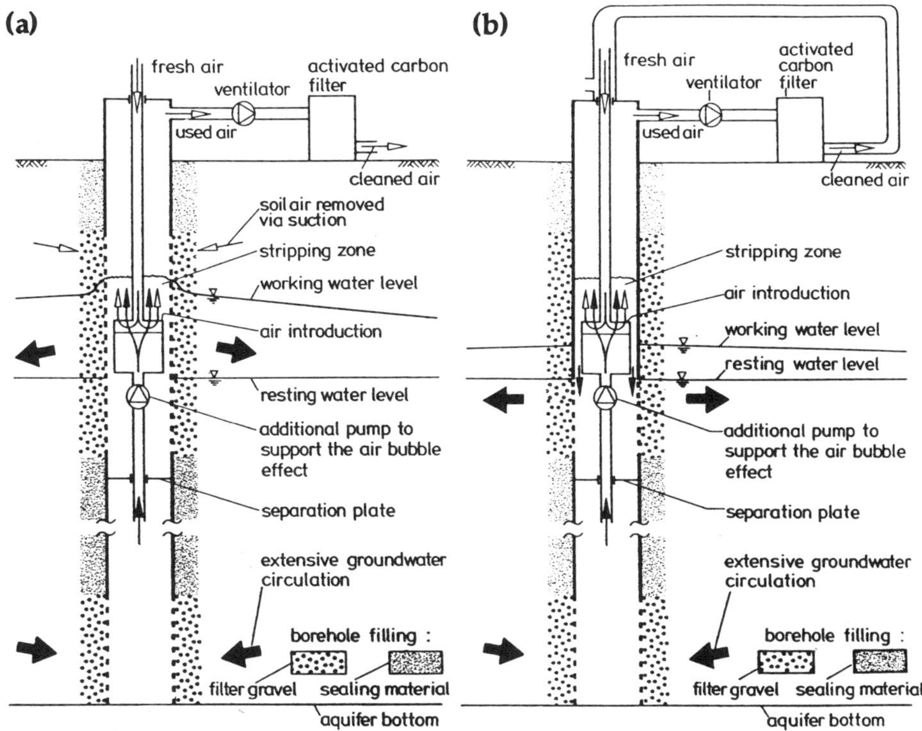

FIGURE 1. Vacuum vaporizer well (UVB): (a) with additional pump and separating plate and (b) with closed air circulation.

The upper, closed part of the well is maintained at below-atmospheric pressure by a ventilator. This lifts the water level within the well casing. The fresh air for the upper part of the well casing is introduced through a fresh air pipe: the upper end is open to the atmosphere, and the lower end terminates in a pinhole plate. The height of the pinhole plate is adjusted such that the water pressure is lower there than the atmospheric pressure. Therefore, the fresh air is drawn into the system. The reach between the pinhole plate and the water surface in the well casing is the stripping zone, in which an air bubble flow develops. The rising air bubbles produce a pump effect, which moves the water up and causes a suction effect at the well bottom. In recent wells, a separating plate and an additional pump (Figure 1a) are used to reinforce the pumping effect of the air bubbles. Additionally, soil air is drawn from the surrounding contaminated unsaturated zone at many sites. Stripped air and possibly soil air are transported through the ventilator and across activated carbon, onto which the contamination is adsorbed. Thus, only clean air escapes into the atmosphere.

The cleaning effect of the well is based on reduced pressure, which reinforces the escape of volatile contamination out of the water and, as a result of the air intermixing, onto the considerable surface area of the air bubbles and onto the concentration gradient. In this sense, the permanent vibration caused by the air bubbles is beneficial to the escape process of the contamination. This vibration is transmitted as compression and shear waves into sediment and fluid and presumably influences the mobility of the contaminants, even outside the well.

The upward-streaming, stripped groundwater leaves the well casing through the upper screen section in the reach of the groundwater surface, which is lifted in a phreatic aquifer by the previously explained pump processes and the below-atmospheric pressure. It then returns in an extensive circulation to the well bottom. In this way, the groundwater surrounding the well is also remediated. The expansion of groundwater circulation is positively influenced by the anisotropy existing in each natural aquifer possessing greater horizontal than vertical hydraulic conductivities. The artificial groundwater circulation determines the sphere of influence of a well and is overlapped with the natural groundwater flow (as described below).

The pinhole plate and all the installations within the well casing are designed as a float so they can adjust automatically to changing groundwater levels.

In the well design represented in Figure 1a with additional pump and separation plate, precipitation of iron, manganese, and calcium may occur depending on the water quality at a remediation site. In general, the water entering the well contains very little oxygen. When it leaves the well after flowing through the stripping zone, it is fully saturated with oxygen. Supersaturation with carbon dioxide at the well influx is decreased in the stripping zone by the exchange with the air. The pH value increases and calcium precipitation occurs. To avoid or to reduce these effects, which may eventually seal the upper screen section, the closed UVB method has been developed (Figure 1b): the cleaned air is recycled as fresh air input in a closed circuit. Thus after only a few circulations, water and air will reach an equilibrium and the unwanted precipitation effect will disappear. Only those pollutants adsorbed by activated carbon are removed from the circulating air. Volatile substances not adsorbed remain in the groundwater and will not contaminate the atmosphere. However, special cleaning methods incorporated into the air circulation are conceivable.

When using the closed UVB method, any contaminated soil air must be separately removed via suction. Efficient technologies exist for this. Further, the upper well casing and the pipe system must be sealed so that no fresh air can enter the closed system of air circulation.

For special contaminants of lower density than water or when only the upper part of the aquifer is contaminated, a special installation within the well is available: the contaminated water enters the well through the upper screen, is stripped there, and with help of the additional pump, leaves the well through the lower screen. Both installations can be used within the same well casing. A special UVB installation allows the operator to alternate from upward to downward operation within the well.

OPERATION OF THE WELL FOR *IN SITU* BIORESTORATION

The hydraulic circulation system outside the well can be used to initiate *in situ* biodegradation. In this case, suitable nutrients (electron donors) and/or electron acceptors, depending on the contaminants, can be added in dosed quantities while the groundwater passes through the well casing. Since the growth of the desired bacteria, activated by the supplied addition, may clog the aquifer behind the outflow screen of the well, the pump system within the well has to be changed from upward to downward operation at specific time intervals. These intervals depend on the biological and chemical situation at a site, the quantity of the added nutrients and/or electron acceptors, and the contamination and its concentration. Each pulse cycle length can be varied for each addition as required for optimal biodegradation at a particular site. Present research attempts to determine suitable additions and time intervals at specific field sites, taking into consideration the amount of aquifer clogging.

Figure 2 displays a well construction in a principle sketch where no stripping is necessary at a site. In principle, the groundwater flowing in the well casing can be pumped to the surface for very special water treatment and returned to the same well, but the energy costs for such a system can be considerable.

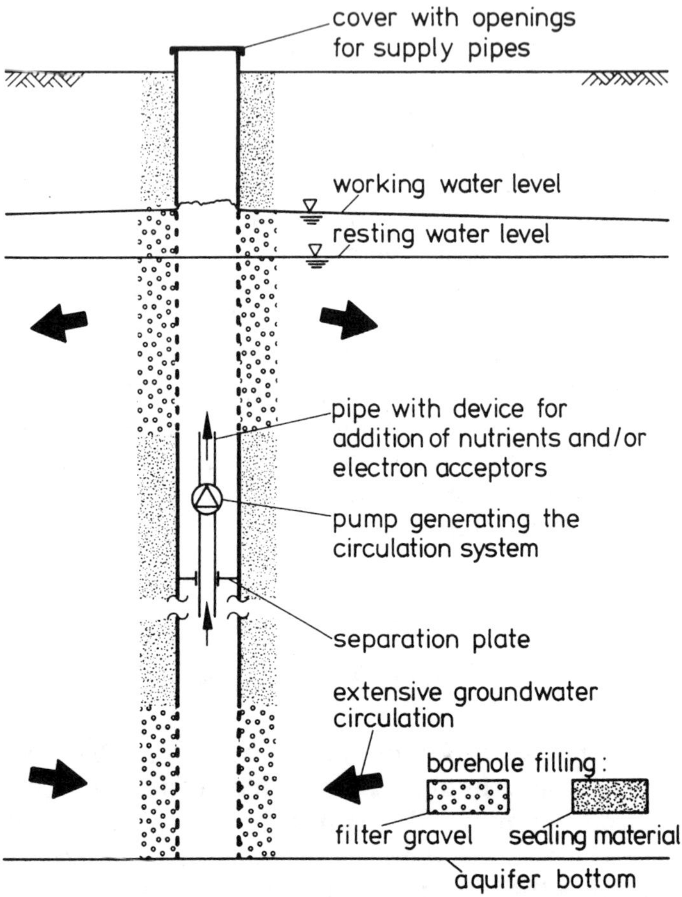

FIGURE 2. An upward-operating UBV installation that includes a circulation system in the area surrounding the well.

In principle, the circulation system in Figure 2, initiated by a pump only, can be combined directly with the above-discussed UVB installation for stripping. It must be determined from case to case whether this combination is suitable at a site.

When using the circulation system of a UVB for biodegradation at a site with natural groundwater flow, three different cases can be differentiated:

1. The contaminants are dissolved in the groundwater and *in situ* biodegradation takes place in the plume.

2. The contaminants, existing in a liquid or solid phase near a source area, can be dissolved into the groundwater where they are biodegraded in the circulation flow of a UVB.

3. The contaminants cannot be dissolved and are trapped in the porous medium at a local site where *in situ* biodegradation is initiated within the sphere of influence of the circulation flow of a UVB.

In the first case, the situation is very clear. The concentrations of the dissolved contaminants, the nutrients, and the electron acceptors are measured in the groundwater entering the UVB. From this data—and perhaps from single investigations of the aquifer material—the desired quantities of the nutrients and/or the electron acceptors can be calculated and are added before the water leaves the well. The estimation of these quantities should consider that a portion of the water will flow directly downstream. The added components should be limited to the amount which is required for the biodegradation of the contaminants down to their desired concentration levels so that no added components or undesirable reactions occur in the far downstream water. Another portion of the water will circulate around the well, and some of the added components may not have been biodegraded when the water enters the well again. By measurement of the inflow concentrations at the UVB, the remaining added components can be considered when new quantities are estimated.

For the second case, the situation is very similar to that of the first case. The only difference is that the groundwater in the upstream capture zone may be clean, but the above-described principles of the process control are the same.

For the third case, the process control for all added components is based on the principle that the desired levels of the concentrations of the added nutrients and electron acceptors are met in the water flowing downstream.

CALCULATION OF THE SPHERE OF INFLUENCE AND OF THE CAPTURE ZONE OF A UBV OR UBV FIELD

The extended circulation field outside the well is of special interest. In principle, two different cases have been considered:

- When there is no (or negligible) natural groundwater flow, the sphere of influence (or the range, R) of a UVB is of interest.

- When natural groundwater flow is significant, the extent of the capture zone has to be determined for locating the well installations at a remediation site.

In this paper the effect of the above-mentioned permanent vibrations, caused by the air bubbles, will not be considered. When a UVB is used without air stripping, this effect does not exist.

The resulting flow field of one or several UVB installations differs from the natural groundwater flow field only in a limited area around the UVB. This is because sinks and sources are located at the bottom and top of the same aquifer, each at places with the same horizontal coordinates. The affected area can, therefore, be limited to the sum of the areas of influence of all the UVBs. When only confined aquifer conditions are considered to reduce the computational effort, the flow field of each UVB can be superimposed onto those of other UVBs and of the natural groundwater flow field.

To estimate the sphere of influence and the capture zone of a UVB, numerical investigations have been performed. To calculate the complex, three-dimensional flow field of a single UVB or a UVB field with minimal effort, the following simplifications and assumptions have been used: (1) The aquifer thickness is constant. (2) Only confined aquifer conditions are considered in the calculation, even if the natural aquifer is phreatic. (3) The aquifer structure is assumed radially homogeneous to hydraulic conductivities. Horizontal layers, each with different conductivities, can be used. The hydraulic conductivities may be anisotropic, but each horizontal layer may have only one vertical and one horizontal conductivity. (4) The local below-atmospheric pressure field near the wells is neglected. (5) Density effects are neglected. (6) The computations assume steady-state conditions. (7) For estimating the capture zone, only convective transport is considered.

The three-dimensional flow field in the above-defined, limited aquifer region is obtained by superimposition of a horizontal, uniform flow field, computed in a vertical cross section and representing the natural groundwater flow, and of radially symmetric flow fields for each UVB. The superimposition of the different flow fields with their own discretization is achieved by interpolating and adding the different flow vectors at the various nodes of a simple rectangular grid with variable grid distances that are independently chosen for each Cartesian coordinate. The rectangular grid can be quickly and simply set up and allows for some refinements near the wells and their screen sections. More details of the numerical computations are given in Herrling and Buermann (1990).

Resulting Flow System

Before going into more detail, the complex flow field near an individual UVB is clarified for a vertical cross section in the direction of the natural groundwater flow (symmetry plane of the flow problem). In Figure 3, the streamlines of three case studies are illustrated with Darcy velocities (v) of natural groundwater flow of 0.0 m/day, 0.3 m/day, and 1.0 m/day. All other parameters remain constant: the discharge (Q) through the well casing is 20.16 m^3/h, the thickness (H) of the aquifer is 10 m, the anisotropic hydraulic conductivities are $K_H = 0.001$ m/sec (horizontal) and $K_V = 0.0001$ m/sec (vertical), and the lengths of the screen sections are $a_B = 1.2$ m at the bottom and $a_T = 2.1$ m at the top.

Figures 3b and 3c show that the groundwater, flowing from the left, dives downward to the lower screen section and is transported upward within the well casing, and that the cleaned water flows out to all sides at the upper screen section. The flow situation can only be calculated and plotted in such a simple way in this cross section, otherwise the complex, three-dimensional flow field has to be considered.

For a deep aquifer contaminated only in the upper groundwater zone, a UVB installation can be used at a hydraulically imperfect well. The resulting flow system is demonstrated in Figure 4, clarified for a vertical cross section in the symmetry plane (Figure 4b). The used parameters are the same as for Figure 3b. The only difference is that the aquifer thickness (H) is 30 m (well length = 10 m, as before).

At most of the UVB installation sites, a natural, nonnegligible groundwater flow will exist. For a normal withdrawal well, a separating streamline can be determined: all the water within this line is captured by the well, and all water outside of it passes the well. In principle, the situation is the same when using a UVB. In contrast to a normal withdrawal well, where the flow can be considered horizontal, the flow around a UVB must be regarded as three-dimensional. Thus the water body, flowing toward the UVB from upstream and being captured by the lower screen section, cannot be delimited by a simple separating streamline, but by a curved separating stream surface. This can be calculated as described in Herrling and Buermann (1990): on the

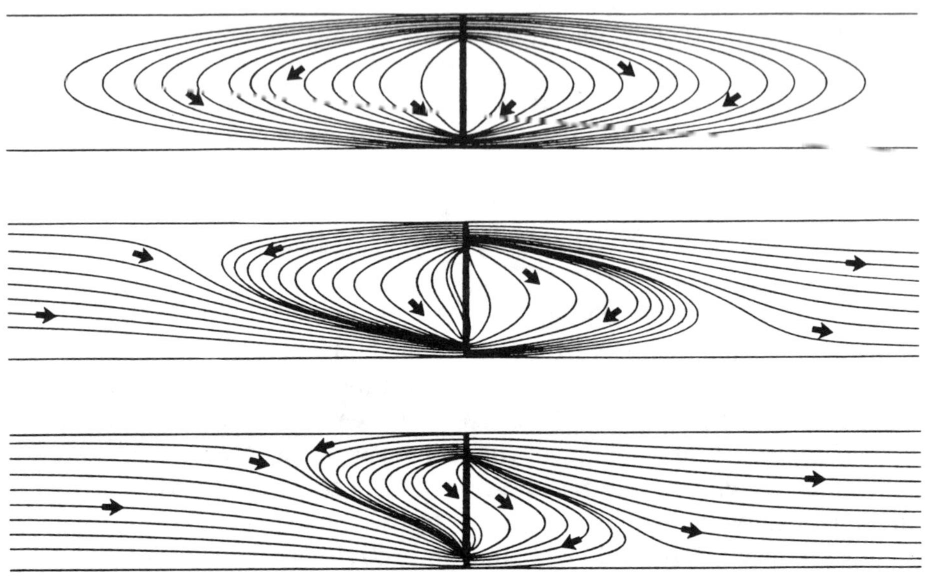

FIGURE 3. Streamlines clarified for a vertical cross section with natural velocities: (a) 0.0 m/day, (b) 0.3 m/day, and (c) 1.0 m/day.

basis of the three-dimensional flow field, a three-dimensional, particle-tracking method is used. The water body within the separating stream surface is captured by the UVB, and that outside of it, which flows from upstream, passes the well.

In Figure 5 the outer surface of the capture zone, calculated numerically, and the surrounding horizontal aquifer bottom and aquifer top are plotted for two situations (the natural groundwater flows from the background at the right side to the UVB, as shown by the vectors). Figures 5a and 5b were calculated for the situation described for Figure 3b; the only difference is that for Figure 5a the vertical hydraulic conductivity is K_V = 0.001 m/sec, which means the calculation is performed for isotropic conditions. The figures have a visible basis area of 50 m × 50 m (Figure 5a) and 100 m × 50 m (Figure 5b).

The captured water is cleaned within the well and leaves it through the upper screen section in all directions (not shown in Figure 5). Parts of it are again captured by the lower screen section, and the rest flows directly downstream.

If a wide plume of contaminated groundwater is to be cleaned, one UVB might not be enough to capture the whole plume. Different UVB installations can be arranged, for example, in one line normal to

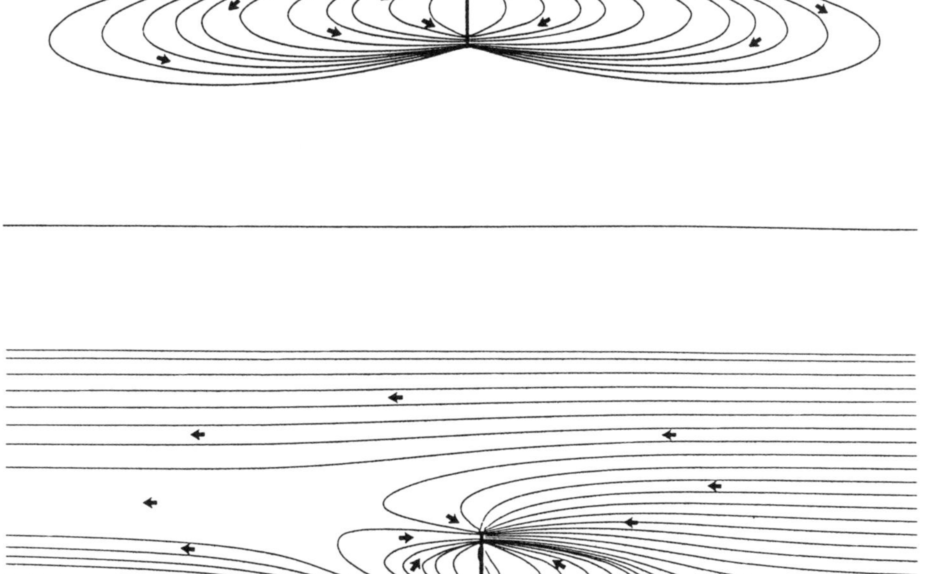

FIGURE 4. Streamlines at a hydraulically imperfect well clarified for a vertical cross section with natural velocities: (a) 0.0 m/day and (b) 0.3 m/day.

the natural flow. An important question concerns the maximum distance that allows no contaminated water to flow between two neighbouring wells without being cleaned. Figure 6 demonstrates such an example for the situation of Figure 5b where the maximum well distance is 46 m. The visible basis area of Figure 6 is to 150 m × 150 m.

Figure 7 presents a view of the separating stream surfaces of all three water bodies in connection with the flow around a UVB. The natural groundwater flow comes from the left side. (In Figure 7b the three water bodies were artificially separated for clarification.)

At the left side of Figure 7, the separating stream surface of the contaminated groundwater captured by the UVB can be seen. In the center a water body is shown which consists of cleaned groundwater and shows the circulation flow around the UVB. At the right side of Figure 7, the separating stream surface of the cleaned groundwater flowing downstream is displayed. The calculation has accounted for the following parameters: $Q/(H^2v) = 30$, $a/H = 0.25$, and $K_H/K_V = 5$. The screen lengths at the bottom and top are the same: $a_T = a_B = a$.

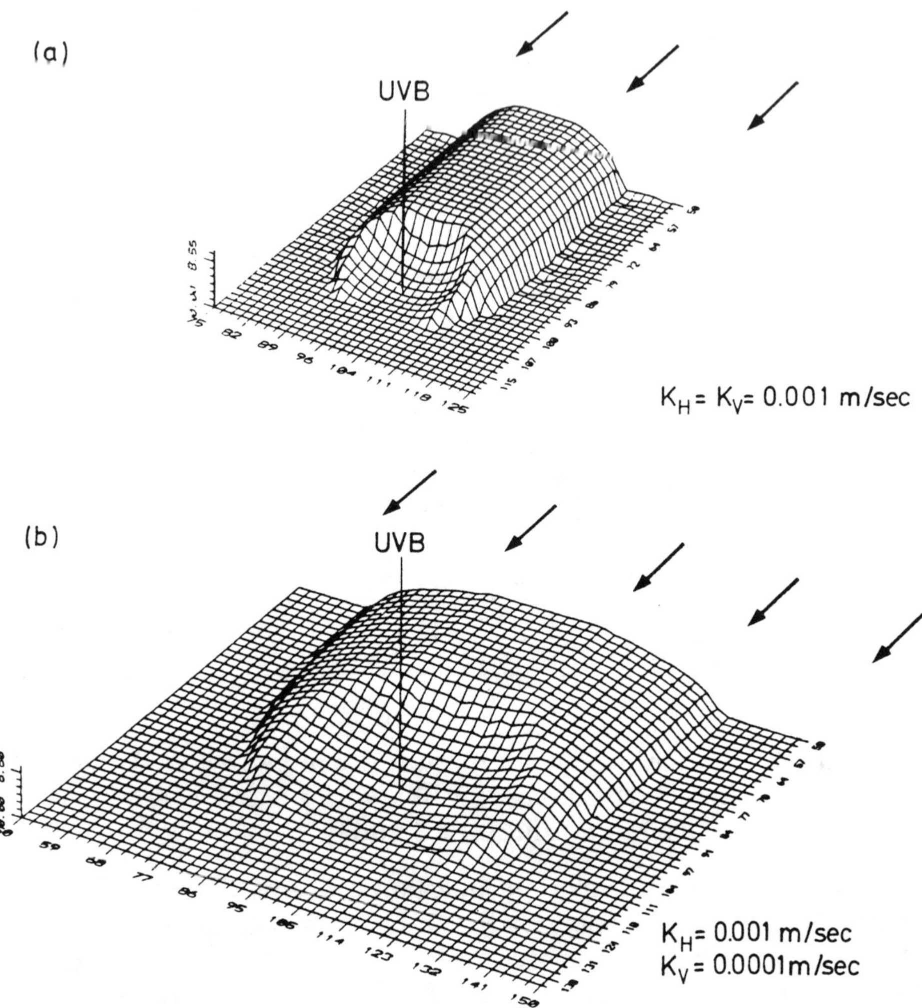

FIGURE 5. Separating stream surface of the capture zone for the situation of Figure 3b: (a) K_H = 0.001 m/sec (isotropic) and (b) anisotropic K_H/K_V = 10.

Diagrams for the Dimensioning of UVB Installations

Absence of Natural Groundwater Flow. At sites without natural groundwater flow, the sphere of influence (R) of a UVB is of special interest. R is dependent on the anisotropy (horizontal over vertical hydraulic conductivity: K_H/K_V), on the thickness (H) of the aquifer, and

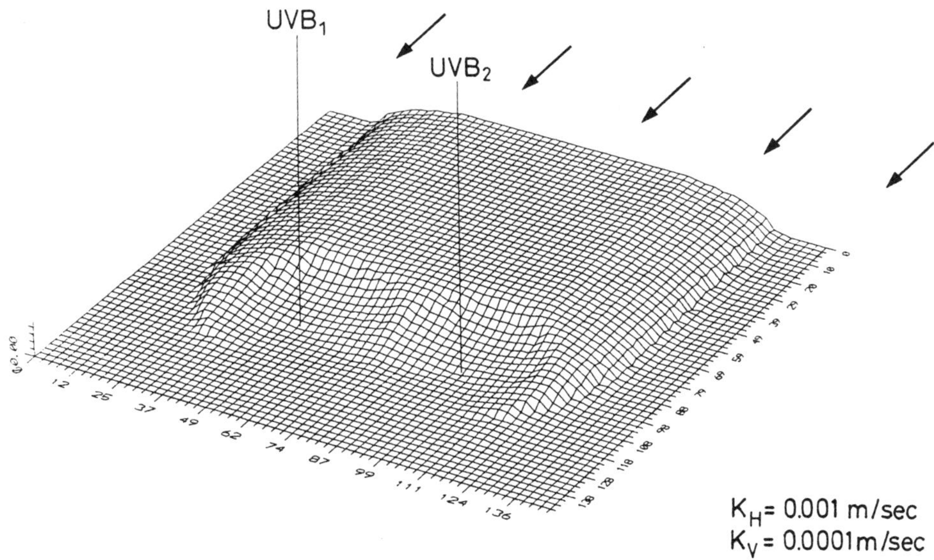

FIGURE 6. Separating stream surface of the capture zone for the situation of Figure 5b, but for two UVB installations at a maximum distance.

on the length of the screen sections a_T and a_B at the top and bottom of the aquifer (see Figure 8) or the ratio a/H (when the same length of the screen section is used for both, then only a is referred to). Although R is mathematically infinite, it is, in practice, defined as the horizontal distance from the well axis to the farthest point at which circulation flow is still significant. In a dimensionless description, R depends on the ratio Q_R/Q, where Q_R is the circulating water quantity, which has already dived downward and flows back toward the lower screen section. The ratio Q_R/Q, which is prescribed for practical reasons, describes the strength of a circulation flow at the distance R from the well.

In Figure 9a, results are presented for ratios $Q_R/Q = 0.98$ and 0.8 and for $a = a_T = a_B$ in a dimensionless diagram. The sphere of influence (R) is independent of the discharge through the well, but strongly dependent on the anisotropy K_H/K_V. Within usual proportions, the length of the screen sections has only a small influence. For a UVB with separating plate and additional pump, a totally screened well casing should be avoided because of hydraulic short-circuiting.

Figure 9b presents a dimensionless diagram that describes the differences (Δh) of the hydraulic heads between the top and bottom of

(a)

(b)

FIGURE 7. Separating stream surfaces of the different water bodies in the outside flow of a UVB: captured, circulating, and flowing downstream water in (a) a real situation and (b) water bodies separated for clarification.

a double-screened well through which a discharge (Q) is pumped. Δh is dependent on the parameter $Q/(H^2K_H)$ and the ratios K_H/K_V and a/H. Abiding by the above-described assumptions, the rise of the hydraulic head at the top of the well amounts to $\Delta h/2$, and the decrease is $-\Delta h/2$ at the bottom (both referring to the position of rest). When using the UVB for stripping, the falling, stripped water in the reactor causes a dynamic effect that will influence the upper hydraulic head within the well.

For the dimensioning or examination of a site, Figure 9b is a valuable expedient. When K_H is known (*e.g.*, by pump test)—along

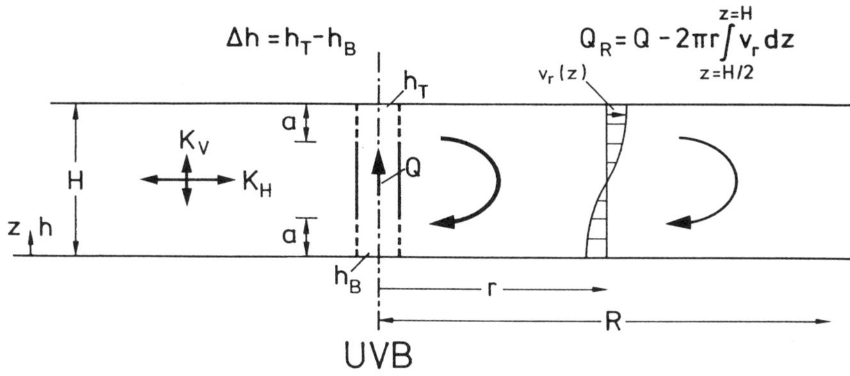

FIGURE 8. Notation in a vertical cross section.

with H, Q, and a—Figure 9b and the measured Δh allow an estimate of the anisotropy at a site.

Presence of Natural Groundwater Flow. At most remediation sites, a natural groundwater flow exists. Figure 10 shows numerical results represented in dimensionless form for the dimensioning of UVB installations under these conditions. Figure 11 introduces the notations for an upstream cross section through the capture zone normal to the natural groundwater flow direction (comparable with the open influx region to the left of the capture zone in Figure 7) for one and two UVB installations. It is often the case when remediating a wide contamination plume that several wells are used in a line normal to the direction of the natural groundwater flow. The length (D) denotes the maximum well distance at which the contaminated groundwater cannot pass between the wells without being cleaned. The results of Figure 10 have been calculated for an upstream distance of 5H from the well and for a constant ratio of $a/H = 0.25$ (screen length over aquifer thickness). The results are discussed for wells that pump upward.

The widths B_T and B_B of the upstream capture zone, measured at the aquifer top and bottom, are shown in Figure 10a. The ratios B_T/H and B_B/H are dependent on the ratios $Q/(H^2v)$, K_H/K_V, and a/H. The Darcy velocity of the natural groundwater flow is denoted as v; all other variables are explained above. For small values of $Q/(H^2v)$, the upper part of the capture zone does not reach the top of the aquifer. This implies that for remediating a plume, a minimum well discharge (Q) is required. Again, the results are quite sensitive to the degree of the anisotropy (see Figure 5, as well).

(a)

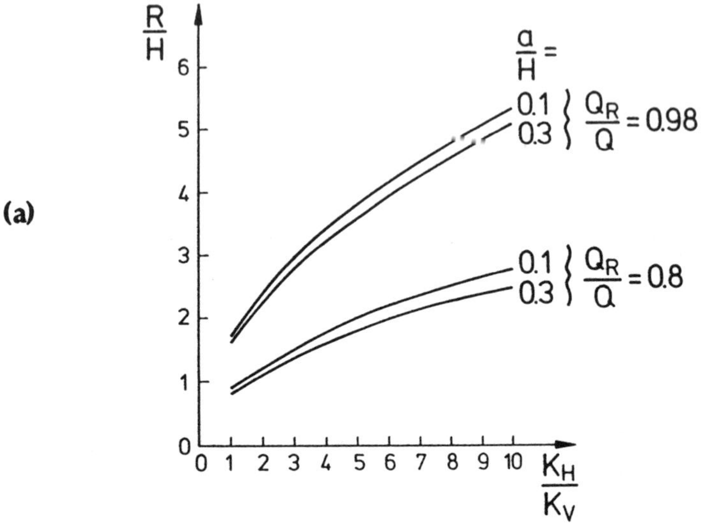

(b)

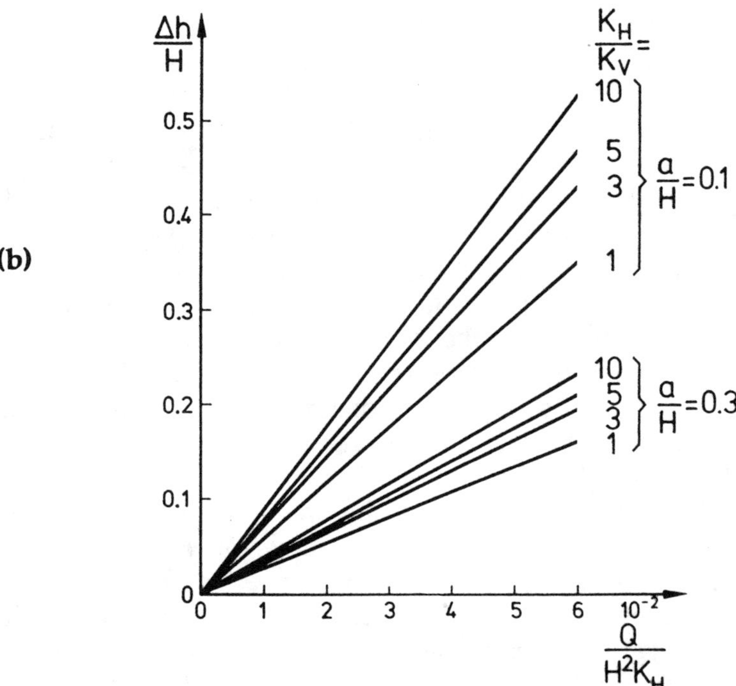

FIGURE 9. (a) Sphere of influence (R) for a site without natural groundwater flow and (b) differences (Δh) of the hydraulic heads between the top and bottom of a well.

(a)

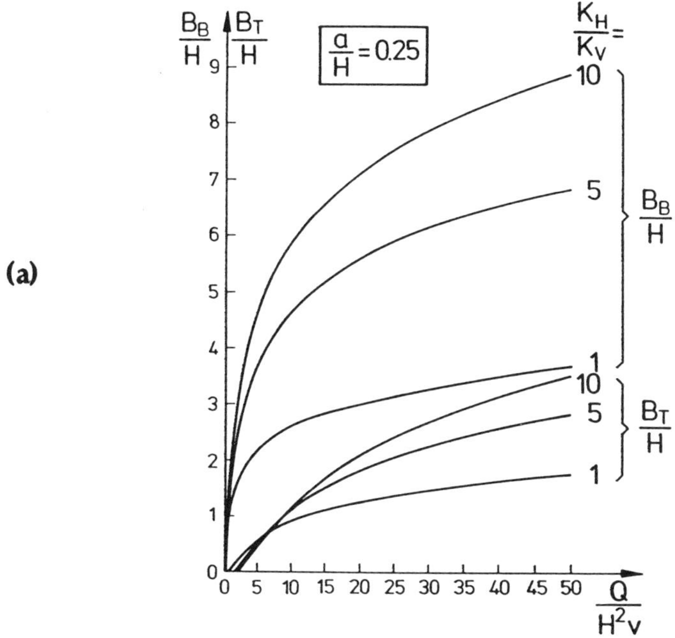

(b)

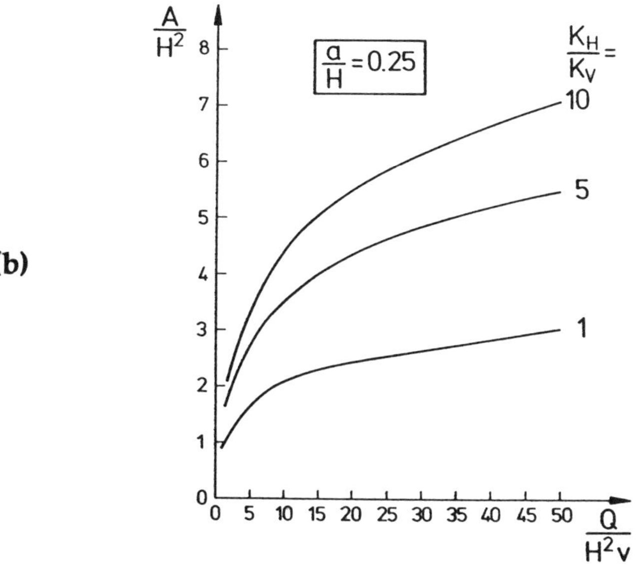

FIGURE 10. (a) Widths B_T and B_B of the upstream capture zone at the aquifer top and bottom. (b) influx area A of the upstream capture zone.

(c)

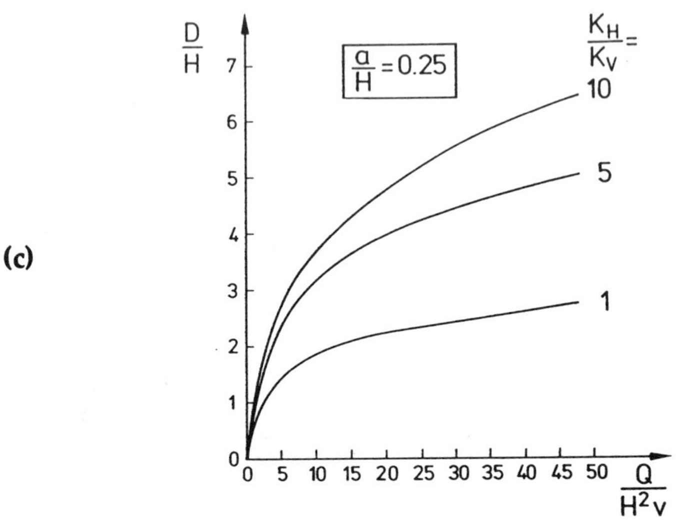

(d)

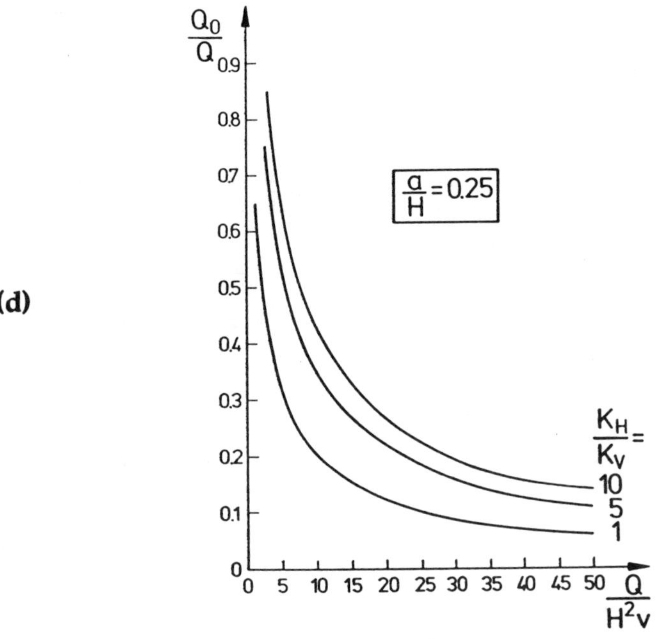

FIGURE 10. **(c) maximum well distance (D) at which the contaminated groundwater cannot pass between the wells without being treated. (d) upstream discharge (Q_o) in the capture zone, which is diluted with the circulating water to the total well discharge (Q).**

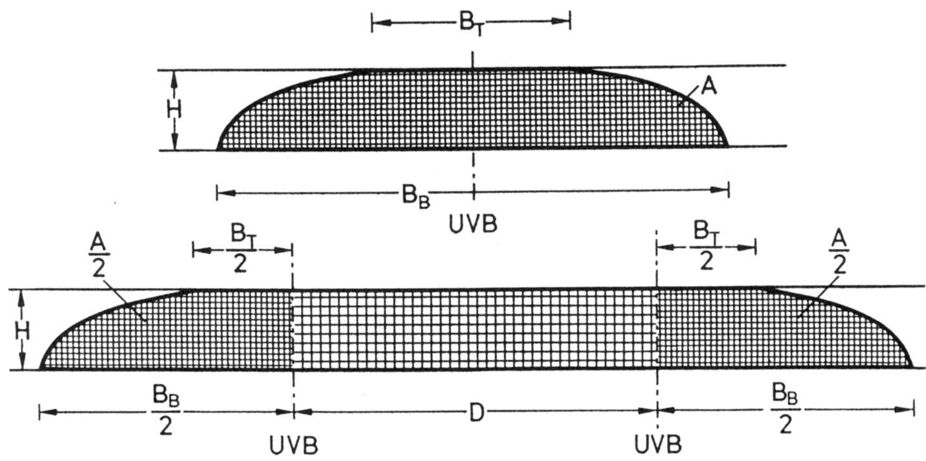

FIGURE 11. Notations in an upstream cross section through the capture zone for one and two UVB installations (for wells pumping upward).

Figure 10b shows the results for the influx area A of the upstream capture zone, and Figure 10c the maximum well distance (D) of two wells between which contaminated groundwater cannot pass without being cleaned by stripping or being treated for biological remediation. The ratios A/H^2 as well as D/H are dependent on the same parameters as the widths B_T and B_B. When a plume of width W is to be cleaned, the number (n) of UVB installations can be estimated by $n = (W - B_T)/D + 1$.

When a plume is remediated, the contaminated water of quantity Q_o, flowing into the capture zone of a UVB from upstream, is diluted with water that has already flowed through the well and circulates around the UVB. Thus, the contaminant concentration of the water within the well casing will be lower than in the upstream plume; near a contamination source the situation is contrary. Figure 10d illustrates the portion Q_o of the total well discharge Q. The ratio Q_o/Q is again dependent on the same parameters as the widths of the upstream capture zone. Figure 10d can be used to estimate the expected concentration value of the water within the well casing for the dimensioning of a UVB installation. It may help to evaluate the progress of remediation at a site when concentration data of the upstream plume and the water within the well are determined, as in case of *in situ* stripping.

In Figure 12 the upstream distance (S) of the stagnation point at the aquifer top from the well axis is described (see Figures 3b and 3c, as well). The ratio S/H is also dependent on the parameters $Q/(H^2v)$, K_H/K_V, and a/H. The location of the stagnation point is highly sensitive to the anisotropy of the aquifer. The length of the screen section is of small importance within usual proportions (as described above). The knowledge of the distance (S) from the stagnation point can be used to determine the positions of measuring equipment. The operation of a UVB can also be supervised using depth-dependent measurements between the stagnation point and the well.

The sphere of influence of the circulation around a UVB at sites with natural groundwater flow is of special interest. This sphere of circulation is limited in a quite different way than at a site with absence of natural flow (Figure 9a) as can be seen in Figures 3b, 3c, 4b, and 7. In the direction of natural groundwater flow, this sphere has a maximum expansion of S (see Figure 12) to the upstream and downstream

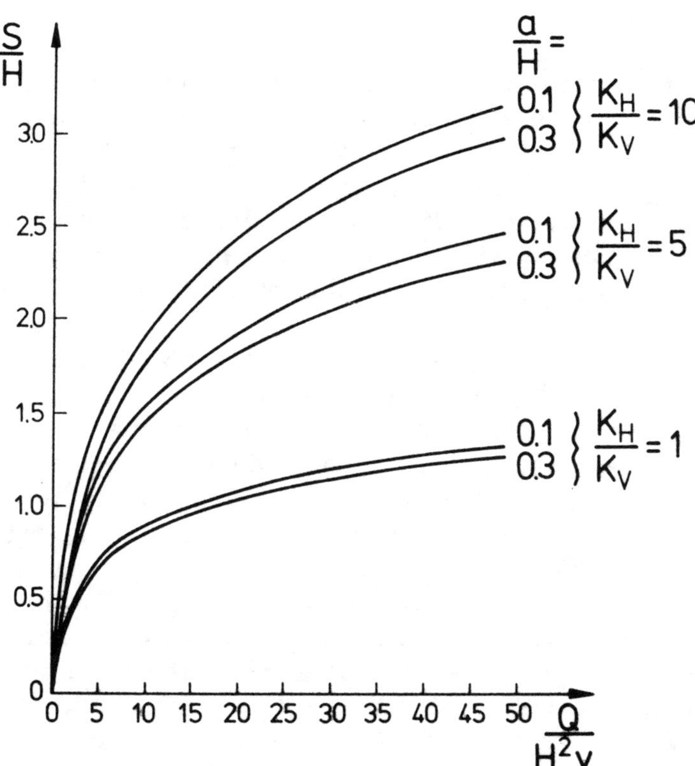

FIGURE 12. Upstream distance (S) of the stagnation point from the well axis.

sides. Normal to this direction, the maximum radius of the sphere of circulation is approximated by $(B_B + B_T)/4$ (Figure 10a), and in the case of several wells in one line, by D/2 (Figure 10c).

Figures 9, 10, and 12 can be used for the dimensioning of a UVB or UVB field when the parameters K_H/K_V and $Q/(H^2v)$ can be estimated, where Q depends on the well size and on the additional pump. For an irregular well field, a layered aquifer, or special critical cases, numerical calculations can be performed.

CONCLUSIONS

The UVB technique can be used for *in situ* stripping of volatile contaminants from the groundwater by itself or in combination with added nutrients and/or electron acceptors for *in situ* biodegradation. Further, the circulation flow around a UVB can be utilized exclusively for *in situ* biorestoration. For all of these cases, the hydraulic system offers many advantages, particularly when compared with a typical hydraulic remediation system of pumping, off-site treatment, and reinfiltration of the groundwater. Such advantages include:

- No lowering of the groundwater level
- No groundwater extraction
- No wastewater
- Less permeable, horizontal layers are penetrated vertically
- Remediation of the groundwater takes place down to the bottom of the aquifer
- Even at low well capacity, remediation operation is continuous
- Soil air extraction is possible at the same time

- Low space requirement

- Investment and operating costs will be considerably lower

When the water discharge through the well casing is directed downward, the hydraulic head is lowered at the well top ($-\Delta h/2$, Figure 9b), but this amount is much smaller than that caused by a normal withdrawal well.

For both *in situ* biodegradation and remediation by *in situ* stripping, the total aquifer is caught by the circulation flow of a UVB. When using different wells for extraction and infiltration, only those areas of an aquifer that are more permeable are penetrated. The other areas are reached mainly by diffusion.

A layered aquifer enlarges the sphere of influence or the distance between the well and the stagnation point. This has been found by numerical simulations and by comparison with field measurements of a tracer test. On the other hand, the positive effect of a layered aquifer is limited when an aquitard is present. Here, several remediation systems must be installed, one for each aquifer.

Acknowledgments

The authors gratefully acknowledge IEG mbH, D-7410 Reutlingen, for supporting this work, and in particular, B. Bernhardt, IEG mbH, D-7410 Reutlingen, inventor and patent holder of the UVB method; E. Alesi, GfS mbH, D-7310 Kirchheim-Teck; P. Brinnel, Hydrodata mbH, D-6370 Oberursel; W. Kaess, D-7801 Umkirch; and H. J. Lochte, UTB mbH, D-4020 Mettmann, for many helpful discussions and contributions to the operation and development of the vacuum vaporizer well.

REFERENCES

Battermann, G.; Werner, P. "Beseitigung einer Untergrundkontamination mit Kohlenwasserstoffen durch mikrobiellen Abbau." *GWF Wasser/Abwasser* **1987**, H.8, 366–373.

Herrling, B.; Buermann, W. "A New Method for *In-Situ* Remediation of Volatile Contaminants in Groundwater—Numerical Simulation of the Flow Regime." In *Computational Methods in Subsurface Hydrology*; Gambolati, G.; Rinaldo, A.; Brebbia, C. A.; Gray, W. G.; and Pinder, G. F.; Eds.; Springer: Berlin, 1990; pp 299–304.

Herrling, B.; Buermann, W.; Stamm, J. "*In-Situ*-Beseitigung leichtflüchtiger Schadstoffe aus dem Grundwasserbereich mit dem UVB-Verfahren." In *Neuer Stand der Sanierungstechniken von Altlasten*; Luehr, H.-P.; Böhnke, B.; Pöppinghaus, K.; Eds.; IWS-Schriftenreihe Bd. 10; E. Schmidt: Berlin, 1990; pp 71–99.

Herrling, B.; Buermann, W.; Stamm, J. "*In-Situ* Remediation of Volatile Contaminants in Groundwater by a New System of 'Vacuum-Vaporizer-Wells.' " In *Subsurface Contamination by Immiscible Fluids*; Weyer, K. U., Ed.; A. A. Balkema: Rotterdam in press.

Lee, M. D.; Thomas, J. M.; Borden, R. C.; Bedient, P. B.; Ward, C. H.; Wilson, J. T. "Biorestauration of Aquifers Contaminated with Organic Compounds." *CRC Crit. Rev. in Environ. Control* **1988**, *18*(1), 29–89.

McCarty, P. L. "Bioengineering Issues Related to *In-Situ* Remediation of Contaminated Soils and Groundwater." In *Environmental Biotechnology*; Omenn, G. S., Ed.; Plenum: New York, 1988.

McCarty, P. L.; Semprini, L.; Roberts, P. V. "*In-Situ* Biotransformation Methodologies." In In-Situ *Aquifer Restauration of Chlorinated Aliphatics by Methanotropic Bacteria*; Roberts, P. V., *et al.*, Eds.; Dept. Civil Eng., Stanford Univ., Stanford, CA, Technical Report No. 310, pp 197–204; 1989.

Raymond, R. L.; Jamison, V. W.; Hudson, J. O. "Beneficial Stimulation of Bacterial Activity in Groundwater Containing Petroleum Products." *AICHE Symposium Series* **1976**, *73*(166), 390–404.

Semprini, L.; Roberts, P. V.; Hopkins, G. D.; McCarty, P. L. "A Field Evaluation of *In-Situ* Biodegradation of Chlorinated Ethenes: Part 2, Results of Biostimulation and Biotransformation Experiments." *Ground Water* **1990**, *28*(5), 715–727.

Wilson, J. T.; Leach, L. F.; Henson, M. J.; Jones, J. N. "*In-Situ* Biorestoration as a Ground-Water Remediation Technique." *Ground Water Monit. Rev. Fall* **1986**, p 56.

Numerical Investigation into the Effects of Aquifer Heterogeneity on *In Situ* Bioremediation

*Wolfgang Schäfer**, *Wolfgang Kinzelbach*
Kassel University

INTRODUCTION

During the last decade a large number of groundwater contamination cases from spills of strongly adsorbing organic pollutants have been discovered in Europe and North America. The presence of these chemicals in the subsurface environment creates a long-term threat to groundwater quality. Therefore, a great effort is being made to find permanent and economic solutions for the removal of these pollutants from underground.

In situ remediation is thought to be one possibility to achieve this goal. This method makes use of the biodegradability of many organic chemicals (Filip *et al.* 1988). Recent investigations show that even halogenated organics can be biologically degraded in the subsurface (*e.g.*, Semprini *et al.* 1990). Limiting substances like electron acceptors or nutrients are delivered to the subsurface to enhance microbial degradation of the pollutant. The numerical investigations presented in this paper assume a theoretically complete microbial degradability of the organic pollutant in question and focus on the effect of aquifer heterogeneities on remediation. A method is suggested to predict the prolongation of an *in situ* remediation caused by spatially

* Department of Civil Engineering, Kassel University, FB 14, Noritzstrasse 21, 3500 Kassel, Federal Republic of Germany

nonuniform pollutant and/or hydraulic conductivity distributions as compared with the duration in the spatially homogeneous case.

THE NUMERICAL MODEL

A two-dimensional finite difference transport model coupled to a model of bacterial action was used for the numerical studies. The transport model computes the movement of mobile chemical species in groundwater flow due to advection and diffusion/dispersion. The chemical model describes heterotrophic growth of aerobic bacteria. It takes into consideration the interaction of a maximum number of six species in three phases: dissolved oxygen and organic carbon in mobile pore water and in an immobile bacterial phase, organic carbon in the aquifer material, and the bacteria that are thought to be attached to the aquifer material. The numerical description of bacterial growth and subsequent consumption of organic carbon and oxygen is based on Monod-type kinetics (Monod 1949).

The two submodels are coupled via an iterative two-step procedure, a method that proved to be computationally efficient. Further, it allows easy modification or replacement of the chemical system equations (Schäfer & Kinzelbach 1990).

The numerical model used here is a reduced version of a more complete model of heterotrophic denitrification described in detail in Kinzelbach and Schäfer (1989). It was successfully applied in modelling denitrification in a natural aquifer (Kinzelbach *et al.* 1989) and in a field case of *in situ* remediation (Kinzelbach & Schäfer 1989). By omitting all terms dealing with respiratory nitrate consumption, the reduced model used here was derived. It is described by the following set of equations:

Steady state isotropic groundwater flow in a confined aquifer:

$$\nabla\left(m \bullet k_f \bullet \nabla h\right) + q = 0 \tag{1}$$

Transport equation for a conservative tracer:

$$\frac{\partial c_T}{\partial t} = -\nabla\left(\vec{v} \bullet c_T\right) + \nabla\left(D \bullet \nabla c_T\right) \tag{2}$$

Coupled transport equations for dissolved oxygen (DO) and dissolved organic pollutant (DOP):

$$\frac{\partial c_{DO}}{\partial t} = -\nabla\left(\vec{v} \bullet c_{DO}\right) + \nabla\left(D \bullet \nabla c_{DO}\right) + S_{DO} \tag{3}$$

$$R\frac{\partial c_{DOP}}{\partial t} = -\nabla\left(\vec{v} \bullet c_{DOP}\right) + \nabla\left(D \bullet \nabla c_{DOP}\right) + S_{DOP}$$

$$\text{with} \quad R = 1 + \frac{k_d \bullet \rho_b}{n_e} \tag{4}$$

Equation for microbial growth and decay:

$$\left[\frac{\partial X}{\partial t}\right]_{aer} = \mu_{max} \bullet \frac{c_{DOP}}{K_{DOP} + c_{DOP}} \bullet \frac{c_{DO}}{K_{DO} + c_{DO}} \bullet X$$

$$\left[\frac{\partial X}{\partial t}\right]_{dec} = -\mu_{dec} \bullet X \tag{5}$$

$$\left[\frac{\partial X}{\partial t}\right]_{gross} = \left[\frac{\partial X}{\partial t}\right]_{aer} + \left[\frac{\partial X}{\partial t}\right]_{dec}$$

Expressions for DO and DOP uptake, respectively:

$$S_{DO} = -\frac{1}{Y_{DO}} \bullet \left[\frac{\partial X}{\partial t}\right]_{aer} \tag{6}$$

$$S_{DOP} = -\frac{1}{Y_{DOP}} \bullet \left[\frac{\partial X}{\partial t}\right]_{aer} + f_{use} \bullet \left[\frac{\partial X}{\partial t}\right]_{dec} \tag{7}$$

Considering aerobic degradation only saves computation time during stochastic analysis, while the main effects on which the numerical studies were focussed can also be observed with a model of aerobic bacterial growth alone.

THE REMEDIATION SETUP

A hypothetical remediation scheme was used to study the effects of aquifer heterogeneities on remediation. The scheme consists of an often-employed arrangement of alternating galleries of injection and pumping wells, with four pumping wells and one injection well forming a so-called five-spot each. One quarter of such a five-spot with an injection well in the upper left-hand corner and a pumping well in the lower right-hand corner is selected as the actual model area (Figure 1). The properties of the homogeneous confined aquifer as well as the discretization and biochemical parameters used are given in Table 1. The discretization fulfills the Courant- and grid-Peclet-number criteria.

It is assumed that the partitioning of the organic pollutant between mobile pore water and aquifer material is governed by a linear adsorption equilibrium of the form

$$c_{AOP} = k_d \bullet c_{DOP} \tag{8}$$

The initial concentration of DOP is assumed to be uniform in the model area, with the concentration of the adsorbed organic pollutant (AOP) being determined by the respective local distribution coefficient.

This procedure implies that the main groundwater flow and the organic pollutant entered the modelled part of the subsurface via different paths in the past. This is the case if, for example, the groundwater is mainly imported from outside the spill area via regional flow and the pollutant reaches the groundwater by vertical movement as an immiscible phase or is dissolved in groundwater recharge.

Spatial Distribution of Model Parameters

The impact of spatial heterogeneity of aquifer and pollutant parameters is studied by means of a Monte Carlo procedure. For each random realization, the outcome of remediation is computed deterministically. The statistics of outcomes over a large number of realizations is analyzed.

Because of the lack of knowledge of both underground and biochemical characteristics, virtually every parameter of the model should be considered to be a random variable.

Here, only the two model parameters thought to be most sensitive were regarded as randomly distributed in space while all other parameters were assumed to be uniformly distributed. The two

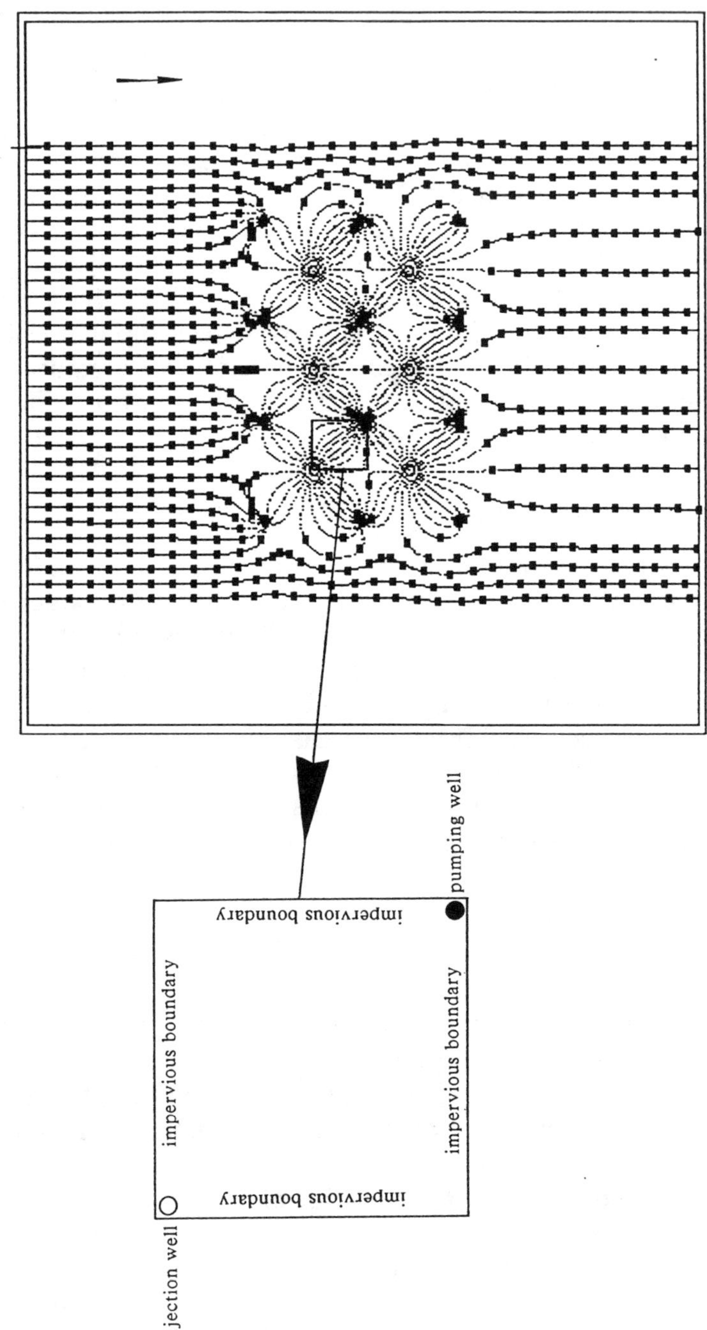

FIGURE 1. Hypothetical remediation scheme with zoomed part used for numerical investigations (so-called quarter five spot). Regional groundwater flow is directed from top to bottom.

TABLE 1. Model aquifer parameters.

	Value	Unit
Aquifer Properties:		
k_f	43	m/d
α_L	0.5	m
α_T	0.1	m
m	30	m
Discretization:		
nx	20	–
ny	20	–
Δx	1	m
Δy	1	m
Δt	0.25	day
Biochemical Parameters:		
μ_{max}	2	1/day
μ_{dec}	0.2	1/day
f_{use}	0.9	–
Y_{DOP}	0.09	mg MC/mg DOP
Y_{DO}	0.03	mg MC/mg DO
K_{DOP}	2	mg/L
K_{DO}	0.2	mg/L

parameters are the hydraulic conductivity (k_f) from equation (1) and the distribution coefficient (k_d) from equation (8). Both were assumed to be log-normally distributed ($Y = \ln(k_f)$, $Z = \ln(k_d)$). For k_f-values this corresponds to field observations (*e.g.*, Freeze 1975). If one derives the k_d-value for an organic pollutant from its octanol-water partition coefficient by an empirical relation as, for example, proposed by Briggs (1981) for aromatic hydrocarbons:

$$k_d = 0.63 \bullet f_{oc} \bullet k_{ow} \tag{9}$$

the assumption of log-normally distributed k_d-values implies a log-normal distribution of f_{oc}. As long as information on the spatial distribution of f_{oc} in aquifer materials is sparse, we use the assumption of log-normal distribution which is conservative for predictive purposes, as a more heterogeneous distribution should require a longer remediation time.

The aquifer realizations were produced using a turning bands generator (Mantoglou & Wilson 1981) with an exponential variogram model. The main characteristics needed for generating the random fields are the first two moments of the respective distributions (i.e., mean value and standard deviation) and the correlation lengths in x and y directions (clx and cly, respectively).

For each realization, the generated mean value and standard deviation were compared with the desired values and corrected by scaling if necessary. As identical geometric mean values for the distributions of k_d in any two realizations do not result in the same initial mass of AOP, the values were scaled to ensure equal initial pollutant mass in each realization. This measure guaranteed comparability of the computer runs for different realizations.

The distribution coefficients were chosen between 0.2 L/kg and 2 L/kg. Following equation (9), these values correspond to those of toluene derived on the basis of the k_{ow} taken from Montgomery and Welkom (1990) and an f_{oc} of the aquifer between 0.1 and 1 percent of dry material weight.

The standard deviation of $\ln(k_f)$ and $\ln(k_d)$, σ_Y and σ_Z respectively, used in the computations varied from 0.23 to 2.3, covering the range from weak to strong heterogeneity. Two different correlation lengths were chosen for the numerical studies, namely 5 m and 2.5 m. These values are on the order of those found by Sudicky (1986). For all realizations, cly was chosen equal to clx resulting in an isotropic correlation structure.

Regarding biodegradability, it was assumed that there is only one single pollutant present and that it will be completely degraded to CO_2 and H_2O. Further, it is assumed at present that microbial growth does not affect groundwater flow by clogging. Based on *in situ* remediations in Germany (*e.g.*, DVWK 1991), there is no evidence of aquifer clogging to date. The biochemical parameters presented in Table 1 are in the range of those given in the literature (*e.g.*, Rittmann & McCarty 1980) and were not varied during the numerical experiments.

RESULTS

The Homogeneous Case

Basic Functioning of the Model. Figures 2 through 4 show the basic functioning of the coupled numerical model in the completely homogeneous case (i.e., for uniform distribution of k_f and k_d).

Figures 2a through 2c show computed concentration isolines for DO, AOP, and bacteria 20 days after the beginning of the remediation. Because of the pattern of the flow field (Figure 3), the center of the remediation cell is better supplied with oxygen than the corner regions. This leads to a faster degradation along the diagonal. At the chosen time the zone of bacterial activity has moved partly through the polluted aquifer, the shape following the isochronal structure of flow. Bacteria consume the DO delivered from the injection well on the left-hand side and simultaneously degrade the organic pollutant. This is possible because DOP is coupled to AOP (Equation [8]) and thus retarded compared with DO movement.

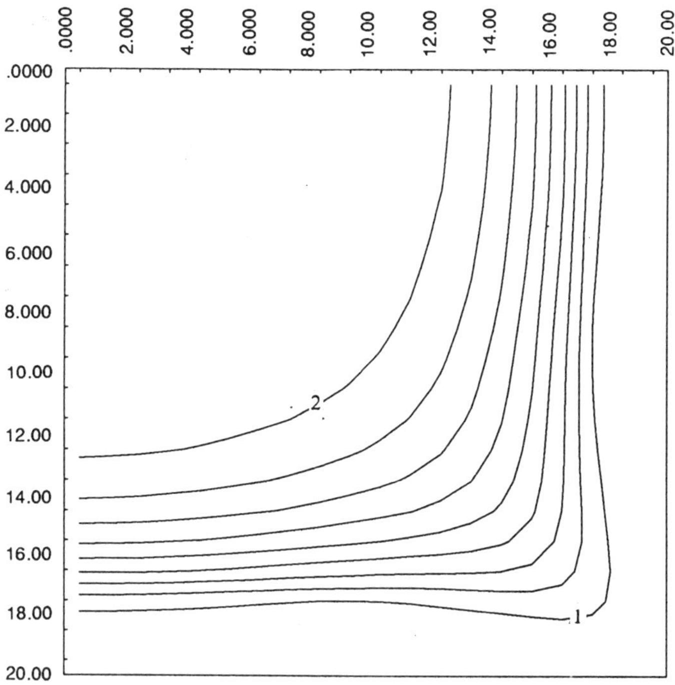

FIGURE 2a. Computed isolines for DO in the homogeneous aquifer ($1 = 5$ mg/L, $2 = 45$ mg/L, $\Delta c = 5$ mg/L).

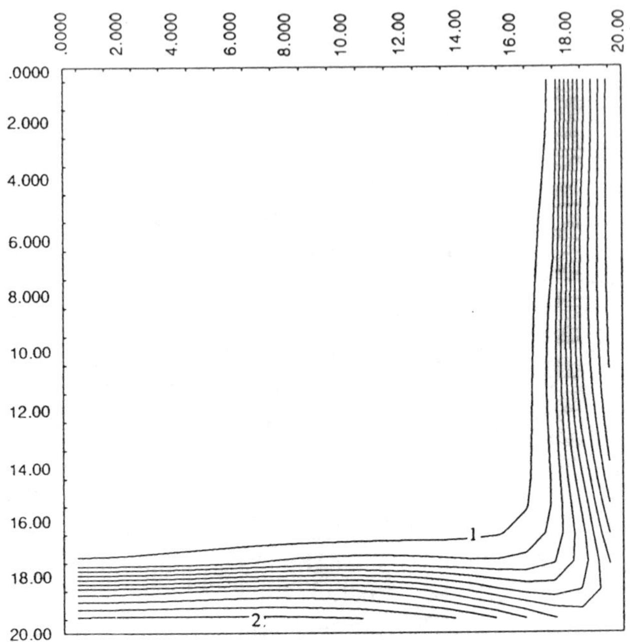

FIGURE 2b. Computed isolines for AOP in the homogeneous aquifer (1 = 0.038 mg/kg, 2 = 0.41 mg/kg, Δc = 0.038 mg/kg).

From the breakthrough curves in the pumping well (Figure 4), it can be seen that DO breakthrough occurs after 20 days. Simultaneously, the concentration of DOP decreases sharply. In the same figure, the breakthrough curve for a conservative tracer is shown for reference. In a field case of *in situ* remediation, the differences between the breakthrough curves of the injected electron acceptor (here DO) and a conservative tracer are an important indicator and measurement device for microbial action in the aquifer.

The total mass of AOP in this case was 12.7 kg, corresponding to a concentration of about 0.6 mg/(kg dry matter). The pumping and injection rates were both 1 L/s, the concentration of injected DO was 50 mg/L, which is roughly the equilibrium concentration for water in contact with a pure oxygen atmosphere. While the total mass of AOP was varied throughout the different model runs, DO concentration and pumping rates always remained constant.

Means to Compare Different Runs. The time in days to achieve 90 percent pollutant removal (90% remediation time [RMT90]) was chosen as the criterion for comparison between different model runs.

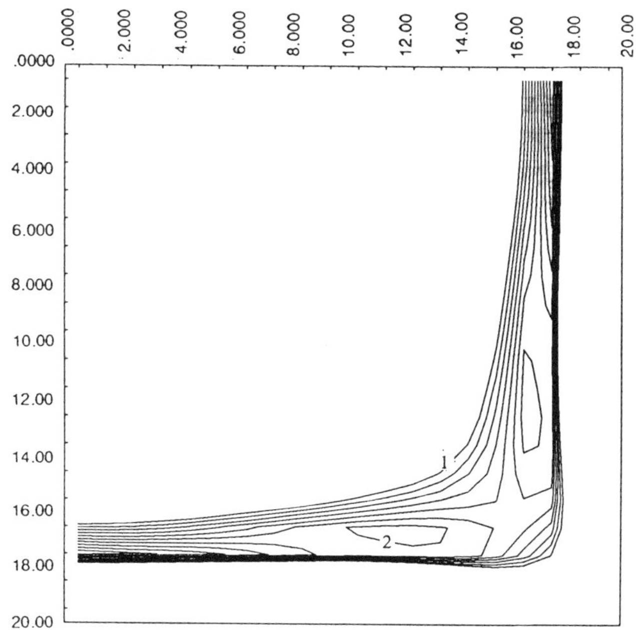

FIGURE 2c. Computed isolines for bacteria in the homogeneous aquifer [1 = 1.5×10^5 cells/(g soil), 2 = 1.9×10^5 cells/(g soil), $\Delta c = 4.5\times10^3$ cells/(g soil)].

This is an arbitrary value, but it is thought to be one that could be used in real remediation cases, where a 100 percent clean-up goal would not be reasonable.

The homogenous-case runs for different values of AOP mass serve as reference for comparison with the results of the computations with heterogeneous k_d and/or k_f distributions. The RMT90 for the homogeneous case with an AOP mass of 12.7 kg was 23 days. This is the reference RMT90 for the following comparable heterogeneous cases.

The Heterogeneous Cases

The Standard Heterogeneous Case (Case 1). The largest number of numerical experiments was carried out for a heterogeneous aquifer with $\sigma_Y = \sigma_Z = 1.84$ and correlation lengths clx = cly = 5 m. Figures 5a and 5b show the areal distributions of $\ln(k_f)$ and $\ln(k_d)$ for one single realization. In Figure 6 the related velocity distribution can be seen, while Figures 7a through 7c show the related isolines for DO, AOP, and bacterial concentrations 20 days after beginning the remediation. Lastly,

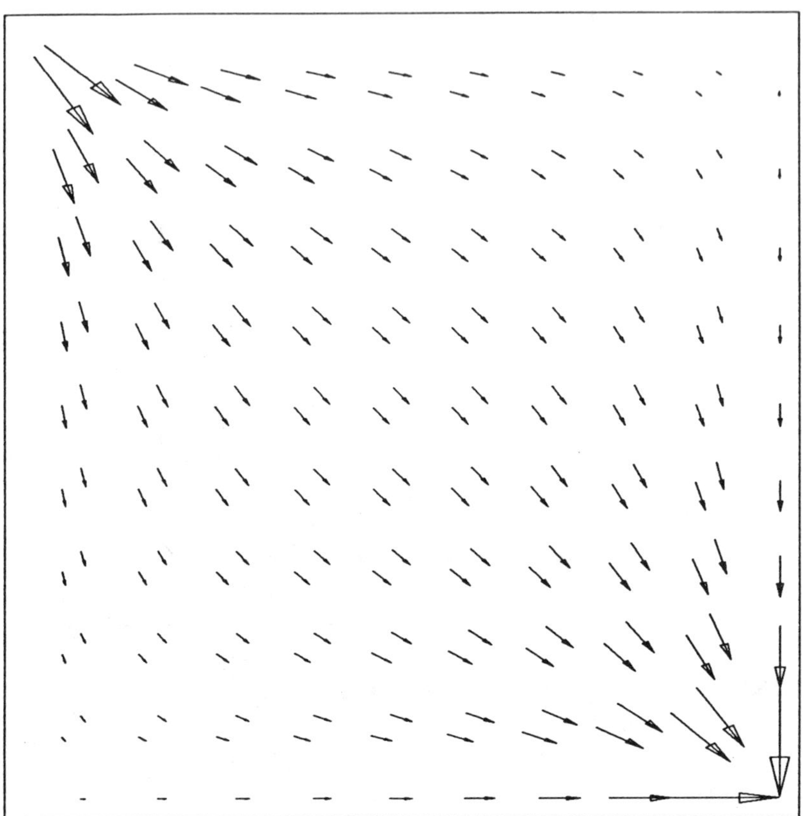

FIGURE 3. Pore velocities in the homogeneous case.

the breakthrough curves for DO, DOP, and a conservative tracer are presented in Figure 8. The mass of AOP was the same as for the homogeneous case shown in Figures 2 through 4.

While in the homogeneous case the concentration distributions were determined by the geometry of the quarter five spot alone, in the heterogeneous case the distributions of the two random parameters disturb the basic pattern decisively.

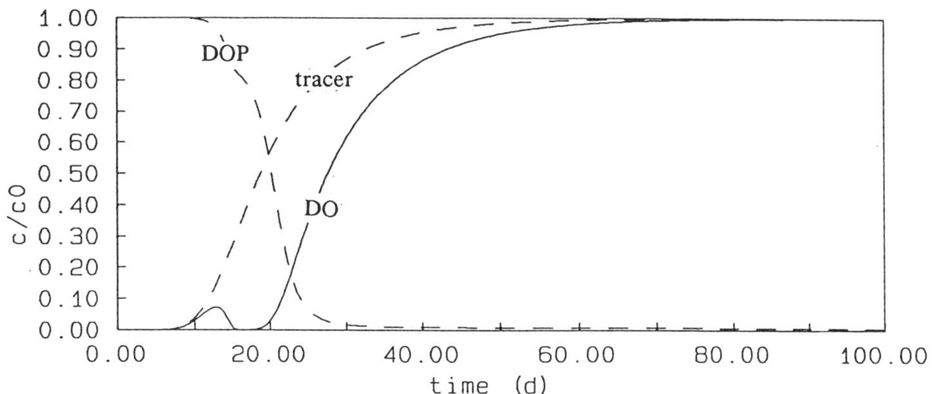

FIGURE 4. Breakthrough curves for DO, DOP, and a conservative tracer in the pumping well of the homogeneous aquifer.

Variations of the Standard Heterogeneous Case. In addition to the standard case, three other cases of heterogeneous aquifers were investigated:

- Case 2: All parameters remained the same as in the standard case, but σ_Y and σ_Z were set to 0.92. This case shows the sensitivity of the results of the standard case with respect to heterogeneity as measured by the standard deviations.

- Case 3: In this case, only the k_f values were distributed heterogeneously with $\sigma_Y = 1.84$, while the value for k_d was set uniform in the whole model area. This case tests the behaviour of a single heterogeneously distributed parameter as compared with the case of two superimposed heterogeneously distributed parameters (Case 1).

- Case 4: The distributions of k_f and k_d remained the same as in Case 1, but the correlation lengths clx and cly were decreased to 2.5 m. This case shows the sensitivity of results of Case 1 with respect to the spatial correlation length.

Comparing Cases 1 through 4 clarifies the impact of various aspects of the heterogeneity structure on the remediation process.

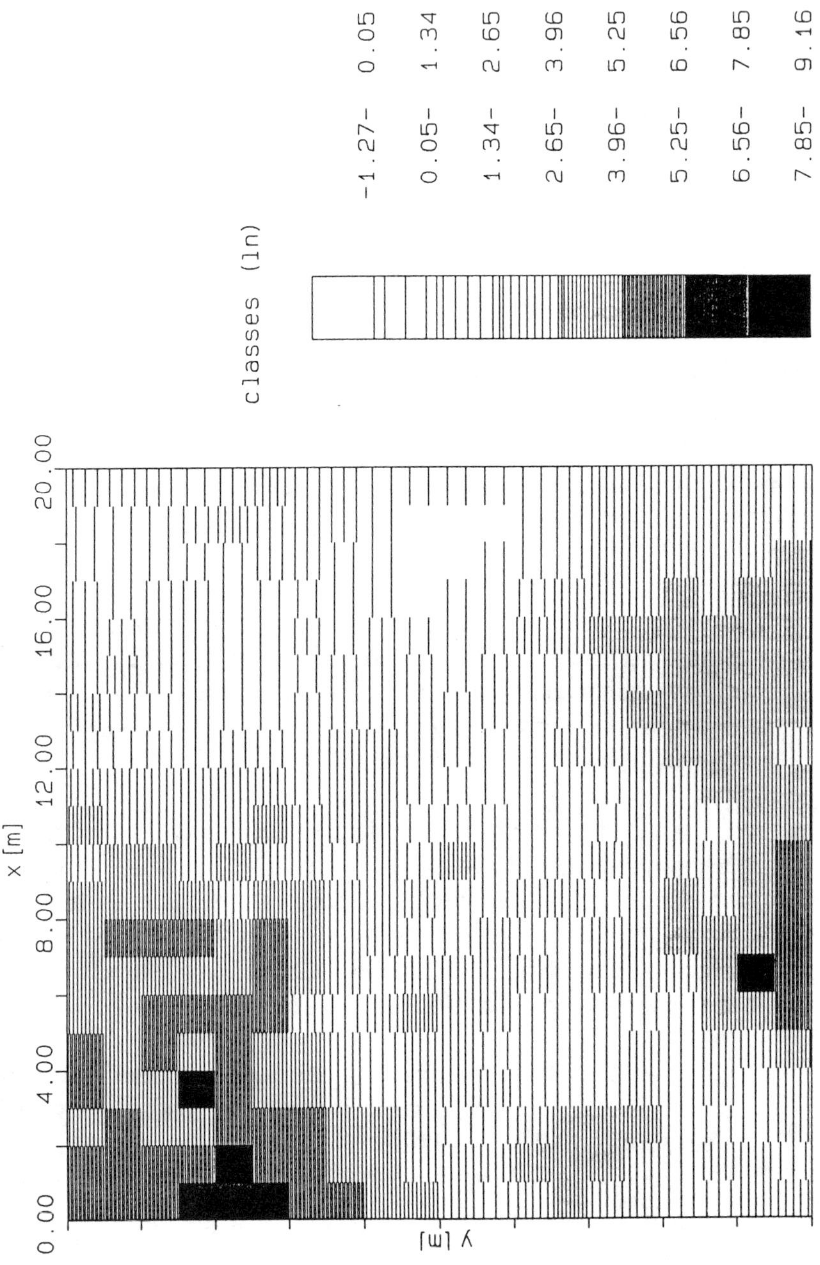

FIGURE 5a. Sample realization for log-k_f distribution of Case 1.

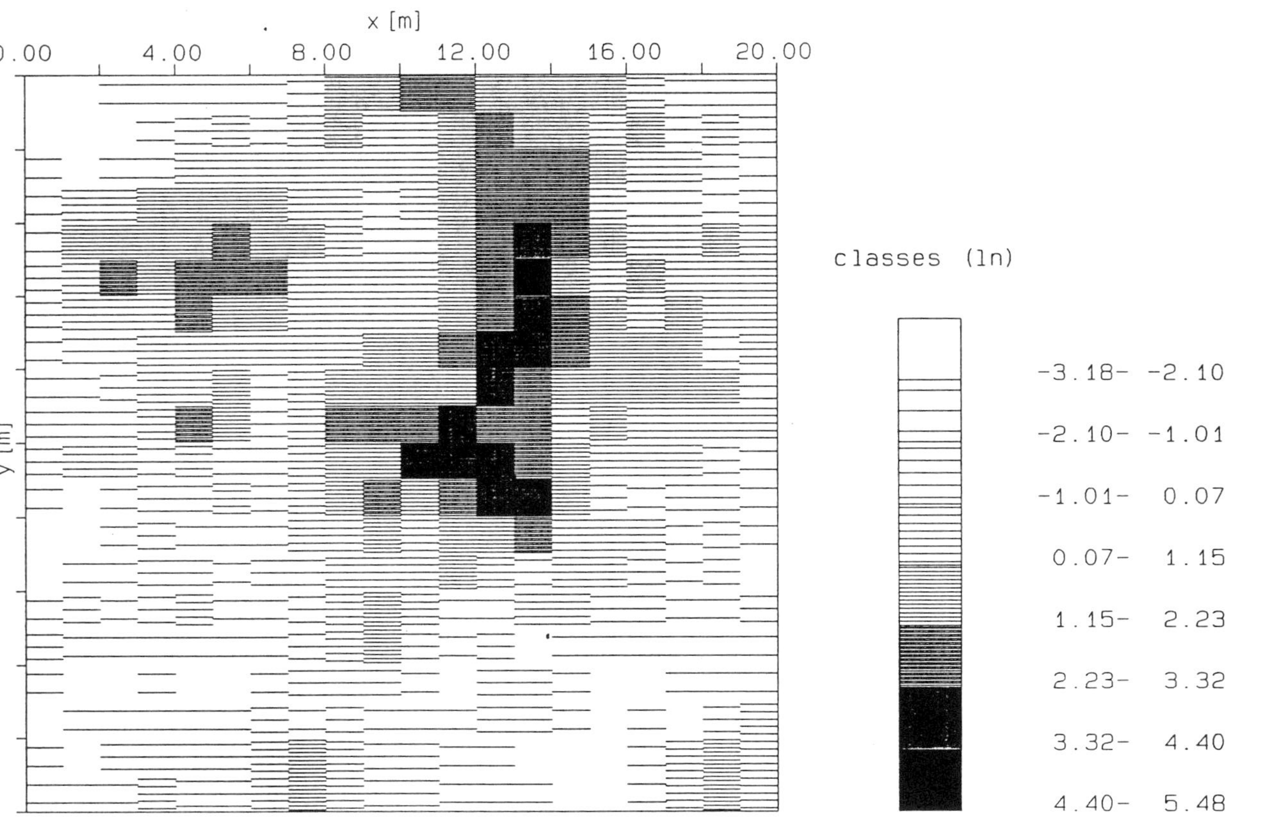

FIGURE 5b. Sample realization for log-k_d distribution of Case 1.

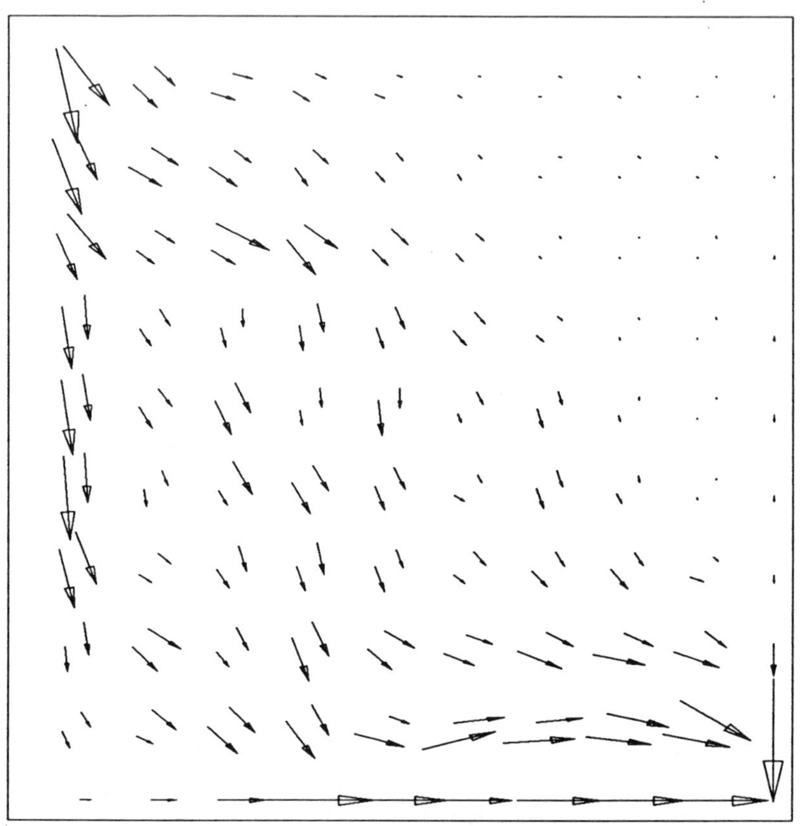

FIGURE 6. Pore velocities in a heterogeneous case (related to the log-k_f distribution of Figure 5a).

In addition, the influence of dispersion and initial mass of AOP on the duration of remediation were tested for Case 1. While the first test allows us to recognize the influence of model dispersivity (numerical and physical), the second test is necessary as a basis for transferring results from Cases 1 through 4 to aquifers with different initial masses of AOP.

Convergence of RMT90 during Monte Carlo Simulation

In any Monte Carlo simulation, the number of runs must be sufficient to yield convergent results. In the study presented, the number of

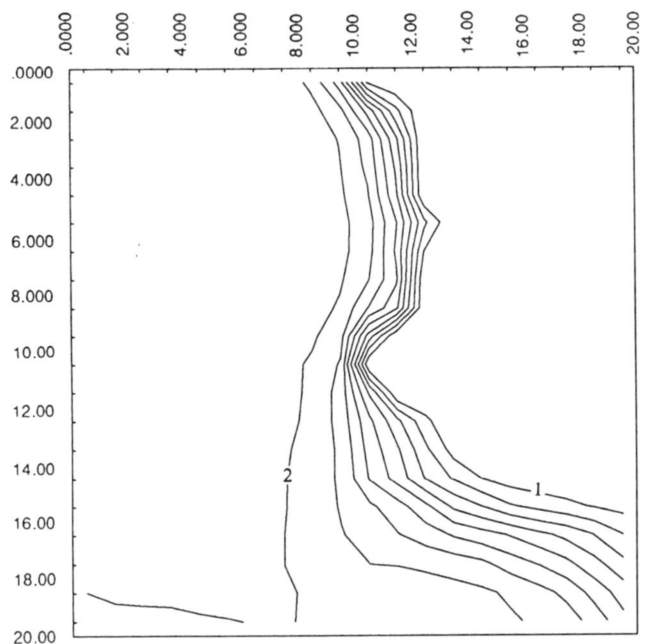

FIGURE 7a. Computed isolines for DO in the heterogeneous aquifer of Figure 5 (1 = 5 mg/L, 2 = 45 mg/L, Δc = 5 mg/L).

Monte Carlo runs needed for convergence of mean RMT90 was largest in Case 1 where more than 200 runs were required. In Cases 2 and 3, convergence was reached after about 40 runs and in Case 4 after 140 runs (Figure 9).

Table 2 lists the mean value, standard deviation, variation coefficient, and skewness for both RMT90 and ln-RMT90 distributions of the four cases.

The mean RMT90 is longest for Case 1, followed by Cases 3, 2, and 4. Case 1 also shows the largest standard deviation resulting from some very long RMT90 for unfavourable combinations of k_f and k_d (*e.g.*, coincidence of small k_f and large k_d values in the corner of the quarter five spot, where DO supply is minimal even in the homogeneous case).

Distribution of the RMT90

While the mean ln-RMT90 are almost equal in Cases 3 and 4 (Table 2), the shapes of their distributions differ markedly. Thus, the RMT90 distributions were studied in more detail in order to describe them with adequate statistical characteristics.

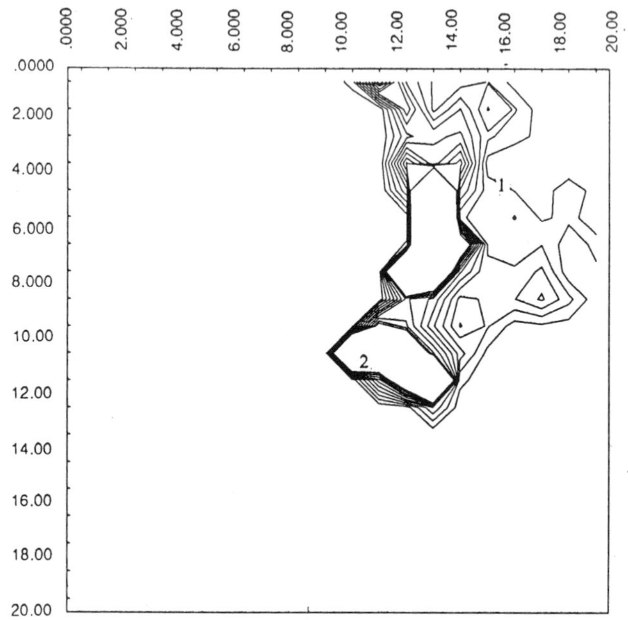

FIGURE 7b. Computed isolines for AOP in the heterogeneous aquifer of Figure 5 (1 = 0.189 mg/kg, 2 = 1.89 mg/kg, Δc = 0.189 mg/kg).

TABLE 2. Mean value x_m, standard deviation s_x, variation coefficient c_v, and skewness c_s of the RMT90 (in days) for all four cases.

	Case 1		Case 2		Case 3		Case 4	
	direct	ln	direct	ln	direct	ln	direct	ln
x_m	69	3.89 (49)	35	3.51 (34)	44	3.70 (40)	51	3.70 (40)
s_x	77	0.76	13	0.33	18	0.39	55	0.61
c_v	1.12	0.48	0.36	0.09	0.41	0.11	1.08	0.17
c_s	3.13	1.00	1.28	0.64	0.94	0.47	4.50	1.45

In all cases, the RMT90 and ln-RMT90 distributions had a non-zero skewness (Table 2). Therefore, skewed theoretical distributions were considered in addition to the widely used normal distribution when comparing model results to a theoretical distribution. Table 3 shows the results of the χ^2 test for normal, Gumbel, Pearson3, and Weibull3 distributions and their respective log-distributions. The best fit expressed by the maximum significance level was obtained for the

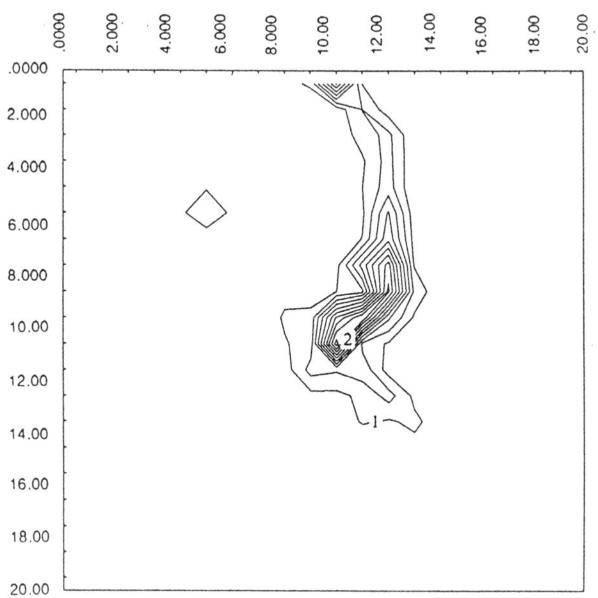

FIGURE 7c. Computed isolines for bacteria in the heterogeneous aquifer of Figure 5 [1 = 1.4×10^5 cells/(g soil), 2 = 1.4×10^6 cells/(g soil), $\Delta c = 1.4\times10^5$ cells/(g soil)].

log-Weibull3 distribution in Cases 1 and 3, the log-Gumbel distribution for Case 2, and the Weibull3 distribution for Case 4. The log-Weibull3 distribution can be used for all four cases to give satisfactory results.

Using the theoretical distribution that fits numerical experiments best, the probability of a remediation event exceeding a given RMT90 can be estimated. Some cumulative probabilities for Cases 1 through 4 are shown in Table 4. These probabilities are a better means to differentiate between the four cases than mean values or standard deviations alone. Cases 3 and 4, for example, have the same geometric mean of RMT90 and thus could be considered very similar. But Case 4 has a markedly larger probability for very long remediation times of more than 100 days, namely 10 percent, while Case 3 has only a 2 percent probability. This means that the risk to exceed a given time for aquifer clean-up is significantly increased in Case 4 compared with Case 3.

Influence of σ_Y and σ_Z

Comparing Cases 1 and 2 shows the strong influence of the standard deviations of k_f and k_d values on RMT90. While in Case 2 about

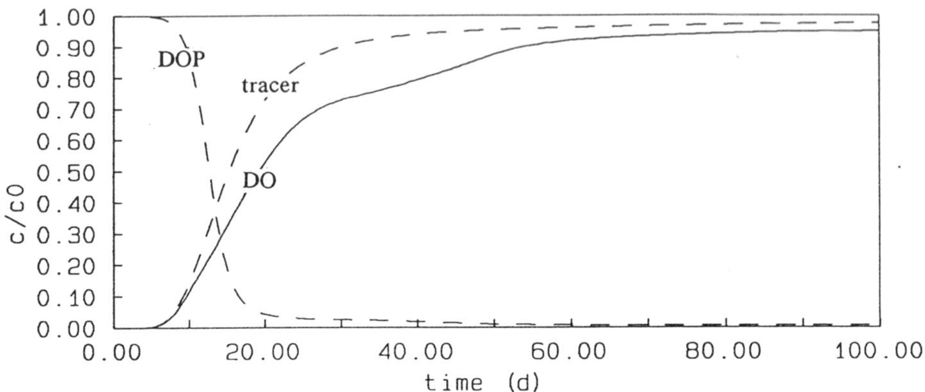

FIGURE 8. Breakthrough curves for DO, DOP, and a conservative tracer in the pumping well of the heterogeneous aquifer of Figure 5.

80 percent of the remediation events were in the range of doubled RMT90 compared with the homogeneous reference case, this was only true for 50 percent in Case 1 (see Table 4). It can also be seen that an increase in σ_Y and σ_Z increases remediation time and especially the probability of extreme events (*e.g.*, events with RMT90 > 100 days).

From a comparison of RMT90 values over a range of σ_Y and σ_Z from 0 to 2.3 in Figure 10, it can be seen that the remediation time grows overproportionately with the standard deviation.

Superposition of Two Random Parameters

If only the k_f values are heterogeneously distributed in the aquifer while the k_d values are uniform (Case 3), the variations and the mean of RMT90 are significantly reduced compared with Case 1 (Table 2).

Contrary to Chiang *et al.* (1990) who observed in their numerical experiments that the spatial pattern of the hydraulic conductivity field dominated the fate of the organic pollutant, we found both k_f and k_d to be equally important. The superposition of the two random variables even had an overproportional effect on the prolongation of remediation as measured by the respective percentiles (Table 4, Cases 1 and 3).

The different conclusions probably arise from the fact that Chiang *et al.* (1990) did not perform systematic Monte Carlo simulations

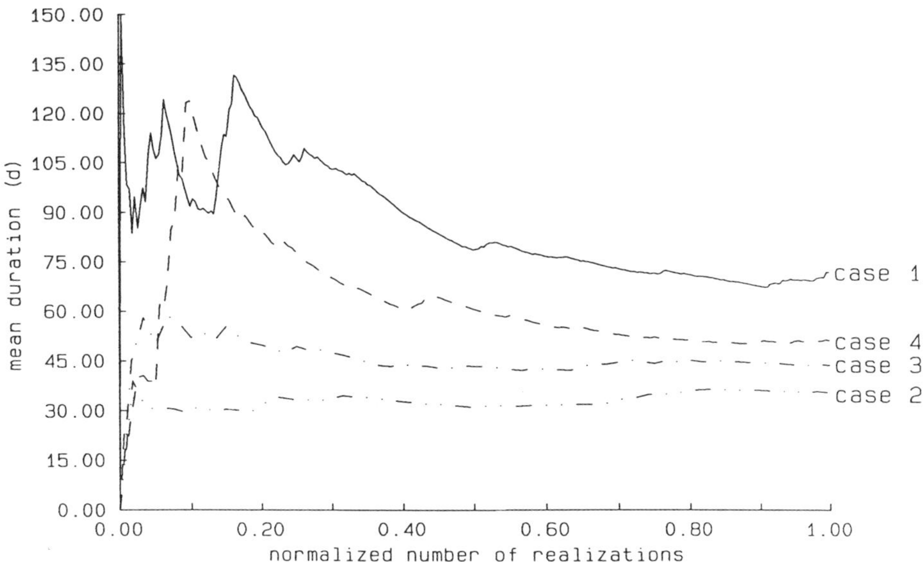

FIGURE 9. Convergence of mean-RMT90 for Cases 1, 2, 3, and 4 with 270, 54, 60, and 180 realizations, respectively.

but considered some special cases only, and that in their studies the degree of heterogeneity was not equal for the two random variables. In our investigations, the standard deviations and correlation structures for k_f and k_d were always the same (apart from Case 3, where $\sigma_Z = 0$).

Influence of Correlation between k_f and k_d. An especially interesting way of superimposing the heterogeneous distributions of k_f and k_d is characterized by a negative linear correlation between the two values:

$$\ln\left(k_d(x,y)\right) = -\ln\left(k_f(x,y)\right) + \ln\left(k_d\right) + \ln\left(k_f\right) \tag{10}$$

A negative correlation in the field would arise if higher distribution coefficients were found in places with lower hydraulic conductivity. There are observations showing that organic carbon is concentrated in the clay fraction (*e.g.*, Matthess 1983) which in turn is larger in regions with small k_f. Until now there is no clear evidence that such a negative correlation generally exists in the field (*e.g.*, Sinclair *et al.* 1990). However, the number of field studies dealing with this subject is still extremely small, so it cannot be ruled out at this time.

Table 5 shows the arithmetic and geometric means of RMT90 for the uncorrelated and the negatively correlated realizations from Cases 1,

TABLE 3. χ^2-test for Cases 1 through 4 (best fits marked with bold letters).

	Case 1			Case 2			Case 3			Case 4		
	χ^2	Degrees of Freedom	s_{max} (%)	χ^2	Degrees of Freedom	s_{max} (%)	χ^2	Degrees of Freedom	s_{max} (%)	χ^2	Degrees of Freedom	s_{max} (%)
normal	> 100	2	< 0.1	21.7	2	< 0.1	> 30.0	2	< 0.1	> 100	2	< 0.1
log-normal	28.3	4	< 0.1	3.5	2	18.0	8.7	2	1.4	92.7	2	< 0.1
Gumbel	41.0	2	< 0.1	7.8	2	2.2	16.7	2	< 0.1	70.1	2	< 0.1
log-Gumbel	9.5	4	5.0	3.4	2	**19.0**	7.6	2	2.3	10.9	2	0.4
Pearson3	–	–	–	5.2	1	2.3	13.1	1	< 0.1	–	–	–
log-Pearson3	8.0	3	5.3	2.2	1	15.5	–	–	–	5.1	1	2.4
Weibull3	2.9	1	9.0	2.5	1	11.8	9.4	1	0.3	2.5	1	**12.4**
log-Weibull3	5.1	3	**13.0**	2.1	1	17.0	4.6	1	**3.5**	5.8	1	1.8

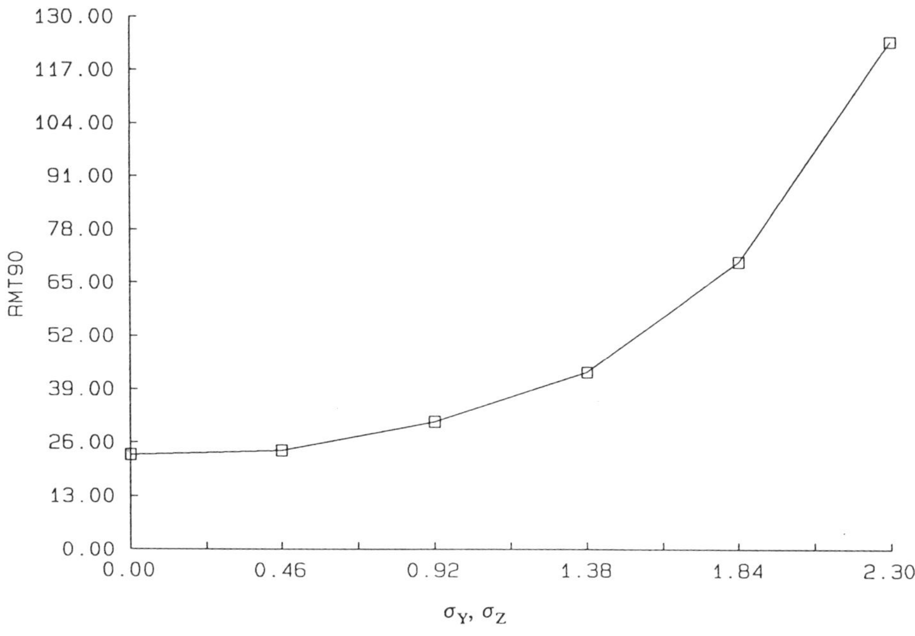

FIGURE 10. Influence of σ_Y and σ_Z on RMT90 for realizations of Case 1.

TABLE 4. Cumulative probabilities of RMT90 (in days) from best fits.

	50%	80%	90%	95%	98%	99%
Case 1	41	86	138	213	364	534
Case 2	32	43	52	62	79	94
Case 3	38	55	69	85	109	128
Case 4	31	68	106	152	223	283

2, and 4. The anticorrelation has a pronounced negative effect on remediation efficiency.

The linear anticorrelations of local k_f and k_d values used here do not necessarily lead to the case with maximum RMT90 for given distribution characteristics, but there were realizations among the uncorrelated cases with markedly longer RMT90.

TABLE 5. Comparison of mean RMT90 (in days) for uncorrelated and anticorrelated realizations.

	Case 1		Case 2		Case 4	
	Not Correlated	Anti-correlated	Not Correlated	Anti-correlated	Not Correlated	Anti-correlated
Arithmetic Mean	69	212	36	53	51	190
Geometric Mean	49	130	34	48	40	101

Influence of Correlation Length

Besides the absolute values of σ_Y and σ_Z, the autocorrelation structure of k_f and k_d also affects the RMT90 distribution. Reducing clx and cly by a factor of 2 led to very short RMT90 for half of the realizations (Case 4 in Table 4), but the probability for remediation events with very long durations is still high. It lies between the corresponding probabilities in Case 1 and Cases 2 and 3.

The overall decrease of RMT90 for smaller correlation lengths is mainly due to the fact that the smaller correlation lengths reduce the probability for the existence of channels with large hydraulic conductivity which connect the injection well with the pumping well. Such a hydraulic "short-circuit" leads to an inefficient distribution of DO in the aquifer with only a small flux reaching the pollutant located outside of the channel.

Influence of Dispersivities and Initial Mass of AOP

The dispersivities used in the model should account for the effects of small-scale mixing only; the dispersive processes caused by large-scale mixing are represented explicitly by the nonuniform velocity distribution resulting from the heterogeneous hydraulic conductivity structure.

An important issue in this investigation was the relative importance of numerical dispersion for RMT90. Therefore, the dispersivities of a given realization were varied from the standard α_L = 0.5 m alternatively to α_L = 2.5 m and α_L = 0 with the ratio of α_L to α_T remaining constant. Figure 11 shows that an increase in α_L leads to a decrease in

RMT90 by enhancing microbial activity because of better mixing of electron acceptor and donator.

Setting α_L to 0 shows the amount of numerical dispersion present. In this case, mean RMT90 is much smaller than in the case with α_L = 0.5 m (and of course in the case with α_L = 2.5 m) so that we can conclude that numerical dispersion does not affect the overall results. Mixing by dispersion/diffusion is a process that strongly influences remediation times, so one should always check for numerical dispersion when modelling *in situ* remediation in heterogeneous aquifers.

The influence of initial masses of AOP different from those in Cases 1 through 4 (12.7 kg) on RMT90 was first investigated for the homogeneous case. Table 6 shows the RMT90 for increase of total mass of organic pollutant by a factor of 1.8, 2.6, 4.1, and 8. A proportionality can hardly be seen neither for the values themselves nor for their logarithms. This may be mainly due to the variable kinetics exhibited by microbial growth when modelled on the basis of Monod-type

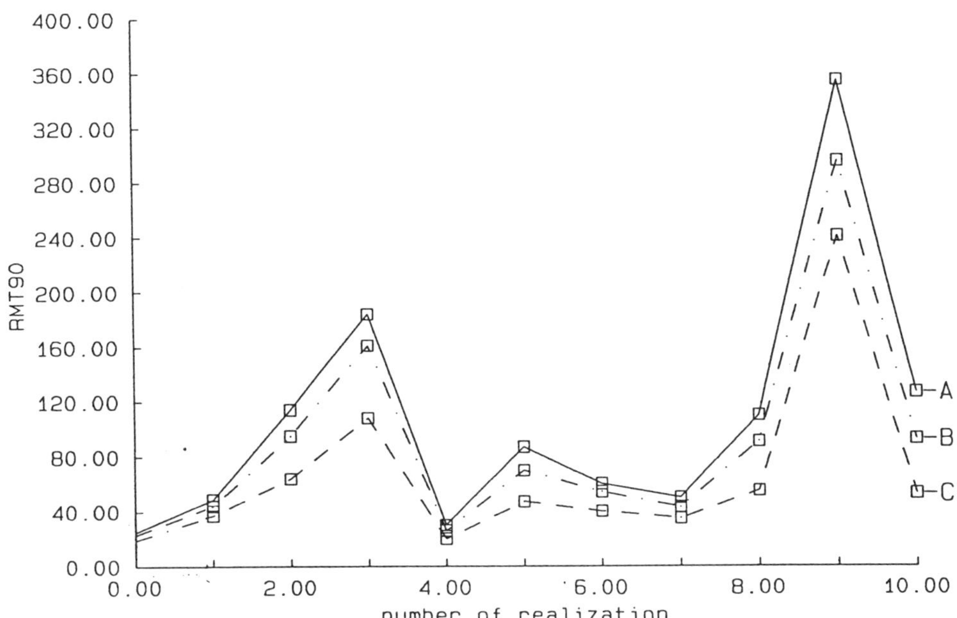

FIGURE 11. Influence of dispersivities on RMT90 for realizations of Case 1 (A: α_L = 0, B: α_L = 0.5 m, C: α_L = 2.5 m, realization 0 indicates homogeneous case).

relations. If pollutant degradation could be described by means of a zero- or first-order degradation term, the RMT90 for different initial masses would be directly proportional to the increase in mass or constant, respectively. The RMT90s obtained from the model of microbially mediated degradation are situated between these two extremes (as are the kinetics), but they are not known *a priori*.

TABLE 6. RMT90 (in days) for various initial masses of organic pollutant and the respective multiplication factors.

Pollutant Mass (kg)	RMT90	Factor		Factor	
		mass	RMT90	log-mass	log-RMT90
16.1	23	1	1	1	1
28.7	32	1.8	1.4	1.06	1.11
41.3	40	2.6	1.7	1.10	1.18
66.5	55	4.1	2.4	1.14	1.28
129.3	93	8.0	4.0	1.21	1.45

These difficulties are even more severe in the heterogeneous cases. Of course, an increase in pollutant mass always leads to an increase in RMT90, but the respective value varies from realization to realization (Figure 12). Only an approximative rule for upscaling the pollutant mass can be given which scales the moments of the distribution and the cumulative probabilities. It consists of two steps:

- Define the scale factor f_s by dividing the ln-RMT90 for the homogeneous case with the new mass by the ln-RMT90 for the homogeneous reference run.

- Multiply the moments of the reference log-distribution and its log-percentiles by the scale factor f_s to obtain the new moments and percentiles.

For mass increases within a factor of 5, the new distributions can be obtained through this scaling of the reference distributions.

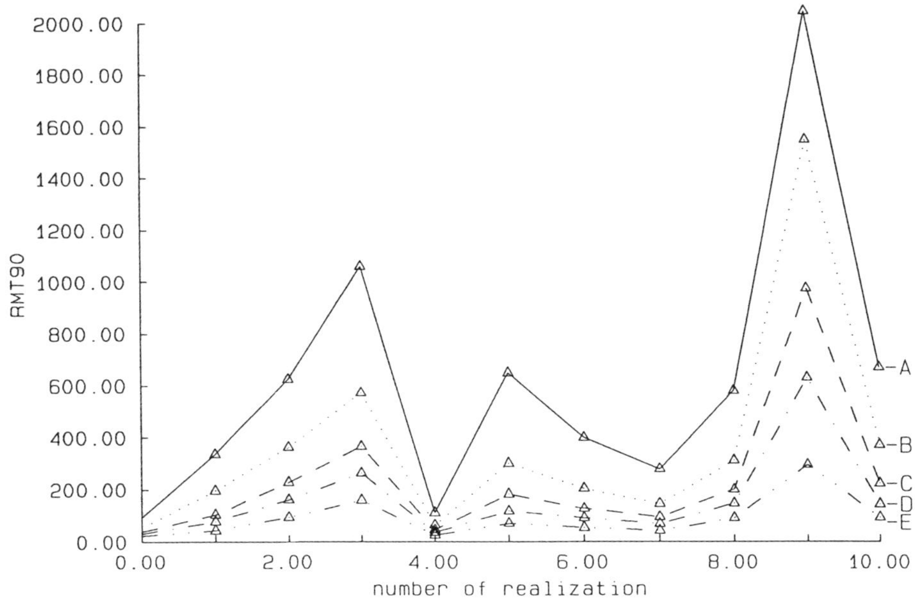

FIGURE 12. Influence of initial pollutant mass (m_i) on RMT90 for realizations of Case 1 (A: m_i = 129.3 kg, B: m_i = 66.5 kg, C: m_i = 41.3 kg, D: m_i = 28.7 kg, E: m_i = 16.1 kg, realization 0 indicates homogeneous case).

Interpretation of Breakthrough Curves

In the homogeneous case, remediation progress can be observed directly by looking at the difference between the breakthrough curves of a conservative tracer and the reactive electron acceptor. From that difference, the mass of degraded pollutant can be estimated via stoichiometric relations and knowledge of groundwater flow rates (*e.g.*, by means of numerical modelling). Furthermore, the difference in curve shapes can give information on the relative importance of the different kinetic processes involved such as microbial kinetics or diffusive exchange kinetics (Schäfer & Kinzelbach 1990).

In the homogeneous case, about 90 percent of the pollutant is already degraded when DO breakthrough reaches 50 percent and DOP is reduced by a factor of 50 (Figure 4). In this case, both breakthrough curves are a good means to monitor the degradation progress, but one must be aware that if the organic pollutant is strongly adsorbing, even in homogeneous aquifers, diffusion-limited pollutant desorption can result in decreased pollutant concentration in the pumped water due to dilution. This decrease in concentration could wrongly be interpreted as efficient remediation.

The DO breakthrough may not be a reliable indicator of remediation progress in the heterogeneous case. For the breakthrough curves of the heterogeneous case presented in Figure 8, a 50 percent breakthrough of DO and a simultaneous almost-complete reduction of DOP is coupled with only 50 percent pollutant degradation (RMT90 for this case was 70 days).

In general, our numerical studies show that breakthrough curves of the dissolved pollutant and the electron acceptor become ambiguous in heterogeneous aquifers and can no longer be easily used to describe the actual state of the remediation process.

DISCUSSION

The numerical investigations showed aquifer heterogeneities to have a generally negative effect on remediation efficiency by prolonging remediation time (Table 4). This result is in contrast to findings from numerical experiments by Sudicky and MacQuarrie (1989) for a natural aquifer situation. They observed an increase of organic carbon degradation with increasing aquifer heterogeneity in their computations. These different findings may be partly due to the fact that Sudicky and MacQuarrie (1989) used a retardation factor much lower (by a factor of 8) than the one used here. Because of the small retardation factor, mixing of electron acceptor and organic pollutant, which is the precondition for microbial degradation, was not very efficient in their homogeneous case. Thus, aquifer heterogeneity enhanced their microbial activities by achieving a better mixing. Considering aquifer remediation, these weakly adsorbing pollutants could be treated by purely hydraulic measures as well. The studies presented here address the less easily recoverable, strongly adsorbing chemicals for which biological remediation actually should be applied. For those substances, sufficient mixing in the homogeneous case is guaranteed by the large difference of retardation factors for DO and DOP. Here the spatially heterogeneous distributions of k_d and/or k_f result in a diminished efficiency of the remediation because part of the DO supplied by injection is soon abstracted without coming into contact with the remaining polluted zones. Furthermore, Sudicky and MacQuarrie (1989) did not perform Monte Carlo simulation but regarded heterogeneity effects using a single realization. Thus, contradictory results may also be due to the particular properties of the realization they used. We also found some realizations to have RMT90 shorter than that of the homogeneous case, while on the average remediation was always prolonged by aquifer heterogeneity.

CONCLUSIONS

The studies revealed that the heterogeneity structure of the key aquifer parameters relevant for remediation, namely spatial distributions of hydraulic conductivity and distribution coefficient of organic pollutants, has a decisive impact on the success of a remedial action. In general all kinds of heterogeneity tested in the hypothetical remediation scheme lead to an increase of remediation time compared to a remediation with equal initial pollutant mass in a homogeneous aquifer. The remediation time estimated on the basis of homogeneous conditions therefore tends to understate time requirements in all real, more or less inhomogeneous aquifers. Even for aquifers with only moderate heterogeneity, this prolongation can be on the order of a factor of two. For aquifers with large heterogeneity (especially with anticorrelation of hydraulic conductivity and distribution coefficient) unfavourable distributions of these quantities can render an *in situ* remediation an expensive failure even if a pollutant is shown to be microbiologically degradable in laboratory column studies.

Regarding the monitoring of remediation progress, the breakthrough curve for the electron acceptor would be a reliable indicator of the actual state of the remediation in homogeneous aquifers. The Monte Carlo simulations presented here, however, showed that in heterogeneous aquifers DO breakthrough becomes ambiguous and is therefore no longer easily interpreted.

Acknowledgments

This research was financially supported by the German Research Foundation (DFG).

REFERENCES

Briggs, G.G.J. *Agric. Food Chem.* **1981**, *29*, 1050–1059.

Chiang, C. Y.; Dawson, C. N.; Wheeler, M. F. To be published in a special issue of *Transport in Porous Media*, 1990.

DVWK (Deutscher Verband für Wasserbau und Kulturwesen). *Sanierungsverfahren für Altlasten und Grundwasserschadensfälle*; Parey Verlag: Hamburg, 1991 (in press).

Filip, Z.; Geller, A.; Schiefer, B.; Schwefer, H.-J.; Weirich, G. *Forschungsbericht 14404567 "Feste Abfallstoffe,"* Bundesministerium für Forschung und Technologie, 1988.

Freeze, R. A. *Water Resour. Res.* **1975**, *11*(5), 725–741, cited in: Gelhar, L. W.; Axness, C. L. "Stochastic analysis of Macrodispersion in Three Dimensionally Heterogeneous Aquifers." Report No. H-8 of the New Mexico Institute of Mining and Technology, 1981.

Kinzelbach, W.; Schäfer, W. IAHS Publication No. 188, 1989; 237–260.

Kinzelbach, W.; Schäfer, W.; Herzer, J. In *Contaminant Transport in Groundwater*; Kobus, H. E.; Kinzelbach, W., Eds.; Proceedings of the International Symposium on Contaminant Transport in Groundwater; A. A. Balkema: Rotterdam, 1989; pp 191–198.

Mantoglou, A.; Wilson, J. L. Technical Report No. 264; Ralph M. Pearsons Lab., MIT: Cambridge, MA, 1981.

Matthess, G. *DVGW Schriftenreihe Wasser* **1983**, *36*, 65–78.

Monod, J. *Annual Review of Microbiology* **1949**, *3*, 373–394.

Montgomery, J. H.; Welkom, L. M. *Groundwater Chemicals Desk Reference*; Lewis Publishers: Chelsea, MI, 1990.

Rittmann, B. E.; McCarty, P. L. *Biotechnology and Bioengineering* **1980**, *XXII*, 2359–2373.

Schäfer, W.; Kinzelbach, W. In *Computational Methods in Subsurface Hydrology*; Gambolati, G.; Rinaldo, A.; Brebbia, C. A.; Gray, W. G.; Pinder, G. F., Eds.; Proceedings of the Eighth International Conference on Computational Methods in Water Resources; Springer-Verlag: Berlin, 1990; 405–412.

Semprini, L.; Roberts, P. V.; Hopkins, G. D.; McCarty, P. L. *Groundwater* **1990**, *28*(5), 715–728.

Sinclair, J. L.; Randtke, S. J.; Denne, J. E.; Hathaway, L. R.; Ghiorse, W. C. *Groundwater* **1990**, *28*(3), 369–377.

Sudicky, E. A. *Water Resour. Res.* **1986**, 22(13), 2069–2082.

Sudicky, E. A.; MacQuarrie, K.T.B. In *Contaminant Transport in Groundwater*; Kobus, H. E.; Kinzelbach, W., Eds.; Proceedings of the International Symposium on Contaminant Transport in Groundwater; A. A. Balkema: Rotterdam, 1989; pp 307–316.

NOMENCLATURE

index aer, dec, gross	: aerobic, decay, gross	[-]
AOP	: adsorbed organic carbon from pollutant	[-]
c_{AOP}	: concentration of AOP	[M/M]

c_{DO}, c_{DOP}, c_T	: concentrations of DO, DOP, and conservative tracer	$[M/L^3]$
clx, cly	: correlation lengths in x and y directions	[L]
D	: tensor of hydrodynamic dispersion	$[L^2/T]$
DO	: dissolved oxygen	[-]
DOP	: dissolved organic carbon from pollutant	[-]
f_{use}	: utilizable portion of dead bacteria	[-]
f_{oc}	: mass fraction of organic carbon in soil	[M/M]
h	: hydraulic head	[L]
k_d	: distribution coefficient	$[L^3/M]$
k_f	: hydraulic conductivity	[L/T]
k_{ow}	: octanol/water partition coefficient	[M/M]
K_{DO}, K_{DOP}	: half-velocity concentrations for DO and DOP	$[M/L^3]$
m	: aquifer thickness	[L]
MC	: microbial carbon	[M]
ne	: effective porosity	[-]
nx,ny	: numbers of cells in x and y directions	[-]
q	: recharge/discharge rate per unit horizontal area	[L/T]
R	: retardation factor	[-]
RMT90	: time in days to achieve 90 percent pollutant removal	[T]
s_{max}	: maximum significance level	[-]
t	: time	[T]
$\vec{v}$	: vector of average pore velocity	[L/T]
μ_{max}	: maximum growth rate of bacteria	[1/T]
μ_{dec}	: rate of bacterial decay	[1/T]
X	: concentration of bacteria	$[M/L^3]$
Y_{DO}, Y_{DOP}	: yield coefficients for DO and DOP	[-]
Y	: $\ln(k_f)$	[-]
Z	: $\ln(k_d)$	[-]
α_L, α_T	: longitudinal and transversal dispersivities	[L]
Δt	: discretization in time	[T]
Δx,Δy	: discretization interval in x and y directions	[L]
∇	: nabla operator $(\partial/\partial x, \partial/\partial y)$	[-]

ρ_b	: bulk mass density	$[M/L^3]$
σ	: standard deviation	[-]

Application of a Numerical Model to the Performance and Analysis of an *In Situ* Bioremediation Project

Mark A. Widdowson[*]
University of South Carolina
C. Marjorie Aelion
U.S. Geological Survey

INTRODUCTION

Numerical flow and transport models often serve as tools in the design of groundwater remediation schemes. Recently, a number of transport models have been developed for the simulation of transport coupled to microbial activity. Multidimensional biodegradation transport models offer hydrologists an additional tool for evaluating the design and performance of bioremediation projects. However, models must also offer an accurate numerical scheme and the flexibility to simulate a range of microbial processes (i.e., multiple respiration and nutrient limitation) in a realistic fashion.

Fate and transport models developed to incorporate microbial activity simulate biodegradation at various levels of sophistication. Borden and Bedient (1986) and Widdowson *et al.* (1988) are representative of contributions to the literature on the subject of model development over the past 5 years. However, very few documented cases exist (*e.g.*, Rifai & Bedient 1989) in which hydrologists have applied models to enhanced *in situ* bioremediation projects.

[*] Department of Civil Engineering, Sumwalt Building #88, University of South Carolina, Columbia SC 29208-0112

This paper describes the application of a two-dimensional model to the cleanup of a hydrocarbon-contaminated aquifer using enhanced *in situ* biorestoration. The model, developed for simulation of natural *in situ* bioremediation, considers two-dimensional advective-dispersive transport and microbial utilization of a hydrocarbon contaminant (or constituency of organic compounds), two electron acceptors (oxygen and nitrate), a mineral nutrient (NH_4^+ or PO_4^{-3}), and a conservative tracer. A governing system of coupled partial differential and algebraic equations are solved simultaneously in a time-stepping solution. This paper describes the study site and remediation plan, outlines numerical model theory and development, and explores the use of the model to assess remediation strategies.

U.S.G.S. HANAHAN SITE

The Defense Fuel Supply Point (DFSP), located near Charleston SC, is a jet fuel storage and transfer facility. The site contains seven 3.3 million-gal aboveground storage tanks. In October 1975, a leak of approximately 83,000 gal of JP-4 aviation turbine fuel was reported. Fuel recovery efforts were undertaken in late 1975 and early 1976, during which time approximately 20,850 gal were recovered.

Site Hydrogeology and Geochemistry

The shallow water table aquifer at the site consists primarily of sediments of unconsolidated fine- to medium-grained sands of the Ladson Formation (Pleistocene age), with interfingering lenses of clay to a depth of approximately 6 to 10.5 m (20 to 35 ft). Occurrence of relatively impermeable clayey beds in this material leads to semiconfined and unconfined conditions (Figure 1), and perched water tables may develop in some locations. These sands and clays are underlain by the Ashley Formation (Oligocene age), which consists of calcareous clay containing silt and fine sand. Sediments of the Ashley Formation are relatively impermeable and function as a lower confining bed to the water table aquifer. Depth to the water table varies seasonally from 1.5 to 3.5 m (5 to 12 ft) below land surface (McClelland 1987).

Several studies have been carried out to assess the contaminant concentrations in the groundwater and sorbed to subsurface sediments. Dissolved concentrations of alkylbenzenes as high as 3 mg/L benzene, 5 mg/L toluene, 1 mg/L ethylbenzene, and 3 mg/L total xylenes have

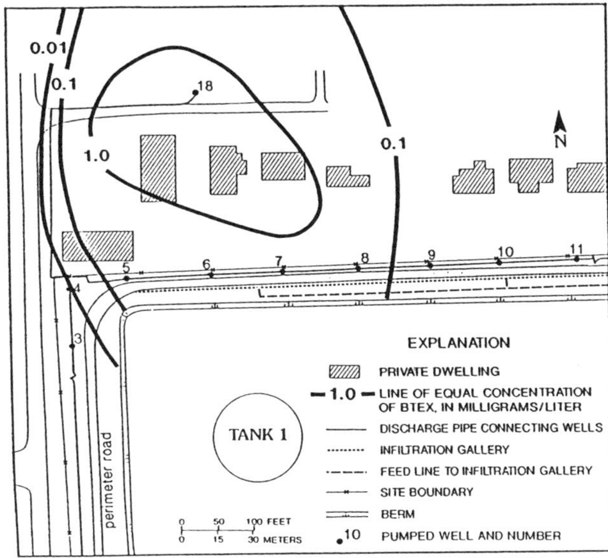

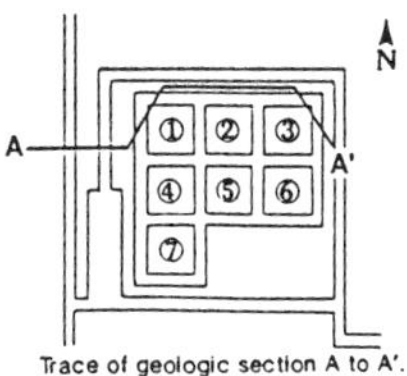

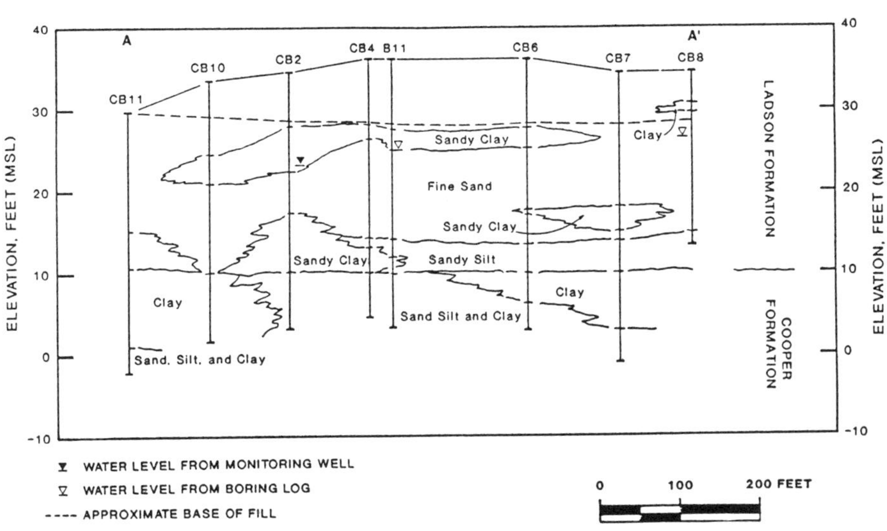

FIGURE 1. Schematic of the infiltration/extraction well bioremediation system and geologic cross section at the Hanahan site showing benzene, toluene, ethylbenzene, and xylene (BTEX) contaminant plume.

been measured (RMT 1988). Sediment contained sorbed concentrations of total petroleum hydrocarbons (TPH) ranging from 11 to 4,487 mg/kg (dry wt).

Groundwater from the site is generally acidic, with pH values ranging from 4.8 to 6.0, and inorganic nutrients in groundwater are present in low concentrations. Although no dissolved oxygen was detected in groundwater from monitoring wells screened at 8 to 10.5 m (27 to 35 ft), 2.9 mg/L oxygen was measured in groundwater from wells screened at 1 to 5.5 m (3 to 18 ft). Significant concentrations of methane have been measured in groundwater from contaminated zones, and laboratory investigations indicate that an active microbial community exists at the site. End products of aerobic, denitrifying, and anaerobic microbial processes (CO_2, N_2O, and CH_4, respectively) have been measured in laboratory incubations using contaminated subsurface sediments from the site. The bioremediation system with added nitrate should promote the activity of the denitrifying bacteria, but it is expected that other microbial processes also will be active (Aelion & Bradley 1991).

Bioremediation Design and Performance Analysis

The enhanced *in situ* bioremediation program at the DFSP will use a series of pumping wells to recover contaminated groundwater and circulate nutrient-enriched (with nitrate and phosphate) water to the subsurface through an infiltration gallery (RMT 1988). Sixty percent of the recovered water will be discharged to the North Charleston sewer district. The infiltration gallery, a perforated pipe installed 0.6 m (2 ft) below land surface, will extend along the length of the remediation area (Figure 1). Nutrients carried by the recirculating water are expected to stimulate indigenous bacteria to degrade JP-4 fuel in the water table aquifer. Wells and recharge galleries will be placed so that contaminated groundwater and nutrient-enriched water are contained within a cone of depression to prevent further spread of the contaminant plume and nutrients. U.S. Geological Survey (U.S.G.S.) hydrologists employed a quasi-three-dimensional flow model to evaluate hydraulic containment within the aquifer for several remediation system designs. The selected system is near completion, and start-up is scheduled for Summer of 1991.

A water quality and hydrologic monitoring program has been implemented by U.S.G.S. hydrologists to assess hydrocarbon contaminant effacement and to project the length of time required for aquifer remediation. This study will include tracer and field testing to quantify hydraulic conductivity distribution within the water table aquifer. Test results and sample data will serve as a basis for verification of simulation results. In addition, microbial kinetic parameters have been derived from laboratory microcosm studies using samples of subsurface materials from the site.

Modeling Approach

At present, simulations of solute transport and biodegradation have been limited to two dimensions. Although it is desirable to simulate the entire region in three dimensions, a two-dimensional vertical profile domain affords several advantages in the early analysis of the bioremediation design. In addition, the hydraulic containment system facilitates unidirectional groundwater flow between the infiltration gallery and the line of pumping wells. Our first approach to modeling bioremediation at the site is to apply the model problem domain at a location where the assumption of horizontal advection is a valid approximation and hydrocarbon contamination in the sediments is concentrated near Tank 1 (Figure 1). This approach allows small-scale simulation of contaminant biodegradation while incorporating the variability in advection rates and source input over the vertical dimension of the aquifer.

MODEL THEORY

Development of the mathematical model has been documented for the one-dimensional case (Widdowson *et al.* 1988). The model herein includes basic processes at the microscale (*e.g.*, microbial utilization), advective-dispersive two-dimensional transport at the macroscale, and processes that govern the transfer of mass between phases at both scales (*e.g.*, diffusion of solute into a microbial colony). Mathematical description of macroscopic advective-dispersive transport is the most straightforward relative to equations describing microbial utilization or interphase transport. The latter are dependent on assumptions applied to microbial mass distribution, generally described by Widdowson (1991) as either continuous growth (i.e., biofilms) or discrete growth (*e.g.*, bacterial cell clusters). The discrete growth model is the basis for the microcolony concept established by Molz *et al.* (1986) in which subsurface biomass is assumed to appear and proliferate in small clusters attached to soil particles. Later versions of the microcolony model (Widdowson *et al.* 1988) make no unwarranted assumptions concerning information on grouping patterns or dimensions to such groups, but require a diffusion surface area parameter for calculating interphase transport and microbial utilization.

Microbial Utilization

However the biomass is configured conceptually, the rate of microbial utilization of each constituent is typically based on equations describing biomass growth and decay (an exception to this statement is the instantaneous reaction concept described by Borden & Bedient, 1986). Assuming that microbial metabolism is limited by several nutrients, utilization rates can be described using modified Monod kinetics in which the rate of growth is dependent on constituent concentrations, specifically, organic carbon substrate, electron acceptor(s), and mineral nutrient. Kinetic expressions in the literature applied to the subsurface employ either pore fluid (macroscopic) or microbial phase (microscale) constituent concentrations in the functional relationships describing growth and decay.

Herein, we choose kinetic expressions based on concentrations associated with the microbial phase as described by Molz *et al.* (1986). Oxygen is the electron acceptor under aerobic conditions, whereas nitrate is the primary electron acceptor under anoxic conditions. This sequence is based on knowledge that the denitrifying enzyme is repressed by oxygen and is quickly derepressed under anoxic conditions. This enzyme inhibition is modeled using a kinetic expression for noncompetitive enzyme inhibition:

$$I(o) = \left(1.0 + \frac{o}{K_c}\right)^{-1} \tag{1}$$

where I(o) is a dimensionless inhibition function, o is the oxygen concentration in the biomass phase (M/L^3), and K_c is the inhibition constant (M/L^3) (Widdowson *et al.* 1988).

Following Herbert's theory (1958) of organic carbon assimilation by aerobic, heterotrophic organisms, the utilization rate for the organic carbon species (r_s; that is, rate of hydrocarbon utilization per unit biomass, M/M/t) is expressed as the sum of the organic carbon substrate utilization rate due to both aerobic (r_{so}) and anaerobic (r_{sn}) respiration:

$$r_s = r_{so} + r_{sn} \tag{2}$$

where

$$r_{so} = \frac{\mu_o}{Y_o}\left(\frac{s}{K_{so} + s}\right)\left(\frac{o}{K_o + o}\right)\left(\frac{a}{K_a + a}\right) \tag{3}$$

$$r_{sn} = \frac{\mu_n}{Y_n}\left(\frac{s}{K_{sn} + s}\right)\left(\frac{n}{K_n + n}\right)\left(\frac{a}{K_a + a}\right)I(o) \tag{4}$$

where s, n, and a are the hydrocarbon substrate, nitrate, and mineral nutrient concentrations in the biomass phase (M/L^3), respectively; μ_o and μ_n are the maximum specific growth rates (M/M/t) under oxygen- and nitrate-based respiration, respectively; Y_o and Y_n are the theoretical yield coefficients under oxygen- and nitrate-based respiration, respectively; K_{so} and K_{sn} are half-saturation constants (M/L^3) for the hydrocarbon compound under aerobic and anaerobic (nitrate) respiration, respectively; and K_o, K_n, and K_a are half-saturation constants (M/L^3) for oxygen, nitrate, and nutrient, respectively.

Similar equations are developed for the utilization rates (M/M/t) for oxygen (r_o), nitrate (r_n), and mineral nutrient (r_a):

$$r_o = Y_o \gamma r_{so} + \alpha_o k_{do}\left(\frac{o}{K_o + o}\right) \tag{5}$$

$$r_n = Y_n \eta r_{sn} + \alpha_n k_{do}\left(\frac{n}{K_n + n}\right)I(o) \tag{6}$$

$$r_a = \psi Y_o r_{so} + \varepsilon Y_n r_{sn} \tag{7}$$

where γ and η are use coefficients for biomass growth and α_o and α_n are use coefficients for biomass maintenance for oxygen and nitrate, respectively; Ψ and ε are use coefficients for the mineral nutrient under oxygen- and nitrate-based respiration, respectively; and k_{do} and k_{dn} are use decay rate coefficients for oxygen- and nitrate-based respiration, respectively.

Interphase Transport

Basic relationships must also be developed to relate macroscale processes to the microbial phase in which utilization of mass occurs. In this stage of model development, it is assumed that mass transport within the attached microbial cell colonies is balanced so that

$$\left(\begin{array}{c} \text{Rate of mass} \\ \text{entering biomass} \end{array} \right) = \left(\begin{array}{c} \text{Rate of mass} \\ \text{utilized in biomass} \end{array} \right) \tag{8}$$

Note that the cell colony approach implies small, discrete growths of biomass; this is in sharp contrast to the biofilm approach in which mass transport within the biofilm by diffusion is a primary component of the model. However, the model developed here is similar to mathematical biofilm models in that mass is transported from the pore fluid to the biomass phase across a diffusional boundary layer of thickness (δ) so that

$$\left(\begin{array}{c} \text{Rate of mass} \\ \text{exiting pore fluid} \end{array} \right) = \left(\begin{array}{c} \text{Rate of mass} \\ \text{entering biomass} \end{array} \right) \tag{9}$$

The rate of mass exiting the pore may vary slowly with time, resulting in a pseudo-steady-state condition within the boundary layer. From the stated assumptions, it follows that diffusion is the primary mechanism for mass transport between the aqueous and microbial phases and that this rate of diffusion is equal to the rate of microbial utilization. As developed in Molz *et al.* (1986) for porous media problems and later modified in Widdowson *et al.* (1988), the diffusion term relates pore fluid concentrations to the microbial phase concentrations

$$D_s \left(\frac{S - s}{\delta} \right) \beta = r_s \tag{10}$$

$$D_o \left(\frac{O - o}{\delta} \right) \beta = r_o \tag{11}$$

$$D_n \left(\frac{N - n}{\delta} \right) \beta = r_n \tag{12}$$

$$D_a \left(\frac{A - a}{\delta} \right) \beta = r_a \tag{13}$$

where S, O, N, and A are the hydrocarbon substrate, oxygen, nitrate, and mineral nutrient concentrations in the pore fluid (M/L^3), respectively; D_s, D_o, D_n, and D_a are molecular diffusion coefficients for the hydrocarbon compound, oxygen, nitrate, and mineral nutrient (L^2/t), respectively; and β is the available surface area for diffusion per unit biomass (L^2/M).

Macroscale Transport

Fundamental equations governing constituent transport (hydrocarbon substrate, oxygen, nitrate, mineral nutrient, and nonconservative tracer) in the pore fluid incorporates the following processes:

- Two-dimensional dispersive/diffusive transport
- Two-dimensional, horizontal advective transport
- Sorption (linear Freundlich isotherm)
- Interphase diffusive transport from the bulk liquid to the attached biomass
- Injection/infiltration of solute from external source (well or infiltration pond)

For any single constituent, the result is one general equation describing the rate of change in pore fluid concentration:

$$f_c \frac{\partial C}{\partial t} = -v_l(x_t) \frac{\partial C}{\partial x_l} + D_l \frac{\partial^2 C}{\partial x_l^2} + D_t \frac{\partial^2 C}{\partial x_t^2} - \frac{MD_c}{\theta} \left(\frac{C-c}{\delta} \right) \beta + R_c \quad (14)$$

where C and c are pore fluid and biomass phase concentrations (M/L^3), respectively; x_l and x_t are defined as coordinate directions longitudinal and transverse to the principal direction of dispersion, respectively; t is time; $v_l(x_t)$ is the pore fluid velocity (L/t); D_l and D_t are hydrodynamic dispersion coefficients (L_2/t) in longitudinal and transverse directions, respectively; f_c is the retardation factor (ratio of the v_l to constituent velocity); D_c is the molecular diffusion coefficient (L^2/t); M is a macroscale variable for the concentration of biomass expressed as biomass per unit bulk volume of porous media (M/L^3); θ is porosity of the porous medium; and R_c is a general term representing mass entering the pore fluid ($M/L^3/t$). The specific representation of R_c will depend on problem geometry and physical configuration of the remediation system as described below. As written, Equation (14) allows for simulation of both vertical profile and horizontal (areal) domain problems.

Microbial Growth Dynamics

The model will account for temporal fluctuations in the microbial population density as concentrations of nutrients increase or decrease. Herbert (1958) states that the rate of growth in biomass is proportional to the rate of hydrocarbon substrate utilization (r_s), and the net rate of biomass production is the difference in biomass production and decay rates. Equation (15) is a statement of mass rate balance for the attached biomass describing the rate of change in the macroscopic variable M as

$$\frac{1}{M}\frac{\partial M}{\partial t} = Y_o r_{so} + Y_n r_{sn} + k_{do}\left(\frac{n}{K_n + n}\right) - k_{dn} I(o) \qquad (15)$$

Thus, the model will account for increased rates of interphase mass transport out of the pore fluid as the biomass concentration increases with time.

Numerical Solution Technique and Verification

The complete set of governing equations consists of four coupled macroscopic transport equations, four coupled algebraic equations describing microbial utilization rates, and an equation describing changes in the biomass concentration (see Widdowson *et al.* 1988). The system of coupled equations is solved in a sequence of steps that comprises an iterative solution to the entire system. The model solves a fifth transport equation for a nonconservative solute so that constituent concentrations can be compared or tracer transport can be simulated.

The transport equations (*e.g.*, Equation [14]) are solved using a split-operator approach in which the advection process and the dispersion/biotransformation processes are solved separately. Advection is solved by adopting a Lagrangian approach using a two-point interpolation scheme along streamlines as detailed by Molz *et al.* (1986)

in one dimension and Widdowson *et al.* (1987) in two dimensions. The resulting versions of the transport equations, with the advection terms removed, are solved by adopting an Eulerian approach using a line-successive overrelaxation finite difference solution technique for the Darcy scale pore fluid concentrations (S, O, N, A, and T, where T is the tracer pore fluid concentration) at the new time level ($t+\Delta t$). The subsystem of Equations (10) through (13) is solved iteratively within the global iteration using a Newton-Raphson-Kantorovich quasilinearization technique for colony concentrations (s, o, n, a) at the new time level. Equation (15), describing time changes in the active microbial population concentration (M), is then solved directly knowing the newly calculated concentrations above. Convergence of all arrays throughout the spatial domain is checked against a tolerance criterion.

Testing of the numerical solution for the two-dimensional case presented in Widdowson *et al.* (1987) shows excellent agreement with the analytical solution to the two-dimensional line boundary-source problem. Careful selection of node spacing (Δx) and time step size (Δt) as input parameters will result in solutions that are devoid of numerical difficulties such as numerical dispersion or oscillations associated with solution of the solute transport equation. Numerical results are compared with the analytical solution for a two-dimensional point-source problem (Figure 2). The continuous-source solution (Hunt 1978) assumes that the velocity field is independent of the rate of injection and is uniform in the two-dimensional domain. Numerical results in dimensionless form are presented in Figure 2 along the centerline of the plume at a simulation time of 500 and 1,000 days. Modification of the grid spacing or time step size did little to improve numerical results using the technique described above (Method 1); however, a favorable comparison with the analytical solution (solid curve) is achieved when the numerical solution technique is modified (Method 2) so that the transport equation with the advection term removed is solved first, followed by the advection calculation in which the front is advanced by a travel distance $v_l \Delta t / f_c$.

REMEDIATION SIMULATION

Figure 3 depicts the vertical profile problem domain and mathematical assumptions at the influent, effluent, and longitudinal (water table and lower confining-layer) boundaries. Pore fluid velocity v_l is directed along the longitudinal axis, but can depart from the depth-average velocity along the axis (x_t) transverse to flow. Contaminant and

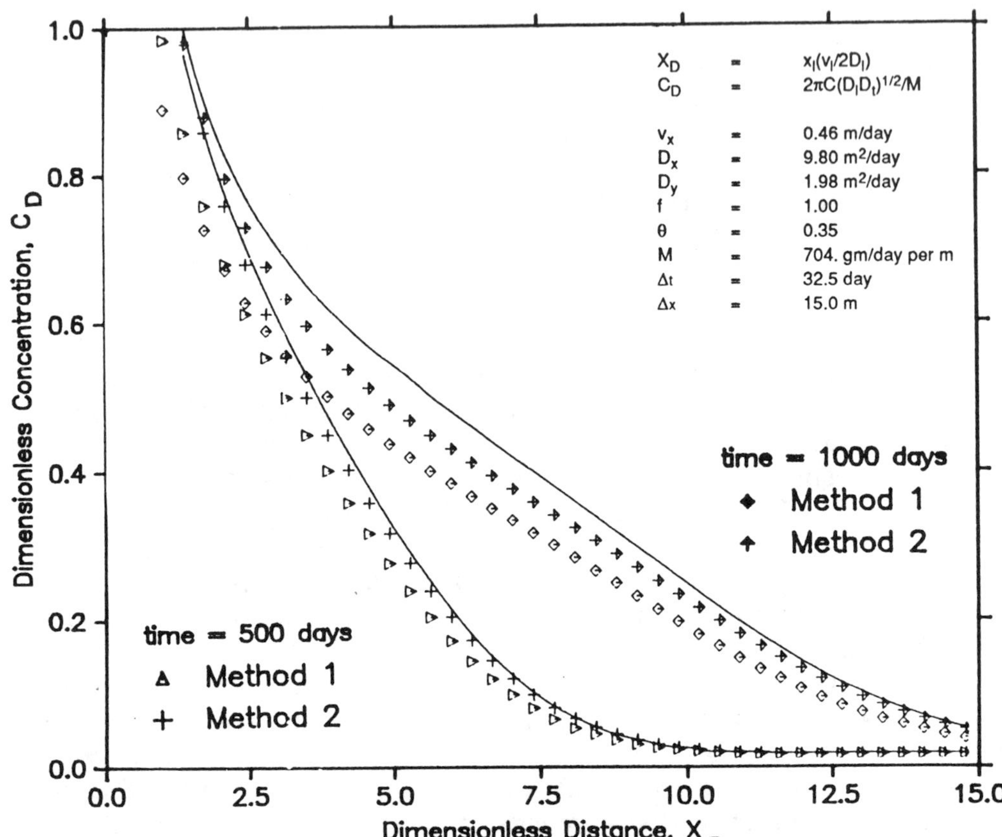

FIGURE 2. Longitudinal profile (y=0) of numerical (individual data points) and analytical (solid lines) results to a point source contamination problem at time = 500 and 1,000 days.

nutrient flux are accounted for using the macroscopic transport equation source term (R_c) as the mass injection rate per unit volume of pore fluid (Huyakorn, P. S., Auburn University, personal communication, 1985). Given a mass injection of C_oQ distributed over n finite difference mesh nodes, the corresponding source term at each of the nodes (domain thickness τ) receiving solute flux is

$$R_c = C_oQ/(n\Delta x_l \Delta x_t \tau\theta) \tag{16}$$

The basis for this preliminary approach has been addressed above. However, experimental data derived from tracer and hydraulic testing within the recirculation system will provide direction on any necessary

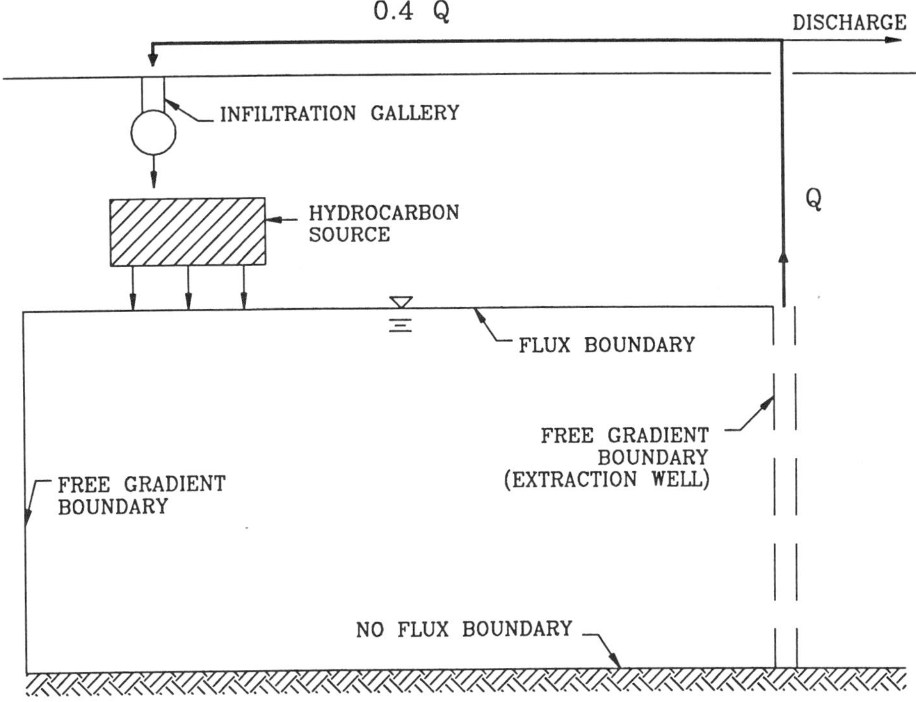

FIGURE 3. Schematic of the numerical problem domain as applied to a 1-m wide vertical section of aquifer. Free-gradient boundary condition allows for a transient concentration gradient as a solute front "exits" the domain at the effluent boundary.

modification to the numerical approach. For example, a velocity distribution, $v_l(x_t)$, will be derived from breakthrough data at multilevel sampling devices. Likewise, the magnitude of vertical flow and its significance to solute transport in the modeled cross section can be evaluated from water level data during the test and through additional flow modeling.

Model input parameters were varied during simulations of the fate and two-dimensional transport of groundwater contamination at the Hanahan site to determine model sensitivity to various parameters. Results indicated that the initial distributions of biomass and oxygen concentrations over the vertical dimension of the domain were factors in the resulting distribution of contaminant. It appears, however, that vertical variation in pore fluid velocity, particularly in the upper region of the domain, was the most significant factor controlling plume size and contaminant distribution. The geologic cross section of the Hanahan site (Figure 1) suggests that hydraulic conductivity variation, $K(x_t)$, within the fine sand is likely; therefore, multilevel field tests remain a necessary step in quantifying $K(x_t)$. A velocity distribution has been generated for model input (2.5 m/day $\leq v_l(x_t) \leq$ 4.0 m/day, 8.2 ft/day $\leq v_l(x_t) \leq$ 13.1 ft/day) to reflect both permeability variability and results of flow modeling (3.0 m/day, 9.8 ft/day average pore fluid velocity) employed in the system design.

Hydrocarbon Distribution and Fate

Based on subsurface sediment sampling, the hydrocarbon mass in the sediments near Tank 1 has been estimated at approximately 725 kg (1,600 lb) of total petroleum hydrocarbons (TPH). Because the numerical model considers only hydrocarbon mass below the water table, an accounting scheme for hydrocarbon mass was developed and incorporated into the computer code. For an initial mass m_o (hatched area of Figure 3), the mass present in the sediments is given as $m(t) = m_o g(t)$, where $g(t)$ is a prescribed function of time (*e.g.*, $\exp[-\lambda t]$). The amount of mass input to the saturated zone is calculated as the difference in mass, $m(t) - m(t+\Delta t)$, over the time step Δt plus any hydrocarbon recirculated through the infiltration gallery. Constituent mass exiting the domain is calculated at the end of each time step, and a fraction (model input, 0.40 in this case) is returned to the domain using the source term. This total input of mass is distributed over several mesh nodes, and a source term (R_s) must be calculated for each time step. Nitrate, oxygen, and mineral nutrient additions are also accounted for through source terms (R_n, R_o, R_a, respectively).

In all cases, the predicted contaminant plume was approximately located within the upper fourth (1.20 m) of the domain. Figure 4 (top) depicts a typical simulated hydrocarbon distribution in the upper half domain. Molz and Widdowson (1988) demonstrated that the numerical method used here can accurately simulate large vertical concentration gradients only if small dispersivity values (in this case, α_l = 1.0 cm and αt = 0.10 cm) are employed. These results are consistent with water quality data at the site in which high TPH concentrations occur at the water table. Vertical transport of hydrocarbons at the site is primarily caused by fluctuation of the water table elevation with recharge events.

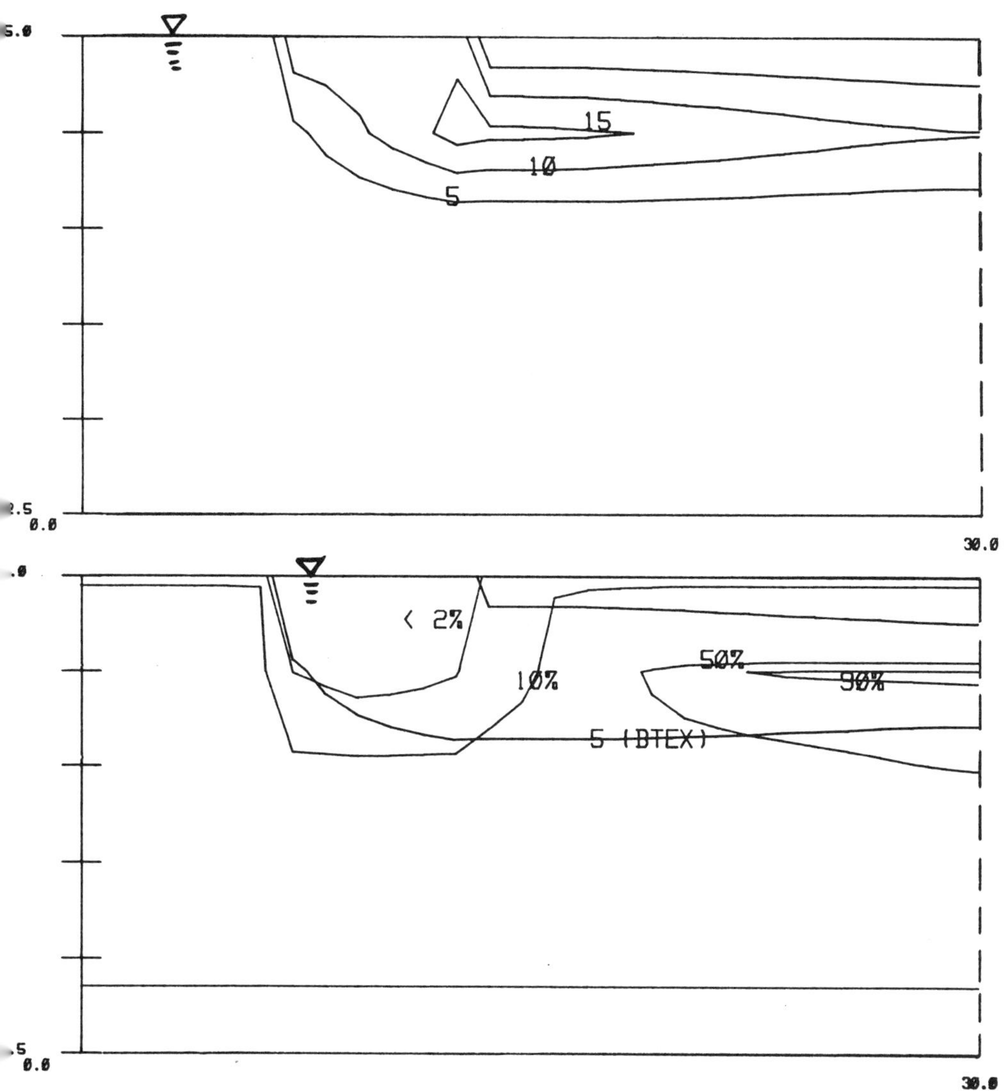

FIGURE 4. Simulated hydrocarbon distribution at time = 60 days (top). Distribution of dissolved oxygen (bottom) reflects varying levels of denitrifying activity.

Electron Acceptor Distribution and Fate

Because some oxygen will be introduced into the aquifer at the recharge gallery, model simulations indicate that oxygen may persist at levels sufficient to inhibit nitrate-based respiration in the first 5 to 10 m near the recharge gallery. Simulations using 2.0 mg/L as the dissolved oxygen (DO) concentration of recharge water always showed the presence of excess nitrate at the extraction well approximately 15 m from the recharge gallery. Thus, system simulations under the residence time as now planned (3.75 to 5 days) imply that nitrate-based respiration, the proposed main mechanism for hydrocarbon reduction, may be effective only in the final one- to two-thirds of the flow path. Figure 4 (bottom) depicts oxygen distribution superimposed over the simulated hydrocarbon plume, and by implication, zones delineated by metabolic respiration type (aerobic or denitrifying). The percentage of denitrifying activity, as determined by Equation (1), I(o) × 100%, is identified in the figure. Oxygen is consumed near the source, and low oxygen concentrations (less than 0.09 mg/L) are present in the pore fluid beyond the line labeled 10 percent. Nitrate-based respiration is only significant within the plume shown as a finger-shaped area labeled 50 percent (or greater). Slower simulated groundwater velocities allow increased contact time between nitrate, hydrocarbon, and microbes and allow better definition of the area along the flowpath where nitrate-based respiration begins. Moreover, simulations involving control of the hydraulic gradient may provide potential optimization approaches that could be effective at the site.

SUMMARY OF RESULTS

Simulation results have been summarized in terms of the cumulative percentage of hydrocarbon removed through aerobic, nitrate-based, anaerobic biodegradation or both (Figure 5). Based on the simulation results, contaminant remediation appears to be primarily limited by the rate at which hydrocarbon mass enters the aquifer treatment zone. Because transfer of hydrocarbon from the solid phase to the aqueous phase is facilitated by flushing from the infiltration galleries alone, the input of hydrocarbon will also be the least controllable process at the site. These preliminary simulations indicate that electron acceptor availability will not be a factor at the Hanahan site due to relatively rapid travel times in the aquifer and the organic carbon substrate limitation.

Mineral nutrient limitation also appears to be a controlling factor, although no definitive conclusions have been reached regarding nutrient supplement to the system. Simulation 10 depicts the case in which an insufficient amount of mineral nutrient is delivered. Simulations 11 and 7 both consider high nutrient (PO_4^{-3}) input, but varied

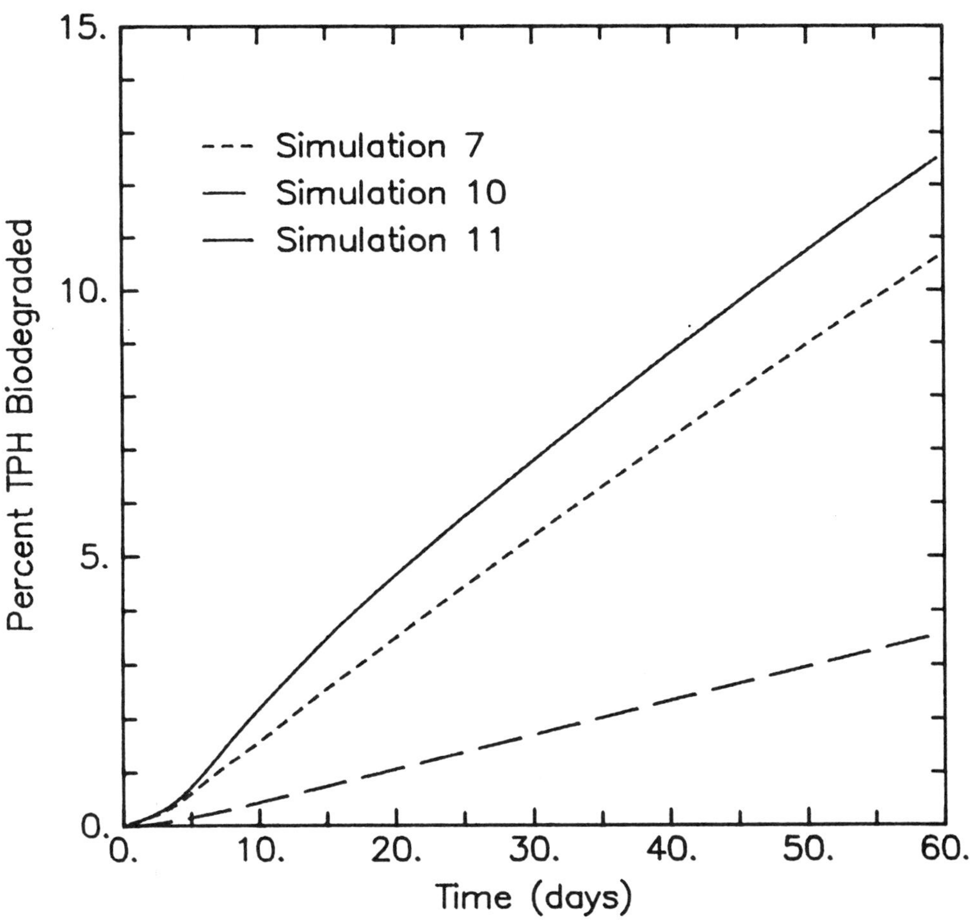

FIGURE 5. Percent of initial TPH in groundwater and sediments biodegraded for three simulations: $C_0(NO_3)$ = 7 mg/L, nutrient excess (sim. 7); $C_0(NO_3)$ = 42 mg/L, nutrient limited (sim. 10); $C_0(NO_3)$ = 42 mg/L, nutrient excess (sim. 11).

with respect to relative nitrate input (42 to 7.0 mg/L, respectively). The small difference in removal is attributed to oxygen inhibition. An approximately equal amount of hydrocarbon mass is simulated as removed from the system through withdrawal and subsequent discharge (10% in simulation 11) as was biodegraded (13%). The percentage of mass remaining in the sediments and pore fluid is 77 percent after 60 days.

Analysis of the remediation system based on model predictions will continue as system start-up approaches. Longer time simulations will be required to forecast an expected time-to-complete remediation. Integrating field data from the water quality and hydrologic monitoring program with the model-based analysis will supply the means to test and verify many of the assumptions made in this simulation study. Analyses of advection patterns, mineral nutrient limitation, and hydrocarbon flushing may result in modification to the modeling approach.

REFERENCES

Aelion, C. M.; Bradley, P. M. *Applied and Environmental Microbiology* **1991**, *57*, 57–63.

Borden, R. C.; Bedient, P. B. *Water Resources Research* **1986**, *22*, 1973–1982.

Herbert, D. In *Some Principles of Continuous Culture. Recent Progress in Microbiology*; Tunevall, G., Ed.; Blackwell Scientific: Oxford, England, 1958.

Hunt, B. *ASCE, Journal of the Hydraulics Division* **1978**, *104*, 75–85.

McClelland Engineering. "Confirmation Study—Characterization Step"; final report to the Defense Fuel Supply Point, Charleston, SC; Houston, TX, 1987.

Molz, F. J.; Widdowson, M. A. *Water Resources Research* **1988**, *24*, 615–619.

Molz, F. J.; Widdowson, M. A.; Benefield, L. D. *Water Resources Research* **1986**, *22*, 1207–1216.

Rifai, H. S.; Bedient, P. B. *Proceedings of the Solving Ground Water Problems with Models Conference and Exposition*, AGWSE/IGWMC, 1989, pp 1187–1203.

RMT Consultants. "Aquifer Evaluation"; final report to the Defense Fuel Supply Point, Charleston, SC; Greenville, SC, 1988.

Widdowson, M. A. Comment to "An Evaluation of Mathematical Models of the Transport of Biologically Reacting Solutes in Saturated Soils and Aquifers," Baveye, P.; Valocchi, A. *Water Resources Research* **1991**, in press.

Widdowson, M. A.; Molz, F. J.; Benefield, L. D. *Proceedings of the Solving Ground Water Problems with Models Conference and Exposition*, AGWSE/IGWMC, 1987, pp 28–51.

Widdowson, M. A.; Molz, F. J.; Benefield, L. D. *Water Resources Research* **1988**, *24*, 1553–1565.

Biodegradation of Volatile Organic Compounds in Porous Media with Natural and Forced Gas-Phase Advection

B. E. Sleep*
University of Toronto
J. F. Sykes
University of Waterloo

INTRODUCTION

Contamination of groundwater resources by hazardous substances has become an issue of increasing importance in North America in the last two decades. Petroleum products and halogenated hydrocarbon solvents, ubiquitous in our industrialized society, are among the most serious threats to groundwater systems. These organic compounds are characterized by their immiscibility with water, and low, but toxicologically significant, air- and water-phase solubilities.

Organic liquids, spilled into the subsurface, flow vertically and horizontally under the influence of capillary, viscous, and buoyancy forces. Liquids less dense than water will pool on the water table, while those more dense than water will continue to move downward past the water table. When the spill is terminated, the organic liquid will continue to move until it is distributed at residual saturation, or is at rest in higher saturation pools of liquid on the water table, or above impenetrable soil layers.

* University of Toronto, Department of Civil Engineering, Galbraith Building, 25 St. George St., Toronto, Ontario, M5S 1A4

The immobilized organic liquids will dissolve into the water phase and volatilize into the gas phase. Subsequent transport in these phases may result in widespread groundwater contamination. The extent of groundwater contamination from immobilized organic liquids is controlled by the rates of dissolution and volatilization of the organic, by the degree of adsorption onto the soil, by groundwater velocities, and by the rate of diffusion of organic vapours to the atmosphere.

A number of numerical models capable of describing the immiscible flow, dissolution, volatilization, and transport of organic compounds have been developed (Abriola & Pinder 1985; Forsyth 1988). All of these models assumed that there was no advection in the gas phase. Sleep and Sykes (1989) developed a numerical model that included gas-phase advection and showed that gas-phase advection due to organic vapour concentration gradients could be significant. The model did not include organic liquid flow.

In addition to dissipation by diffusion to the atmosphere, the amount of a volatile organic compound in the subsurface may be reduced by chemical and biological reactions. Biodegradation of organic compounds occurs naturally under typical subsurface conditions (Borden & Bedient 1986). Under aerobic conditions, the rate of biodegradation of a compound is dependent on microbial population, substrate concentration, electron acceptor (oxygen) concentration, as well as the presence of cosubstrates, inorganic nutrients, inhibiting substances, moisture content, and temperature.

Several numerical models of biodegradation in subsurface systems have been developed. Sykes *et al.* (1982) developed a model for the anaerobic biodegradation of organic leachate from landfills. Biodegradation rate expressions were based on Monod kinetics. Borden and Bedient (1986), MacQuarrie and Sudicky (1990) and Molz *et al.* (1986), developed models for the aerobic biodegradation of organic compounds in saturated systems. In all of these models biodegradation rates were represented by Monod kinetics, using the dual Monod formulation.

All of the models presented in the literature have been for single-phase, water-saturated systems. Many of the most common organic contaminants of groundwater systems are less dense than water and will not move below the water table. In addition to dissolution and volatilization, vadose zone biodegradation may account for a significant mechanism for the removal of these organic compounds. Oxygen availability for aerobic biodegradation may be much greater in the unsaturated zone than in the saturated zone if diffusion of oxygen from the atmosphere occurs, because gas-phase diffusion coefficients are 4 orders of magnitude greater than water-phase diffusion coefficients.

The application of forced gas venting to increase organic volatilization rates has become a popular method for remediating spills of volatile organic compounds (Johnson *et al.* 1990). In inducing high gas-phase flow rates the availability of oxygen in the subsurface is also increased, as it is when oxygen sparging is used to enhance biodegradation (Thomas *et al.* 1987). In designing forced gas venting and/or enhanced biodegradation systems, one must consider the dynamics of the dissolution, volatilization, and biodegradation processes, and the availability of electron acceptors such as oxygen. This requires a multiphase model that includes a mobile gas phase, with the capability to model the transport of the organic species in the liquid organic, water, and gas phases, and the transport of oxygen in the water and gas phases.

In this study a numerical model is developed to simulate three-phase (air-organic-water) flow and transport, with interphase mass transfer and biodegradation. The flow equations include the effects of gravity, capillarity, compressibility, and variable gas-phase pressure. In addition to movement of air, water, and organic phases, the simulator can be used to predict the transport and transformation of an arbitrary number of additional components, such as oxygen, that are soluble in the water and/or soil gas phases. Microbial growth and oxygen and substrate consumption due to biodegradation are described by Monod kinetics.

Simulations of the dissolution, volatilization, and aerobic biodegradation of common volatile organic compounds in a variably saturated subsurface system demonstrate the relative importance of the various processes. The imposition of forced gas-phase advection leads to both increased volatilization and increased biodegradation from the greater availability of oxygen.

GOVERNING EQUATIONS

The macroscopically averaged equation describing the movement of a species α in phase β is (Abriola & Pinder 1985)

$$\frac{\partial}{\partial t}\left[\rho_\beta x_{\alpha\beta}\left(\phi S_\beta + K_{\alpha\beta,d}\rho_b\right)\right] + \nabla \bullet \left[\rho_\beta x_{\alpha\beta} v_\beta\right] - \nabla \bullet \left[\phi S_\beta D_{\alpha\beta} \nabla\left(\rho_\beta x_{\alpha\beta}\right)\right] - r_{\alpha\beta} - q_{\alpha\beta} = 0 \tag{1}$$

where ϕ is porosity, S_β is phase saturation, ρ_β is molar density, $x_{\alpha\beta}$ is species mole fraction, υ_β is Darcy velocity vector, $D_{\alpha\beta}$ is the dispersion tensor, $r_{\alpha\beta}$ represents interphase transfer of species α to or from phase β, and $q_{\alpha\beta}$ represents sinks and sources of species α to phase β. $q_{\alpha\beta}$ can include processes such as pumping or injection and chemical and biological reactions.

In Equation 1 it is assumed that partitioning of a species between a fluid phase (β) and the soil phase is described by a linear adsorption isotherm. $K_{\alpha\beta,d}$ is the ratio of mass concentration of species in the soil phase to mass concentration of species in the β phase, and ρ_b is the bulk mass density of the soil phase. If it is assumed that water saturations are sufficient to coat soil grains with water and are not in direct contact with the gas and organic phases, then $K_{\alpha\beta,d}$ will be zero for the gas and organic phases.

The Darcy velocity vector for a phase β is given by

$$\upsilon_\beta = -\frac{kk_{r\beta}}{\mu_\beta}(\nabla p_\beta + \gamma_{\beta g}\nabla z) \tag{2}$$

where k is intrinsic permeability tensor, $k_{r\beta}$ is relative permeability, μ_β is the viscosity, p_β is the phase pressure, g is gravitational acceleration, γ_β is mass density, and z is elevation.

The following constraints apply to Equations 1 and 2:

$$\sum_{\beta=1}^{3} S_\beta = 1.0 \tag{3}$$

$$\sum_{\alpha=1}^{N} x_{\alpha\beta} = 1.0 \tag{4}$$

The capillary pressures associated with interfaces between phases are given by

$$P_{cgw} = p_g - p_w \tag{5}$$

$$p_{cow} = p_o - p_w \tag{6}$$

$$p_{cgo} = p_g - p_o \tag{7}$$

where p_{cgw} is the capillary pressure between the water (w) and gas (g) phases in a two-phase, gas-water system; p_{cow} is the capillary pressure between organic (o) and water phases in a two-phase, organic-water or three-phase, organic-water-gas system; and p_{cgo} is the capillary pressure between organic and gas phases in a two-phase, organic-gas system or a three-phase, organic-water-gas system. It is commonly assumed that water has the highest wettability, gas has the lowest wettability, and the organic phase is of intermediate wettability.

$D_{\alpha\beta}$ is the dispersion tensor, defined by Bear (1972) as

$$D_{\alpha\beta ij} = \alpha_T |\upsilon| \delta_{ij} + (\alpha_L - \alpha_T) \frac{\upsilon_i \upsilon_j}{|\upsilon|} - D^m_{\alpha\beta} \tau_\beta \tag{8}$$

where α_L and α_T are the longitudinal and transverse dispersivities (L), respectively, δ_{ij} is the Kronecker delta, $D^m_{\alpha\beta}$ is the molecular diffusion coefficient, and τ_β is the tortuosity factor.

The Millington and Quirk (1961) model for tortuosities in partially saturated porous media is

$$\tau_g = \frac{S_g^{7/3}}{\phi^2} \tag{9}$$

where S_g is gas-filled porosity. The same form can be used for the tortuosity in the water phase with by replacing S_g with S_w.

Summing the equation for a species α over all the phases gives the component molar balance

$$\sum_{\beta=1}^{3} \frac{\partial}{\partial t} [\rho_\beta x_{\alpha\beta} (\phi S_\beta + K_{\alpha\beta,d} \rho_b)] + \nabla \bullet [\rho_\beta x_{\alpha\beta} \upsilon_\beta] - \nabla \bullet [\phi S_\beta D_{\alpha\beta} \nabla (\rho_\beta x_{\alpha\beta})] - q_{\alpha\beta} = 0 \tag{10}$$

$K_{\alpha\beta,d}$ is zero for the gas and organic phases.

Interphase mass transfer of organic species occurs by dissolution and volatilization of the organic phase and partitioning of organic species between the gas and water phases. When water, gas, and organic phases are present, the organic vapour concentration will be equal to the saturated vapour concentration determined from the organic liquid vapour pressure by

$$x_{\alpha g} = \frac{P_{vp,\alpha}}{P_g} \tag{11}$$

If the organic compound is sparingly soluble in the water phase, Henry's law may be used to relate concentrations in the water and gas phases. Henry's law is

$$Hx_{\alpha w} = x_{\alpha g} P_g \tag{12}$$

Biodegradation may be included as a sink or source term ($q_{\alpha\beta}$) in Equation 10. If the dual Monod formulation (Borden & Bedient 1986; MacQuarrie & Sudicky 1990) is used, the sink term for the organic substrate will be

$$q_{ow}^{bio} = -\, kc_{mt} \left(\frac{x_{ow}}{K_o + x_{ow}} \right) \left(\frac{x_{aw}}{K_a + x_{aw}} \right) \tag{13}$$

where q_{ow}^{bio} is the rate of biodegradation of organic compound, c_{mt} is the total molar concentration of microorganisms, k is the maximum organic utilization rate, x_{ow} is the mole fraction of organic compound in the water phase, x_{aw} is the mole fraction of electron acceptor (oxygen) in the water phase, K_o is the organic compound half-saturation constant, and K_a is the electron acceptor half-saturation constant.

In most aquifers, more than 95 percent of the microorganisms are attached to the soil matrix (Borden & Bedient 1986). Borden and Bedient (1986) and MacQuarrie and Sudicky (1990) assumed that the movement of the microorganisms could be described by the advective-dispersive transport equation, with a linear adsorption isotherm to represent the distribution of the microorganisms between the soil and water phases. In the present study this approach is also used. It is assumed that all microorganisms occur only in the water and soil phases. The total mole fraction of microorganisms, c_{mt}, is given by

$$c_{mt} = \rho_w x_{mw} \left(\phi S_w + K_{mw,d} \rho_b \right) \tag{14}$$

where ρ_w is the molar density of water, x_{mw} is the mole fraction of microorganisms in the water phase, $K_{mw,d}$ is the linear adsorption coefficient for microorganisms in the water and soil phases, and ρ_b is the bulk mass density of the soil.

Mass concentrations of microorganisms are converted to mole fraction using an arbitrary molecular weight and molar density for the microorganisms. The half-saturation constants are also expressed as mole fractions. Half-saturation constants expressed on a mass basis can be converted to molar basis using the molecular weight of the appropriate compound and the density of water.

The rate of consumption of oxygen in the water phase, q_{aw}^{bio}, is given by

$$q_{aw}^{bio} = -kFc_{mt}\left(\frac{x_{ow}}{K_o + x_{ow}}\right)\left(\frac{x_{aw}}{K_a + x_{aw}}\right) \tag{15}$$

where F is the ratio of moles of oxygen consumed to moles of organic compound utilized.

The rate of production of microorganisms, q_m^{bio}, is

$$q_m^{bio} = -kYc_{mt}\left(\frac{x_{ow}}{K_o + x_{ow}}\right)\left(\frac{x_{aw}}{K_a + x_{aw}}\right) \tag{16}$$

where Y is the ratio of moles of microorganisms consumed to moles of organic compound biodegraded (microbial yield coefficient). Y is converted from a mass basis to a molar basis using the arbitrary molecular weight assigned to the microorganisms.

NUMERICAL IMPLEMENTATION

It is assumed that the soil gas phase is composed of nitrogen and oxygen only. Component mole balance equations are required for water, nitrogen, oxygen, organic compounds and microorganisms. In the simulations the production of carbon dioxide is not included, although this is possible with the present model formulation. The component mole balance equations are solved using block-centred finite differences. To avoid serious mass balance errors the biodegradation terms are evaluated implicitly. The advective and dispersive flux terms are calculated explicitly with respect to saturations and concentrations, as is the practice with compositional reservoir models. This allows the combination of all five component mole balances into one equation that is implicit in water-phase pressure only.

The system of pressure equations representing the discretized system is solved using incomplete Gaussian elimination with Orthomin acceleration (Forsyth & Sammon 1986). The saturations and concentrations are then calculated by back-substitution of pressures in the component balance equations. The explicit evaluation of saturation- and concentration-dependent quantities in the flux terms imposes a time step limitation but avoids the necessity of solving a matrix system $5n \times 5n$ (n grid blocks) associated with fully implicit methods.

The relative permeability and concentrations in the advective flux terms are evaluated using two-point upstream weighting (Todd *et al.* 1972). The two-point upstream method is second order accurate in space. The majority of reservoir models use one-point upstream weighting, but one-point upstream weighting is prone to excessive numerical dispersion. The computational cost of using two-point upstream weighting in an explicit formulation is not significantly greater than that of one-point upstream weighting, but the reduction in numerical dispersion is significant.

Model Verification

If the half-saturation constant for the organic compound is very large and the oxygen half-saturation constant is very small, the rate of biodegradation of organic compound will be approximated by

$$q_{ow}^{bio} = - k\rho_w x_{mt} \left(\frac{x_{ow}}{K_o} \right) \tag{17}$$

If the oxygen and microorganism concentrations are kept constant, the transport of the organic compound in one-phase water systems can be described by the single-phase, advective-dispersive transport equation with linear adsorption and linear decay (MacQuarrie & Sudicky 1990).

The two-dimensional analytical solution of Cleary and Ungs (1978) was used to verify the model and to examine the degree of numerical dispersion present in the solution at coarse levels of discretization.

A domain 60 m wide by 10 m deep was discretized into 400 (20 × 20) grid blocks. Water-phase boundary conditions were prescribed to produce constant velocity (0.09 m/day linear porewater velocity) flow parallel to the X-axis. Organic contaminant concentrations were prescribed along the inflow vertical boundary ($x_{ow} = 1 \times 10^{-6}$ from y = 1.0 to y = 3.5 m., x_{ow} = 0.0 along the rest of the boundary).

The longitudinal and transverse dispersivities were 0.6 m and 0.005 m as used by MacQuarrie & Sudicky (1990). Microorganism concentrations and the substrate utilization constant, *k*, were chosen to give a linear decay rate of 0.007/day. The retardation factor $(1 + K_{ow,d}\rho_b/\phi)$ from adsorption was 1.2. The agreement of the numerical and analytical models shown in Figure 1 is satisfactory.

VOLATILIZATION AND BIODEGRADATION OF TOLUENE

Simulations were performed to examine the rate of dissipation of an underground leak of toluene under natural conditions with and without biodegradation and with enhanced volatilization and biodegradation from air and oxygen sparging.

The simulation domain was 90 m long by 20 m deep, with a 2 percent inclination below the horizontal in the direction of groundwater flow. Hydrostatic water-phase pressures were prescribed on vertical boundaries to produce linear porewater velocities of approximately 0.1 m/day with the water table approximately 10 m below the ground surface boundaries. Vertical infiltration rates at the ground surface were approximately 0.3 m^3/m^2/year. Gas-phase pressures at the ground surface were prescribed as atmospheric. No gas-phase flux was allowed across the bottom of the domain, or the vertical boundaries.

Horizontal and vertical soil permeabilities were 2×10^{-12} and 1×10^{-12} m^2, respectively, and soil porosity was 0.3. The longitudinal and transverse dispersivities were 0.6 m and 0.005 m, respectively. The low transverse dispersivity is characteristic of sandy aquifers (MacQuarrie & Sudicky 1990). Relative permeability functions were prescribed to produce a residual organic liquid saturation of 0.1. The retardation factor for toluene adsorption was 1.4. It was assumed that oxygen was not subject to adsorption. The properties of toluene are given in Table 1. It was assumed that the subsurface temperature was 15 C. The air phase was assumed to be composed initially of 80 percent nitrogen and 20 percent oxygen. A Henry's constant of 3.64×10^5 atm was used for oxygen (Perry & Chilton 1973).

Approximately 1,500 kg of toluene were injected into the domain over a period of 10 days, at a position about 4 m below the ground

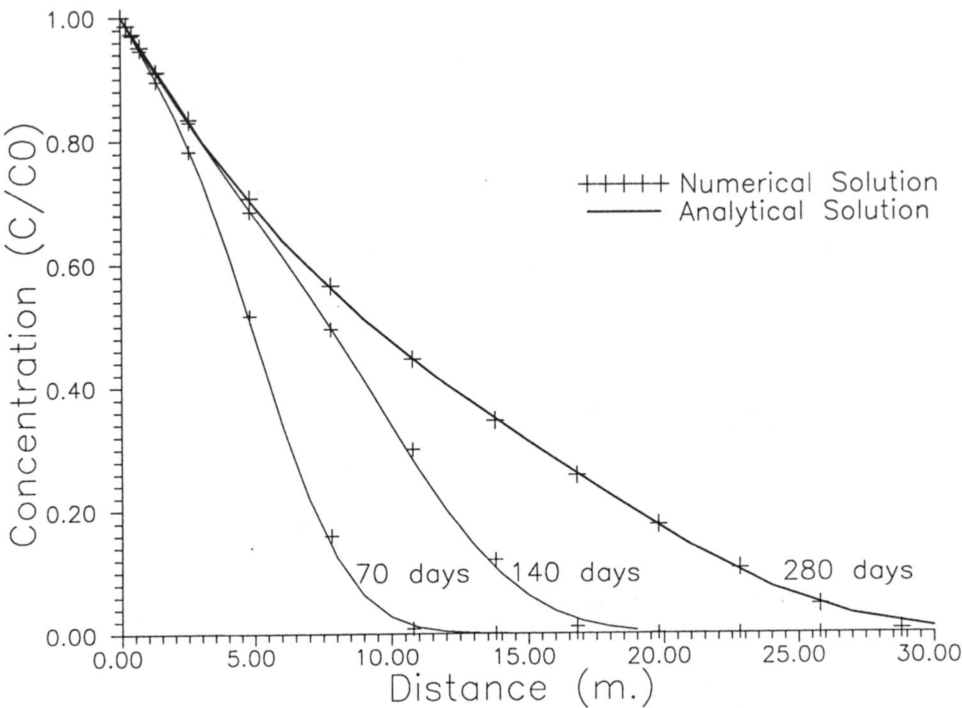

FIGURE 1. Verification of numerical model for one-phase advective-dispersive transport with linear adsorption and linear decay.

surface. The toluene flowed downward, pooled on the water table, and partitioned into the water and gas phases. The concentrations of toluene in the water-phase after 340 days with no biodegradation are shown in Figure 2. At the ground surface gas-phase toluene concentrations were prescribed as zero to simulate the effect of diffusion to the atmosphere.

The effects of biodegradation were simulated using the data on the biodegradation of toluene provided by MacQuarrie & Sudicky (1990). The data are summarized in Table 2. The concentrations of toluene in the water phase are shown in Figure 3, and the corresponding dissolved oxygen concentrations are shown in Figure 4. All concentrations are plotted as the ratio of actual mole fraction to the mole fraction corresponding to solubility limits.

Biodegradation has significantly reduced the extent of contamination in the vadose zone, particularly close to the ground surface. The horizontal extent of the dissolved organic plume below the water

TABLE 1. Properties of toluene used in simulations.

Molecular weight	92.41 g/gmol
Liquid phase density	867.0 kg/m^3
Liquid phase viscosity	0.9 cP
Liquid surface tension, σ_{ao}	23.8 dynes/cm
Aqueous solubility	0.515 kg/m^3
Water phase diffusion coefficient	1. × 10^{-4} m^2/day
Vapour pressure at 15 C	2190 Pa
Gaseous dissufion coefficient	0.6 m^2/day

Source: Reid *et al.* (1977).

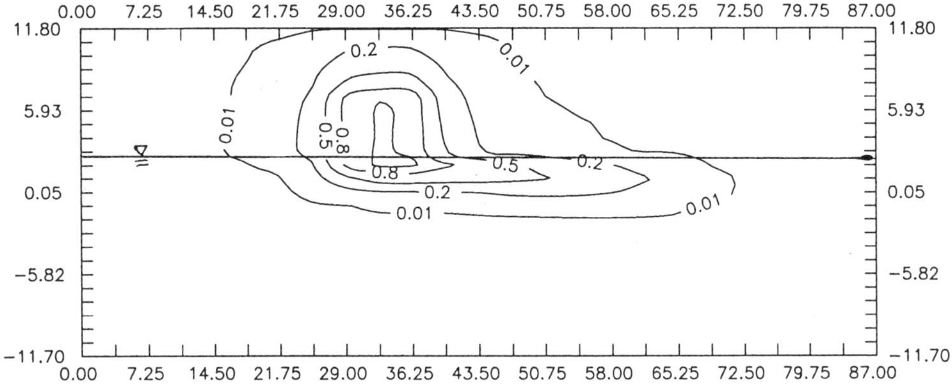

FIGURE 2. Toluene concentrations in water phase at 340 days without biodegradation. Concentrations expressed as c/c_0.

table is essentially the same with and without biodegradation. The oxygen concentrations in the contaminated region below the water table are essentially zero. The rate of diffusion of oxygen from the unsaturated zone to the saturated zone and the rate of supply of

TABLE 2. Biodegradation parameters.

Maximum utilization rate, *k*	0.493/day
Hydrocarbon half saturation constant, K_o	654.6 µg/L
Yield coefficient, *Y*	0.426 g cell/g organic
Decay coefficient	0.0
Initial total microbial population	230 µg/L
Microbial retardation factor	100
Oxygen half saturation constant:	0.1 µg/L
Oxygen utilization ratio, *F*	3.13 g oxygen/g organic

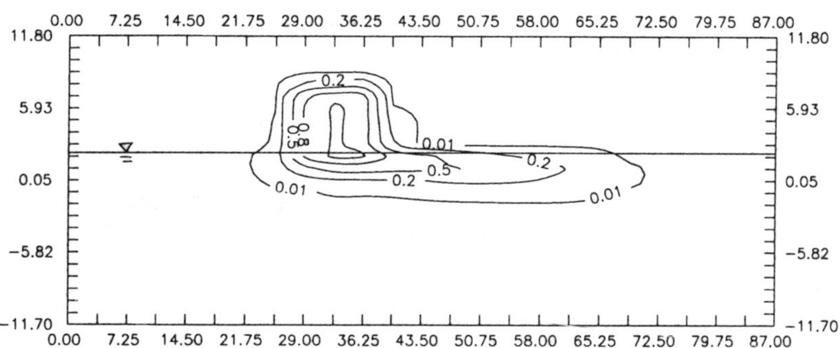

FIGURE 3. Toluene concentrations in water phase at 340 days with biodegradation. Concentrations expressed as c/c_0.

oxygen by horizontal flow of oxygenated groundwater is not sufficient to maintain oxygenated conditions in the contaminant plume below the water table.

The total rates of loss of toluene from the system are shown in Figure 5. The plume had not reached the right boundary at 340 days, so the only losses resulted from diffusion to the atmosphere in the first case and atmospheric diffusion and biodegradation in the second. The rate of loss of toluene is about twice as great with biodegradation as without biodegradation.

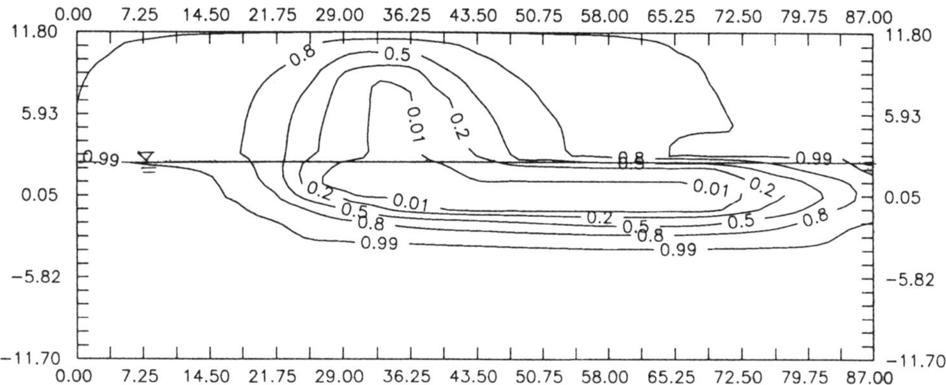

FIGURE 4. Oxygen concentrations in water phase at 340 days with biodegradation. Concentrations expressed as c/c_0.

AIR AND OXYGEN SPARGING

To increase the supply of oxygen to the microorganisms and to increase volatilization, air (20 percent oxygen) was injected into the domain at four points (indicated in Figure 6) just above the water table to the left and right of the toluene leak. At each injection point the gas injection rate was 2×10^{-6} m^3/sec for a section 1 m deep in the third dimension.

The injection of air produced gas flow upward to the ground surface. The organic concentrations and oxygen concentrations are shown in Figures 6 and 7, respectively. Air sparging did not greatly change the position of the organic plume below the water table. The oxygen concentrations at the periphery of the contaminated zone above the water table were slightly greater. The rate of dissipation of toluene (Figure 5) was increased slightly, but transfer of oxygen to the saturated zone was still inadequate. If greater injection rates were used; more significant air sparging effects would be expected.

To increase the rate of biodegradation pure oxygen was injected instead of air. The results are plotted in Figures 8 and 9. The contaminated area was reduced in size from the case with air sparging because of the greater supply of oxygen for biodegradation. The supply of oxygen to the saturated zone was not great enough to prevent oxygen limitation of biodegradation. The rate of dissipation of toluene was substantially increased by the much greater biodegradation rates in the region of residual organic liquid above the water table.

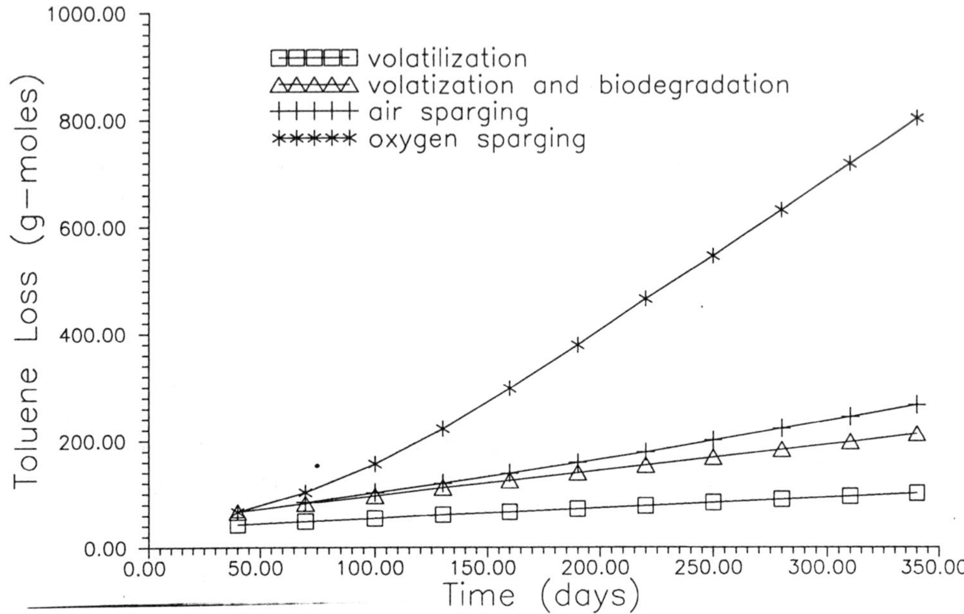

FIGURE 5. Loss of toluene from system with time.

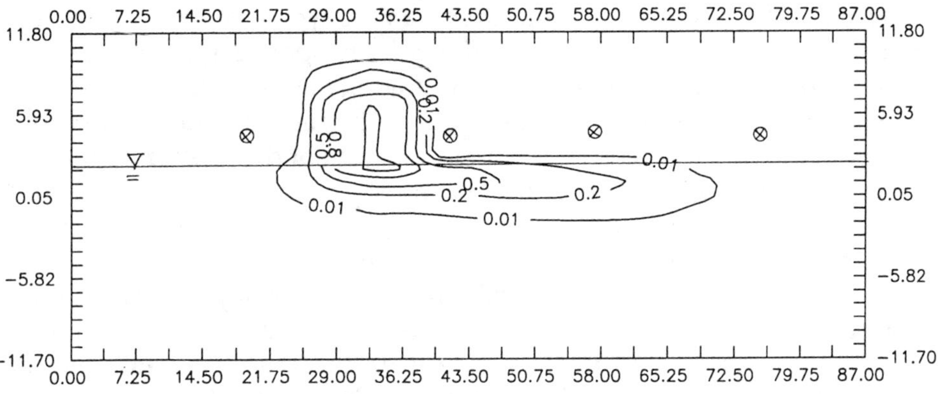

FIGURE 6. Toluene concentrations in water phase at 340 days with air sparging. Concentrations expressed as c/c_0.

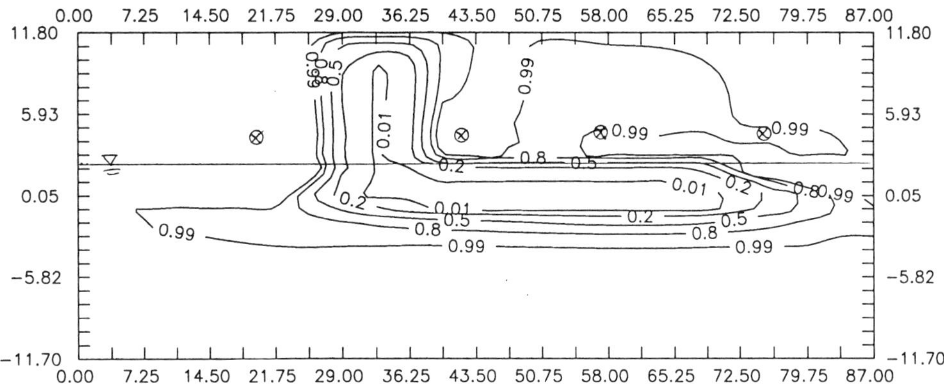

FIGURE 7. Oxygen concentrations in water phase at 340 days with air sparging. Concentrations expressed as c/c_0.

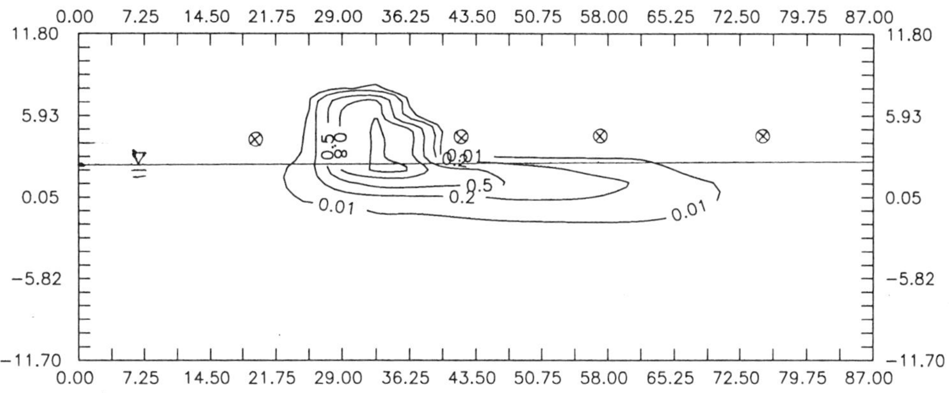

FIGURE 8. Toluene concentrations in water phase at 340 days with oxygen sparging. Concentrations expressed as c/c_0.

CONCLUSIONS

Naturally occurring biodegradation may substantially increase the rate of dissipation of organic compounds above the rate of dissipation from volatilization and diffusion to the atmosphere. The rate of transfer of oxygen to the saturated zone is inadequate to prevent significant contaminant migration below the water table.

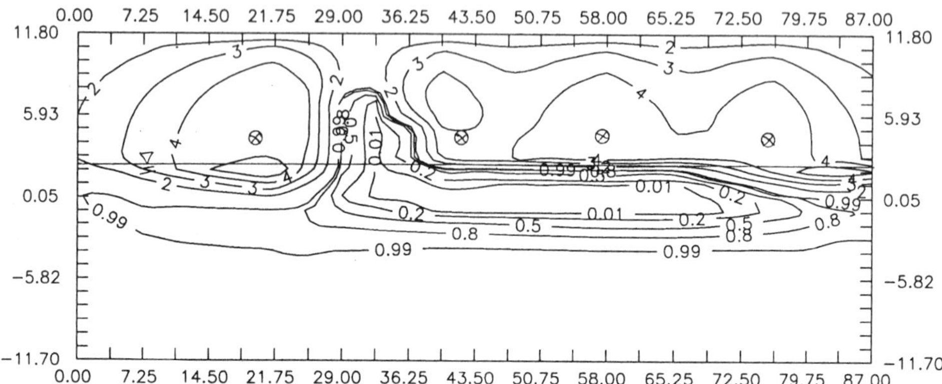

FIGURE 9. Oxygen concentrations in water phase at 340 days with oxygen sparging. Concentrations expressed as c/c_0.

Injection of air or oxygen increases the rate of organic liquid volatilization and biotransformation, but migration of organic contaminants dissolved in the water phase in the saturated zone still occurs.

REFERENCES

Abriola, L. M.; Pinder, G. F. "A Multiphase Approach to the Modeling of Porous Media Contamination by Organic Compounds, Numerical Simulation." *Water Resour. Res.* **1985**, *21*(1), 19–26.

Bear, J. *Dynamics of Fluids in Porous Media*; American Elsevier: New York, 1972.

Borden, R. C.; Bedient, P. B. "Transport of Dissolved Hydrocarbons Influenced by Reaeration and Oxygen Limited Biodegradation, 1: Theoretical Development." *Water Resour. Res.* **1986**, 22, 1973–1982.

Cleary, R. W.; Ungs, M. J. "Groundwater Pollution and Hydrology, Mathematical Models and Computer Programs"; Report 78-WR-15; Water Resources Program, Princeton Univ.: Princeton, NJ, 1978.

Forsyth, P. A. "Simulation of Nonaqueous Phase Groundwater Contamination." *Adv. Water Resour.* **1988**, *11*(2), 74–83.

Forsyth, P. A.; Sammon, P. H. "Practical Considerations for Adaptive Implicit Methods in Reservoir Simulation." *J. Comp. Phys.* **1986**, *62*, 265–281.

Johnson, P. C.; Kemblowski, M. W.; Colthart, J. D. "Quantitative Analysis for the Cleanup of Hydrocarbon-Contaminated Soils by *In-Situ* Venting." *Groundwater* **1990**, *28*(3), 413–429.

MacQuarrie, K. T.; Sudicky, E. A. "Simulation of Biodegradable Organic Contaminants in Groundwater, 1, Numerical Formulation in Principal Directions." *Water Resour. Res.* **1990** *26*(2), 207–222.

Millington, R. J.; Quirk, J. M. "Permeability of Porous Solids." *Trans. Faraday Soc.* **1961**, *57*, 1200–1207.

Molz, F. J.; Widdowson, M. A.; Benefield, L. D. "Simulation of Microbial Growth Dynamics Coupled to Nutrient and Oxygen Transport in Porous Media." *Water Resour. Res.* **1986**, 22(8), 1207–1216.

Perry, R. H.; Chilton, C. H. *Chemical Engineers' Handbook*, 5th ed.; McGraw-Hill: New York, 1973.

Reid, R. C.; Prausnitz, J. M.; Sherwood, T. K. *The Properties of Gases and Liquids*, 3rd ed.; McGraw-Hill: New York, 1977.

Sleep, B.; Sykes, J. "Modeling the Transport of Volatile Organics in Variably Saturated Media." *Water Res. Res.* **1989**, 25(1), 81–92.

Sykes, J. F.; Soyupak, S.; Farquhar, G. J. "Modeling of Leachate Organic Migration and Attenuation in Ground Waters Below Sanitary Landfills." *Water Resour. Res.* **1982**, *18*, 135.

Thomas, J. M.; Lee, M. D.; Bedient, P. B.; Borden, R. C.; Canter, L. W.; Ward, C. H. "Leaking Underground Storage Tanks: Remediation With Emphasis on *In-Situ* Biorestoration"; R. S. Kerr Environmental Research Laboratory: Ada, OK, 1987; EPA/600/2-87/008.

Todd, M. R.; O'Dell, P. M.; Hirasaki, G. J. Methods for Increased Accuracy in Numerical Reservoir Simulators." *SPEJ*, Dec. **1972**, 515–529.

Assessment of *In Situ* Bioremediation Potential and the Application of Bioventing at a Fuel-Contaminated Site

*R. Ryan Dupont**, *William J. Doucette*
Utah State University
Robert E. Hinchee
Battelle

INTRODUCTION

Soil vacuum extraction (SVE) has found wide application in the in-place treatment of fuels and solvent-contaminated soils. SVE has not generally been applied to heavier fuels, i.e., diesel or JP-4 jet fuel, and fuel oils because of the large fraction of high boiling point, high molecular weight compounds these fuels contain. Although these high molecular weight constituents are not volatile, they are amenable to biodegradation. In this research the potential for use of the gas transfer capability of vacuum extraction systems was assessed, in conjunction with nutrient and moisture management, for the optimization of *in situ* biodegradation to yield an enhanced bioventing system capable of degrading residual fuel organics not amenable to SVE treatment alone.

Soil Vacuum Extraction Systems

Soil vacuum extraction is the process of applying a vacuum to an unsaturated zone to collect vapor-phase contaminants and stimulate their removal from the soil by encouraging their volatilization. SVE systems

* Utah Water Research Laboratory, Utah State University, UMC-8200, Logan UT 84322-8200

consist of small-diameter vertical wells and/or lateral trenches connected via a piping network to a blower or vacuum pump. Hutzler *et al.* (1988) provides an excellent review of the current state of technology of SVE systems. These systems are attractive for soil remediation because of their low cost, ease of implementation, and apparently favorable performance in many instances of gasoline contamination in relatively permeable soils. Their costs increase significantly, however, when treatment is required for the off-gas. Their application has also generally been limited to volatile contaminants, i.e., gasoline, solvents, etc., as conventional SVE systems are not effective in removing residual contaminants that do not readily partition into the soil vapor phase.

Bioventing

The biodegradation of petroleum hydrocarbons and related compounds in soil environments has been extensively described in the technical literature, and details of the pathways and populations responsible for their biotransformation have been summarized in a number of textbooks and reviews on soil microbial ecology (*e.g.*, Alexander 1977; Atlas 1981; Dragun 1988). Initial application of these principles of biodegradation have been applied to the aerobic bioremediation of contaminated groundwater, as groundwater has historically been the human exposure route of most concern. In conventional, *in situ* bioremediation, water is used to carry oxygen or alternative electron acceptors to the site of this subsurface contamination. Oxygen has been preferred over other terminal electron acceptors (i.e., NO_3^{-2}, SO_4^{-2}, and CO_2) because of the accelerated degradation rates generally possible under aerobic conditions. These efforts have met with limited success, however, because of what many authors have cited as the inability of remediation schemes to transfer oxygen adequately to areas of subsurface contamination using water as the carrier fluid (Downey *et al.* 1988; Hinchee & Downey 1988; Hinchee *et al.* 1989; Lee *et al.* 1988; Wetzel *et al.* 1987).

This oxygen transfer limitation might well be anticipated when one considers the data summarized in Table 1. These values represent the mass of fluid required to transfer a unit mass of oxygen under the stated conditions, and indicate the large amounts of oxygen-saturated water necessary to transfer oxygen to the subsurface. This oxygen supply limitation is exacerbated by the high oxygen demand of hydrocarbon contaminants as indicated by the simple stoichiometric reactions

TABLE 1. Carrier fluid oxygen supply requirements.

	g Carrier/g O_2
Water	
Air Saturated	110,000
Pure O_2 Saturated	22,000
500 mg/L H_2O_2 (100% Utilization)	2,000
Air (20.9% O_2)	13

for benzene and hexane oxidation shown below, assuming no incorporation of substrate carbon into cell material:

$$C_6H_6 + 7.5\ O_2 \rightarrow 6\ CO_2 + 3\ H_2O \qquad (1)$$
$$3.1\ g\ O_2/g\ C_6H_6$$

$$C_6H_{14} + 9.5\ O_2 \rightarrow 6\ CO_2 + 7\ H_2O \qquad (2)$$
$$3.5\ g\ O_2/g\ C_6H_{14}$$

With greater than 3 g O_2/g hydrocarbon required for the mineralization of hydrocarbon contaminants, hydraulic limitations to massive water movement have been the primary cause of the limited success of saturated phase bioremediation.

Consideration of soil vacuum extraction for the efficient transfer of oxygen to the subsurface was proposed in 1988 by Wilson and Ward, who noted that systems designed for the removal of volatiles from the subsurface could also be used to transport oxygen. A number of other authors have postulated the potential improvement in *in situ,* aerobic, subsurface bioremediation using SVE for oxygen transfer (Bennedsen *et al.* 1987; Ely & Heffner 1988; Ostendorf & Kampbell 1989; Riser 1988; Staps 1989), but little supporting data validating the claims have been published to date.

The U.S. Air Force has recently been interested in the application of SVE to JP-4 sites and has encouraged the investigation of enhanced biodegradation associated with SVE systems. Hinchee *et al.* (1991) reported on findings of field measurements of *in situ* respiration during conventional SVE of a JP-4 site at Hill AFB, UT, indicating that active microbial respiration was occurring at this site, and that from 15 to

25 percent of the observed hydrocarbon removal was attributable to biodegradation. These findings led to the study described in this paper, which investigated further the bioremediation potential and feasible engineering management options for enhancement of bioventing at this fuel site.

Site Description

The site at Hill AFB, UT, was the location of a JP-4 jet fuel spill that occurred in January 1985, after the failure of an automatic shut-off valve. Failure of the valve resulted in the release of approximately 100,000 L (27,000 gal) of JP-4, some 7,500 L (2,000 gal) of which were recovered as free product. The balance of the release fuel migrated away from the tank and contaminated an area around it of approximately 0.4 hectares (1 acre) to a depth of approximately 15 m (50 ft). The soil at the site was formed as a delta outwash of the Weber River and consists of mixed coarse sand and gravel deposits with interspersed, discontinuous clay stringers to a confined groundwater table located approximately 180 m (600 ft) below ground surface (bgs). JP-4 contamination resulted in soil total petroleum hydrocarbon (TPH) concentrations at the site as high as 20,000 mg/kg, with most of the contaminated soils ranging from 1,000 to 5,000 mg/kg. Prior to initiation of the full-scale venting system, the fuel tanks were excavated, refurbished, and installed in a concrete cradle above ground.

A venting system consisting of 15 vertical wells and 10 lateral wells in the excavated soil pile and under the tanks was installed by Oak Ridge National Laboratory to provide access to the contaminated soil and allow flexibility in the operation of the venting system (Figure 1). The vertical wells were placed at 12-m (40-ft) intervals to a depth of 15 m (50 ft) bgs and were slotted over an interval from 3 to 15 m (10 to 50 ft) bgs. Twenty-one pressure monitoring points (PMP) were installed at various depths throughout the site to provide point measurements of subsurface pressure and soil gas conditions (Figure 2), and a background well was placed approximately 210 m (700 ft) north of the site in the same geological unit and at the same depth as the vent wells to provide a control for basal soil respiration levels during the study.

Prior to initiation of bioventing studies at the Hill site, the SVE system was operated under a conventional mode to maximize the recovery of volatile components of the JP-4 through volatilization. Venting was initiated on December 18, 1988, at a rate of 36 m^3/h

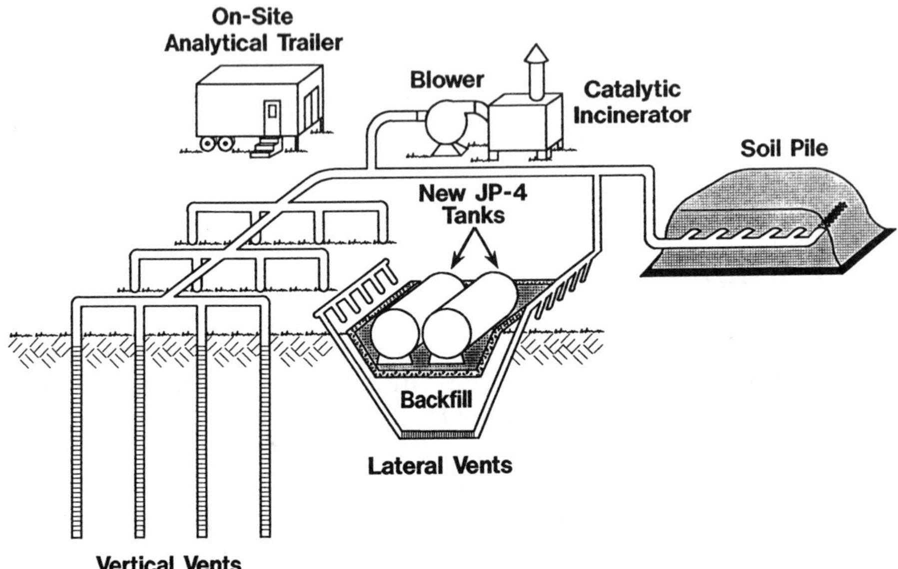

FIGURE 1. Conceptual drawing of the Hill AFB, UT, field soil-venting site.

(26 acfm, approximately 0.11 pore volumes/day), and gradually increased to approximately 2,100 m^3/h (1,500 acfm, approximately 6.5 pore volumes/day) as the hydrocarbon levels in the vent gas decreased over time. Vent gas was collected through Wells V5 to V11 (Figure 2), where the bulk of the soil contamination was located. The venting rate during the start-up period was limited by the operating conditions of the catalytic incinerator used to treat the collected vent gas. This high-rate operating mode was maintained from December 18, 1988, through September 15, 1989, when approximately 7,000,000 m^3 (300,000,000 acf, approximately 900 pore volumes) of soil gas were extracted from the site.

Three *in situ* respiration tests were conducted during the high-rate operating period and have been described in detail elsewhere (Hinchee *et al.* 1991). These tests were conducted following vented air volumes of 970 m^3 (42,000 acf), 13,000 m^3 (540,000 acf), and 1,000,000 m^3 (45,000,000 acf), and showed first order oxygen uptake rates that ranged from 0.85/day to nondetectable in the pressure monitoring points throughout the field site. Comparison of results from specific monitoring points over time also indicated the incremental removal of residual hydrocarbons as respiration rates declined throughout the treatment period. Hinchee *et al.* (1991) concluded that significant respiration was

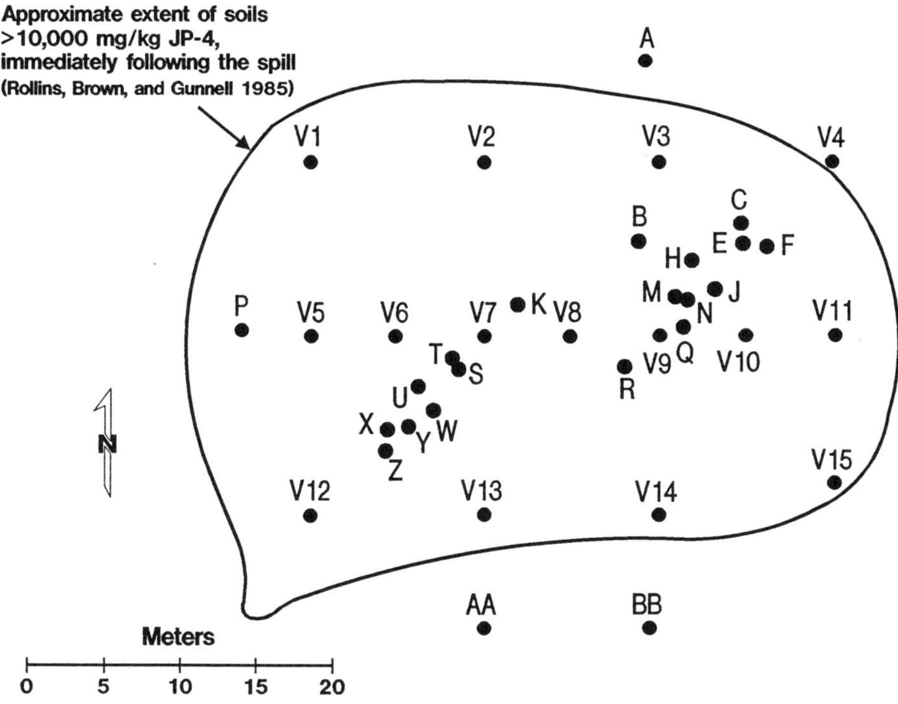

FIGURE 2. Site map of the Hill AFB, UT, field soil-venting site indicating vent well and pressure monitoring point locations.

occurring during conventional SVE without nutrient or moisture addition, and that enhancement of biodegradation could be possible under modified site management conditions.

BIOTREATABILITY STUDIES

To assess the potential for enhancing the biodegradation reactions observed at the Hill site, a series of laboratory and field biotreatability studies were conducted from September 1989 to November 1990. In addition, operational changes were made to the SVE system to evaluate their effect on *in situ* bioremediation of residual fuels at the Hill site.

Laboratory Studies

Laboratory studies were conducted by Battelle, in Columbus OH, on soil collected from the Hill site in May 1989 using standard hollow-stem auger drilling procedures. These soil samples were used to evaluate the potential for increasing biodegradation rates via moisture and/or nutrient addition. Results of these laboratory studies, detailed elsewhere (Hinchee & Arthur 1991), are summarized here for background purposes.

A total of 15 soil columns, 3.8 cm (1.5 in.) in diameter and 30.5 cm (12 in.) long, were prepared in the study, 12 being treatment reactors and three being dead controls (500 mg/kg each of cadmium and mercuric chloride). The columns were packed using approximately 500 g of composited subsurface soils from the Hill site, and JP-4 was added to the soil to yield 5,000 mg/kg total petroleum hydrocarbon concentrations at the beginning of the study. Three soil moisture levels were investigated (6, 12, and 18% corresponding to 25, 50, and 75% field capacity), with two columns at each moisture level amended with 2 percent (by weight) Restore® 375 (FMC nutrient formulation containing 50% ammonium chloride, 20% sodium phosphate, 17.5% sodium tripolyphosphate, and 12.5% monosodium phosphate). Each soil column was vented at 100 mL/min using a water-saturated hydrocarbon and CO_2-free air stream. Moisture was added on a weekly basis as required to maintain each reactor at its desired soil moisture level. Biodegradation rates were quantified during the 48-day test period by measuring CO_2 collected in alkali scrubbers over time. In addition, soil total hydrocarbon levels were determined at the end of the test period to quantify hydrocarbon removal in each reactor.

As indicated in Figure 3, although moisture was shown to be significant in improving soil respiration rates based on CO_2 evolution, the greatest impact on biological activity was nutrient addition. It is interesting to note, however, that the authors found that moisture addition actually had a more significant effect on activity than did nutrients when activity was based on total hydrocarbon removal in the reactors (Hinchee & Arthur 1991). From these laboratory treatability studies it was determined that both moisture and nutrient addition had a potentially positive effect on soil biodegradation rates, and it was this basis upon which further field studies were undertaken.

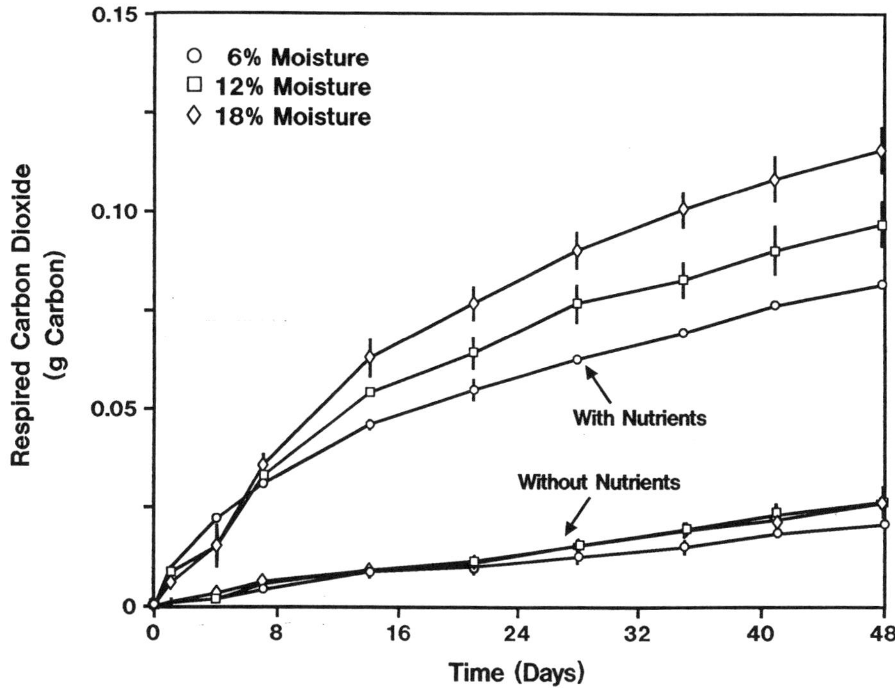

FIGURE 3. Results from bench-scale biotreatability experiments for the Hill AFB, UT, field soil-venting site indicating the effects of moisture and nutrient addition on cumulative CO_2 production rates.

Field Studies

To quantitate biodegradation and volatilization reactions taking place in the field studies, all vent gas components were converted to an equivalent carbon basis. Total hydrocarbon concentration in the vent gas was measured in ppm using a continuously reading, total hydrocarbon analyzer calibrated to hexane. The carbon equivalent of this measurement was calculated based on a carbon-to-hexane ratio of 72 g C/86 g C_6H_{14}. Vent gas CO_2 and O_2 measurements were measured weekly and at specific time intervals at pressure monitoring points and vent wells during shutdown tests. Evacuated stainless steel canisters were utilized to collect samples and GC/Thermal Conductivity Detector analysis was utilized for laboratory quantitation of CO_2 and O_2 during routine monitoring activities. A Gastechtor Model 32520X (Gastech Inc., Newark CA) was used to monitor CO_2 and O_2 throughout the field site during shutdown tests. This instrument provides real-time analysis of

CO_2 and O_2 over the range of soil gas concentrations from 0 to 5 percent and 0 to 25 percent, respectively, within ±0.5 percent accuracy for both gases. Equivalent C values for CO_2 and O_2 were based on the carbon ratio of CO_2 (12 g C/44 g CO_2) and the oxygen equivalent of hexane (3.5 g O_2/g hexane) expressed in carbon equivalents, as shown in equation (2) above.

Accumulation of total hydrocarbons measured in the vent gas over time expressed on a carbon basis was used to quantify the recovery of JP-4 by volatilization. Biodegradation reactions were estimated based on cumulative oxygen consumption and carbon dioxide production, expressed on an equivalent carbon basis with respect to background well measurements. All biodegradation calculations were normalized to background CO_2 and O_2 concentrations so that the effects of field management techniques on biodegradation reactions could be isolated from changes in background respiration taking place during the study throughout the field site.

Modified Operating Conditions. Based on results of soil gas measurements taken from pressure monitoring points during the initial, high-rate venting period at Hill, it was determined that at high extraction rates, i.e., 2,100 m^3/h (1,500 acfm), the entire contaminated zone was aerated to near atmospheric O_2 levels. In addition, with extraction of vapors from the areas of maximum contamination at the interior of the site, elevated hydrocarbon levels were being found in the vent gas. To maximize biodegradation and minimize volatilization, operating flow rates were reduced to 490 to 970 m^3/h (350 to 700 acfm), and vent gas was drawn from wells on the periphery of the site (Wells V12 to V15 in Figure 2) to maximize the flow path and retention time of vapors in the contaminated zone.

Field *In Situ* Respiration Tests. A number of field-scale *in situ* respiration tests were conducted during this study to assess the changes in *in situ* respiration as engineering management options were applied at the site. Based on results of the laboratory-scale treatability studies described above, a total of three tests were conducted from September 1989 to November 1990 representing different levels of management at the site, i.e., following flow rate and operating configuration modifications at the site, following moisture addition, and following moisture and nutrient addition. All tests were conducted by shutting down the venting system and monitoring changes in soil gas CO_2 and O_2 composition in all pressure monitoring points and vent wells over a 10- to 14-day period. Soil gas samples were analyzed by first evacuating three volumes of the monitoring points and vent wells using a portable sampling pump prior to connecting the Gastechtor instrument.

The moisture addition phase of the field treatability study consisted of the addition of culinary water to the field site to yield soil moisture levels throughout the site of approximately 8 to 12 percent (30 to 50% field capacity). This moisture was added via surface spray irrigation at rates of approximately 110 L/min (30 gpm), 8 h/day, 7 days/week, until approximately 3,800,000 L (1,000,000 gal) were applied. Three neutron probe access tubes were used to monitor soil moisture throughout the site during the course of water addition. Data in Figure 4 indicate that soil moisture was successfully increased from pre-irrigation conditions and maintained between 8 and 12 percent over the entire contaminated depth.

Nutrients were added to the site in the form of ammonium nitrate and sodium triphosphate at a C:N:P ratio of 100:10:10, based on soil hydrocarbon analyses in September 1989, which indicated residual hydrocarbon levels throughout the site of approximately 100 mg/kg. A total of 140 kg (300 lb) of N (applied as fertilizer grade ammonium nitrate, 34.5% N) and 14 kg (30 lb) of P (applied as fertilizer grade treble superphosphate, 20% P) were added in three equal increments, 3 weeks apart, by surface-applying the dry mix, tilling it into the upper 15 cm (6 in.) soil horizon, and continuing surface spray irrigation at 110 L/min (30 gpm), 8 h/day, 2 days/week, during this phase of the study.

All field data were analyzed assuming a first order reaction law through linear regression of the natural log transform of CO_2 and O_2 concentrations normalized to the background well at each sampling interval, i.e., $\ln(C_{\text{vent well}}/C_{\text{background well}})_{\text{time } i}$, versus time. Each regression line was tested for the significance of its slope, i.e., the probability of the slope not equaling zero being ≥0.05. In addition, an evaluation of overlapping 95 percent confidence intervals of each regression slope was used to test for significant differences among treatments.

RESULTS AND DISCUSSION

Field Studies

Modified Operating Conditions. Figure 5 summarizes the percent of total fuel recovery that could be attributed to biodegradation, expressed on an oxygen consumption basis, during both the conventional

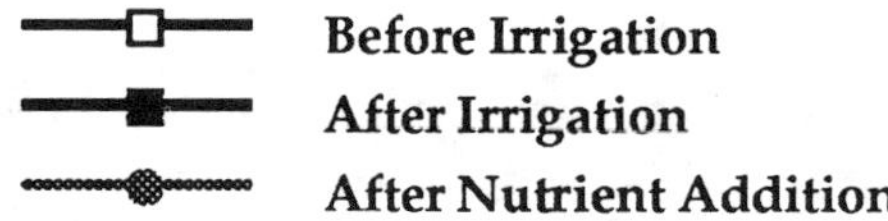

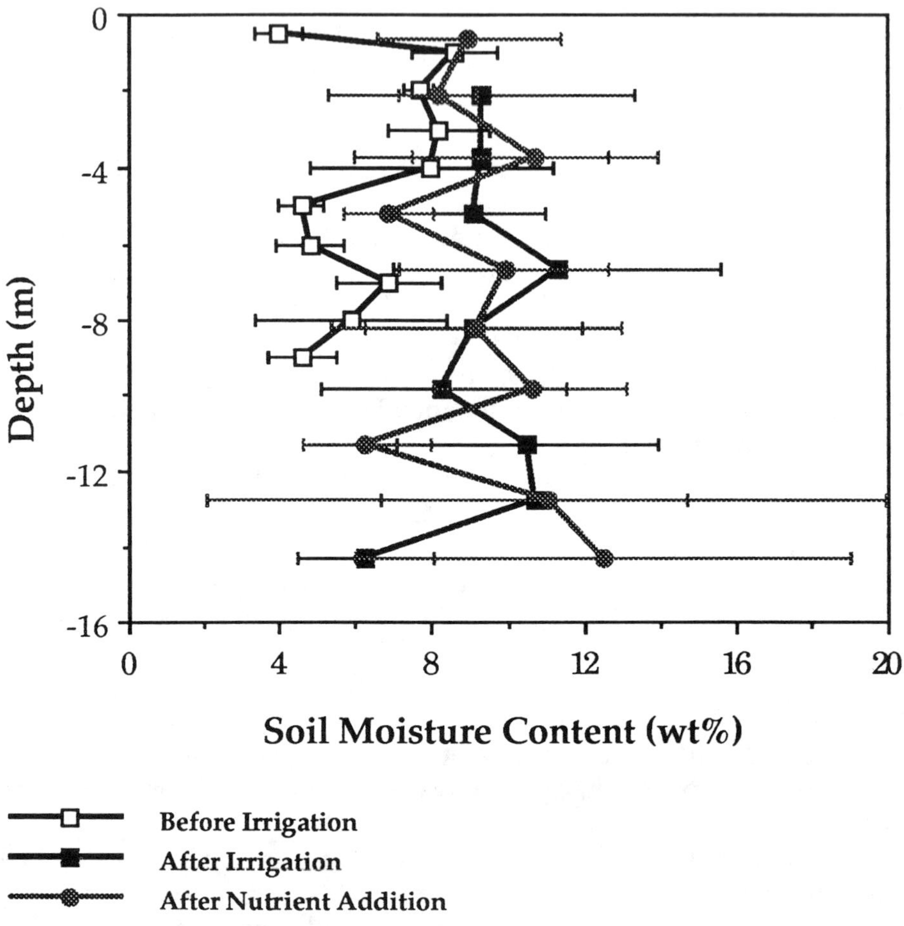

FIGURE 4. Mean soil moisture content at Hill AFB, UT, before and after irrigation and nutrient application at the field site.

high-rate and modified bioventing phases of the study. As indicated earlier, biodegradation accounted for 15 to 25 percent of the recovered hydrocarbon even during high-rate venting. This rate was drastically altered in September 1989, when volatilization was reduced from 90 to 180 kg/day (200 to 400 lb/day) to less than 9 kg/day (20 lb/day) by reducing vacuum flow rates and making changes to the system flow configuration. These operating modifications had no detrimental effect on biodegradation reactions, however, as initial hydrocarbon-C removal rates of 30 kg/day (70 lb/day) were maintained at an average rate of greater than 45 kg/d (100 lb/day) following system operating modifications.

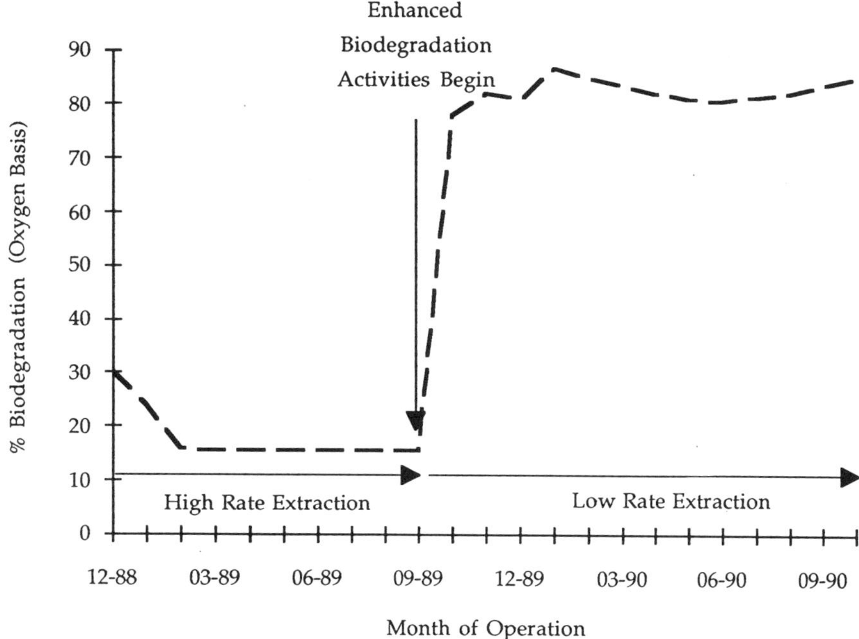

FIGURE 5. Percent of recovered hydrocarbon attributed to biodegradation reactions at the Hill AFB, UT, field soil-venting site based on oxygen consumption measured in the vent gas.

Field *In Situ* Respiration Tests. Figure 6 shows results of a typical oxygen uptake rate determination obtained during *in situ* respiration tests conducted in this study. Detailed results from the three field respiration tests are presented in Tables 2 and 3, and are summarized in Table 4 along with results from the high-rate venting phase of the study. As indicated in Table 4, oxygen uptake was found to be more consistent and more sensitive than carbon dioxide production rates in detecting effects of treatments on microbial activity at the site. This was particularly true for the moisture addition cases, where the interaction of CO_2 and added water greatly affected observed CO_2 production rates within the field site.

Oxygen uptake rates observed during the high-rate venting period of the study were shown to be significantly greater than respiration rates determined approximately one year later following operating system modifications. This difference was likely caused by drying of the soil over one year of high-rate venting as well as to the removal and degradation of intermediate molecular weight, highly degradable JP-4

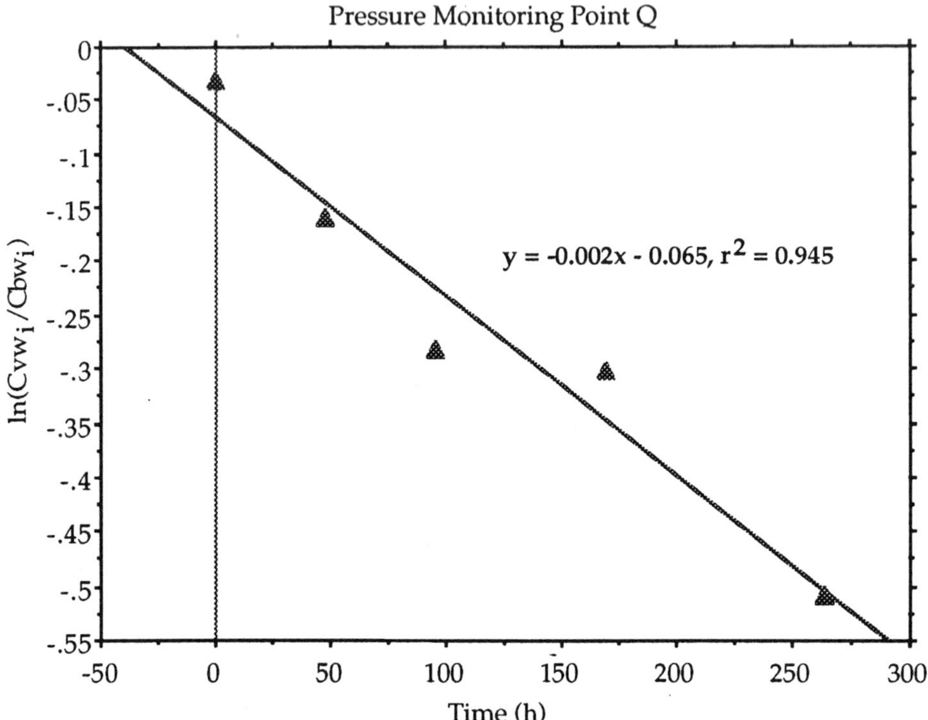

FIGURE 6. A sample first order regression analysis of oxygen uptake rate data obtained during *in situ* respiration measurements.

constituents during the first year of vent system operation. As venting progresses and residual hydrocarbons become enriched in higher molecular weight, lower vapor pressure components, the degradation rate of these residual organics would be expected to decrease as shown in the "base case" labeled "Low Rate Venting" in Table 4. This base case respiration rate was nearly 50 times lower than that observed a year earlier, and was the only treatment period that showed a greater number of significant CO_2 production rate than O_2 uptake rate regression relationships.

The addition of moisture to the field site yielded a significant increase in oxygen uptake rate in 12 of 34 vent wells and monitoring points. A total of 14 significant regressions were observed following moisture treatment. The mean oxygen uptake rate increased by a factor of two, while maximum rates increased by a factor of six with water amendment. CO_2 production rate was not found to be significant at

TABLE 2. Field *in situ* respiration test results—O_2 uptake rate data.

	Low Rate Venting				Moisture Addition				Nutrient & Moisture Addition			
Well	Slope (1/d)	r^2	UCI	LCI	Slope (1/d)	r^2	UCI	LCI	Slope (1/d)	r^2	UCI	LCI
1	−0.001[a]	0.909	−0.003	0.000	−0.003	0.234			−0.004	0.478		
2	−0.003	0.855			−0.024[a]	0.751	−0.024	−0.003	0.001	0.028		
3	−0.003	0.876			−0.011[a]	0.718	−0.024	−0.002	0.000	0.001		
4	−0.007	0.727			−0.011[a]	0.803	−0.024	−0.004	−0.005[a]	0.819	−0.008	−0.001
5	−0.001	0.408			−0.009	0.511			0.001	0.035		
6												
7	0.002	0.143			−0.005[a]	0.983	−0.007	−0.003	0.000	0.000		
8	0.000	0.007			−0.001	0.029			0.001	0.012		
9	0.000	0.011			−0.011	0.539			0.010	0.045		
10	−0.004	0.439			−0.024	0.535			0.000	0.001		
11	−0.006	0.636			−0.005	0.425			−0.001	0.007		
12	0.071	0.296			0.001	0.053			−0.001	0.070		
13	−0.001	0.066			0.001	0.085			−0.003[a]	0.854	−0.006	−0.001
14	−0.002	0.784			0.002	0.116			−0.001	0.018		
15	−0.008	0.784			−0.011[a]	0.902	−0.017	−0.004	0.003	0.325		
PMP												
A	0.000	0.051			−0.001	0.019			−0.003	0.556		
B	−0.020[a]	0.977	−0.029	−0.010	−0.072	0.605			−0.056[a]	0.971	−0.070	−0.043
C	−0.003	0.368			−0.024[a]	0.804	−0.048	−0.008	−0.006[a]	0.735	−0.012	−0.001
E	−0.001	0.074			0.004	0.346			−0.004	0.570		
F	−0.015[a]	0.997	−0.017	−0.012	−0.168[a]	0.995	−0.189	−0.144	−0.034[a]	0.955	−0.045	−0.024
H	−0.026[a]	0.962	−0.042	−0.010	−0.072	0.598			−0.061[a]	0.918	−0.087	−0.036
K	−0.009[a]	0.984	−0.013	−0.006	−0.010	0.366			−0.007	0.139		
M	−0.023[a]	0.957	−0.037	−0.008	−0.048[a]	0.699	−0.096	−0.004	−0.032[a]	0.950	−0.042	−0.022
N					−0.048	0.385						
P	0.000	0.000			−0.006	0.639			−0.008[a]	0.830	−0.014	−0.003
Q	−0.021[a]	0.985	−0.029	−0.013	−0.048[a]	0.945	−0.058	−0.022	−0.020[a]	0.941	−0.028	−0.014
R	−0.019[a]	0.991	−0.025	−0.014	−0.024[a]	0.846	−0.036	−0.044	−0.013[a]	0.974	−0.016	−0.010
S	0.000	0.000			−0.008	0.488			−0.006[a]	0.950	−0.008	−0.004
T	−0.002	0.143			−0.009	0.603			−0.018[a]	0.845	−0.024	−0.004
U	0.000	0.485			−0.009[a]	0.837	−0.024	−0.004	−0.004[a]	0.739	−0.007	−0.001
W	−0.001	0.488			−0.024[a]	0.825	−0.024	−0.008	−0.010[a]	0.906	−0.015	−0.006
X	0.000	0.177			−0.005[a]	0.777	−0.009	−0.001	−0.004[a]	0.870	−0.005	−0.002
Y	−0.005	0.835			−0.011[a]	0.790	−0.024	−0.003	−0.007[a]	0.846	−0.011	−0.003
Z	0.000	0.177			−0.004	0.474			−0.004[a]	0.785	−0.007	−0.001
AA	0.000	0.012			0.001	0.042			0.000	0.010		
BB	0.001	0.357			0.002	0.373			−0.005[a]	0.863	−0.008	−0.001

(a) Significant regression at 95% confidence level.

any monitoring point or vent well under these conditions, however, indicating the sensitivity of the CO_2 production measurement to changes in environmental conditions that affect CO_2 distribution in the subsurface. Nutrient addition increased the number of significant oxygen uptake relationships to 18 of 34 sites, but actually resulted in a net decrease in both maximum and mean uptake rates compared to the previous moisture only treatment.

An analysis of overlapping 95 percent confidence intervals of the slopes of significant regression relationships for the three treatment

TABLE 3. Field *in situ* respiration test results—CO_2 production rate data.

	Low Rate Venting				Moisture Addition				Nutrient & Moisture Addition			
Well	Slope (1/d)	r^2	UCI	LCI	Slope (1/d)	r^2	UCI	LCI	Slope (1/d)	r^2	UCI	LCI
1	0.084	0.627			–0.216	0.431			–0.140	0.197		
2	0.190[a]	0.919	0.362	0.018	0.024	0.076			–0.321[a]	0.970		
3	0.123[a]	0.943	0.215	0.031	–0.384	0.826			0.071	0.031		
4	0.155	0.647			0.096	0.848			0.249	0.826		
5	0.089	0.517			0.120	0.816			–0.461	0.698		
6												
7	0.040	0.092			–0.600[a]	0.996			–0.296	0.652		
8	0.100[a]	0.698	0.155	0.045	–0.480[a]	0.996			–0.044	0.052		
9	0.072	0.260			–0.144	0.897			–0.139	0.619		
10	0.123	0.437			0.120	0.707			0.179[a]	0.844		
11	0.158	0.706			–0.432	0.715			–0.128[a]	0.951		
12	0.000	0.003			–0.384	0.989			–0.260	0.868		
13	0.098	0.496			–0.072	0.830			–0.366	0.990		
14	0.151[a]	0.965	0.238	0.063	–0.144	0.520			0.030	0.023		
15	0.155	0.845			–0.072	0.747			–0.248	0.630		
PMP												
A	0.061[a]	0.967	0.095	0.027	0.288	0.836			–0.045	0.200		
B	0.346[a]	0.952	0.582	0.109	0.504	0.982			0.164	0.572		
C	0.075	0.465			0.432	0.980			–0.018	0.079		
E	0.090	0.789			–	–			–0.044	0.334		
F	0.305[a]	0.983	0.427	0.182	0.768	0.859			0.264[a]	0.666	0.523	0.005
H	0.417[a]	0.937	0.747	0.087	0.600	0.899			0.263[a]	0.999	0.302	0.205
K	0.147[a]	0.990	0.192	0.102	0.288	0.660			0.026	0.007		
M	0.369[a]	0.948	0.633	0.106	0.456	0.967			0.223[a]	0.768	0.393	0.053
N					–0.480	0.973						
P	0.077[a]	0.954	0.129	0.026	–0.168[a]	1.000			0.132[a]	0.698	0.253	0.012
Q	0.301[a]	0.979	0.435	0.167	0.456	0.885			0.174[a]	0.996	0.207	0.142
R	0.294[a]	0.981	0.417	0.171	0.384	0.938			0.051	0.288		
S	0.101	0.724			–0.288[a]	0.997			–0.075	0.327		
T	0.110	0.549			0.216	0.930			0.020	0.056		
U	0.102	0.766			–0.024	0.060			–0.128	0.305		
W	0.080[a]	0.997	0.094	0.066	0.240	0.933			0.047	0.093		
X	0.080	0.669			0.024	0.201			–0.134[a]	0.975		
Y	0.162	0.860			0.072	0.767			0.005	0.001		
Z	0.069[a]	0.910	0.136	0.003	–0.192[a]	0.999			–0.390	0.249		
AA	0.064	0.424			–0.048	0.956			0.188	0.516		
BB	0.054	0.160			0.024	0.151			0.064[a]	0.933	0.117	0.012

(a) Significant regression at 95% confidence level.

cases during the bioventing study indicated that statistically significant increases in rates were observed in 19 of the 35 wells following some form of treatment (Table 2). Moisture addition accounted for 12 of these occurrences of enhanced microbial activity, but only seven cases displayed increased activity attributable to the nutrient amendment. In no cases did nutrient addition significantly increase respiration rates above statistically significant levels following moisture addition alone (Table 2).

TABLE 4. Summarized biodegradation results obtained from field *in situ* respiration studies.

Treatment	Number of Significant Regressions	Maximum O_2 Uptake Rate (1/day)	Maximum CO_2 Production Rate (1/day)	Mean O_2 Uptake Rate (1/day)	Mean CO_2 Production Rate (1/day)
High Rate Venting	7 (O_2) – (CO_2)	–1.11		–0.367	
Low Rate Venting	8 (CO_2) 15 (CO_2)	–0.026	0.417	–0.016	0.202
Moisture Addition	14 (O_2) 0 (CO_2)	–0.168	–	–0.030	–
Nutrient & Moisture Addition	18 (O_2) 6 (CO_2)	–0.060	0.260	–0.016	0.185

Soil Sampling

TPH concentrations in soils were determined before and after site treatment. The results of those analyses are illustrated in Figure 7. The initial sampling was conducted by Oak Ridge National Laboratory in October 1988, prior to any venting (DePaoli *et al.* 1990). The final sampling was conducted after cessation of all venting in November 1990. Each bar in Figure 7 illustrates the average of 14 or more analyses of soil collected in the specified depth interval. It should be noted that both before and after, a substantial number of samples had no detectable TPH; this was due to a highly heterogenous distribution of TPH within the spill area. All averages reflect both contaminated and uncontaminated soil samples. The initial soil borings were utilized for vent well placement. To assure representative after sampling, soils were collected from locations approximately 3 m (10 ft) from the initial boring locations. Initially, most of the TPH contaminated soils were in the 1,000 to 5,000 mg/kg TPH range with a high of 20,400 mg/kg. The initial average for all soil samples with TPH was 410 mg/kg. In the final sampling of 158 soil samples collected, only 18 had TPH above the 5 mg/kg detection limit, and the highest concentration was 38 mg/kg. The final average for all soil samples was 3.8 mg/kg. For the purpose of averaging, concentrations below the detection limit were assumed to be one-half of the detection limit.

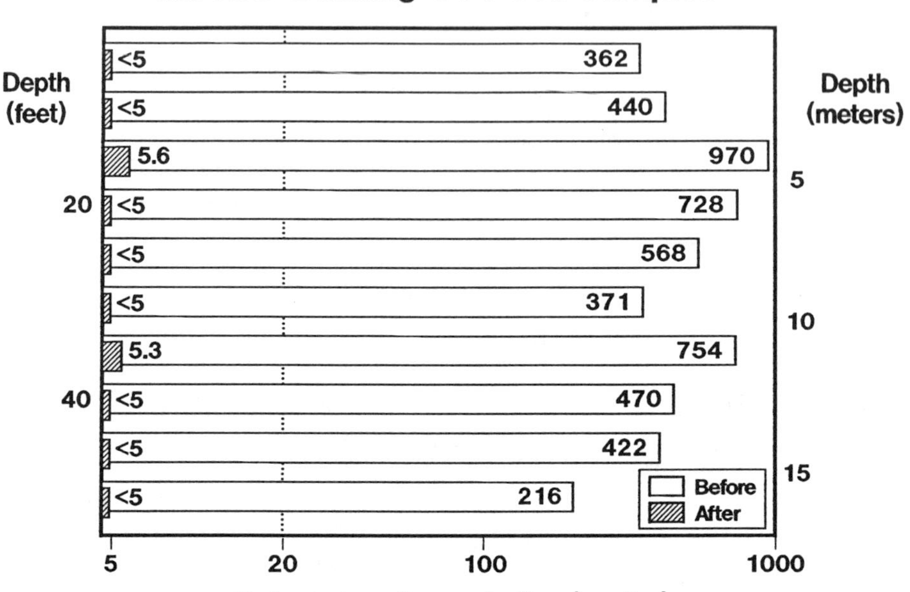

FIGURE 7. Soil hydrocarbon concentrations before and after treatment. Each column represents 14 or more averaged analytical results from the specified depth interval.

SUMMARY AND CONCLUSIONS

In situ measurements of aerobic biodegradation reactions taking place at a JP-4 jet fuel spill site being remediated via SVE indicated the importance of these reactions even under conventional venting operating modes. Laboratory treatability studies using field soils spiked with JP-4 suggested that moisture, and particularly nutrient addition, could be used to significantly increase respiration rates and to enhance biodegradation reactions taking place under unmanaged conditions. Field studies were used to assess these findings at full scale and involved modifications of operating flow rates and flow configuration, moisture addition, and nutrient plus moisture phases. Based on the results of these field studies, the following conclusions can be reached:

1. Conclusive evidence was provided to indicate significant biological activity at the field site. Without enhancement, a total of 15 to 25 percent of the recovered JP-4 could be attributed to biodegradation reactions. With enhancement this proportion increased to greater than 80 percent, resulting in substantial additional total petroleum hydrocarbons being degraded during the bioventing portion of the study.

2. Modification of operating conditions, i.e., reduced flow rates and maximized flow path distances, resulted in significantly reduced volatilization rates, allowing the direct discharge of vent gas without off-gas treatment, while still being below the regulatory limit of 50 ppm total petroleum hydrocarbons. These operating changes had no effect on biodegradation reactions occurring throughout the site.

3. *In situ* field respiration studies indicated that O_2 uptake rate measurements were better indicators of biological activity at the site than were CO_2 production rate determinations. CO_2 measurement sensitivity was susceptible to varying soil environmental conditions, notably soil water content. These CO_2 measurements did not consistently detect respiration changes during the study.

4. Care must be taken in directly interpreting results of laboratory treatability studies when they are to be extrapolated to the field scale. The impact of nutrient addition in this field study was overwhelmed by the effects of moisture amendment, and field data did not generally reflect the outcome expected based on laboratory findings. It should be noted that laboratory conclusions were primarily based on CO_2 production rates following relatively high-level JP-4 spiking with "uniform" nutrient application to the test soils. Field conclusions were driven, however, by O_2 uptake rate changes over time at an aged JP-4 site where the uniformity of JP-4, nutrients, and soil moisture could differ considerably from laboratory column studies.

5. Based on analyses of overlapping 95 percent confidence intervals of statistically significant first order regression relationships for O_2 uptake rates, it was found that moisture addition (35 to 50 percent field capacity) statistically accelerated *in situ* respiration at the site. However, nutrient addition generally did not statistically increase the degradation rates of residual JP-4 constituents.

6. Hydrocarbon concentrations before and after treatment appear to support the conclusion that significant remediation was achieved.

7. Bioventing appears to be a feasible option for the *in situ* degradation of residual fuel contaminants not amenable to recovery by SVE alone. Methods to reduce vapor extraction rates and maximize vapor retention times in the soil are compatible with enhancing biodegradation reactions through moisture management. These procedures result in minimizing volatilization, potentially eliminate the need for vent gas treatment, maximize the utilization of oxygen *in situ,* and provide a framework for the development of truly optimized *in situ* biological treatment systems in the future.

Acknowledgments

This research was funded by Hill Air Force Base's Environmental Management directorate and the Environics Division of the U.S. Air Force's Engineering and Services Center. The authors would like to extend their appreciation to Robert Elliot and Edward Heyse of Hill Air Force Base for invaluable support in this project.

REFERENCES

Alexander, M. *Introduction to Soil Microbiology,* 2nd ed.; John Wiley & Sons: New York, 1977.

Atlas, R. M. "Microbial Degradation of Petroleum Hydrocarbons: An Environmental Perspective." *Microbiological Review* **1981,** *45*(1), 185–209.

Bennedsen, M. B.; Scott, J. P.; Hartley, J. D. "Use of Vapor Extraction Systems for *In Situ* Removal of Volatile Organic Compounds from Soil." *Proceedings,* National Conference on Hazardous Waste and Hazardous Materials, Washington DC, 1987; pp 92–95.

DePaoli, D. W.; Herbes, S. E.; Hylton, T. D.; Jennings, H. L.; Nyquist, J. E.; Solomon, D. K.; Wilson, J. H. "Field Demonstration of *In-Situ* Soil Venting of JP-4 Jet Fuel Spill Site at Hill Air Force Base." Draft Final Report Volume III. ESL-TR-90-21, AFESC, Tyndall AFB, FL, 1990.

Downey, D. C.; Hinchee, R. E.; Westray, M. S.; Slaughter, J. K. "Combined Biological and Physical Treatment of a Jet Fuel-Contaminated Aquifer." *Proceedings,* Petroleum Hydrocarbons and Organic Chemicals in Ground Water: Prevention Detection and Restoration; National Water Well Association: Dublin OH, 1988; pp 627–645.

Dragun, J. "Microbial Degradation of Petroleum Products in Soil." In *Soils Contaminated by Petroleum—Environmental and Public Health Effects*; Calabrese, E. J.; Kostecki, P. T., Eds.; John Wiley & Sons: New York, 1988; pp 289–300.

Ely, D. L.; Heffner, D. A. Process for *In-Situ* Biodegradation of Hydrocarbon Contaminated Soil. U.S. Patent 4,765,902, 1988.

Hinchee, R. E.; Arthur, M. "Bench Scale Studies of the Soil Aeration Process for Bioremediation of Petroleum Hydrocarbons." *Journal of Applied Biochemistry and Biotechnology* **1991**, *28/29*, 901–906.

Hinchee, R. E.; Downey, D. C. "The Role of Hydrogen Peroxide in Enhanced Bioreclamation." *Proceedings,* Petroleum Hydrocarbons and Organic Chemicals in Ground Water: Prevention Detection and Restoration; National Water Well Association: Dublin OH, 1988; pp 715–722.

Hinchee, R. E.; Downey, D. C.; Slaughter, J. K.; Selby, D. A.; Westray, M. S.; Long, G. M. "Enhanced Bioreclamation of Jet Fuels—A Full-Scale Test at Eglin AFB, FL"; Final Report ESL-TR-88-78; Headquarters Air Force Engineering Services Center, Tyndall AFB, FL, 1989.

Hinchee, R. E.; Downey, D. C.; Dupont, R. R.; Aggarwal, P.; Miller, R. E. "Enhancing Biodegradation of Petroleum Hydrocarbons Through Soil Venting." *Journal of Hazardous Materials* **1991**, *28*(3).

Hutzler, N. J.; Murphy, B. E.; Gierke, J. S. "State of Technology Review: Soil Vapor Extraction Systems"; final report to the U.S. Environmental Protection Agency, Hazardous Waste Engineering Research Laboratory: Cincinnati OH, 1988.

Lee, M. D.; Thomas, J. M.; Borden, R. C.; Bedient, P. B.; Wilson, J. T.; Ward, C. H. "Biorestoration of Aquifers Contaminated with Organic Compounds." *CRC Critical Review In Environmental Control* **1988**, *18*(1), 29–89.

Ostendorf, D. W.; Kampbell, D. H. "Vertical Profiles and Near Surface Traps for Field Measurement of Volatile Pollution in the Subsurface Environment." *Proceedings,* New Field Techniques for Quantifying the Physical and Chemical Properties of Heterogeneous Aquifers; American Water Well Association: Dallas TX, 1989.

Riser, E. "Technology Review—*In Situ*/On-Site Biodegradation of Refined Oils and Fuel"; Final Report PO No. N68305-6317-7115; Naval Civil Engineering Laboratory, Port Hueneme CA, 1988.

Rollins, Brown, and Gunnel, Inc. "JP-4 Spill Substance Investigation and Remedial Action"; unpublished report submitted to Environmental Management, Hill AFB, UT, 1985.

Staps, J.J.M. "International Evaluation of *In-Situ* Biorestoration of Contaminated Soil and Groundwater"; RIVM-Report No. 738708006; National Institute of Public Health and Environmental Protection (RIVM): Den Hague, The Netherlands, 1989.

Wetzel, R. S.; Darst, C. M.; Davidson, D. H.; Sarno, D. J. "*In Situ* Biological Treatment Test at Kelly Air Force Base, Volume II—Field Test Results and Cost Model;" Final Report TR-85-52; Headquarters Air Force Engineering Services Center, Tyndall AFB, FL, 1987.

Wilson, J. T.; Ward, C. H. "Opportunities for Bioremediation of Aquifers Contaminated with Petroleum Hydrocarbons." *Journal of Industrial Microbiology* **1988**, *27*, 109–116.

A Field-Scale Investigation of Petroleum Hydrocarbon Biodegradation in the Vadose Zone Enhanced by Soil Venting at Tyndall AFB, Florida

*Ross N. Miller**, *Catherine C. Vogel*
U.S. Air Force
Robert E. Hinchee
Battelle

INTRODUCTION

Soil venting is effective for the physical removal of volatile hydrocarbons from unsaturated soils and as a source of oxygen for biological degradation of the volatile and nonvolatile fractions of hydrocarbons in contaminated soil. Treatment of soil venting off-gas is expensive, constituting at least 50 percent of soil venting remediation costs (Miller 1990). In this research, methods for enhancing biodegradation through soil venting were investigated, with the goal of eliminating the need for expensive off-gas treatment.

A 7-month field investigation (October 1989 to April 1990) was conducted at Tyndall Air Force Base (AFB) FL, where past jet fuel storage had resulted in contamination of a sandy soil. The contaminated area was dewatered to maintain approximately 1.6 m of

* U.S. Air Force, HSO/YAQE, Brooks AFB TX 78235

unsaturated soil. Soil hydrocarbon concentrations ranged from 30 to 23,000 mg/kg. Contaminated and uncontaminated test plots were vented for 188 days. Venting was interrupted five times during operation to allow for measurement of biological activity (CO_2 production and O_2 consumption) under varying moisture and nutrient conditions.

Moisture addition had no significant effect on soil moisture content or biodegradation rate. Soil moisture content ranged from 6.5 to 9.8 percent (by weight) throughout the field test. Nutrient addition was also shown to have no statistically significant effect on biodegradation rate. Initial soil sampling results indicated that naturally occurring nutrients were adequate for the amount of biodegradation observed. Biodegradation rate constants were shown to be affected by soil temperature and followed predicted values based on the van't Hoff-Arrhenius equation.

In one treatment cell, approximately 26 kg of hydrocarbons volatilized and 32 kg biodegraded over the 7-month field test. Although this equates to 55 percent removal attributed to biodegradation, a series of flow rate tests showed that biodegradation could be increased to 85 percent by managing air flow rate.

This research indicates that air flow management is an important factor in influencing total biodegradation of jet fuel, eliminating or substantially reducing the remediation costs associated with treatment of soil venting off-gas.

MATERIALS AND METHODS

Site Description. An *in situ* field demonstration of enhanced biodegradation through soil venting was conducted at the site of an abandoned tank farm located on Tyndall Air Force Base (AFB). The site is contaminated with fuel, primarily JP-4, and free product has been observed floating on the shallow groundwater table. Tyndall AFB is located on a peninsula that extends along the shoreline of the Gulf of Mexico in the central part of the Florida Panhandle. The highest ground on the peninsula is 7.6 to 9.1 m (25 to 30 ft) above mean sea level. The uppermost sediments at Tyndall AFB are sands and gravels of Pleistocene to Holocene age (ESE 1988). Soils at the site are best described by the Mandarin series consisting of somewhat poorly drained, moderately permeable soils that formed in thick beds of sandy material (USDA 1984).

The climate at the site is subtropical with an annual average temperature of 20.5 C (69 F). Average daily maximum and minimum temperatures are 25 and 16 C (77 and 61 F), respectively. Temperatures of 32 C (90 F) or higher are frequently reached during summer months, but temperatures above 38 C (100 F) are reached only rarely. Average annual rainfall is 140 cm (55.2 in.) with approximately 125 days of recordable precipitation during the year. The depth to groundwater varies from about 0.3 to 3.0 m (1 to 10 ft). The water table elevation rises during periods of heavy rainfall and declines during periods of low rainfall. Yearly fluctuations in groundwater elevations of approximately 1.5 m (5 ft) are typical (ESE 1988). Prior to dewatering at the site, the water table was observed to be as shallow as 46 cm (1.5 ft).

Field Testing Objectives. A 7-month field study (October 1989 to April 1990) was designed to address the following areas:

1. Does soil venting enhance biodegradation of JP-4 at this site?

2. Does moisture addition coupled with soil venting enhance biodegradation at this site?

3. Does nutrient addition coupled with soil venting and moisture addition enhance biodegradation at this site?

4. Evaluate ventilation rate manipulation to maximize biodegradation and minimize volatilization.

5. Calculate specific biodegradation rate constants from a series of respiration tests conducted during shutdown of the air extraction system.

Test Plot Design and Operation. To accomplish project objectives, two treatment plots (Figure 1) and two background plots were constructed and operated in the following manner:

- Contaminated treatment plot 1 (V1)—Venting only for approximately 8 weeks, followed by moisture addition for approximately 14 weeks, followed by moisture and nutrient addition for approximately 7 weeks.

- Contaminated treatment plot 2 (V2)—Venting coupled with moisture and nutrient addition for 29 weeks.

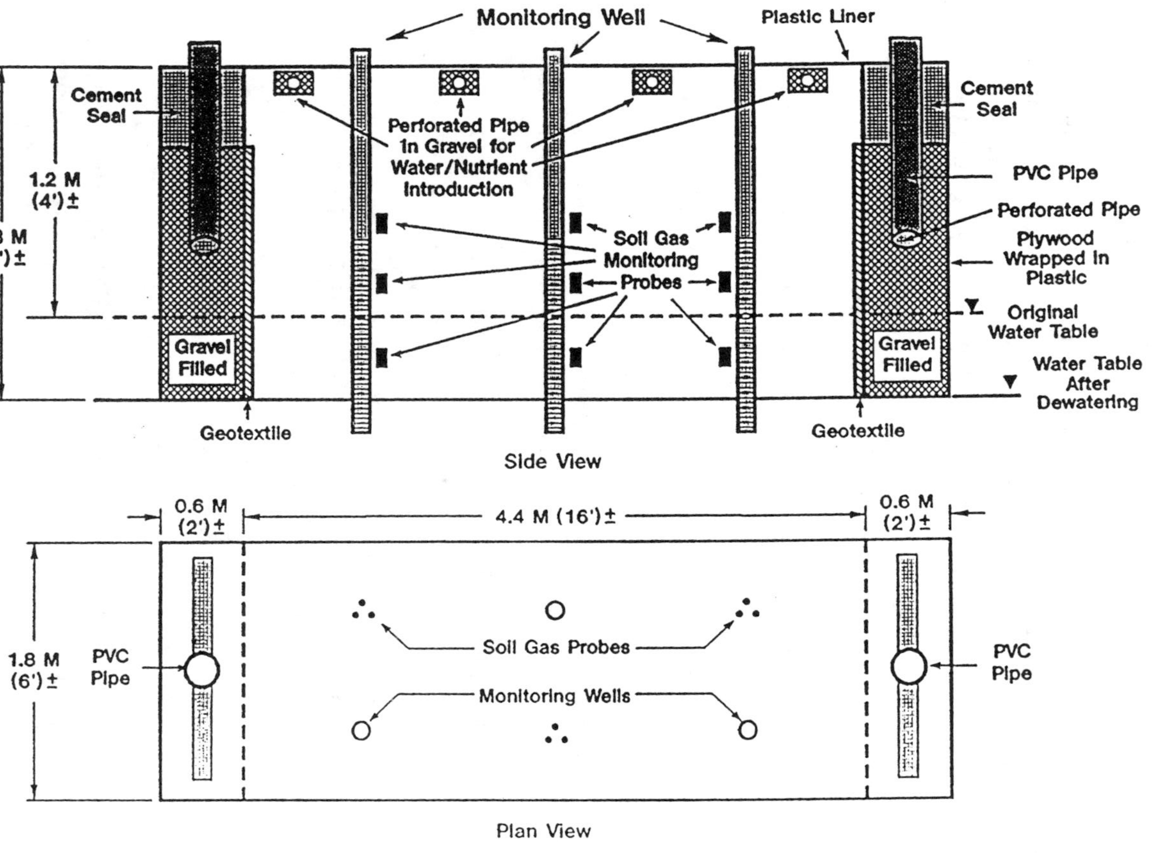

FIGURE 1. Design of contaminated test plots installed at Tyndall AFB, FL.

- Background Plot 3 (V3)—Venting with moisture and nutrient addition at rates similar to V2, with injection of hydrocarbon contaminated off-gas from V1.

- Background Plot 4 (V4)—Venting with moisture and nutrient addition at rates similar to V2.

Air Flow. Air flow was maintained throughout the field test duration except during *in situ* respiration tests. Flow rates were adjusted to maintain aerobic conditions in treatment plots, and background plots were operated at similar air retention times. Off-gas treatment experiments in one background plot (V3) involved operation at a series of flow rates and retention times.

Soil gas was withdrawn from the center monitoring well in V1 and V2 (Figure 1) and from the only monitoring well in V3 and V4. This configuration was selected to minimize leakage of soil gas from outside the test plot, which was observed when air was withdrawn from the ends of the plots. In all but one plot, V3, atmospheric air was allowed to enter passively at both ends. Off-gas from Vl was pumped back to the upstream ends of V3. Flow rates through all test plots were measured with calibrated rotameters.

Water Flow. To allow control of soil moisture, tap water was applied to the surface of the treatment plots through a series of leach lines. The design water flow rates allowed variation from 10 to 100 mL/min in the contaminated treatment plots and 2.5 to 25 mL/min in the background vents. This corresponds to an average annual surface application rate of 43 to 430 cm (17 to 170 in.). Based on vacuum and oxygen measurements in the soil gas monitoring probes, it was determined that a flow rate of 100 mL/min in the treatment plots inhibited air flow and oxygen transfer. Using the same technique, a flow rate of 50 mL/min (215 cm/year, surface application rate) was selected as the final water application rate. This application rate did not appear to inhibit air flow rate as confirmed by vacuum and oxygen measurements at the soil gas monitoring points.

Nutrient Addition. The objective of nutrient addition was to apply sufficient inorganic nitrogen (N), phosphorus (P), and potassium (K) at C:N:P ratios less than 100:10:1 to ensure, as far as possible, that these nutrients would not become limiting during the biodegradation of fuel hydrocarbons in the test plots. Optimizing nutrient addition rates was not the primary objective of this phase of the study. Sodium trimetaphosphate (Na-TMP), ammonium chloride (NH_4Cl), and potassium nitrate (KNO_3) were used as sources of P, N, and K, respectively.

RESULTS AND DISCUSSION

Operational Monitoring of Treatment Plots

Treatment plots were operated for 188 days between October 4, 1989 and April 24, 1990. Operation was interrupted only for scheduled respiration tests. Discharge gases were monitored for oxygen, carbon dioxide, and total hydrocarbons throughout the operational period. Oxygen and carbon dioxide were measured with a single field calibrated instrument, and total hydrocarbons were measured with a portable gas chromatograph with FID detector calibrated to hexane. The biodegradation component was calculated using the stoichiometric oxidation of hexane (Equation 1); the total volume of air extracted from the contaminated soil; and observed oxygen consumption

$$C_6H_{14} + 9.5O_2 \rightarrow 6CO_2 + 7H_2O \quad (1)$$

3.5 g O_2 consumed per g C_6H_{14} biodegraded

Oxygen consumption was calculated as the difference between oxygen in background plot V4 and oxygen in the treatment plots. Using the oxygen concentration in the background plot, rather than the atmospheric oxygen concentration, the natural biodegradation of organic carbon in uncontaminated soil was accounted for. This method ensured that the biodegradation of fuel hydrocarbons was not overestimated. Biodegradation rates based on carbon dioxide production were similarly calculated using Equation 1. Hydrocarbon removal rates attributed to volatilization and biodegradation are presented in Figures 2 and 3, respectively, for treatment plots V1 and V2. Removal rates are expressed in mg/(kg/day) and are based on an estimated soil bulk density of 1,440 kg/m^3 (90 lb/ft^3) and a treatment volume of 20 m^3 (704 ft^3).

As the more volatile compounds are stripped from the soil, biodegradation becomes increasingly important over time as the primary hydrocarbon removal mechanism, as illustrated in Figures 4 and 5 for

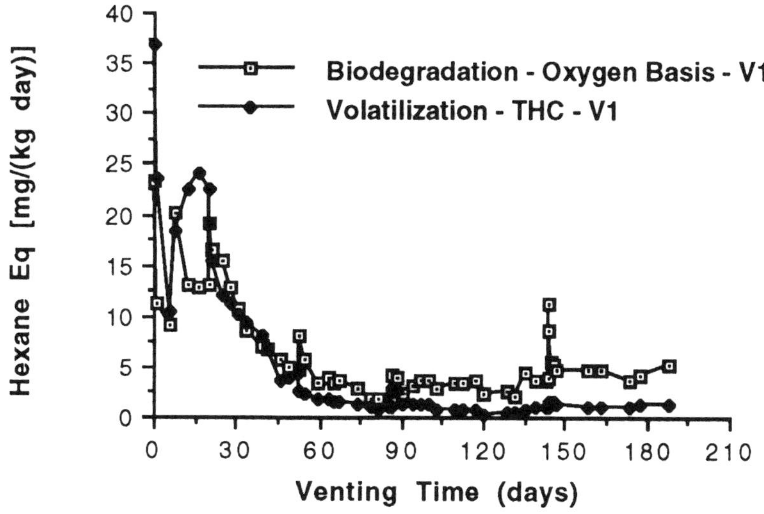

FIGURE 2. Hydrocarbon removal rate attributed to volatilization and biodegradation (oxygen basis) in treatment plot V1 during the field study.

treatment plots V1 and V2, respectively. The percent of hydrocarbons removed by biodegradation relative to the total hydrocarbons removed both by biodegradation and volatilizaton are compared in Figure 6 for treatment plots V1 and V2.

Operational data for the treatment plots are remarkably similar considering that treatment plot V2 received moisture and nutrients throughout the experimental period and treatment plot V1 received moisture after 8 weeks of operation and nutrients after 22 weeks of operation. The relationships demonstrated above indicate that moisture and nutrients were not limiting factors in hydrocarbon biodegradation removal rate for this particular site. Other studies associated with this project but not reported in this paper are cited as potential explanations for this observation. Initial soil samples (Miller 1990) indicated that naturally available nitrogen and phosphorus were adequate for the amount of biodegradation measured. Acetylene reduction studies (Miller 1990) revealed an organic nitrogen fixation potential capable of fixing the observed organic nitrogen, under anaerobic conditions, in 5 to 8 years. Soil moisture levels did not significantly change during the field study (Miller 1990). Soil moisture levels ranged from 6.5 to 7.4 percent, and 8.5 to 9.8 percent, by weight, respectively, in treatment plots V1 and V2. Neither venting nor moisture addition had a

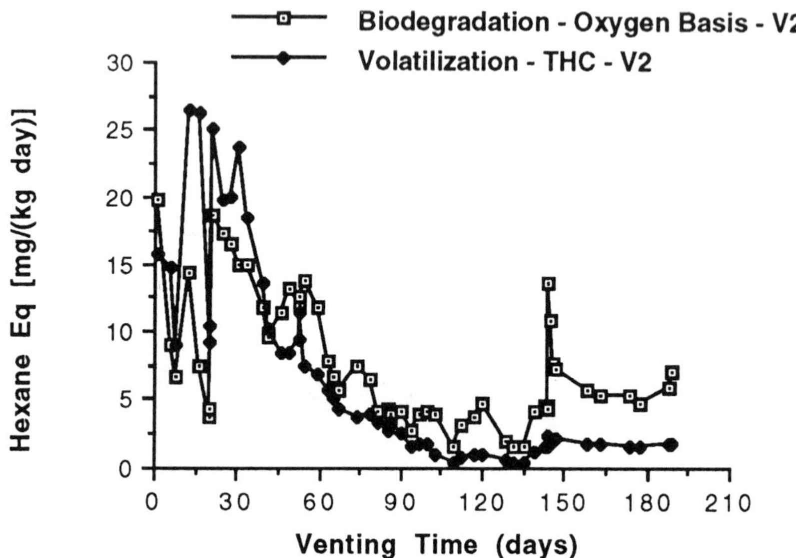

FIGURE 3. Hydrocarbon removal rate attributed to volatilization and biodegradation (oxygen basis) in treatment plot V2 during the field study.

statistically significant effect on soil moisture at this site even though water application rates far exceeded average annual precipitation.

Flow Rate *versus* Total Hydrocarbon Removal Rate and Percent Biodegradation

An experiment was conducted to evaluate the relationship between air flow rate, total hydrocarbon removal rate, and percent of total removal attributed to biodegradation following the period of high volatilization removal (after approximately 75 days of venting). The objective of this experiment was to reduce the amount of hydrocarbons volatilized by reducing air flow rates (i.e., reducing off-gas treatment cost) but yet providing sufficient oxygen to sustain biodegradation.

Rate constants (k) for oxygen consumption and carbon dioxide production have been shown, through respiration tests in this research (Miller 1990), to follow zero order kinetics for oxygen concentrations above 1 percent. Therefore, lower flow rates resulting in longer retention times should result in higher percentages of hydrocarbon removal by biodegradation relative to the total hydrocarbons removed.

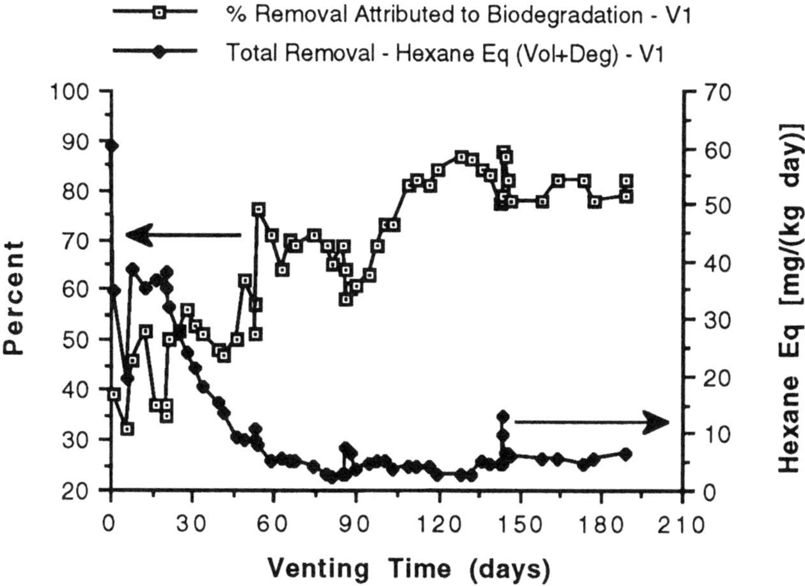

FIGURE 4. Comparison of the combined volatilization and biodegradation removal rates and the percent of removal rate attributed to biodegradation (oxygen basis) in treatment plot V1 during the field study.

Beginning at Day 89, air flow rates in treatment plots V1 and V2 were varied over a 7-week period from January 8, 1990 to February 28, 1990. Flow rates were approximately 8, 4, 2, and 1 L/min, which equate to approximately 2, 1, 0.5, and 0.25 air-filled void volumes per day, respectively. Oxygen, carbon dioxide, and hydrocarbon concentrations were allowed to stabilize at each air flow rate. Figures 7 and 8 illustrate the data for treatment plots V1 and V2, respectively, using oxygen and total hydrocarbon concentrations in the discharge gas streams as the basis for calculating percent removal by biodegradation. Figures 7 and 8 reveal a trade-off between maximizing the percent of hydrocarbon removed by biodegradation and maximizing the combined hydrocarbon removal rate due to biodegradation and volatilization, thereby minimizing the operational time required to remediate a contaminated site.

Using the data illustrated in Figure 7 for treatment plot V1 and assuming that 100,000 g (3,500 mg/kg) of hydrocarbons must be removed, a hypothetical case can be evaluated. If 62 percent biodegradation is desired, then 8 L/min (two air void volumes per day) would be selected with an expected operational time of 571 days. However,

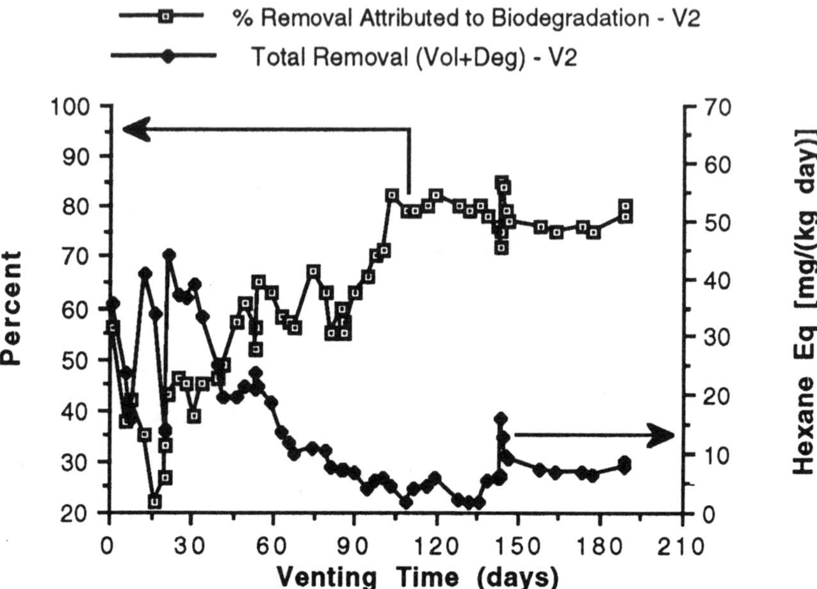

FIGURE 5. Comparison of the combined volatilization and biodegradation removal rates and the percent of removal rate attributed to biodegradation (oxygen basis) in treatment plot V2 during the field study.

if 85 percent biodegradation were desired, then 1 L/min (0.25 air void volumes per day) would be selected with an expected operational time of 1,370 days. Although operational time is increased by a factor of 2.4, total air requirement actually decreases from 6.6 to 2.2 million L. Optimal air flow conditions in V1 appear to be 2 L/min (0.5 air void volumes per day) where 82 percent biodegradation is achieved. Although 85 percent biodegradation is achieved at 1 L/min in V1, hydrocarbon removal rate from combined volatilization and biodegradation is greatly reduced. Operating at 2 L/min in V1, expected operation time is 820 days (1.4 times that required at 8 L/min) and the total air requirement is only 2.3 million L. It is emphasized that operational times in this case are merely illustrative since relationships between air flow and removal rate are applicable only over the 7-week field test period. However, it is likely that similar relationships would exist throughout the remediation period, although the magnitude of removal rates varies widely.

This research has documented that decreasing air flow rates will increase the percent of hydrocarbon removal by biodegradation and

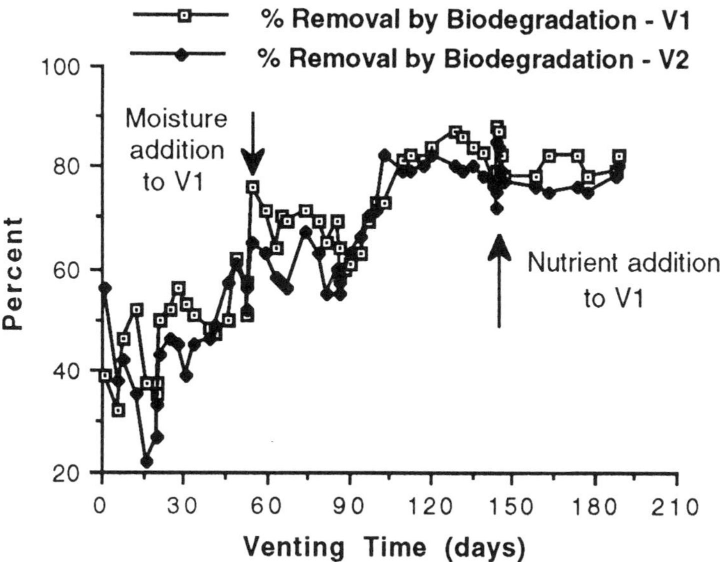

FIGURE 6. Comparison of the percent of combined volatilization and biodegradation hydrocarbon removal rates attributed to biodegradation (oxygen basis) in treatment plots V1 and V2 during the field study.

decrease the percent of hydrocarbon removal by volatilization relative to the total hydrocarbons removed.

Respiration Tests

Respiration tests 1 through 5 were conducted October 24–26, 1989; November 28 through December 1, 1989; January 3–8, 1990; March 3–11, 1990; and April 24–26, 1990, respectively. In addition, two limited respiration tests, 3A and 4A, were conducted from January 25–26 and March 9–12, 1990. The respiration tests were designed to determine the order and rate of hydrocarbon biodegradation kinetics under varying conditions of moisture and nutrient addition. Treatment plot V2 received moisture and nutrients throughout the experimental period and therefore served as a control for kinetic changes resulting from soil temperature and other factors not related to moisture and nutrients. The respiration tests were conducted by first shutting down the air delivery system to both the treatment and background plots, followed by measuring oxygen consumption and carbon dioxide production over time. Biological respiration in treatment plots V1 and V2 was most

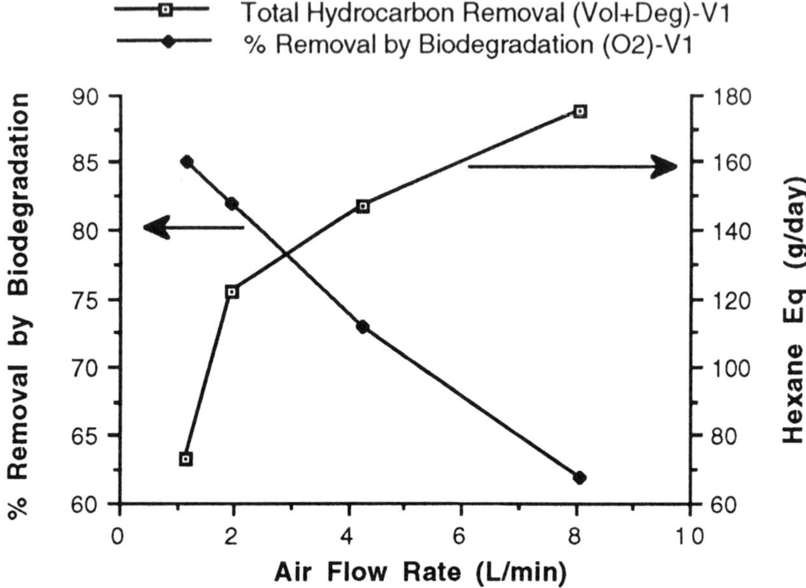

FIGURE 7. Comparison of air flow rate versus total hydrocarbon removal and percent of total removal attributed to biodegradation in treatment plot V1 (O_2 basis) during the variable air flow rate study.

consistently modeled by zero order kinetics during all respiration tests (Miller 1990). In a system not limited by substrate, such as fuel-contaminated soil, biodegradation is likely to be best modeled by zero order kinetics (Riser 1988).

Oxygen and carbon dioxide concentrations, measured in the treatment plot vapor-monitoring wells prior to initiating the respiration tests, were highly variable. Regardless of initial concentration, however, oxygen consumption and carbon dioxide production rates were relatively consistent. For this reason, the data were normalized by dividing oxygen concentration data measured in each vapor-monitoring well by the initial oxygen concentration at each location. A regression of the normalized data versus time for each plot and each respiration test yielded a normalized zero order rate constant that, when multiplied by the initial average oxygen concentration in the plot, yielded the actual zero order rate constant (k = %/min). The normalized regression and 95 percent confidence interval band for treatment plot V1 is illustrated in Figure 9 for respiration test 4. Figure 9 is typical of all respiration tests conducted in treatment plots V1 and V2.

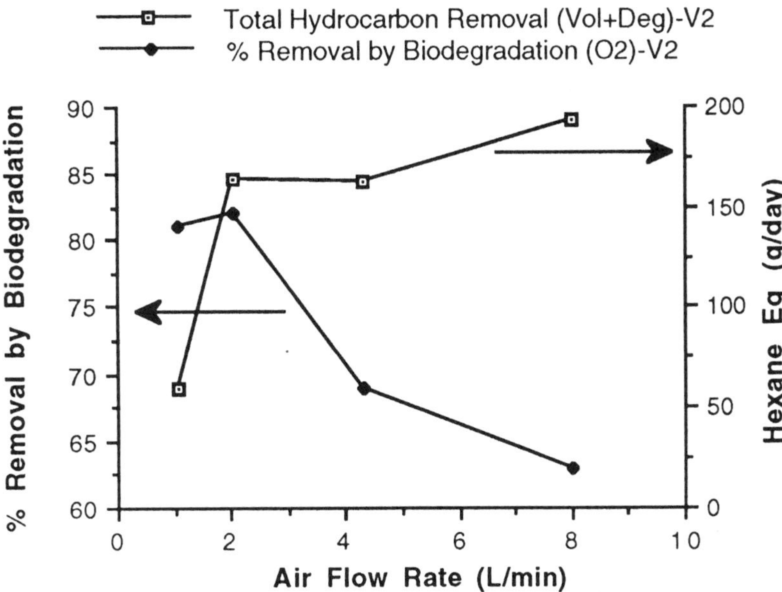

FIGURE 8. Comparison of air flow rate versus total hydrocarbon removal and percent of total removal attributed to biodegradation in treatment plot V2 (O_2 basis) during the variable air flow rate study.

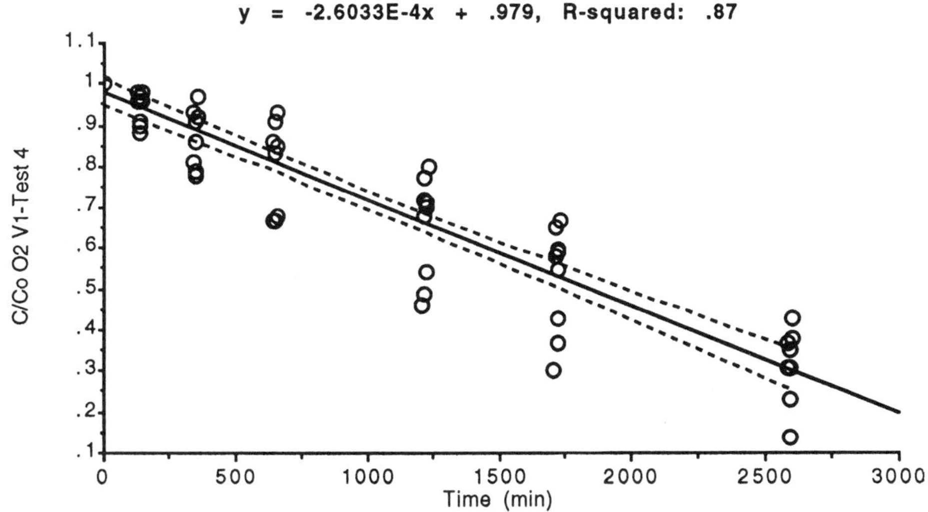

FIGURE 9. Regression of normalized data and 95 percent confidence band for treatment plot V1 and respiration test 4.

A summary of the zero order rate constant data obtained from the respiration tests is graphically illustrated in Figure 10. In treatment plot V1, the rate constant showed a significant (95% confidence interval) drop between tests 1 and 2, and between tests 2 and 3. The rate constant significantly increased between tests 3 and 4 in treatment plot V1, but did not significantly increase between tests 4 and 5. Since moisture was added to treatment plot V1 after test 2 and nutrients after test 4, their addition would seem, without further analysis, to be of no benefit and even detrimental, in the case of moisture addition. In treatment plot V2, there was a statistically significant drop in the rate constant from tests 2 to 3 and a statistically significant increase in the rate constant between tests 3 and 4. Although a depression appears in the rate constant data, there were no other statistically significant differences in treatment plot V2 rate constants.

Statistically significant differences in respiration rate between treatment plots V1 and V2, and the background plot V4, on all tests, and between off-gas treatment plot V3 and background plot V4 on tests 3, 4A, and 5 are illustrated in Figure 10. From the data presented, it was concluded that biodegradation both of jet fuel in contaminated soil and of hydrocarbon off-gas resulted in statistically significant increases in respiration over that observed in uncontaminated soil.

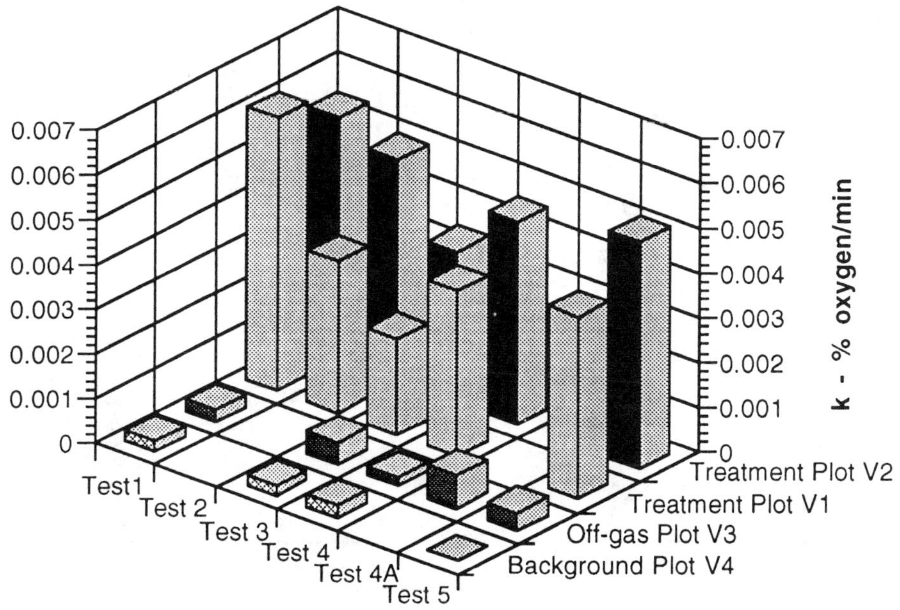

FIGURE 10. Average zero order rate constants determined by respiration tests.

Potential Temperature Effects on Respiration Tests

As described above, hydrocarbon biodegradation rates appear to have been unaffected by moisture and nutrient addition. This conclusion was based on insignificant differences in biodegradation rates in treatment plots V1 and V2 (Miller 1990), even though the treatment plots were operated under different moisture and nutrient conditions. Although biodegradation rates in the treatment plots were similar, there was a general decline in hydrocarbon removal rates from initiation of the field study, reaching minimum values near the middle of the experimental period, followed by a general increase in hydrocarbon removal rates through the completion of the field study. Since the treatment plots appeared unaffected by moisture and nutrient addition, soil temperature was investigated as the potential cause of the depression in biodegradation rates.

Soil temperature at this field site was related to ambient air temperature because air was continually drawn through the soil. More importantly, the moisture provided to the treatment plots affected soil temperature as the applied water temperature was a function of air temperature because this water was temporarily stored in the site building prior to delivery to the treatment plots.

Local ambient temperature data were obtained from a weather station located near the field site. The 10-day moving average, aboveground air temperature data are compared with soil temperature in Figure 11. Soil temperature data before January 5, 1990 were not collected at the field site. Therefore, the relationship between ambient temperature and soil temperature was used to estimate soil temperatures prior to this date. Comparison of average soil temperature (Figure 11) to oxygen consumption rate (Figure 10) during respiration tests in treatment plots V1 and V2 imply a relationship between soil temperature and biological activity. It appears from the respiration data presented that soil temperature had a much more significant effect on the rate of hydrocarbon biodegradation than moisture and nutrient addition.

To evaluate the effect of moisture and nutrient addition on biological activity in treatment plot V1, the effect of temperature must be understood. Treatment plot V2 received moisture and nutrients throughout the experimental period and should be a control on temperature and other unmeasured variables. Therefore, a model that eliminates the effect of temperature on the oxygen consumption rate constants (Figure 10) in treatment plot V2 should be adequate for

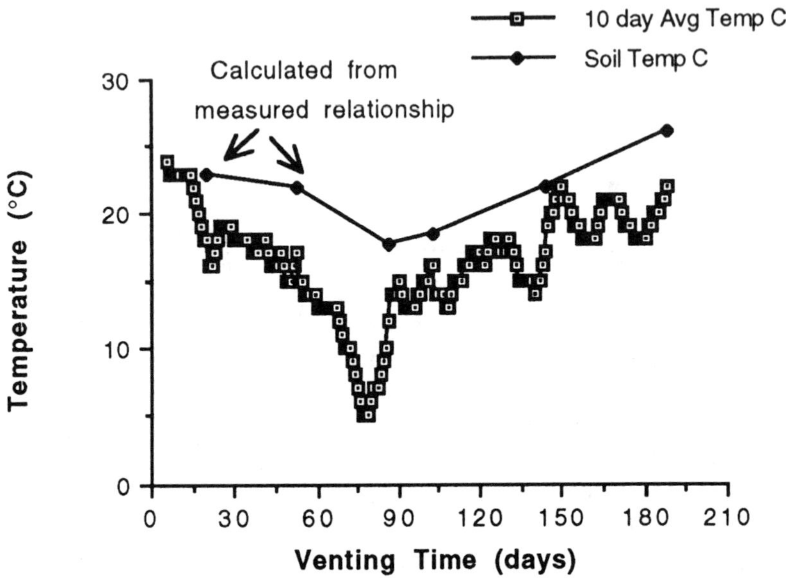

FIGURE 11. Comparison of the 10-day moving average of the mean ambient aboveground air temperature and corresponding estimated and measured soil temperature.

temperature correcting rate constants measured in treatment plot V1, thereby allowing an assessment of the effect of moisture and nutrient addition in treatment plot V1.

In aquatic systems, the van't Hoff-Arrhenius equation predicts a doubling of the rate constant with each temperature increase of 10 C, assuming typical activation energy values (Benefield & Randall 1980). Figure 12 is the Arrhenius plot for determining activation energy using measured soil temperature and rate constant relationships from tests 3, 4, and 5 for treatment plots V1 and V2. Using the Arrhenius constants determined from the plots in Figure 12, the rate constants for treatment plots V1 and V2 were corrected to 23 C, the soil temperature of test 1. The Arrhenius model for temperature correction resulted in insignificant rate constant differences between tests 2, 3, 4, and 5 in treatment plot V2 (Figure 13). Therefore, the Arrhenius equation adequately modeled the effects of temperature on hydrocarbon biodegradation rate. Using the same model, the oxygen consumption rate constants in treatment plot V1 were corrected for temperature (Figure 14). Although statistically significant differences in rate constants remained between test 3 and tests 2 and 4 in treatment plot V1, the magnitude of the differences are not important from a practical application standpoint.

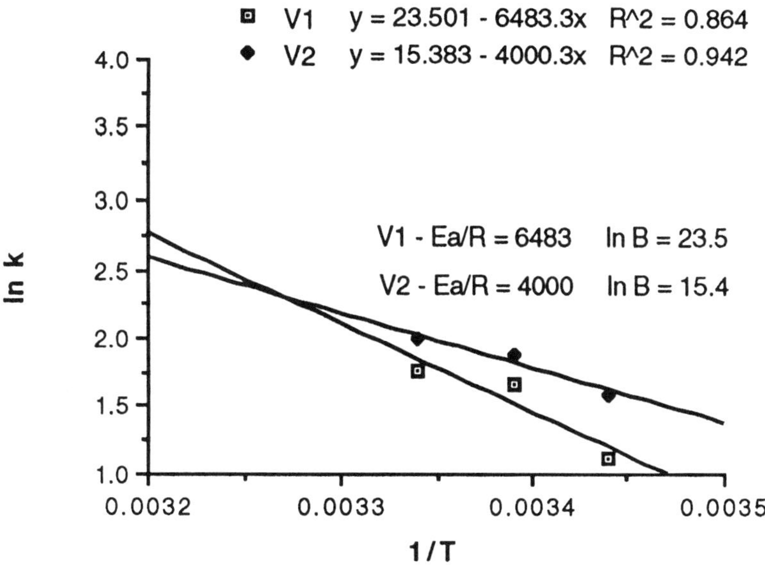

FIGURE 12. Arrhenius plot for determining activation energy using measured soil temperature and rate constant relationships from tests 3, 4, and 5 for treatment plots V1 and V2.

Test 1 in both treatment plots was not considered because it was conducted when hydrocarbon concentrations in the soil gas were still very high.

Moisture was added to treatment plot V1 following respiration test 2, and nutrients were added following respiration test 4. Temperature corrected rate constants (Figure 14) were not significantly increased between tests 2 and 3 and between tests 4 and 5. Therefore, it can be concluded that moisture and nutrient addition were of insignificant benefit to the rate of hydrocarbon biodegradation in treatment plot V1 for this particular site. Although moisture and nutrient addition did not affect biodegradation rates, the data indicate that soil temperature likely did.

CONCLUSIONS

This field-scale investigation has demonstrated that soil venting is an effective source of oxygen for enhanced aerobic biodegradation of

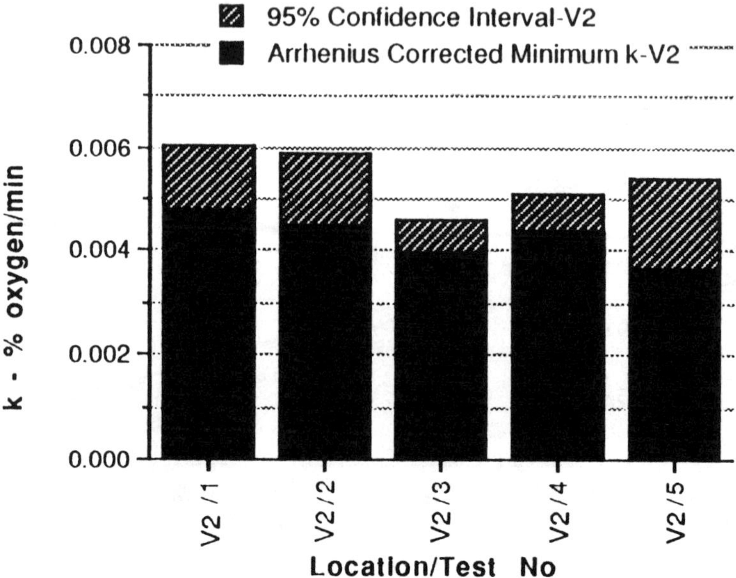

FIGURE 13. Temperature-corrected (23 C based on Arrhenius plot) oxygen consumption rate constants determined by respiration tests for treatment plot V2. Mean k is at the center of the 95 percent confidence level.

petroleum hydrocarbons (jet fuel) in the vadose zone. Specific conclusions are

1. Biodegradation removal rates during venting ranged from approximately 2 to 20 mg/(kg day), but stabilized values averaged about 5 mg/(kg day). The effect of soil temperature on biodegradation rates was shown to approximate effects predicted by the van't Hoff-Arrhenius equation.

2. Operational data and respiration tests indicated that soil moisture (6.5 to 9.8 percent by weight) and nutrients were not limiting factors in hydrocarbon biodegradation for this site.

3. Air flow tests documented that decreasing flow rates increased the percent of hydrocarbon removal by biodegradation and decreased the percent of hydrocarbon removal by volatilization relative to the total mass of

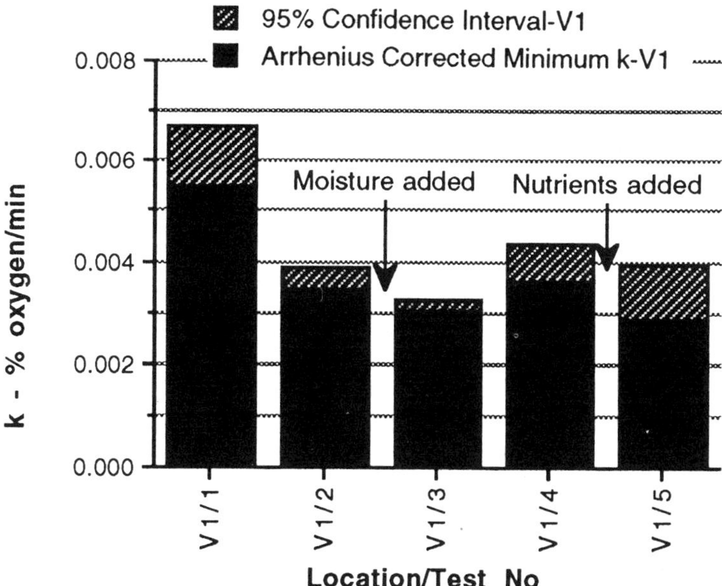

FIGURE 14. Temperature-corrected (23 C based on Arrhenius plot) oxygen consumption rate constants (k) determined by respiration tests for treatment plot V1. Mean k is at the center of the 95 percent confidence level.

hydrocarbon removed. Under optimal air flow conditions (0.5 air void volumes per day), 82 percent of hydrocarbon removal was biodegraded and 18 percent volatilized.

4. Respiration tests documented that oxygen consumption rates followed zero order kinetics and that rates were linear down to about 2 to 4 percent oxygen. Therefore, air flow rates can be minimized to maintain oxygen levels between 2 and 4 percent without inhibiting biodegradation of fuel, with the added benefit that lower air flow rates will increase the percent of removal by biodegradation and decrease the percent of removal by volatilization.

RECOMMENDATIONS FOR FUTURE STUDY

To pursue further the development of technology to enhance biodegradation of petroleum hydrocarbons through soil venting (bioventing), the following studies are recommended:

1. Further investigate the relationship between soil temperature and hydrocarbon biodegradation rate.

2. Investigate methods to increase hydrocarbon biodegradation rate by increasing soil temperature with heated air, heated water, or low-level radio frequency radiation.

3. Investigate the effect of soil moisture content on biodegradation rate in different soils with and without nutrient addition.

4. Investigate nutrient recycling to determine maximum C:N:P ratios that do not limit biodegradation rates.

5. Investigate different types of uncontaminated soil for use as a reactor for biodegradation of generated hydrocarbon off-gas and determine off-gas biodegradation rates.

6. Investigate gas transport in the vadose zone to allow adequate design of air delivery systems.

REFERENCES

Benefield, L. D; Randall, C. W. *Biological process design for wastewater treatment*; Prentice-Hall: Englewood Cliffs, NJ, 1980; pp 11–13.

ESE (Environmental Science and Engineering, Inc.) "Installation Restoration Program Confirmation/Quantification Stage 2 Volume 1 Tyndall AFB, FL"; final report to Headquarters Tactical Air Command, Command Surgeon's Office (HQTAC/SGPB); Bioenvironmental Engineering Division; Langley AFB, VA, 1988.

Miller, R. N. Ph.D. Dissertation, Utah State University, 1990.

Riser, E. "Technology Review *In Situ*/On-site Biodegradation of Refined Oils and Fuel." PO No. N683056317-7115 to the Naval Civil Engineering Laboratory; Port Hueneme, CA, 1988.

USDA (U.S. Department of Agriculture). *Soil Survey of Bay County Florida*; Soil Conservation Service. U.S. Government Printing Office: Washington, DC, 1984.

In Situ and On-Site Subsoil and Aquifer Restoration at a Retail Gasoline Station

J. van Eyk[*], *C. Vreeken*
Delft Geotechnics

INTRODUCTION

The technique of venting to remove petroleum vapours and volatile petroleum products from the vadose zone has been well documented (Hutzler *et al.* 1988). In addition to the possibility of enhancing the evaporation of volatile organics, venting can also be used to enhance convective air transport in soil to stimulate the oxygen-dependent petroleum biodegradation in the vadose zone (Anon. 1986; Staatsuitgererij Den Haag 1986).

The possibilities for bioventing were indicated by lysimeter studies on a microbially mediated gas-oil mixture (gas-oil) removal from sand. In the absence of bioventing, an almost 100 percent removal was achieved in a period of 8 years (Tibbetts 1982). A feasibility study subsequently carried out confirmed the possibilitites for the enhancement of biodegradation by bioventing (van Eyk & Vreeken 1989a). A large-scale field experiment was started in 1984 to validate the feasibility study (van Eyk & Vreeken 1988, 1989b).

The next step is to develop this methodology further in a real spill situation. After careful consideration and screening, the contaminated area surrounding a retail gasoline station was selected as a test case to develop and demonstrate the feasibility of the approach. The demonstration project was financially supported by the Department for the Environment (Grant No. MJZ 20 D 8037).

* Delft Geotechnics, P.O. Box 69, 2600 AB Delft, The Netherlands

The first priority after a mineral oil spill has occurred is to prevent the spreading of the spilled products. In cases of groundwater contamination, pumping wells can be used to control contaminant plume movement. When large quantities of petroleum products have been spilled into the ground, a skimming pump for the removal of liquid product allows bulk removal of most of the spilled product in a relatively short period of time. Soil venting can subsequently be used as a medium- to long-term remedial strategy. According to the guidelines outlined by Dutch law on soil remediation, mineral oil levels for soil with an organic matter content of 10 percent should be reduced to 50 mg/kg dry soil and for groundwater to 50 µg/L. In addition, individual values for benzene, toluene, and the xylenes should be equal to or smaller than 0.2 µg/L.

The purpose of this demonstration project is to demonstrate that it is possible to comply with these guidelines using an *in situ* methodology.

SITE SELECTION AND CHARACTERISATION

The site selected for this demonstration project was based on specific criteria. The hydraulic permeability of the soil had to be equal to or greater than 10^{-5} m/s because smaller values would result in exceedingly long clean-up times. In addition, the groundwater table had to be at least 2 m below ground surface (BGS) because the economics of this *in situ* technique, when compared to excavation only, become progressively favourable with increasing depth of the groundwater table.

The site eventually selected for this project is shown in Figure 1. It is a plan view showing the central area with fuel supply pumps, car shop, and garage.

Degree and Extent of Contamination

Soil. Analysis results for borings B1 to B4 are shown in Table 1 (B5 to B10 are only slightly contaminated). A considerable spillage appears to have occurred in the area of B2, and concentration levels decrease toward B3 and B4. Although neither the cashier in the shop nor

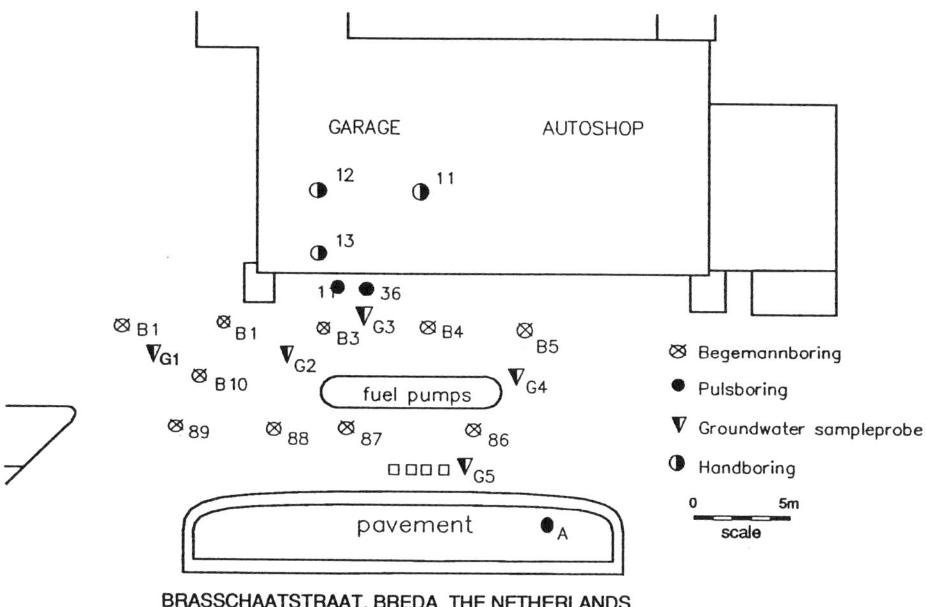

FIGURE 1. Plan view of retail gasoline station showing the locations for soil and groundwater sampling.

TABLE 1. Results of borings B1 to B4 with respect to mineral oil and BTX (mg/kg).

Component	B1 0.0–3.0	B2 0.0–3.0	B3 0.0–3.0	B4 1.0–3.0
Benzene	<.5	130	39	5.4
Toluene	<.5	860	250	130
Ethylbenzene	<.5	280	96	60
Xylenes	<.5	2,000	740	430
CH-total	<50	6,900	2,300	1,300
Mineral oil	<100	$15 \cdot 10^3$	1,300	320

personnel in the car shop have complained about foul smells, some extra handborings have been carried out in the garage (11, 12, and 13).

The analysis results are shown in Table 2. The results show that only boring 13, which is close to B3 (compare with Figure 1 and Table 1), has significantly elevated levels.

TABLE 2. Analysis results of handborings carried out under the garage floor (mg/kg).

	Boring Number and Depth Below Ground Surface		
Component	No. 11 1.0–1.5	No. 12 1.5–2.0	No. 13 1.0–1.3
Benzene	0.15	3.7	28
Toluene	0.56	5.6	270
Ethylbenzene	0.09	1.1	81
Xylenes	0.75	4.9	460
Mineral oil	<20	<20	850

Groundwater. Analyses carried out on groundwater samples obtained from approximately 15 monitoring wells generally confirmed the contaminant situation that was found via soil sample analysis. The greatly elevated levels in monitoring well 11 confirmed the elevated levels found in borings B2 to B4 and indicated the presence of mobile mineral oil on the capillary fringe.

Sampling from some monitoring wells indicated that the aquifer had been contaminated up to approximately 10 m BGS. Contamination of groundwater up to 10 m BGS was confirmed by samples obtained with the groundwater probe (Table 3). This device is more reliable than conventional observation wells because groundwater samples are obtained directly. After having reached the required depth of penetration, the probe allows the *in situ* removal of the outer casing. This exposes the filter and allows groundwater to enter the sample probe. This device excludes the possibility for contamination from soil horizons located above the testing level.

TABLE 3. Results of BTEX analyses for groundwater samples.

Component[a] (μg/L)	Monitoring Well No. 11[b]	Monitoring Well No. 36	Groundwater Sample Probe Number and Depth BGS GW2 9.0	GW3 9.5	GW4 7.5	GW5 9.5
Benzene	40 • 10^3	280	3.2	<0.2	<0.2	<0.2
Toluene	85 • 10^3	1,400	1.7	<0.5	<0.5	<0.5
Ethylbenzene	8.4 • 10^3	150	<0.5	<0.5	<0.5	<0.5
Xylenes	58 • 10^3	980	0.8	<0.5	<0.5	<0.5
CH—total	200 • 10^3	3,300	<20	<20	<20	<20

(a) Filter at 1.60–4.10 m BGS.
(b) Filter at 9.10–11.10 m BGS.

Soil Structure and Composition

The phreatic zone up to 12 m BGS is composed of medium-fine to medium-coarse sand, above fine silty sands and clay deposits (Twente Formation). A pulse boring to 20 m BGS showed the presence of a 1-m-thick loamy layer at 6 m BGS. This layer did not show up, however, in other borings carried out farther than 6 m BGS. From observation wells it appeared that the downward vertical hydraulic gradient is very small. Because of these observations, the phreatic zone has been considered as one and the same deposit. The permeability results obtained in the laboratory for a handboring carried out in the vicinity of the fuel supply pumps (see Figure 1), to evaluate the possibilities for *in situ* remediation by venting, are shown in Table 4.

The different porosities reflect the different degrees of compaction applied in the laboratory. In addition, a number of grain size distribution curves have been obtained (Table 5) from Begemann* borings Bl to B10 (Figure 1), which were also used to calculate hydraulic permeabilities. The hydraulic permeabilities calculated on the basis of the data of Table 5, according to Seelheim (S) and Beyer (B), are shown in Table 6. The value of c in Beyer's equation is based on the natural compaction (second graph in Figure 2, Beyer 1964).

A comparison of Tables 4 and 6 shows that experimental values obtained in the laboratory (Table 4) are higher than the calculated permeabilities based on *in situ* grain size distribution. To calculate air

* A method patented by Delft Geotechnics to take undisturbed soil samples.

TABLE 4. Permeability results obtained in the laboratory from a handboring 1.2 to 1.5 m BGS.

Porosity (%)	Permeability (m/sec)
45.3	$2.14 \cdot 10^{-4}$
42.7	$1.92 \cdot 10^{-4}$
40.8	$1.52 \cdot 10^{-4}$

TABLE 5. Results of grain size distribution curves.

Boring number	Metres below ground surface	d_{10} µm	d_{50} µm	d_{60} µm
1	0.00–1.80	36	160	180
2	1.00–1.80	80	175	200
3	1.00–1.80	75	165	185
4	0.00–1.80	53	150	175
5	0.00–1.80	75	160	185
6	0.00–1.55	63	160	185
7	0.00–1.80	75	140	160
8	1.00–1.80	80	165	185
9	1.00–1.80	75	160	180
10	0.00–1.80	36	160	190

transport in the vadose zone, a saturated water permeability of $1.0 \cdot 10^{-4}$ m/s has been selected as a middle value between the calculated and measured values.

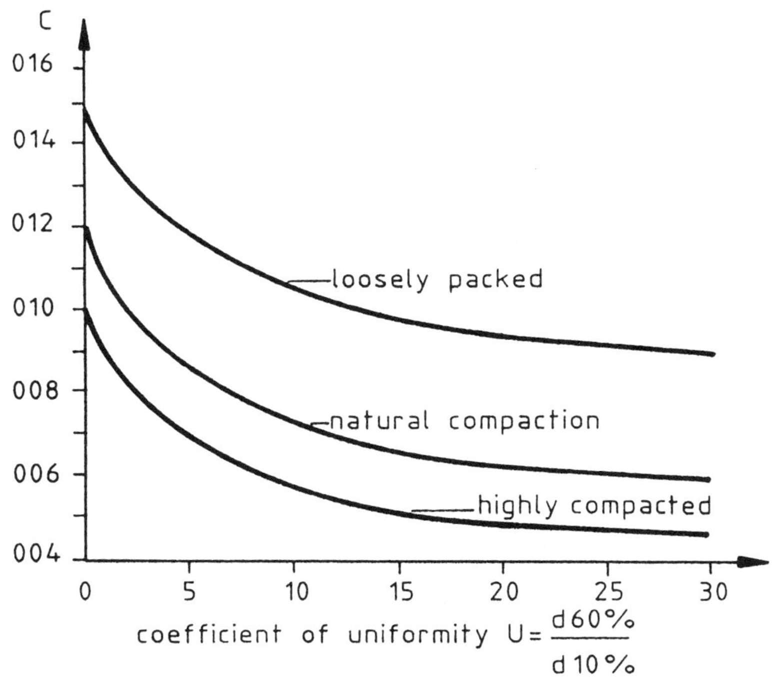

FIGURE 2. Value of c in Beyer's equation as a function of uniformity and grain compaction.

TABLE 6. Calculated water permeabilities based on grain size distribution.

Boring depth below ground surface	d_{10} µm	d_{50} µm	$K_w(S)$ m/sec	$K_w(B)$ m/sec
0.00–1.80	36	160	$8.0 \bullet 10^{-5}$	$1.2 \bullet 10^{-5}$
aver. 1–10	65	156	$8.7 \bullet 10^{-5}$	$4.2 \bullet 10^{-5}$
1.00–1.80	80	175	$1.1 \bullet 10^{-4}$	$6.4 \bullet 10^{-5}$

SYSTEM DESIGN

Model Calculations

Groundwater Withdrawal. To prevent contaminated plume movement and to clean up contaminated groundwater, pumping rates had to be calculated. Analysis of groundwater from observation wells showed that contaminated groundwater had migrated to a distance of 27 m away from the planned location of the pumping well(s) (Figure 3). In other words, 27 m is critical with respect to the radius of influence of the pumping well, and the withdrawal area, therefore, should contain at least this point source. The radius of influence on static water levels was calculated with the Single Layer Analytic Element Model (SLAEM) computer programme (Strack 1989). The model was verified with measured data from both groundwater and surface water levels.

Numerical calculations with SLAEM have been carried out under the following assumptions: the phreatic aquifer is confined to the impermeable base at 11 m BGS and the net rainwater infiltration rate amounts up to 0.15 m/year and a k_d value of 250 m^2/day. These parameters were used to calculate the effect of pumping on static water levels. The diameter of the well, including backfilling, was 0.2 m. Several scenarios were calculated using pumping well rates of up to 5 m^3/h. The calculations showed that a pumping rate of 5 m^3 was sufficient to control plume movement.

Air Transport Calculations. Air transport was calculated using the Hydrology Contaminant Transport Model (Vreeken & Sman 1989). The calculations are required to determine the spacing of the air extraction wells.

Grain size distribution curves show that the soil is composed of medium-fine sand. Stroosnijder's data (1976) were used to quantify the variation in water content as a function of depth to the water table. With respect to medium-fine sand, the following assumptions have been made:

- Porosity = 0.35
- Air-filled porosity at 2.0 m above groundwater table = 0.27

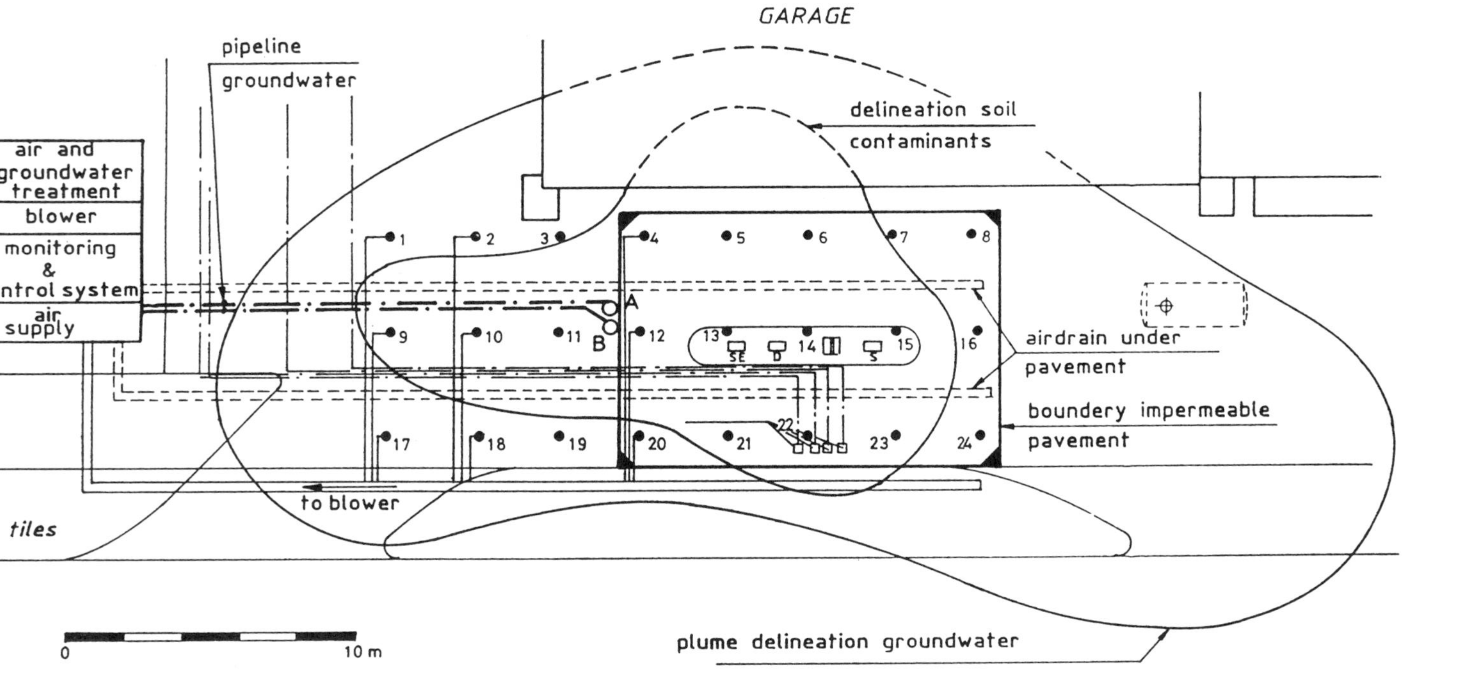

A○ pumping well filter Ø0,2m(9–11m -gs)
B○ pumping well filter Ø0,2m(1,5–4m -gs)
1t/m24 ● air pumping wells

FIGURE 3. Plan view of retail gasoline station outlining the installation of the venting system, pumping wells, and cleanup and monitoring systems.

- Air-filled porosity at 1.0 m above groundwater table = 0.18
- Air-filled porosity at 0.5 m above groundwater table = 0.05

It has been assumed that the soil up to 0.50 m above the groundwater table is completely impermeable to air. Above 0.5 m, an air-filled porosity of 0.2 has been assumed.

On the basis of the hydraulic permeabilities presented before (site characterisation, Table 6), a value of $8.7 \bullet 10^{-4}$ m/s was calculated for soil air permeability. Permeability of moist soil was calculated with the Carman Kozeny equation. These calculations have been described before (Vreeken & Sman 1989).

The schematization for the calculations is shown in Figure 4. To prevent flow of air along the air well, the soil surface was assumed to be partially covered. The flow lines are shown in Figure 5, which shows the flow of air just beneath the cover on the ground surface. Based on a distance between the air wells of 3 m, an air flux of 68 m^3/day was calculated at a negative pressure of 0.5 m water column (–0.05 m bar or 0.95 bara).

Evaluation of Maximum Attainable Biodegradation Rates. Although the obligatory participation of atmospheric oxygen may not be axiomatic (Vogel & Grbić-Galić 1986; Zeyer *et al.* 1986), petroleum biodegradation can take place only in the presence of atmospheric oxygen (van Eyk & Vreeken 1989b). In addition, since hydrocarbons are devoid of any oxygen, aerobic mineralisation of petroleum products requires more oxygen than primary products of photosynthesis such as sugars.

Calculations carried out for the design of the venting system (compare with Figure 4) show that a venting ratio of 6 can be accomplished under the circumstances outlined above. This means that an air flux of 68 m^3/day will purge the air-filled soil pores six times per day. One m^3 of air contains approximately 0.29 kg oxygen. With a net pore volume of 20 percent, the oxygen supply amounts to $6 \bullet 0.29 \bullet 0.20$ or 0.34 kg of oxygen/m^3/day. Assuming complete mineralisation of hydrocarbons $(CH_2)_n$ to carbon dioxyde and water, 0.34 kg oxygen could sustain complete oxydation of 100 g of hydrocarbons/m^3/day (63 mg/kg/day). Since biodegradation rates in the subsoil are far from optimal (because of the absence of mixing, availability of nutrients, and suboptimal temperatures), such rates are unlikely to be realized. The biodegradation rate of approximately 10 mg/kg/day, determined in a

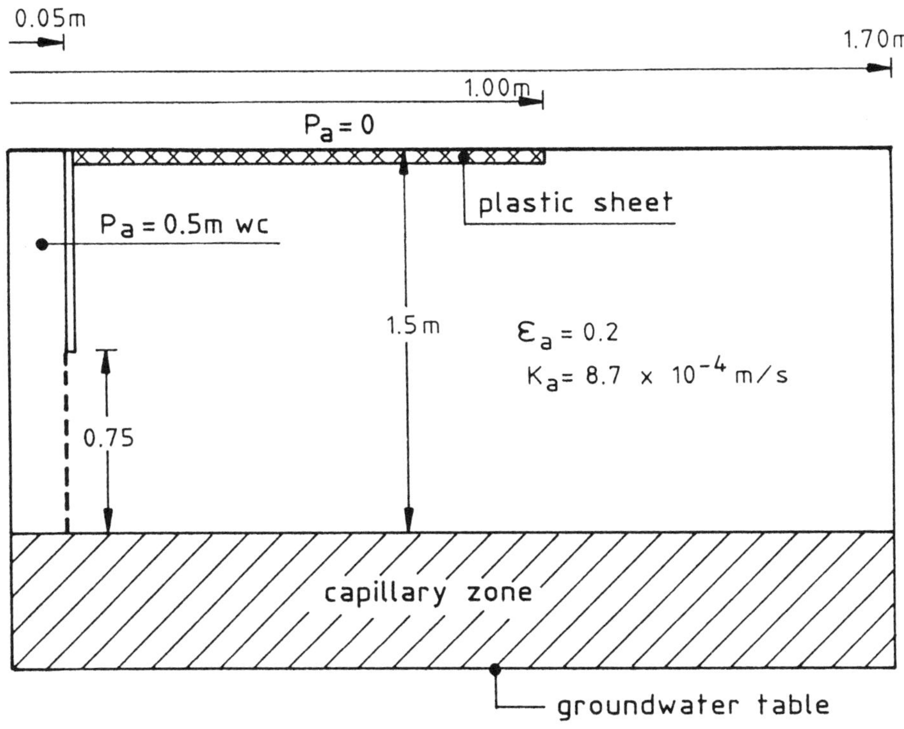

FIGURE 4. Computational scheme to calculate air transport in soil.

large-scale field experiment, is significantly lower than this figure (van Eyk & Vreeken 1989b).

Design Considerations

Extraction of Groundwater. From the investigations of degree and extent of contamination, it appeared that the concentration of contaminants in the upper sandy layer of the phreatic zone is much larger than in the lower parts of the aquifer. It was decided, therefore, to install two pumping wells (Figure 3). The first well has a filter between 1.60 m and 4.10 m BGS; the filter of the second pumping well was installed between 9.10 m and 11.10 m BGS. The shallow water is strongly contaminated; the deeper groundwater is only slightly contaminated. The flow of each pumping well can be controlled, which makes it possible to control the loading of the water treatment system.

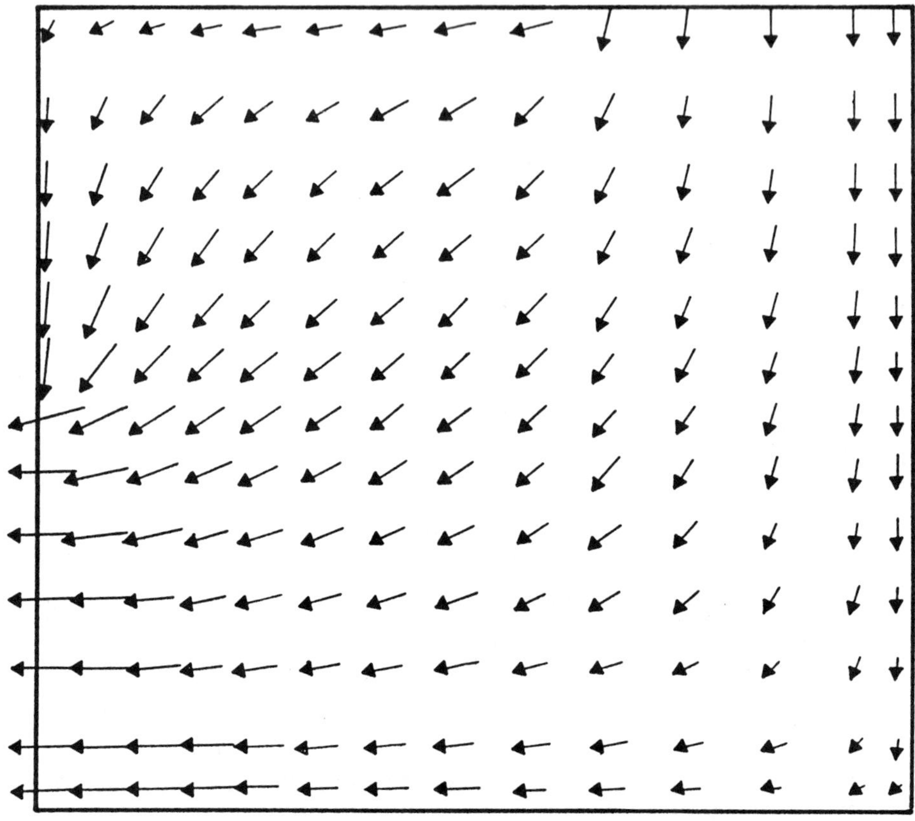

FIGURE 5. Schematic representation of calculated flow pattern toward an air extraction well.

Venting System. Pumping air wells were installed according to the results of the calculated design specifications. Air wells were connected to a pipeline that, in turn, was connected to a blower. Monitoring equipment was installed to control and measure gas flux and gas pressure. Because the ground surface has an impermeable paving (see below), air access is obtained via two horizontal drains installed below the paving (Figure 6).

Impermeable Paving. An impermeable paving is an integral part of a venting system for *in situ* remediation for several reasons. The venting system has to be protected from any form of adverse effect or damage, such as that caused by heavy trucks.

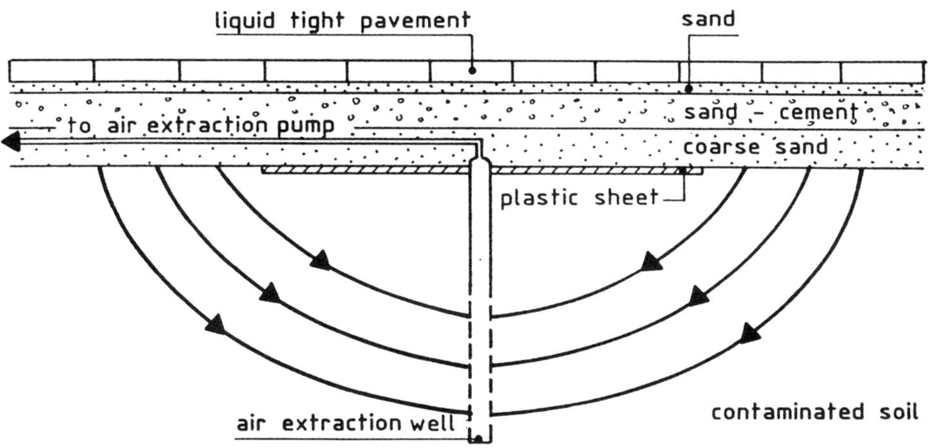

FIGURE 6. Outline of the design of air extraction wells under the liquid tight pavement.

In addition, because the retail gasoline station has ongoing business during the period of remediation, migration of fresh spilled products around the fuel supply pumps into the treated zone must be excluded.

Air and Groundwater Clean-up System. The clean-up system is shown in Figure 7. The groundwater, which is mainly contaminated with volatile hydrocarbons, is cleaned in a stripping tower. An oil-water separator was installed because analysis of boring B2 (Table 1, Figure 1) indicated the presence of free product floating on the groundwater table.

Preliminary investigations indicated that the soil contained a considerable amount of iron. As a result of aeration, this will produce iron hydroxyde. A sand filter was therefore installed after the cascade aerator.

The biofilter, installed to remove gaseous hydrocarbons from air extracted from soil, is also used for cleaning the air from the stripping tower.

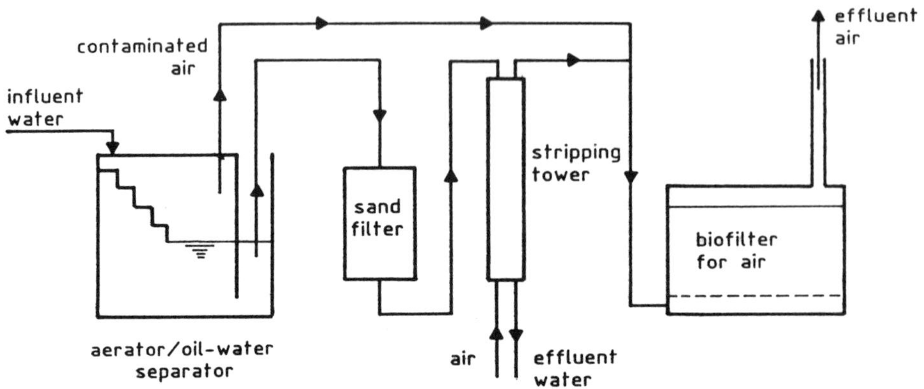

FIGURE 7. Schematic overview of the system for the treatment of polluted air and groundwater.

Monitoring and Control

Initial progress of the remediation process is monitored via vented air samples and groundwater samples. Soil samples are taken in the final stages of the clean-up process.

PRELIMINARY RESULTS

Air Extraction. To avoid the possibility of overloading the biofilter, only the extracted air from the soil was passed over the biofilter when the *in situ* clean-up operation was started in August 1990.

After several weeks of operation, air sampling showed the biofilter to be working effectively. The pumping wells were then activated, and air produced by the air stripper was passed into the biofilter. At the start of the operation, the air flux amounted to approximately 25 m^3/h. After 2 weeks it was increased to 38 m^3/h. Air transport calculations predicted a value of 68 m^3/h.

The removal of gaseous hydrocarbons from soil by venting is shown in Table 7. Over a period of 3 months, approximately 600 kg of hydrocarbon vapours were extracted from the subsurface. The results confirm results obtained earlier (van Eyk & Vreeken 1988), which showed that components like benzene are removed quickly and effectively. After 3 months, the benzene concentration is reduced to approximately 4 percent of its original value. For toluene and the xylenes, the figures are 8 and 23 percent, respectively.

TABLE 7. Hydrocarbons removed by the soil vacuum extraction system.

Date	Air Flux (m^3/h)	Concentration in Air (mg/L)				Total Hydrocarbons Removed (kg)
		Benzene	Toluene	Xylenes	CH—tot	
1990-08-22	23.2	2.09	4.14	1.65	29.32	124
1990-08-29	29.8	0.38	1.44	1.08	8.16	199
1990-09-04	38.4	0.14	0.88	0.77	4.50	239
1990-09-11	42.0	0.11	0.56	0.44	3.41	266
1990-10-11	35.8	0.08	0.32	0.46	2.60	352
1990-11-29	42.1	0.08	0.34	0.38	2.40	583

Groundwater Clean-up. Table 8 shows the hydrocarbon levels analysed in the two influents discharged into the clean-up system. The data show that groundwater extracted from the deep pumping well is only marginally polluted. Groundwater pumped from the shallow well, in contrast, appears to be considerably polluted.

Approximately 3 months after the start of the demonstration project in August 1990, hydrocarbon levels start to rise significantly. The rise appeared to be caused by free product entering the pumping well. Two months later, concentrations start to fall again, particulary those for benzene, toluene, and xylene. Hydrocarbon concentrations in the effluent were determined as well as maximum allowable levels for oil and BTEX (10,000 and 100 µg/L, respectively), prior to discharge into the sewer. Analyses show that these targets are being met by the groundwater clean-up system.

The Biofilter. Unfortunately, it is not possible to produce actual data on biofilter performance because local environmental authorities do not, as yet, impose quantitative restrictions on the amount of hydrocarbons that can be vented off into the atmosphere. All that is presently required is that the air vented off is odourless.

To get some idea of the performance of the biofilter, air samples were taken before entering and after passing through the filter with the help of so-called Dräger tubes. Because these tubes are designed to determine BTX compounds singly and not in combination, the results are not conclusive. They show that BTX compounds can no longer be detected after 3 weeks of operation. The biofilter volume amounted to 13.8 m^3. Assuming a 75 percent removal rate during the first 3 weeks,

TABLE 8. Deep groundwater clean-up (influent concentrations in µg/L; filter at 4 m BGS) and shallow groundwater clean-up (influent concentrations in µg/L; filter at 11 m BGS).

Date	Q[(a)]	COD[(b)]	Mineral Oil	Benzene	Toluene	Xylene
			Deep groundwater clean-up			
1990-10-12	25	66	6,900	1,600	2,100	2,100
1990-11-13	164	189	91,000	2,600	7,800	11,000
1990-12-11	104	116	72,000	1,400	5,300	8,000
1991-01-14	112	-	88,000	510	1,200	1,750
			Shallow groundwater clean-up			
1990-10-12	27	28	53	<0.2	<0.5	<0.5
1990-11-13	43	35	<50	<0.2	<0.5	<0.5
1990-12-11	80	24	<50	<0.2	<0.5	<0.5
1991-01-14	87	-	97	<0.2	<0.5	<0.5

(a) Flux in m^3/day.
(b) COD = chemical oxygen demand.

or 179 kg, and 100 percent removal for the remaining period, an average removal rate of 0.3 kg/m^3/day can be calculated (Table 7).

DISCUSSION

The demonstration project described in this paper aims to demonstrate for a real spill situation the feasibility of a two-pronged attack, venting and bioventing, for the *in situ* removal of volatile and semivolatile hydrocarbons from subsurface soil strata. The reported *in situ* and on-site subsoil and aquifer restoration is part of a wider concept, which we have proposed, namely to combine the Prevention, Isolation, and Sanitation (PRISAN) methodology into one integrated approach for *in situ* soil and on-site groundwater clean-up.

The basic idea of the concept is that, when all possible environmental risks and possible risks for exposure have been taken care of, it will be possible to apply relatively inexpensive *in situ* remedial techniques, which require considerable longer clean-up times than conventional techniques (excavation).

The bioventing approach was validated in a large-scale field experiment in 1984 and first reported in 1986 (Anon. 1986). From this experiment it transpired that carbon dioxyde production rates are not always a reliable indicator of biodegradation rates, as carbon dioxyde can react with carbonates present in soil to produce bicarbonates that disappear into the groundwater (van Eyk & Vreeken 1988). Because of the way in which that particular field experiment was designed and because we knew exactly the carbonate content of the soil prior to bioventing, it was possible to evaluate the contribution of biodegradation to total hydrocarbon removal.

In the present demonstration project, however, such an approach, both for practical and economic reasons, was not feasible. On the other hand, carbon dioxyde production rates may indicate the onset and course of biodegradation. For that reason, carbon dioxyde fluxes have been and will be measured at intervals. Carbon dioxyde levels, determined 3 months after the start, were approximately 1.0 ppm and subsided in the (very) cold winter to around 500 ppm (difference between the control and production wells). Based on an extraction rate of 40 m^3/h, these figures amount to the removal of approximately 0.4 to 0.2 kg carbon/day. These are fairly low figures, but we do not yet know how far carbon dioxyde is absorbed into the soil.

The clean-up system shown in Figure 7 was selected on the basis of cost-effectiveness, and on that basis, the use of activated carbon for air clean-up was rejected. It is anticipated that the integrated groundwater clean-up system will have to be in operation for a period of approximately 1 year. As soon as groundwater analyses show that the effluent complies with the standards set by the authorities, the clean-up system can be removed. If the targets have been met after 2 to 3 years, the air extraction system can be deactivated and remain installed as a standby and early warning system.

CONCLUSIONS

After all possible risks resulting from a spill were eliminated, results so far indicate that an *in situ* clean-up technique as a medium- to long-term strategy is a feasible methodology.

REFERENCES

Anonymous. "*In Situ* Reclamation of Petroleum Contaminated Sub-soil by Subsurface Venting and Enhanced Biodegradation." *Research Disclosure* **February 1986,** No. 26233, 92–93.

Beyer, W. "Zur Bestimmung der Wasserdurchlässigkeit von Kiesen und Sanden aus der Kornverteilungskurve." Berlin, *Wasserwirtsch Wassertechn.* **1964,** H6.

Hutzler, N. J.; Murphy, B. E.; Gierke, J. S. "State of Technology Review: Soil Vapor Extraction Systems"; final report to the U.S. EPA, Hazardous Waste Engineering Research: Cincinnati, OH, 1988.

Staatsuitgeverij Den Haag. Proceedings of a workshop, 20–21 March, 1986, Bodembeschermingsreeeks No. 9: *Biotechnologische Bodemsanering,* pp 31–33, Rapportnr. 851105002, ISBN 90-12-054133, Ordernr. 250-154-59; Staatsuitgeverij Den Haag: The Netherlands, 1986.

Strack, O.T.L. *Groundwater Mechanics;* Prentice-Hall: Englewood Cliffs, NJ, 1989.

Stroosnijder, L. *Infiltratie en herverdeling van water in grond* (English summary); ISBN 90-220-0596-8; PUDOC Centrum voor landbouwpublicaties en landbouwdocumentatie Wageningen: The Netherlands, 1976.

Tibbetts, P.J.C. "The Analysis of Oil in Sand from Four Lysimeters in Katwijk, The Netherlands"; a COOW-CONCAWE draft report, No. 2870/227/1/925; 1982.

van Eyk, J.; Vreeken, C. "Venting-Mediated Removal of Petrol from Subsurface Soil Strata as a Result of Stimulated Evaporation and Enhanced Biodegradation." *Med. Fac. Landbouww. Riiksuniv. Gent* **1988,** *53*(4b), 1873–1884.

van Eyk, J.; Vreeken, C. "Model of Petroleum Mineralisation Response to Soil Aeration to Aid in Site-Specific, *In Situ* Biological Remediation." In *Groundwater Contamination: Use of Models in Decision-Making,* Proceedings of an International Conference on Groundwater Contamination; Jousma *et al.,* Eds.; Kluwer: Boston/London, 1989a; pp 365–371.

van Eyk, J.; Vreeken, C. "Venting-Mediated Removal of Diesel Oil from Subsurface Soil strata as a Result of Stimulated Evaporation and Enhanced Biodegradation." In *Hazardous Waste and Contaminated Sites,* Envirotech Vienna, Vol. 2, Session 3, ISBN 389432-009-5; Westarp Wiss: Essen, 1989b; pp 475–485.

Vogel, T.H.M.; Grbić-Galić, D. "Incorporation of Oxygen from Water into Toluene and Benzene during Anaerobic Fermentative Transformation." *Applied Environ. Microbiol.* **1986,** *52,* 200–202.

Vreeken, C.; Sman, H. T. "The Use of A Hydrology Contaminant Transport Model for the Prediction of the Effect of Air-Stripping on the *In Situ* Cleaning of Contaminated Soil." In *Groundwater Contamination: Use of Models in Decision-Making,* Proceedings of an International Conference on Groundwater Contamination; Jousma *et al.,* Eds.; Kluwer: Boston/London, 1989; pp 329–327.

Zeyer, J.; *et al.* "Rapid Microbiol. Mineralisation of Toluene and 1,3-Dimethyl-benzene in the Absence of Molecular Oxygen." *Applied Environ. Microbiol.* **1986,** *52,* 944–947.

Soil Vapour Extraction of Hydrocarbons: *In Situ* and On-Site Biological Treatment

L.G.C.M. Urlings[*], *F. Spuy, S. Coffa, H.B.R.J. van Vree*
TAUW Infra Consult B.V.

INTRODUCTION

In industrial countries the extent of soil contamination is enormous, as indicated in the two examples presented in this paper. The overall cost of soil remediation in the Netherlands is estimated to exceed $30 billion; this corresponds to 150 million m^3 of excavated soil from more than 100,000 polluted sites ("Ten-Year Scenario" 1990). Based on studies by the U.S. Environmental Protection Agency (EPA), Miller *et al.* 1990 estimate more than 1 million leaky underground fuel tanks in the United States.

Certainly, contamination of soil and groundwater by hydrocarbons is the most important contamination, based on quantity. Miller *et al.* (1990) describe, for the United States, that 90 percent of the transported hazardous material consists of gasoline, fuel oil, and jet fuel. In this article the remedial action approach is focused on hydrocarbons, particularly toluene (Site 1) and gasoline (Site 2).

***In Situ* Techniques Versus Excavation.** Excavation of contaminated soil is a very effective method of removing pollution. Treatment or disposal of excavated soil is necessary, and in most cases additional

[*] TAUW Infra Consult B.V., P.O. Box 479, 7400 AL Deventer, The Netherlands

groundwater treatment is also needed. Nevertheless, excavation can be difficult or even impossible under certain circumstances:

- The presence of buildings and civil engineering works (roads, bridges, etc.)
- Certain cables, power lines, and pipelines in the subsurface
- Very deep contamination (i.e., >4 m)
- Shortage of space and traffic problems (city centres)
- Interference with an irreplaceable function of the site (i.e., railway station)

Generally, the criteria favourable for applying *in situ* techniques are the following:

- Only one contaminant (*e.g.*, toluene) or comparable components (*e.g.*, gasoline) are present.
- The quantity of contaminated soil is substantial.
- The contaminant can be biodegraded.
- The contaminant can be leached and/or volatilized.
- The soil is reasonably permeable.
- Less disturbed layers of clay or peat appear in the subsoil.
- The contamination is infiltrated (i.e., not buried).

Above- and underground infrastructure often makes a site unsuitable for conventional excavating techniques; this occurs especially in city centres and industrial areas. Therefore, more effort has to be put into developing innovative remedial techniques.

This article focuses on the soil vacuum extraction technique in general and air-based biodegradation in particular. A newly developed treatment attending to both soil gas and groundwater will also be discussed.

REMEDIAL TECHNOLOGY

Soil Vapour Extraction (SVE). The mechanism by which SVE operates is relatively simple. By creating negative pressure gradients in a series of zones within the unsaturated soil, a subsurface airflow is induced (Figure 1). This flow volatilizes the contaminants present in the unsaturated soil. This process, in theory, continues until all volatile components are removed. The extraction wells are individually connected to the transfer pipes, then manifolded to a vacuum unit, and the soil vapour is transported to a soil vapour treatment system. Figure 1 gives an outline of three different SVE applications.

The withdrawn soil vapour is often treated by charcoal adsorption or catalytic incineration. The groundwater is usually treated by stripping and/or charcoal adsorption. To minimize treatment costs of both groundwater and soil vapour, TAUW Infra Consult B.V. has applied a biological system for combined aerobic treatment since 1989.

SVE and Related New Technologies. Experience with SVE is so widespread that it can be considered a proven technology. A review of SVE techniques is given by Hutzler *et al.* (1989). The treatment costs of withdrawn soil vapour are substantial, usually more than 50 percent of the total remediation costs (Miller *et al.* 1990). Air-based biodegradation is applied with the aim of reducing SVE costs. This is a new and innovative technique, as indicated by an international review of *in situ* bioreclamation practice (Staps 1990). Only two of the 23 studied sites used air-based technology; most of the applied bioreclamation was water-based. In recent literature (Hinchee & Miller 1990; Miller *et al.* 1990), air-enhanced biodegradation is described.

Apart from the mentioned advantage of lower hydrocarbon concentration in the withdrawn soil gas, substantially less carrier medium is needed when air is used as an oxygen carrier, as indicated in Table 1. An additional advantage of using air as an oxygen source in the gas-phase vadose zone is the fact that diffusion rates are much higher in air than in the water-filled saturated zone.

Two sites equipped with SVE systems will be discussed in this paper. Site 1 is an industrial site contaminated with toluene. The gasoline contamination in Site 2 is situated under a road.

Estimated Duration of SVE. To select the most suitable remedial action technique for a specific site, it is necessary to estimate the costs of several techniques and to know the required duration of SVE treatment. For this purpose, a relatively simple spreadsheet model was

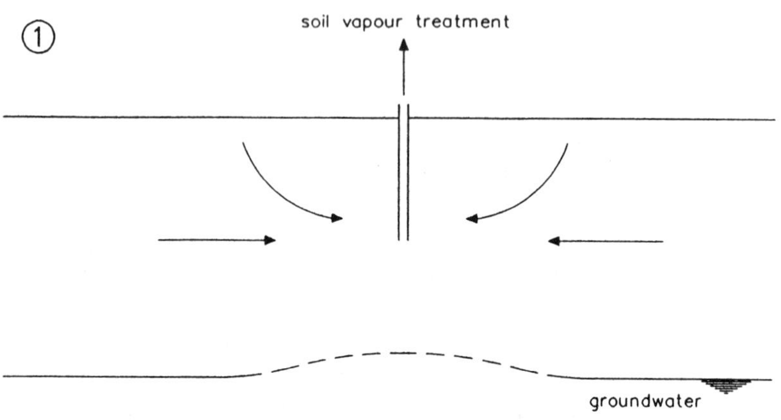

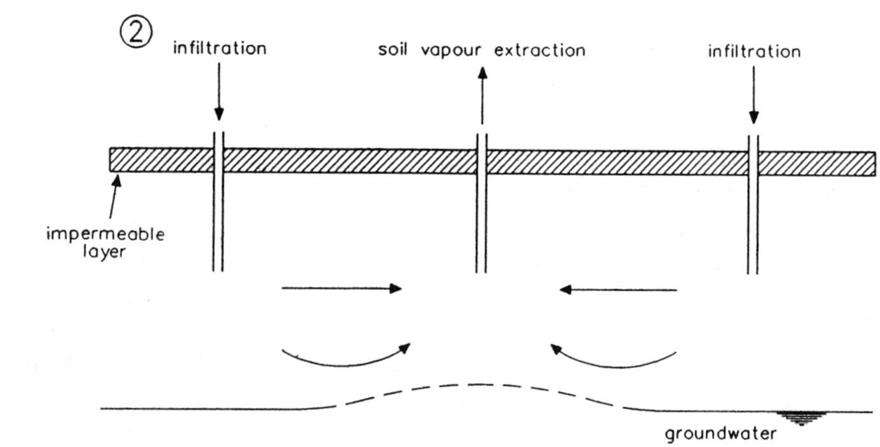

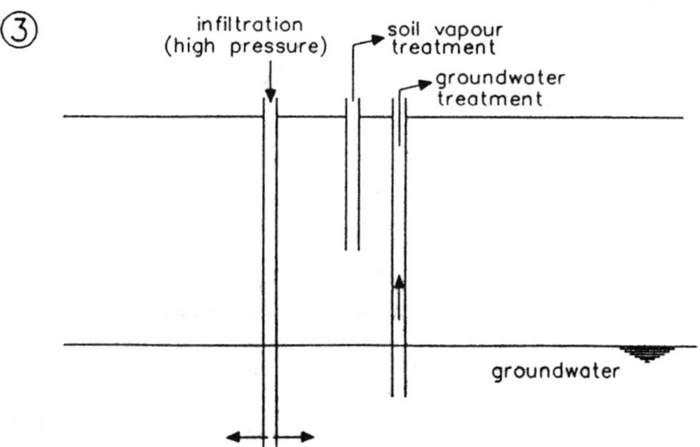

FIGURE 1. Three different applications of SVE.

TABLE 1. Mass requirements to deliver sufficient oxygen or nitrate for biodegradation to mineralization.

Electron Acceptor	Carrier Medium	Mass requirements (kg carrier/kg hydrocarbon)
Oxygen (in air)	8.0 mg/L in water	400,000
Oxygen (pure)	40.0 mg/L in water	80,000
Oxygen (in H_2O_2)	100.0 mg/L H_2O_2 in water	65,000
Oxygen (in H_2O_2)	500.0 mg/L H_2O_2 in water	13,000
Nitrate	50.0 mg/L in water	90,000
Nitrate	300.0 mg/L in water	15,000
Oxygen	20.9 percent in air	13

Source: Hinchee & Miller 1990.

developed. The most important input parameters fall into several major categories:

Contaminants

- Estimated amount
- Molecular mass, vapour pressure, and solubility (three maximum)
- Adsorption/desorption coefficients (k_{oc})

Soil

- Volume of contaminated soil
- Density, porosity, and moisture
- Fraction of organic carbon in the soil

Biodegradation

- Zero order biodegradation rate

Application of SVE

- Estimated effective airflow in the subsoil

The model can be validated by column testing with contaminated soil from the site. Column testing can give detailed information on volatilization, biodegradation, and possible remedial action limit concentrations.

SITE 1 CHARACTERISTICS

The contamination was caused by the spillage of solvents (mainly toluene) from a paint factory, probably between 1959 and 1978. The groundwater was heavily contaminated with toluene. Because the site is situated close to drinking water extraction wells, this project received a high priority in 1988. Remedial action was initiated in the beginning of December 1989.

Approximately half of the contaminated area is situated underneath a building. The groundwater table is about 7 m below ground level. The unsaturated zone of the soil consists of fine sand to gravel. The aerodynamic conductivity of this layer in a vertical gradient was estimated at 70 to 90 m/day. The soil is heavily contaminated with toluene and includes minor amounts of other aromatic hydrocarbons, such as benzene and xylenes. The highest toluene concentration appeared to be 2,200 mg/kg (dw).

Three remedial action techniques were studied: excavation, soil vapour extraction, and soil flushing. For financial and practical reasons, soil vapour extraction combined with groundwater sanitation was considered the best method. The location is shown in Figures 2 and 3 in both horizontal and cross-sectional views.

Under the building next to the extraction wells (perforated 2 to 3.5 m and 5 to 6.5 m below ground level), infiltration wells were installed (perforated 2 to 6.5 m below ground level). Outside the building only extraction wells were installed (perforated 5 to 6.5 m below ground level). All the wells could be separately monitored for both concentration and pressure. Pressure was applied at approximately 30 to 80 millibar with an airflow of approximately 150 m^3/h. (Hydraulic residence time in the soil vapour is approximately 0.5 h.) The soil vapour is treated using activated carbon filtration. The average groundwater withdrawal was 10 m^3/h. After stripping, the exhaust gas was treated in the activated carbon unit.

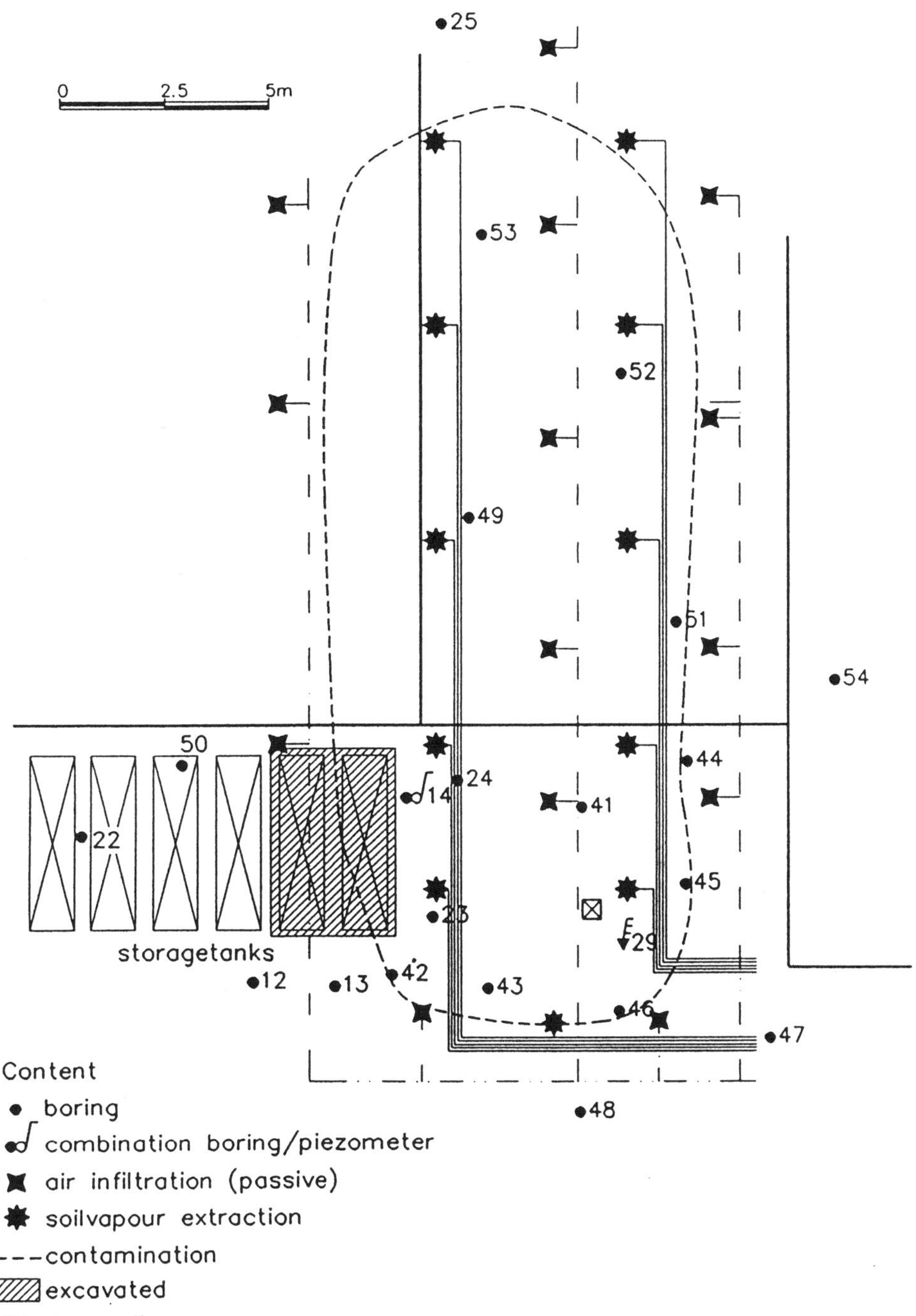

FIGURE 2. Horizontal view of Site 1.

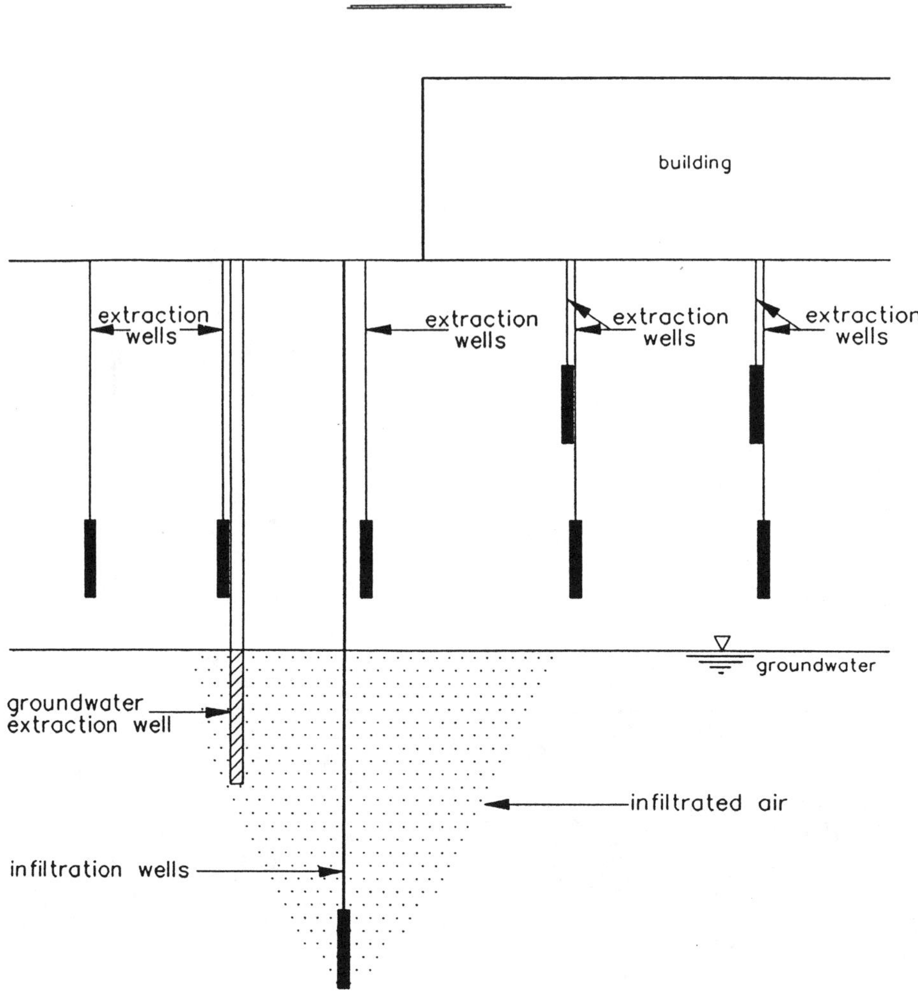

FIGURE 3. Cross section of Site 1.

Results and Discussion. The results of soil vapour extraction are given in Figures 4 and 5. Within 4 months, approximately 580 kg of toluene was withdrawn using the SVE system. Concentrations of up to 8,000 mg toluene/m^3 were measured (up to 40 g/m^3 in a specific extraction well) in the withdrawn soil vapour. At 3 and 6 months, the soil was sampled at almost exactly the same spot prior to remediation. The results are given in Table 2.

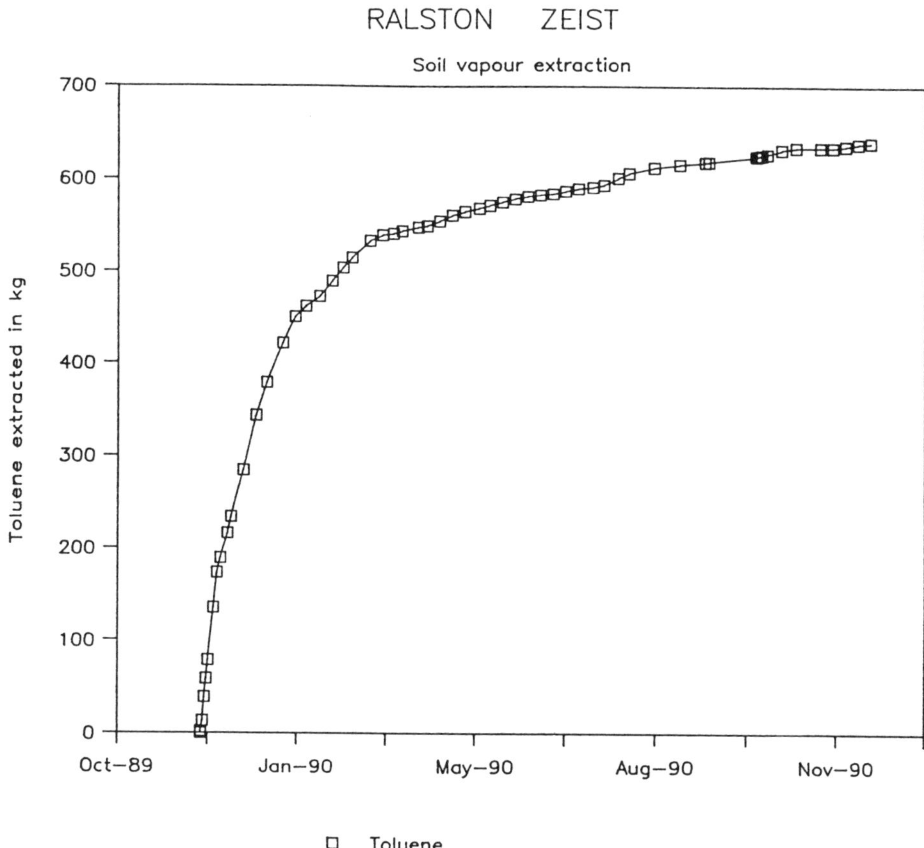

FIGURE 4. Total amount of toluene withdrawn.

Additional Investigations during Remedial Action. Apart from the usual analytical and supervisory activities during remedial action, special attention was focused on three areas:

- Modelling of SVE remedial action duration. As previously mentioned, several factors determine SVE duration. The input parameters were evaluated during remediation. Toluene removal flow was more or less as expected for the first months. Thereafter, the toluene extraction was influenced mainly by mineral oil contamination (floating layer).

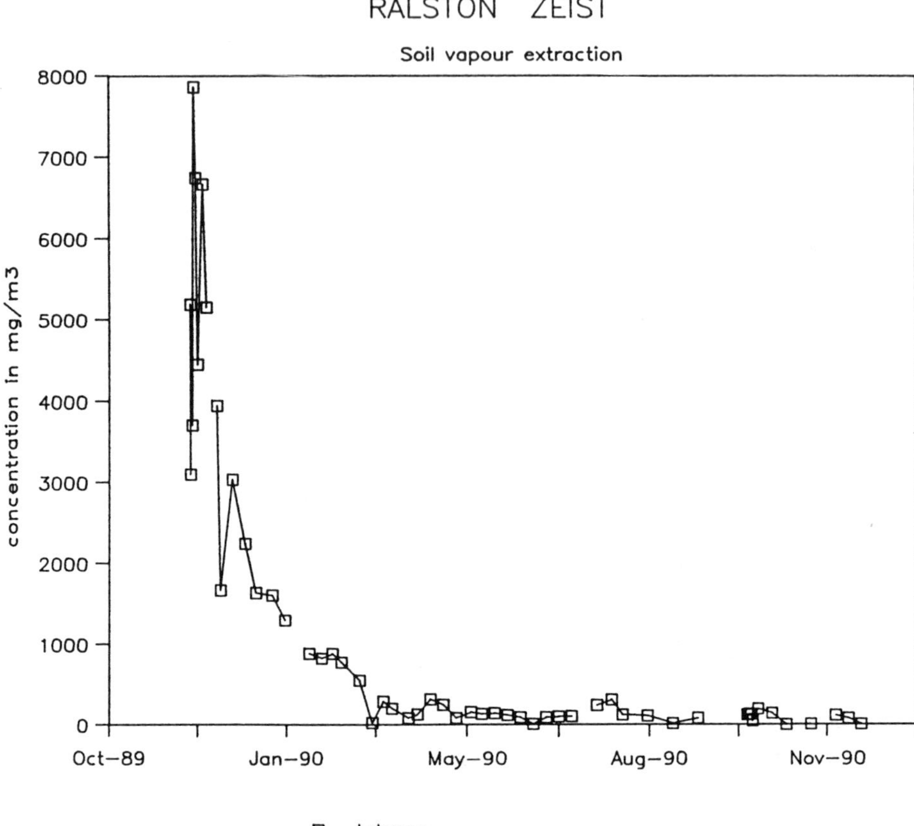

FIGURE 5. Concentration of toluene in the withdrawn soil vapour.

- Horizontal versus vertical aerodynamical conductivity of the soil. The aerodynamic conductivity of the contaminated layer in vertical gradient was about 70 to 90 m/day, compared with 150 m/day in the horizonal gradient. These numerical values result from airflow and tracer (helium) velocity measurements in relation to the applied negative pressure gradient in the subsoil.

- Measurement of bacterial activity (*e.g.*, counting, oxygen and carbon dioxide sampling). The volume of the contaminated layer is about 900 m^3 (about 1,500 tons). The oxygen consumption amounts to 0.3 to 0.5 kg/h, and the carbon dioxide production was 0.3 to 0.4 kg/h. Consequently, the

TABLE 2. Toluene concentrations in the soil.

Location	Depth (m below ground level)	t = 0 (mg/kg dw)	t = 13 weeks (mg/kg dw)	t = 31 weeks (mg/kg dw)
215–12	5.0–5.5	–	–	<0.05
215–12	5.5–6.0	530	1.5	<1.00
215–13	6.0–6.5	2,200	3.8	<1.00
	6.5–7.0	–	6.4	7.00
207–12	5.5–6.0	310	<1.0	<0.05
207–13	6.0–6.5	1,100	<1.0	<1.0
	6.5–7.0	<1		

rate of biodegradation of toluene was estimated to be approximately 2 mg C/kg/day.

Progress. In October 1990, an air injection system was installed at 14 m below ground level to enhance groundwater sanitation. The toluene concentration in the groundwater was lowered from 14,000 µg/L (November 1989) to 3,000 µg/L (October 1990). The injection flow was 50 Nm^3/h, and the pressure was 3 bar. An increase in toluene concentration in the vapour gas could not be observed (within the sampling error; a decrease of 40 µg/L/h corresponds to an increase of 10 mg/m^3).

In December 1990, the air injection was stopped mainly because of decreased efficiency and problems related to another contaminant present in the groundwater from a different origin. Air injection performance was studied successfully by the tracer experiments.

SITE 2 CHARACTERISTICS

During soil sanitation at a petrol station, contaminants were found underneath a provincial road. Excavation of the contaminated soil was not feasible for financial and technical (traffic) reasons. The most favourable solution was an SVE system in combination with biostimulation. This system not only had to remove the volatile compounds from the gasoline, but also had to stimulate biodegradation of (particularly) the nonvolatile components caused by the (passive) infiltration of air (oxygen). The unsaturated zone of the soil consisted of fine sand to gravel. The groundwater surface had to be lowered from 2 to 3 m below ground level to enlarge the unsaturated zone and make the smear zone available for SVE. Figure 6 shows the location.

Seven soil vapour extraction wells (perforation 2 to 2.75 m below ground level) were installed on one side of the road and seven infiltration wells (passive) on the other side. To prevent direct air infiltration at the extraction side of the road, a plastic liner was placed between the road and the sheet pile wall.

Results and Discussion. In Figure 7 the removal results are compiled based on soil vapour analyses. Based upon the speed of oxygen consumption and carbon dioxide production, the biodegradation rate of gasoline can be estimated at 7 mg/kg soil/day (3 sampling days, weeks 30 to 50). The *in situ* biodegradation seems to take place by zero order kinetics.

Hydrocarbon concentrations in the withdrawn soil vapour varied enormously. Initial concentrations of up to 80 g/m^3 were measured. After 20 weeks, the concentration dropped to 3 g/m^3 and gradually decreased until 240 mg/m^3 (week 60). The dimensions of the soil vapour treatment system allowed the withdrawn flow of the soil vapour to be gradually increased from 25 Nm^3/h to 50 Nm^3/h in week 36 and to 63 Nm^3/h in week 46. The hydraulic residence time of (ambient) air in the soil is approximately 1 h.

Progress. After being operational for more than a year, some soil samples show relatively high gasoline concentrations. This might result from soil heterogeneities or aerodynamically stagnant areas. To hasten the remedial action, the groundwater level will be discontinuously lowered and the flow rate of the separate soil vapour extraction wells will be varied. Additionally, heated air (30 to 40 C) will be introduced on the other side of the road.

COMBINED ON-SITE AIR AND WATER TREATMENT

A sludge or carrier type of bioreactor can mineralize the vapours in the withdrawn soil vapour as well as the dissolved contaminants in the pumped groundwater. The first treatment unit with an efficient volume of 10 m^3 was tested at site 2 and functions well. Retention time is

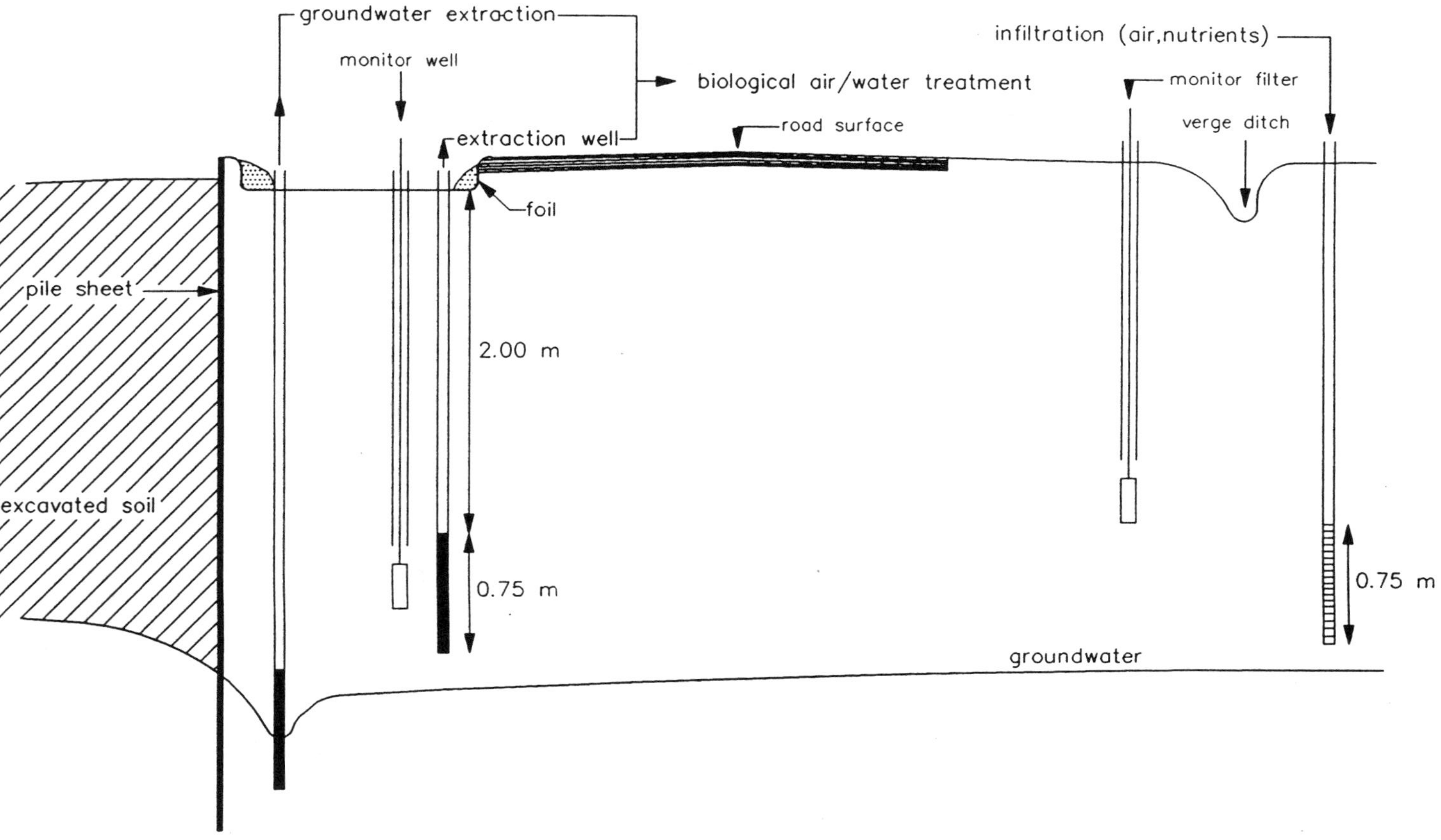

FIGURE 6. Cross section of Site 2.

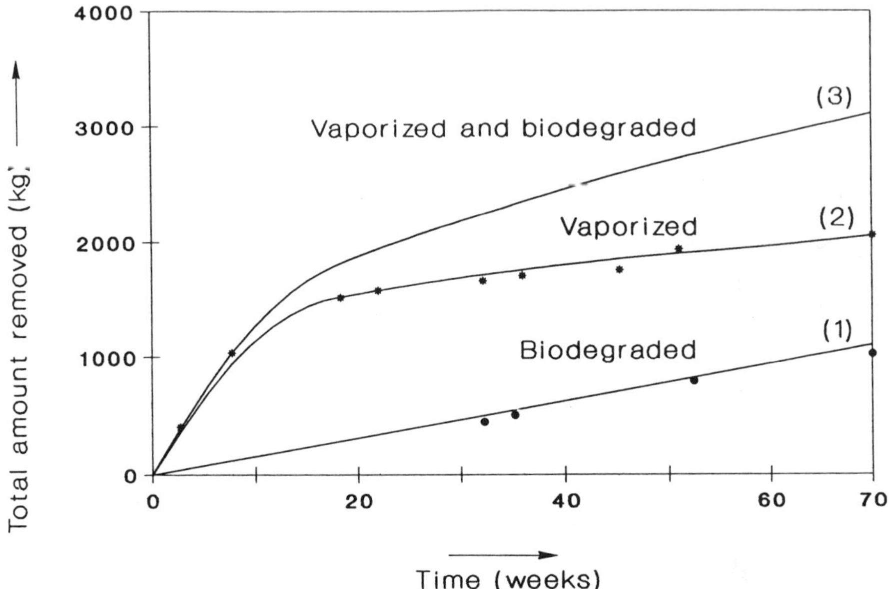

FIGURE 7. Cumulative amounts of gasoline removed during soil vapour extraction.

approximately 15 min for groundwater and less than 10 min for soil vapour. The addition of N and P nutrients takes place in the influent stream. In different compartments of the reactor, ambient air is supplied. Mixing in the reactor can be described as plug flow. The treatment efficiency of the total load (soil, air, and groundwater) exceeds 98 percent. All the requirements regarding groundwater discharge are met: 100 μg aromatics/L and 1 mg mineral oil/L. In the search for metabolites with coupled gas chromatography/mass spectrometry techniques, no components could be detected at the μg/L level.

In the purified soil vapour (exhaust gas), no aromatics or other volatile carbon dioxides were detected (detection limit 0.1 ppm). In Figure 8 the input and output of the on-site treatment unit are outlined. Figure 9 presents chromatograms of GC analyses. During 1 year the combined air and groundwater treatment showed that the biotreatment system is reliable, even in winter (–10 C). Maintenance requirements were negligible. A real advantage for the environment is the breakdown, rather than concentration or replacement, of contaminants. The latter often occurs in physiochemical techniques.

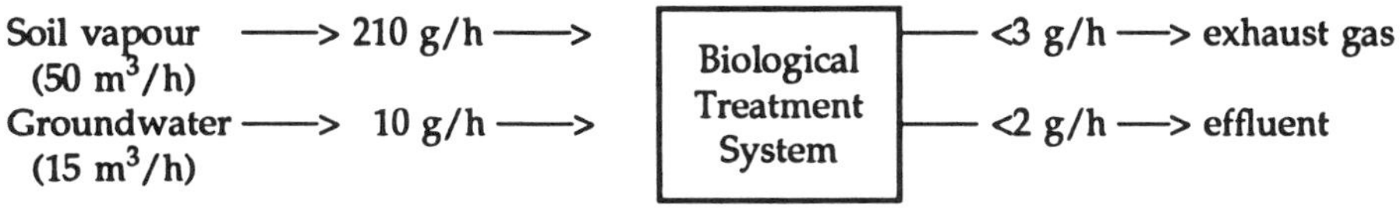

Soil vapour ——> 210 g/h ——> | Biological Treatment System | —— <3 g/h ——> exhaust gas
(50 m^3/h)
Groundwater ——> 10 g/h ——> | | —— <2 g/h ——> effluent
(15 m^3/h)

FIGURE 8. Gasoline mass balance at steady state.

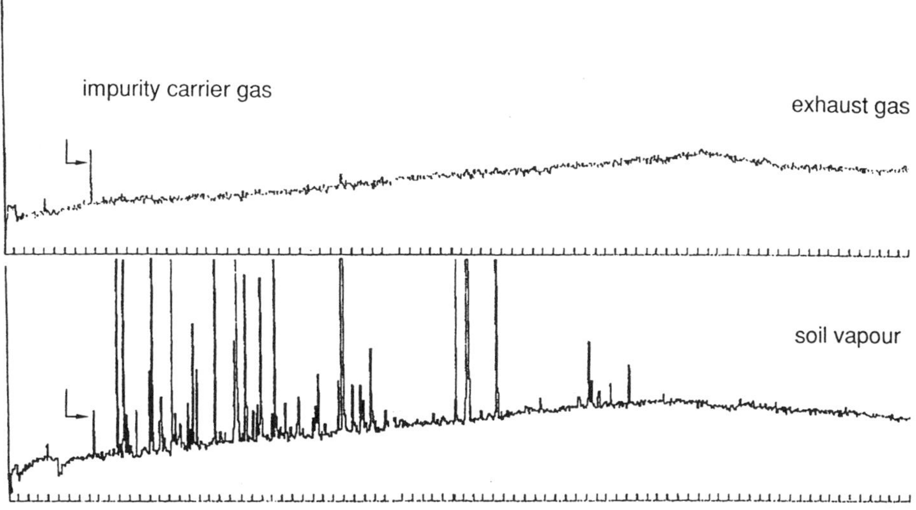

FIGURE 9. Chromatogram of soil vapour (untreated) and exhaust gas (after biological treatment).

RECOMMENDATIONS

In soil remediation the application of the *in situ* SVE technique can often be a good alternative to conventional remedial action techniques. The removal of volatile contaminants can be done effectively even in less permeable soils, such as loamy sands. The SVE technique becomes even more feasible when air-enhanced biodegradation is applied. Important advantages of *in situ* SVE biodegradation are the reduced costs for on-site treatment for withdrawn soil vapour and the breakdown of less volatile (hydrocarbon) contaminants in the subsoil.

For an accurate estimate of the costs involved for the SVE, it is important to calculate tretment duration. The application of computer simulations are necessary, and more attention should be paid to the further development of these models. Column testing with contaminated soil from the site is recommended to validate the simulation.

Treatment costs of the withdrawn soil vapour are substantial; therefore, the search for alternative techniques must continue. TAUW Infra Consult B.V. has successfully applied a combined soil vapour and groundwater treatment based on biodegradation at a gasoline-polluted site. For soil contaminated with hydrocarbons, SVE combined with on-site air and water treatment can be an efficient and effective remedial action strategy.

REFERENCES

Hinchee, R. E.; Miller, R. N. "Bioreclamation of Hydrocarbons in the Unsaturated Zone." In *Hazardous Waste Management of Contaminated Sites and Industrial Risk Assessment*; Pillman, W.; Zirm, K., Eds.; Vienna, 1990; pp 641–650.

Hutzel N. J.; Murphy, B. E.; Gierke, J. S. *State of Technology Review, Soil Vapour Extraction System*; U.S. Environmental Protection Agency, 1989; EPA/600/2-89/024.

Miller, R. N.; Hinchee, R. E.; Vogel, C. M.; Dupont, R. R.; Downey, D. C. "A Field Scale Investigation of Enhanced Petrol Hydrocarbon Biodegradation in the Vadose Zone at Tyndall AFB, Florida," *Proceedings*, NATO/CCMS meeting, France, December 1990.

Staps, J. J. "International Evaluation of *In Situ* Biorestoration of Contaminated Groundwater"; Report 738708006; National Institute of Public Health and Environmental Protection; The Netherlands, 1990.

"Ten-Year Scenario, Soil Remediation"; Stuurgroep Tien Jaren-Scenario Bodemsanering, Ministerie van VROM, 1989.

Potential Applications for Monitoring Remediations in Australia Using Geoelectric and Soil Gas Techniques

G. B. Davis[*], *C. Barber*
CSIRO Division of Water Resources
G. Buselli
CSIRO Division of Exploration Geoscience
A. Sheehy
Canberra University

INTRODUCTION

Contamination of soil and groundwater by industrial, agricultural, and urban development in Australia has been recognised for some time (Jacobsen & Lau 1988). With increasing public interest in the environment, and particularly with the need for expansion and redevelopment of major cities, remediation of contamination is fast becoming of major significance.

In Western Australia, Perth relies on groundwater resources in an unconfined aquifer on the Swan Coastal Plain for 35 percent of its domestic water supplies (Cargeeg *et al.* 1987). There are also at least 60,000 private wells within the metropolitan area, which are mainly used for irrigation. A wide range of wetland habitats also rely on groundwater for their existence. The city and its suburbs are located directly over the aquifer, which consists of coastal sands and limestones same 30 to 40 m thick bordering the Indian Ocean.

---

[*] CSIRO Division of Water Resources, Private Bag P.O., Wembley, Western Australia 6014, Australia

Expansion of the urban area across the coastal plain has led to further moves to protect the groundwater resource. The need for remediation of contamination from past industrial and domestic waste disposal and from leaking underground petroleum fuel tanks is becoming increasingly apparent. Also recognised generally in Australia (Swinton 1989), as overseas (Grady 1990), is a need for research on remediation technologies, particularly *in situ* bioremediation of groundwater polluted by petroleum fuels.

Effective monitoring is crucial to the success of *in situ* groundwater treatment methodologies. Invariably, this involves the use of extensive networks of observation boreholes, which are expensive to install and operate. Our recent research has evaluated the use of remote (surface) techniques of monitoring groundwater quality, using geoelectric sounding and soil gas sampling and analysis. Such techniques potentially provide a qualitative measure of the extent of residual groundwater contamination during remediation. These would be appropriate where remediation changes the formation resistivity, for example with abstraction of polluted water or following mineralisation of organic pollutants, which increases the electrical conductivity (EC) of groundwater.

GEOELECTRIC SOUNDINGS

Geoelectric techniques are often used to identify plumes in aquifers (Davis *et al.* 1988; Evans & Schweitzer 1984; Mazac *et al.* 1987). Invariably, contamination shows up as a contrast in groundwater conductivity affecting overall formation resistivity. Transient electromagnetic (TEM) sounding using the early-time variant of SIROTEM a TEM instrument which was developed by CSIRO in Australia, has been evaluated for measuring changes in formation resistivity in shallow aquifers (Buselli *et al.* 1990). Although used for identification of groundwater contamination, rarely have electromagnetic techniques been used for monitoring contamination changes in the subsurface over time, such as during remediation.

Early-time TEM and direct current (DC) resistivity soundings have been used to monitor remotely the changes with time of a groundwater pollution plume at the Morley landfill site. The Morley landfill overlies the sand aquifer on the Swan Coastal Plain north of the Swan River in Western Australia. Solid domestic refuse has been disposed at the site since 1980, and leachate has produced a plume of contamination extending nearly one kilometre from the site in the direction of groundwater flow. The site topography varies with natural dune development and sand excavations. The sand aquifer consists of an unsaturated zone 10 to 40 m thick and a saturated zone 25 to 40 m thick (Salama *et al.* 1989). The sand aquifer is underlain by Cretaceous clays of the Osborne Formation.

TEM and DC surveys (Figure 1) have been conducted at six-month intervals over a four-year period, with TEM soundings taken along transects across the direction of groundwater flow downgradient from the site. A range of techniques for interpretation of TEM sounding data have been evaluated (Barber *et al.* 1991). Early efforts modelled the underlying aquifer as a two-layered stratigraphic system; layer 1 being the unsaturated and saturated sand zones, and layer 2 the basement clays of the aquifer. Two-layer numerical inversion of the TEM sounding data gave good resolution of the clay basement (low resistivity layer 2) and showed a resistivity low for the aquifer (layer 1) extending approximately from borehole M4 southwest through M10 (see Figure 1). The position of the low resistivity coincides with the centre of the plume.

Some inconsistencies were evident for the two-layer inversions. Close to the wastes, estimated resistivities from the model inversions are much higher than expected from the electrical conductivity of sampled groundwater (*e.g.*, at borehole M5). Also, low resistivity values from the two-layer inversions appear to be associated with topographic low points as well as the pollution plume. The unsaturated zone therefore appears to have a significant effect on the TEM response.

To account for the effects of topographic variations on inverted resistivity results for the saturated aquifer, three-layer numerical inversion of the TEM sounding data was undertaken. Layer 1 was taken to correspond to the unsaturated zone, layer 2 to the saturated sand and layer 3 to the basement clay. To reduce further the error in the inverted results, the layer 1 thickness was fixed to the thickness of the unsaturated zone. The depth was determined by surveying in each of the sounding locations and measuring borehole water levels. Additionally, the resistivity of the unsaturated zone (layer 1) was assumed to be that obtained from DC-resistivity soundings.

Figure 2 shows variation with time in resistivities of the saturated zone of the aquifer (layer 2) after three-layer inversion for surveys carried out in October 1988, May 1989, and November 1989. The formation resistivities show no obvious association with topography. A gradual decrease in the resistivities can be seen over time along the G, U, and L lines with an expansion of the areal extent of the low resistivity zone towards the southwest. Groundwater sampled from M1O,

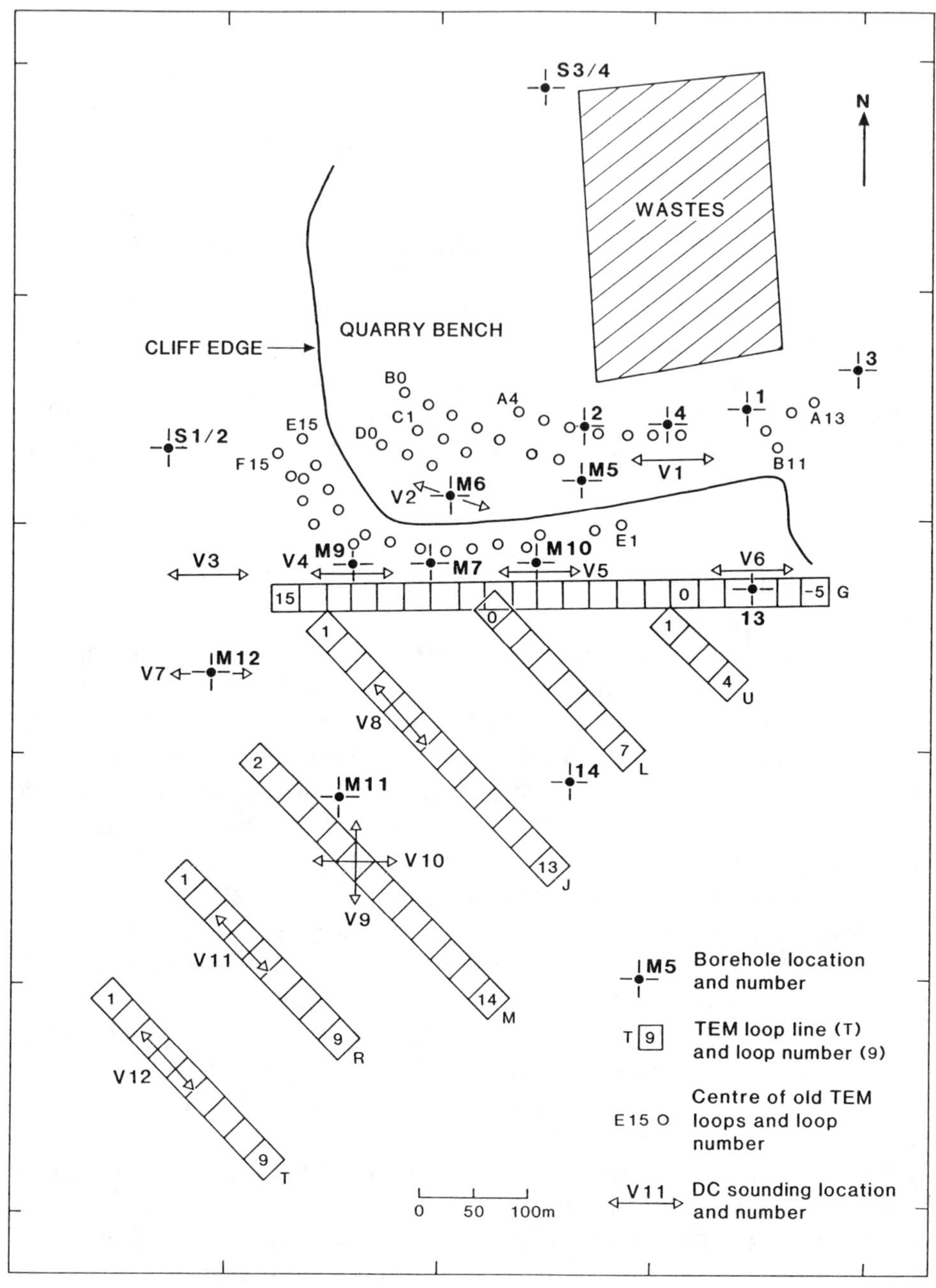

FIGURE 1. The Morley sand quarry and landfill showing the location of boreholes and DC-resistivity and TEM soundings.

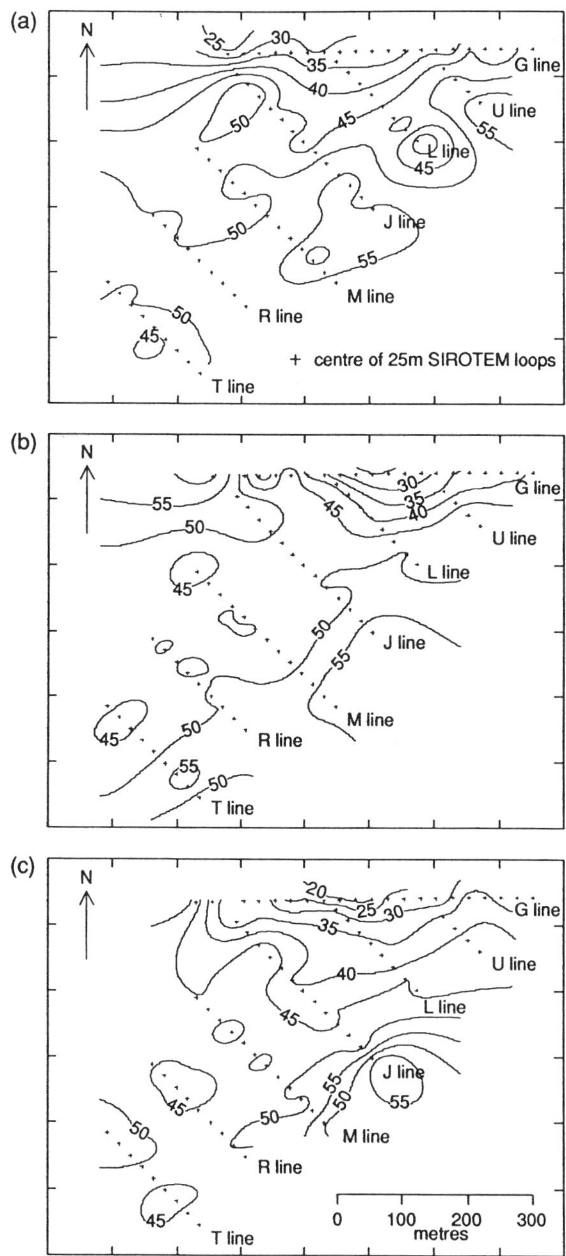

FIGURE 2. Formation resistivities (in ohm.m) of the saturated zone of the sand aquifer at Morley, given by a three-layer inversion model of TEM soundings taken in (a) October 1988, (b) May 1989, and (c) November 1989.

close to the G line, has shown a marked increase in the EC over this same period of time, from 240 to 340 mS/m (mmho/m) in September 1988 to 600 to 800 mS/m in November 1989. Conversion of these EC values into formation resistivities via Archie's law yields resistivities comparable to those obtained via TEM soundings (Barber *et al.* 1991). No increase in groundwater EC was observed at borehole M11, close to the M line where little decrease with time in resistivity was observed (Figure 2).

The results for November 1989 suggest the extent of the plume is limited to resistivity contours of less than 45 ohm.m. There are, however, significant problems in accurately defining the plume boundaries. This is thought to result from the large volume of the aquifer contributing to the response measured at the ground surface. Each sounding was thus an integrated response from a large volume of the plume. Near the plume edges, both polluted and unpolluted portions of the aquifer contributed to the response.

SOIL GAS SURVEYING

Recently, soil gas monitoring techniques have been used to assess remotely the extent of groundwater pollution by volatile organic contaminants (Marrin & Thompson 1984; Thompson & Marrin 1987) and by waste leachates (Barber & Davis 1986; Barber *et al.* 1990). The technique has also been used successfully to identify leakages of petroleum fuel from service stations (Glaccum *et al.* 1983). The technique consists of sampling and analysis of soil gas for volatile constituents or trace gases at shallow depths (usually 1.5 to 2 m below surface) and inferring the pollution of underlying soil or groundwater. These techniques have shown some success, although the processes controlling the loss of volatiles from groundwater and movement in the unsaturated zone are still poorly understood. Here we give details of two field studies that were carried out in Perth, Western Australia, to determine the usefulness of this approach as a remote monitoring technique for detecting contaminated leachate movement (at the Morley site) and fuel hydrocarbon contamination of groundwater.

At the Morley landfill (Figure 1) variations in methane concentrations in groundwater and soil gas were determined in detail over a two-year period. An initial investigation consisted of a survey of methane concentrations in soil gas at shallow depth using a spear probe (Barber & Davis 1986). This was carried out in an area adjacent to the landfill within the excavated sandpit 10 m above the groundwater, and on an undisturbed dune rise nearly 35 m above the water table. Results are shown in Figure 3 from Barber *et al.* 1990. These show extreme ranges in concentration, from 50 percent by volume close to the wastes to below parts per million by volume (ppmv) levels on the higher ground. The distribution of methane at 2 m depth successfully indicated the general direction of leachate movement in groundwater, but did not identify the full extent of contamination. The latter was determined by sampling from boreholes, using SIROTEM soundings and from pollution plume modelling (Figure 4 and Davis *et al.* 1988).

Following this initial survey, permanent gas sampling lines (3 mm o.d. nylon tubing) were installed at ten sampling stations to determine concentration-at-depth profiles of methane in the unsaturated zone. Additionally, to relate methane profiles in the unsaturated zone to dissolved concentrations in groundwater near the water table, two stations (M4 and M7 in Figure 4) were chosen for detailed monitoring of groundwater. Here, distribution of dissolved methane at the water table and in the capillary fringe was determined by using diffusion cells (Barber & Briegel 1987).

Figure 4 summarizes the variation in methane concentrations in soil gas taken from close to the water table for all ten sampling stations. Background concentrations of methane in soil gas are between 0.2 and 1.7 ppmv. Above-background concentrations occur to the southwest of the site and immediately upgradient to the northeast. These correlate closely with the determined extent of polluted groundwater to the southwest of the site in the general direction of groundwater flow. Lateral advective fluxes of methane can also be significant near the landfill. However, there was a general correspondence between high methane concentrations determined in soil gas and contaminated groundwater. Most of the variability with time was considered to be the result of seasonal changes in soil moisture content. Nevertheless, the use of soil gas techniques for monitoring temporal changes in groundwater contamination is thus of dubious value. Soil gas surveying can provide an effective indicator of the likely occurrence of leakage of hydrocarbons or other volatiles (*e.g.*, PCE or TCE) from underground storage tanks (Glaccum *et al.* 1983). As part of a wider study of the effects of urbanisation in Perth on groundwater quality, forty service station sites were surveyed by sampling soil gas to assess the extent of hydrocarbon leakage.

To obtain a soil gas sample, a stainless steel probe is inserted to a depth of 1.5 m below the ground surface into available grassed areas around the service stations. A 50-mL sample of soil gas is taken from the spear probe, following purging, and organic compounds in the gas

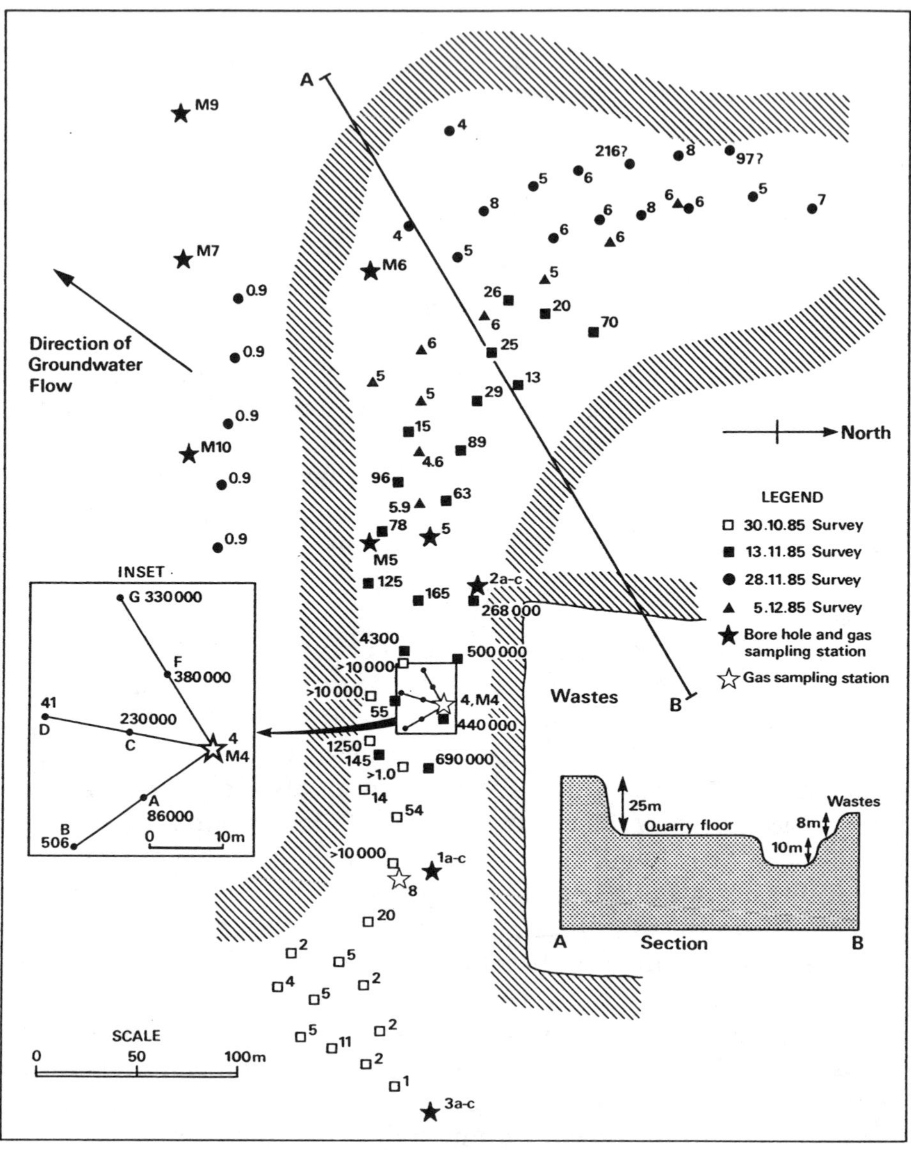

FIGURE 3. Concentration of methane in ppmv at a depth of 2 m below surface adjacent to the Morley landfill.

are concentrated on Tenax Traps. The traps are thermally desorbed onto a chromatographic column for separation of organics and quantification using flame ionisation detection. Analyses are generally carried out onsite in a mobile laboratory.

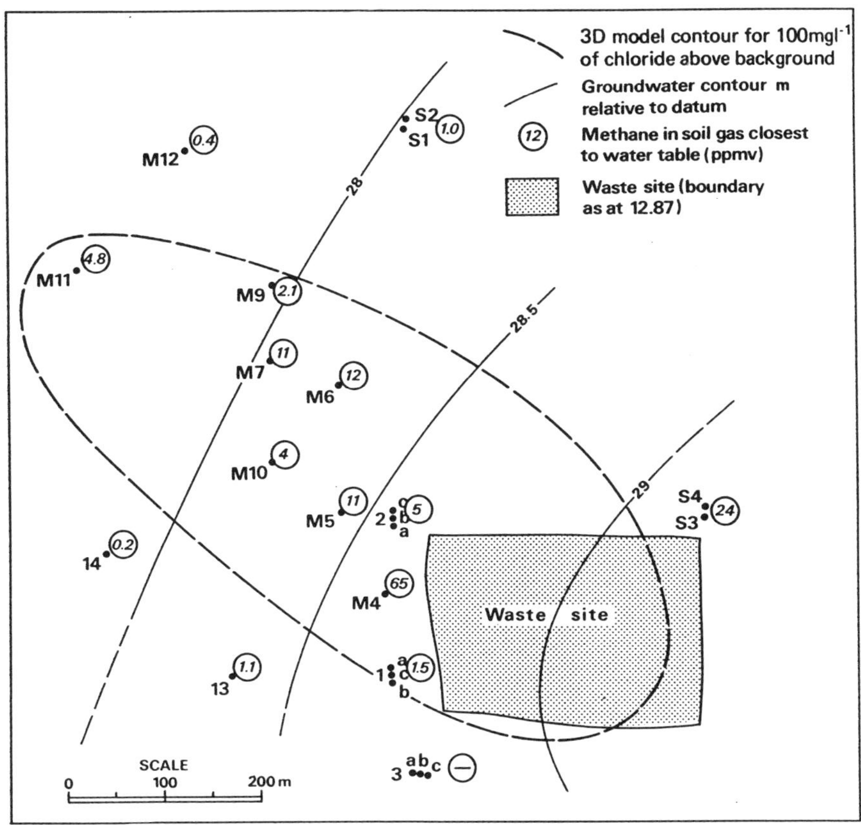

FIGURE 4. Distribution of groundwater contamination (as chloride from Davis *et al.* 1988) in relation to the concentration of methane (ppmv) in soil gas close to the water table (circled figures) at the Morley site.

Of the 40 service stations sampled, eight showed elevated levels of BTX monoaromatics and total hydrocarbons. At the most polluted site, all sampling locations close to the site recorded higher than background concentrations of total hydrocarbons and BTX aromatics (Figure 5). Concentrations of hydrocarbons were highest on the southern edge of the site: total hydrocarbons ranged up to 1,911 ppmv, benzene up to 16.2 µg/L, toluene up to 120 µg/L and xylene up to 37 µg/L). The lowest concentrations were found to the east of the site where total hydrocarbons ranged between 7 and 62 ppmv.

Based on the initial soil gas sampling, five shallow bores were drilled in the vicinity of the site. This drilling and sampling confirmed the presence of dissolved petroleum product in the groundwater, but

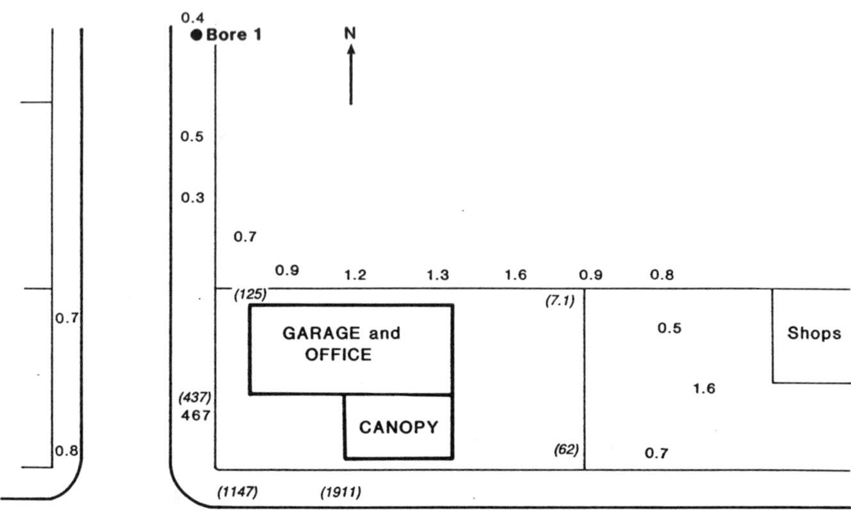

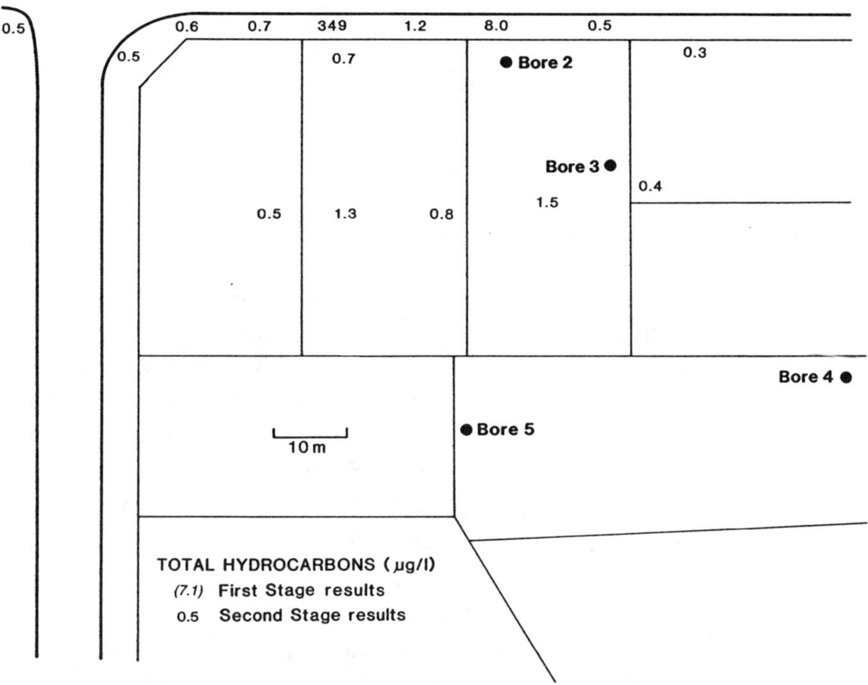

FIGURE 5. Service station site in Perth, Western Australia showing the location of groundwater boreholes and total hydrocarbon concentrations (μg/L) from shallow soil gas sampling.

no free product was identified. Figure 5 shows the location of the boreholes and Table 1 gives dissolved BTX constituents for each of the bores. All boreholes show above-background concentrations of the BTX compounds. However, highest concentrations were found in boreholes 2, 3, and 4. Boreholes 1 and 5 have BTX concentrations 2 to 4 orders of magnitude lower. The data suggest that the plume of dissolved hydrocarbon contamination is extensive, having migrated beyond 100 m (and possibly much further distances) toward the southeast consistent with the regional and local direction of groundwater flow. The boreholes also indicate that free product, if present, is confined within the service station boundaries.

TABLE 1. BTX concentrations determined from groundwater samples at the service station site.

	Benzene	Toluene (μg/L)	Xylene
Borehole 1	17	8	1
Borehole 2	11,300	23,600	3,090
Borehole 3	2,700	1,660	380
Borehole 4	8,390	5,200	970
Borehole 5	17	6	0.9

As with the landfill study, volatile concentrations in soil gas were found to be sensitive to soil heterogeneity, particularly to moisture content. Increasing the soil moisture content significantly reduces effective soil porosity and gas permeability, thus limiting the mobility of gases (Kerfoot 1990). In repeat soil gas surveys at five of the eight sites where evidence of leakage was found, significant differences in concentrations were found as shown in the examples in Figure 6. This was attributed to the effects of heavy winter rainfall immediately prior to the repeat survey. Initially, soil gas was investigated in late summer when the soil profiles were at their driest. These results again indicate difficulties with using soil gas techniques for monitoring changes with time in subsurface contamination. However, at the site shown in Figure 5, there are indications that soil-gas gives useful information on

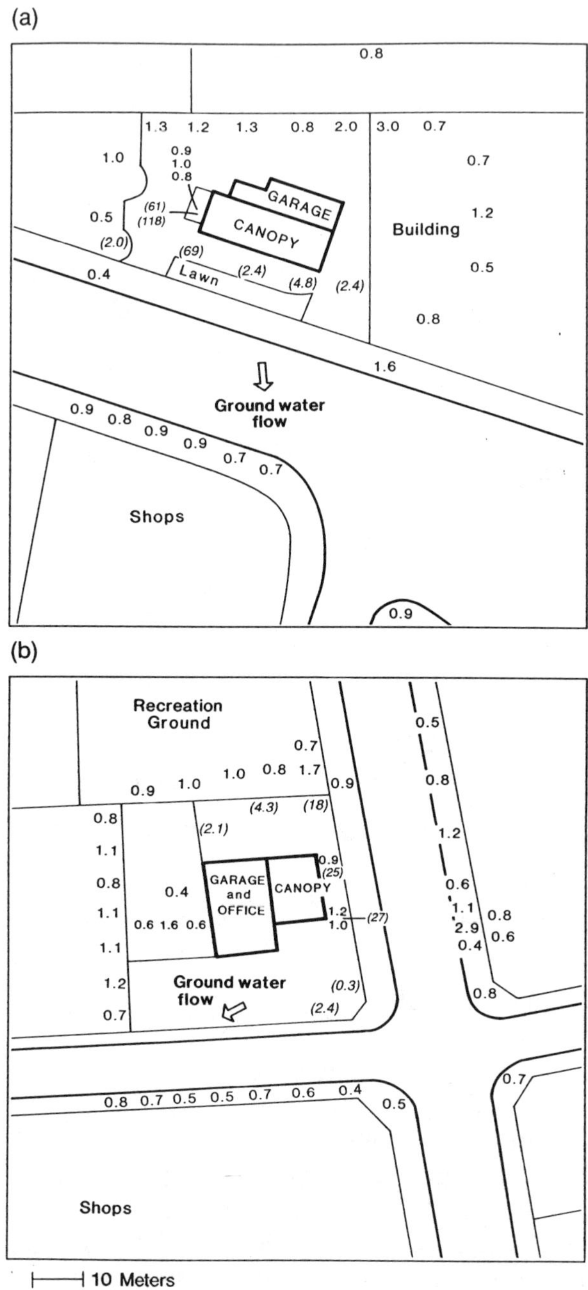

FIGURE 6. Examples of investigation of shallow soil-gas around two petrol service stations, showing variability with time in concentrations of total hydrocarbons (µg/L) at 1.5 m below surface.

the distribution of free product. This is potentially useful for monitoring remediation.

CONCLUSIONS

Remote monitoring from the ground surface using electramagnetic techniques and/or shallow soil gas sampling offers a cost-effective means of determining the distribution of contaminants in groundwater. For the Morley landfill, initial soil gas sampling indicated correctly the general direction of leachate plume movement, although transient changes in soil gas were not related to changes in groundwater pollution. Transient EM surveys at the Morley landfill did successfully identify plume development with time.

Soil gas sampling and analysis has also been successful in identifying a site of hydrocarbon leakage and in indicating the likely direction of plume movement, but not its extent. Concentration of volatiles in soil gas, however, are sensitive to changes in soil porosity and therefore shallow soil gas surveys have little advantage for monitoring (relative) changes with time in subsurface contamination.

Remote techniques alone cannot be used to monitor the effectiveness of a remediation strategy. However, in combination with conventional borehole monitoring, the techniques can reduce the need for drilling extensive networks of monitoring bores. Further refinement of these techniques will be undertaken as part of a large research program on bioremediation which has recently cammenced. The research will entail further work at the service station site to determine the distribution of dissolved hydrocarbons in the shallow aquifer system. The use of TEM and other surface geophysical techniques, such as ground penetrating radar (Mellett 1990) for mapping hydrocarbons, will be investigated. The use of soil gas monitoring as an indicator of presence of nonaqueous phase liquids will be evaluated. It is also proposed to implement and test bioremediation strategies at this field site over the next three years. Hydrogeological data for the site will be integrated with complex coupled solute transport and chemical kinetics modelling, and with microbiological studies.

REFERENCES

Barber, C.; Briegel, D. *J. Contaminant Hydrology* **1987**, *2*, 51–60.

Barber, C.; Davis, G. B. In *Monitoring to Detect Changes in Water Quality Series;* IAHS Public. No. 157; **1986**, pp 3–12.

Barber, C.; Davis, G. B.; Briegel, D.; Ward, J. K. *J. Contaminant Hydrology* **1990**, *5*, 155–169.

Barber, C.; Davis, G. B.; Buselli, G.; Height, M. *Int. J. of Environment and Pollution* **1991**, *1*, in press.

Buselli, G.; Barber, C.; Davis, G. B.; Salama, R. B. In *Geotechnical an Environmental Geophysics: Volume II: Environmental and Groundwater*; Ward, S. H., Ed; SEG: Tulsa, 1990; pp 27–39.

Cargeeg, G. C.; Boughton, G. N.; Townley, L. R.; Smith, G. R.; Appleyard, S. J.; Smith, R. A. "Perth Urban Water Balance Study, Vol. 1 Findings"; report to the Water Authority of Western Australia; Perth, 1987.

Davis, G. B.; Barber, C.; Buselli, G. In *Proceedings*, 3rd International Minewater Congress, Melbourne, Australia, 1988; pp 261–270.

Evans, R. B.; Schweitzer, G. E. *Environ. Sci. Technol.* **1984**, *18(11)*, 330A–339A.

Glaccum, R.; Noel, M.; Evans, R.; McMillion, L. In *Proceedings*, 3rd National Symposium on Aquifer Restoration and Groundwater Monitoring, National Water Well Assoc., Columbus, OH, 1983; pp 421–427.

Grady, C.P.L. *J. Environmental Engineering* **1990**, *116*, 805–828.

Jacobsen, G.; Lau, J. E. "Groundwater Contamination Incidents in Australia, An Initial Survey"; Geology and Geophysics Report 287; Bureau of Mineral Resources: Canberra, 1988.

Kerfoot, H. B. *Trends in Analytical Chemistry* **1990**, *9(5)*, 157–163.

Marrin, D. L.; Thompson, G. M. In *Proceedings*, NWWA/API Conference on Petroleum Hydrocarbons and Organic Chemicals in Groundwater, National Water Well Assoc., Houston, TX, 1984; pp 172–187.

Mazac, O.; Kelly, W. E.; Landa, I. *J. Hydrology* **1987**, *93*, 277–294.

Mellett, J. S. *Geotimes* **1990**, September, 12–14.

Salama, R. B.; Davis, G. B.; Barber, C. In *Groundwater Management, Quantity and Quality*; IAHS Public. No. 188; 1989; pp 215–226.

Swinton, R. *Water* **1989**, December, 12–17.

Thompson, G. M.; Marrin, D. L. *Groundwater Monitoring Review* **1987**, 88–93.

Stratification of Anoxic BTEX-Degrading Bacteria at Three Petroleum-Contaminated Sites

Mark D. Mikesell, Ronald H. Olsen, Jerome J. Kukor*
The University of Michigan Medical School

INTRODUCTION

Among the most important considerations in evaluating the potential for *in situ* bioremediation of a hydrocarbon plume is the limitation imposed by the availability of an electron acceptor. The rates of benzene, toluene, ethylbenzene, and xylene(s) (BTEX) biodegradation, for example, have been shown in some laboratory studies to be greater with oxygen than with nitrate (Major *et al.* 1988), but other studies have shown similar hydrocarbon degradation rates under denitrifying and aerobic conditions (Kuhn *et al.* 1985, 1988). In most cases of bioremediation, oxygen availability is thought to limit hydrocarbon removal, even when oxygen (as H_2O_2 or in air) is directed to the plume (Hinchee & Miller 1990). The use of nitrate is considered a viable alternative to oxygen (Battermann & Werner 1984) because, although degradation rates are slower, the water solubility of NO_3^- is much greater than that of oxygen (O_2 solubility is ca. 10 to 11 mg/L at typical groundwater temperatures, while nitrate salts can dissolve fairly freely in water). In addition to the availability of an electron acceptor, the stratification of microorganisms and the contaminants within aquifers, and perhaps within the plumes themselves, must also be considered.

* Department of Microbiology and Immunology, The University of Michigan Medical School, Ann Arbor MI 48109-0620, (313) 764-4380

In this work we demonstrate relationships among these parameters in samples from three contaminated aquifers.

EXPERIMENTAL METHODS

Three sites were selected in which the groundwater had been contaminated by petroleum products including B, T, E, and X. Our analysis of the samples taken from these sites consisted of total viable bacterial counts, counts of aerobic and anoxic BTEX-utilizing bacteria, and anoxic enrichment for BTEX-degrading, denitrifying bacteria during which we monitored BTEX disappearance by high-performance liquid chromatography (HPLC), sampled for isolation of anoxic BTEX degraders, and assayed samples for nitrate reductase activity. This protocol is illustrated schematically in Figure 1.

Aquifer Sampling and Characteristics

The aquifers sampled for this study (designated A, B, and C) are composed of fine to medium sands with varying amounts of gravel. The depths to the water table are 1.2, 8.5, and 20.7 m below ground level for sites A, B, and C, respectively. The contamination at these sites originated with leaks or spills of unknown volume from pipelines or tanks at petroleum processing installations.

Aquifer core samples were obtained from the three sites using a Waterloo Cohesionless-Aquifer Core Barrel sampling device (Zapico *et al.* 1987). This sampling method permitted collection of aquifer material in a 1.5-m core barrel from which a portion was removed from each end, giving approximately 0.9 m of intact core suitable for microbiological analyses. The core samples were sealed and stored in an upright position at 4 C until the enrichment cultures were prepared.

At each site several monitoring wells were established so that hydrocarbon (BTEX) concentrations and the extent of the plume could be determined. Also, groundwater samples were collected during screened-auger drilling for the measurement of aqueous-phase BTEX concentrations. In addition to BTEX concentrations, groundwater samples were analyzed by a commercial laboratory for a variety of chemical parameters for which the results are summarized in Table 1. Dissolved oxygen was measured in the field using a portable DO meter and the method described by Franson (1976).

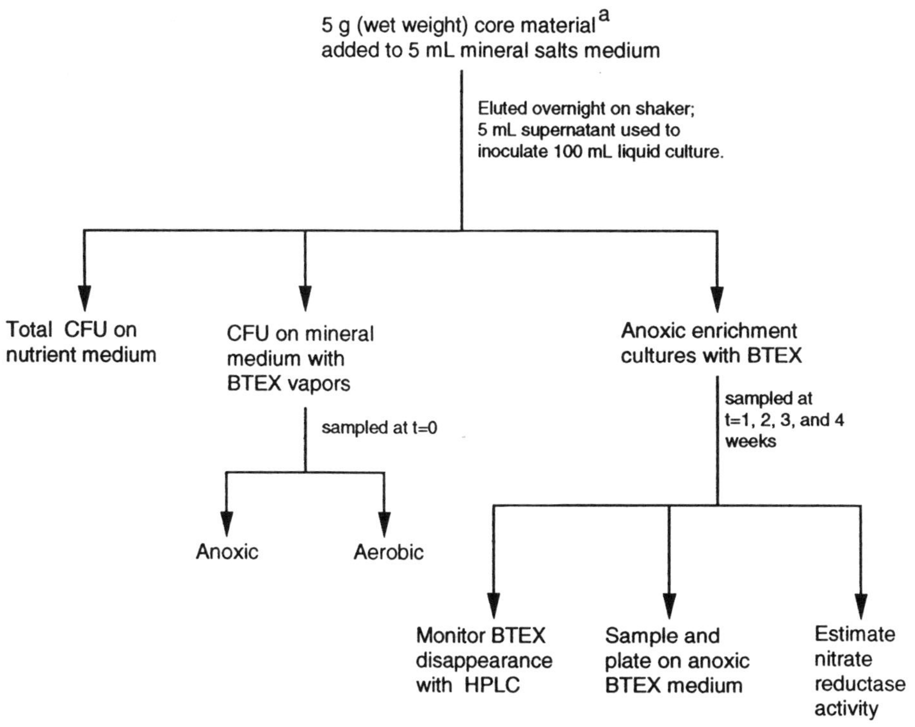

[a]Three sites (A, B, and C); five 5 cm x 1.5 meter cores taken from saturated sands at each site.

FIGURE 1. Protocol for the estimation of BTEX-degrading bacteria in saturated sand samples.

Enrichment Culture Preparation

Aquifer samples spanning a range of depths at each of the three sites were used to prepare the anoxic enrichment cultures. As shown in Figure 1, from each sample a slurry (1:1 wt:vol) was prepared in sterile aqueous medium and shaken overnight at 30 C. After allowing the coarse fraction to settle, 5 mL of the slurry was added to 100 mL of sterile gas-flushed (ca. 1.5 L/min, 10 min) liquid medium. The culture was flushed again with N_2:H_2:CO_2 (85:10:5), and the substrates (BTEX) were added to a final concentration 250 µM each. The bottles were sealed with Teflon®-lined septa and incubated at 30 C. At the start of the incubations, the O_2 content of the medium was approximately

TABLE 1. Characteristics of groundwater from the aquifer at study sites A, B, and C.

		Parameters Measured[a]								
Core Site	Location	pH	DO	TOC	NO_3^-N	NO_2^-N	B	T	E	X
A	plume	6.7	1.9	34	1.4	<0.02	<.001	.10	nd	1.35
A	pristine	7.7	4.2	<5	1.9	<0.02	<.001	<.001	nd	<.001
B	plume	7.5	nd	15	<0.1	<0.02	.62	.001	<.001	<.001
B	pristine	7.5	nd	1	1.1	<0.02	<.001	<.001	<.001	<.001
C	plume	7.0	1.1	280	<0.1	<0.02	2.50	3.00	.27	1.40
C	pristine	7.0	3.1	1	5.6	0.17	<.001	<.001	<.001	<.001

(a) Parameters (except pH) are expressed in mg/L.
Abbreviations: nd, not determined; DO, dissolved oxygen; TOC, total organic carbon; B, benzene; T, toluene; E, ethylbenzene; X, xylenes.

2 mg/L, as determined polarographically using a Clark-type electrode. The medium used (Page & Sadoff 1975) contains (per liter) 0.02 g $MgSO_4 \cdot 7H_2O$, 0.01 g $CaCl_2$, 5 mg $FeSO_4 \cdot 7H_20$, and 25 mg $NaMoO_4 \cdot 2H_2O$, and is buffered at pH 7.1 with 5 mM potassium phosphate buffer; anaerobic techniques are described in Kaspar and Tiedje (1982). Nitrate was added to a final concentration of 10 mM (140 mg/L nitrate-N). Sterile controls were established by the addition of sodium azide to 40 mM in the soil slurries and the enrichment cultures.

Viable Counting Procedures

Viable cell counts for soil slurries and for enrichment cultures were determined by plating serial dilutions on relevant media. Total counts were determined by plating on tryptone/yeast extract medium (each liter contains 2.5 g tryptone, 1.25 g yeast extract, 0.5 g dextrose, 4.5 g

NaCl, and 0.07 g $CaCl_2$) and counting colonies after 3 days incubation at 30 C. Enumeration of BTEX-degrading bacteria was performed by plating on minimal medium containing 10 mM nitrate. Plates were incubated in glass desiccators containing a small piece (ca. 2 cm x 2 cm) of BTEX-saturated filter paper; BTEX was replenished after 1 day and colonies were counted after the second day of incubation. Anoxic conditions were established in an anaerobic jar (BBL Gas-Pak) containing a H_2-CO_2 atmosphere from which oxygen was removed with a palladium catalyst. BTEX was supplied as in the aerobic incubations and plates were counted after 7 days at 30 C.

Analytical Procedures

The concentrations of benzene, toluene, ethylbenzene, and p-xylene were determined by HPLC employing a 25 cm × 4 mm C_{18} column from which the analytes were eluted with 75 percent methanol and 25 percent water at a flow rate of 1.5 mL/min. Samples were withdrawn from the bottles using a sterile syringe and immediately mixed with an equal volume of methanol. BTEX concentrations were determined by the external standards method. The detection limit of the method was 5 to 10 µM.

Nitrite levels in the enrichment cultures were determined using a colorimetric method consisting of diazotization with sulfanilamide and derivatization with N-naphthyl-ethylenediamine (Hanson & Phillips 1981).

RESULTS

Table 1 shows some of the characteristics of the three test sites designated A, B, and C. Groundwater was obtained from the top of the water table at these sites and analyzed for its salient features. Although all three sites initially were contaminated with B, T, E, and X, their content of these compounds varied at the time of sampling. Perhaps this variance reflects ongoing biodegradation from the time of spill to the time of sampling and/or relative amounts of B, T, E, or X in the original contamination. At site A, toluene and xylene were the major contaminants; at site B, benzene was the major contaminant; and at site C, all four compounds were present at significant concentrations. Pristine samples were obtained adjacent to and up-gradient of the plume area. These samples did not show significant B, T, E, or X contamination and, furthermore, showed low TOC. Also, TOC was found in inverse proportion to nitrate concentration in both plume and pristine samples: lower nitrate content was observed for plume samples with higher TOC. This diminution of nitrate content may have resulted from active denitrification associated with TOC degradation. Correlating with this also are low DO values associated with high TOC for the plume samples. Collectively, then, these observations suggest biological activity occurring within the plume.

Enrichment cultures derived from various depths in the plume were scored for their degradative activity towards B, T, E, or X and microbial content at the beginning and at the end of 1, 2, 3, and 4 weeks of incubation for cultures from each site. To estimate loss of BTEX into the head space during the incubation period, sterile aqueous suspensions containing the same liquid volume in identical vessels were incubated with the enrichment cultures and also sampled. In Table 2 are shown the data resulting from analysis of the sterile control suspensions of BTEX, and a score of zero was assigned to all enrichment cultures showing B, T, E or X concentrations similar to the abiotic controls. Scores of 1 or 2 were assigned to the enrichment cultures for samples in which the concentrations of remaining B, T, E, or X were less than those observed for the controls. For example, a score of 1 was assigned to a culture that showed degradation of B greater than 15 to 30 percent, and a score of 2 might be assigned to a culture showing degradation of T greater than 42 percent from the end of week 1 until the end of week 4. This approach allowed a comparison of the degradative activity against each of the four compounds for samples from all the sites. Scores achieved by cultures from all three sites are shown in Table 3.

The data in Table 3 show stratification for B, T, E, or X degradation under anoxic incubation conditions at various depths. For site A, activity obtains maximally at 3.4 m, but is also observed for inocula taken from adjacent depths. For site B, degradation was observed at 13.9 m. The scores derived from site C samples, on the other hand, suggest two zones of degradative activity at 24.5 m and 26.4 m, although insignificant degradation of *p*-xylene occurred for the 26.4 m inoculum. As shown in Figure 1, other microbial characteristics of the samples and their enrichment cultures were also ascertained. These characteristics in relation to the BTEX degradation scores for the various samples are shown in Tables 4, 5, and 6.

Table 4 shows data developed from site A. These data suggest that the total number of colony forming units (CFU) is uniform throughout the cores and that similar counts were obtained for bacteria

TABLE 2. Estimation of BTEX degradation in anoxic liquid culture.

Score	Percent BTEX degradation during weeks 1 through 4 per compound(a)(b)			
	B	T	E	X
0	<15	<27	<50	<54
1	15–30	27–42	50–65	54–69
2	>30	>42	>65	>69

(a) (Amount at end of week 1 minus amount at end of week 4)/(Amount present at end of week 1) × 100.
(b) Average loss during week 1 was <16 percent.
Abbreviations: B, benzene; T, toluene; E, ethylbenzene; X, *p*-xylene.

TABLE 3. Scoring of BTEX degradation in anoxic liquid culture inoculated with eluted core samples from the saturated zone of sites A, B, and C.

Site A[a]						Site B[a]						Site C[a]					
	Score per cpd.						Score per cpd.						Score per cpd.				
Sample Depth	B	T	E	X	Total Score	Sample Depth	B	T	E	X	Total Score	Sample Depth	B	T	E	X	Total Score
1.7	0	0	0	0	0	9.3	0	0	0	0	0	24.5	2	1	1	1	5
2.7	1	1	0	0	2	12.3	0	0	0	0	0	25.8	1	0	0	0	1
3.4	1	1	1	1	4	13.9	2	1	0	0	3	26.4	1	1	1	0	3
4.7	1	1	0	0	2	15.4	0	0	0	0	0	27.6	1	0	0	0	1
6.3	0	0	0	0	0												
7.8	0	0	0	0	0												

(a) All depths in meters; depth to the water table: Site A, 1.2 m; Site B, 8.5 m; Site C, 20.7 m.
Abbreviations: B, benzene; T, toluene; E, ethylbenzene; X, *p*-xylene.

able to use BTEX as a sole carbon source for aerobic growth. This does not imply, however, that identical isolates occur on both nutrient agar (for total count) and mineral salts-BTEX medium since many microorganisms growing on the latter often show suboptimal or no growth on complex nutrient medium (unpublished data). The distinctive feature, however, of the microbial counts shown in Table 4 is suggested by the development of significant populations able to grow anoxically on the mineral salts-BTEX medium during the 4-week incubation period. Such microorganisms also uniquely developed from core samples that scored for BTEX degradation during the incubation, suggesting that BTEX degradation was contingent on the presence of such microorganisms. Furthermore, denitrifying activity corresponded to BTEX degradation and the emergence of BTEX degrading bacteria under anoxic conditions. Three correlates, then, were obtained from such analysis of anoxic enrichment cultures: BTEX degradation, denitrifying activity, and enrichment of CFUs growing anoxically on the mineral salts-BTEX medium.

Tables 5 and 6, derived from sites B and C, respectively, follow the trends observed for site A (Table 4). Their variances from those trends may reflect variance in local conditions. Data in Table 5, for example, show a uniform distribution of denitrification activity for all core samples. This unique result may reflect the influence of fertile crop land adjacent to site B, unlike the other two sites. With respect to anoxic BTEX degradation, however, the development of CFUs growing anoxically on BTEX correlates with the score for BTEX degradation in the enrichment culture during the 4-week incubation period.

Enrichment cultures derived from site C show the same correlations as observed for site A. However, in this instance, at two depths within the region sampled, the score for anoxic BTEX degradation by the enrichment cultures correlated with the development of CFUs growing anoxically on BTEX and with denitrifying activity.

DISCUSSION

The data presented here clearly show that, within the hydrocarbon plumes at the three study sites, microorganisms capable of utilizing BTEX under anoxic conditions are stratified, and that their occurrence is generally correlated with hydrocarbon degradation and denitrification in enrichment cultures. The stratification of BTEX degradation, BTEX CFU's, and denitrification activity we observed could be caused by a number of factors. Variations in the concentration of BTEX as well as

TABLE 4. Denitrification and viable counts for site A.

Depth, (m)	Total BTEX Score[(a)]	Viable Count (CFU/mL enrichment culture): Aerobic Total t=0wk	Aerobic BTEX t=0wk	Anoxic with BTEX t=0wk	t=1wk	t=2wk	t=4wk	Denitrification[(b)] t=1wk	t=2wk	t=3wk	t=4wk
1.7	0	3×10^6	3×10^6	<10	<10	1×10^2	8×10^3	–	–	–	–
2.7	2	2×10^6	2×10^6	<10	<10	1×10^4	2×10^4	+	+	+	+
3.4	4	1×10^6	8×10^5	<10	2×10^3	6×10^4	4×10^5	++	++	++	++
4.7	2	6×10^5	1×10^6	<10	<10	6×10^3	7×10^4	+	+	+	+
6.3	0	1×10^6	8×10^5	<10	<10	<10	<10	–	–	–	–
7.8	0	4×10^5	7×10^5	<10	<10	<10	<10	–	–	–	–

(a) Total BTEX score = sum of scores on B, T, E, and X for that sample depth.
(b) –, no apparent nitrate disappearance, no nitrite produced; +, ~0.5 mg/L nitrite-N; ++, ~1.0 mg/L nitrite-N; +++, ~5.0 mg/L nitrite-N; ++++, ~15 mg/L nitrite-N (initial nitrate concentration, 140 mg nitrate-N/L).

TABLE 5. Denitrification and viable counts for site B.

Depth, (m)	Total BTEX Score[(a)]	Viable Count (CFU/mL enrichment culture): Aerobic Total t=0wk	Aerobic BTEX t=0wk	Anoxic with BTEX t=0wk	t=1wk	t=2wk	t=4wk	Denitrification[(b)] t=1wk	t=2wk	t=3wk	t=4wk
9.3	0	1×10^6	9×10^5	<10	<10	<10	6×10^1	+	+	+	+
12.3	0	1×10^6	1×10^6	<10	<10	<10	<10	+	+	+	+
13.9	3	2×10^6	4×10^5	<10	3×10^2	2×10^3	5×10^4	+	+	+	+
15.4	0	2×10^6	2×10^6	<10	<10	1×10^3	9×10^3	+	+	+	+

(a) Total BTEX score = sum of scores on B, T, E, and X for that sample depth.
(b) –, no apparent nitrate disappearance, no nitrite produced; +, ~0.5 mg/L nitrite-N; ++, ~1.0 mg/L nitrite-N; +++, ~5.0 mg/L nitrite-N; ++++, ~15 mg/L nitrite-N (initial nitrate concentration, 140 mg nitrate-N/L).

TABLE 6. Denitrification and viable counts for site C.

Depth, (m)	Total BTEX Score[a]	Viable Count (CFU/mL enrichment culture) Aerobic Total t=0wk	Aerobic BTEX t=0wk	Anoxic with BTEX t=0wk	t=1wk	t=2wk	t=4wk	Denitrification[b] t=1wk	t=2wk	t=3wk	t=4wk
24.5	5	1×10^6	1×10^6	<10	<10	8×10^4	3×10^5	–	++	++++	+++++
25.8	1	7×10^5	1×10^6	<10	<10	<10	<10	+/–	–	–	–
26.4	3	2×10^6	9×10^5	<10	1×10^2	5×10^3	3×10^4	+/–	+	++	+++
27.6	1	8×10^5	4×10^5	<10	4×10^2	5×10^3	8×10^3	+	+	+	+

(a) Total BTEX score = sum of scores on B, T, E, and X for that sample depth.
(b) –, no apparent nitrate disappearance, no nitrite produced; +, ~0.5 mg/L nitrite-N; ++, ~1.0 mg/L nitrite-N; +++, ~5.0 mg/L nitrite-N; ++++, ~15 mg/L nitrite-N (initial nitrate concentration, 140 mg nitrate-N/L).

other significant parameters may be caused by structural discontinuities within the plume. Clearly, the developing biological activity will reflect such discontinuities insofar as they affect the availability of substrates and electron acceptors.

The increase in anoxic CFUs, however, is not in all cases accompanied by a corresponding increase in nitrate reductase activity (*e.g.*, Table 4, 2.7 and 4.7 m; Table 5, 13.9 and 15.4 m; Table 6, 27.6 m). This may be attributable to the insensitivity of our denitrification assay, which relies on the accumulation of nitrite, a situation that may not always occur. It is known, however, that nitrite accumulation is more likely when either the available carbon sources are present at insufficient concentrations, or the incubation period is short relative to the time required for the process, or toxic intermediates interfere with the completion of the reaction (Tiedje 1988).

Within the plume at each site, dissolved oxygen and nitrate-N are depleted, while total organic carbon is elevated. This is the expected situation, since with the addition of carbon (BTEX or other petroleum compounds), dissolved oxygen and NO_3^- will be used up and residual C (BTEX or otherwise) will be recalcitrant.

Benzene degradation was most common in samples from all three sites, occurring in eight of the fourteen samples; degradation of ethylbenzene and xylene is comparatively rare, with three and two occurrences, respectively. The difference in the hydrocarbons present in the plume at site B, compared to those at sites A and C may help explain the site differences apparent in Table 3. Since benzene is the only aromatic hydrocarbon present at appreciable concentrations at site B (Table 1), the data of Table 3 may be explainable in terms of current exposure to limited substrate range (benzene only, no toluene). The absence of *p*-xylene and ethylbenzene degradation in site B enrichments may result from the fact that only benzene is present. That is, our ability to enrich for denitrifying BTEX-degrading bacteria is affected by the exposure history of the inocula. The fact that BTEX-utilizing bacteria could be isolated from samples at site B does not contradict this possibility, since the isolations were carried out in the presence of all four hydrocarbons simultaneously, and therefore, no substrate differences can be derived from the numbers of isolates.

CONCLUSIONS

We have shown that bacteria able to utilize BTEX under anoxic, denitrifying conditions can be isolated from hydrocarbon-contaminated aquifers, and that the occurrence of these bacteria is correlated with anoxic BTEX degradation and denitrifying activity. We also observed inverse relationships between nitrate and organic carbon in both contaminated and pristine groundwater, implying active denitrification *in situ*. The high organic carbon levels within the plumes at the three sites correlate also with low levels of dissolved oxygen. Finally, whereas aerobic BTEX degraders were distributed uniformly in the sampled profile, anoxic BTEX-utilizing bacteria appear to be stratified in zones within the plume at each site.

ACKNOWLEDGMENTS

This work was partially supported by grants from the National Institute of Environmental Health Sciences Superfund Research and Education Grant No. ES-04911, a round table research agreement with the Michigan Oil and Gas Association (MOGA-CoBioRem, Inc.), and the Office of Research and Development, U.S. Environmental Protection Agency under grant R-81570-01-0 to the Great Lakes and Mid-Atlantic Hazardous Substance Research Center. Partial funding of the research activities of the Center is also provided by the State of Michigan Department of Natural Resources.

REFERENCES

Battermann, G.; Werner, P. *Grundwasserforschung-Wasser/Abwasser* **1984**, *125*, 366–373.

Franson, M. A., Ed; *Standard Methods for the Examination of Water and Wastewater*, 14th edition; American Public Health Association: Washington, DC, 1976.

Hanson, R. S.; Phillips, J. A. In *Manual of Methods for General Bacteriology*; Gerhardt, P., *et al.*, Eds.; American Society for Microbiology: Washington, DC, 1981; p 355.

Hinchee, R. E.; Miller, R. N. *Haz. Mater. Contr.* **1990**, *5*, 30–34.

Kaspar, H. F.; Tiedje, J. M. In *Methods of Soil Analysis*; Page, A. L., Ed.; American Society of Agronomy: Madison, WI, 1982; Part 2, Chapter 45.

Kuhn. E. P.; Colberg, P. J.; Schnoor, J. L.; Wanner, O.; Zehnder, A.J.B.; Schwarzenbach, R. P. *Environ. Sci. Technol.* **1985**, *19*, 961–968.

Kuhn, E. P.; Zeyer, J.; Eicher, P.; Schwarzenbach, R. P. *Appl. Environ. Microbiol.* **1988**, *54*, 490–496.

Major, D. W. ; Mayfield, C. I.; Barker, J. F. *Ground Water* **1988**, *27*, 7–14.

Page, W. J.; Sadoff, H. L. *J. Bacteriol.* **1975**, *125*, 1080–1087.

Tiedje, J. M. In *Biology of Anaerobic Microorganisms*; Zehnder, A.J.B., Ed.; Wiley: New York, 1988; Chapter 4.

Zapico, M. M.; Vales, S.; Cherry, J. A. *Ground Water Monit. Rev.* **1987**, *7*, 74–82.

Redox Zones Downgradient of a Landfill and Implications for Biodegradation of Organic Compounds

J. Lyngkilde[*], *T. H. Christensen, B. Skov, A. Foverskov*
Technical University of Denmark

INTRODUCTION

Leachates from municipal as well as hazardous landfills contain a broad spectrum of xenobiotic compounds (Beckerath 1985; Först *et al.* 1989; Harkov *et al.* 1985) that may migrate into the surrounding aquifer if the leachate collection system is insufficient or, sometimes for older landfills, completely missing. For hazardous landfills, several reports in the literature document transport of xenobiotic contaminants over long distances (*e.g.*, Jackson & Patterson 1989; Johnson *et al.* 1985; Lesage *et al.* 1990; Williams *et al.* 1984), whereas the reports for municipal landfills are fewer and more ambiguous with respect to the extent of organic contaminant migration (*e.g.*, Barker *et al.* 1986; Reinhard *et al.* 1984). As well as the xenobiotic organic compounds, municipal landfill leachate typically contains high concentrations of organic matter originating from the degradation of the landfilled organic waste. This organic matter, for example, may include organic fatty acids and humic- and fulvic-like compounds that may act as a large pool of substrate for microbial activities in the aquifer, potentially also affecting the fate of the xenobiotic organic compounds. This migration of reduced leachate

[*] **Department of Environmental Engineering, Groundwater Research Centre, Technical University of Denmark, Building 115, DK-2800 Lyngby, Denmark**

into the aquifer may theoretically, through the microbial utilization of the organic carbon in the leachate, lead to the development of a redox zone sequence that ranges from methanogenic conditions close to the landfill, over sulphate-reducing, iron-reducing, manganese-reducing, and denitrifying conditions, to aerobic conditions at the outskirts of the leachate pollution plume. A redox zonation caused by landfill leachate migration has been proposed by Champ *et al.* (1979), but to our knowledge has never been documented in any detail.

The presence of various redox zones downgradient from a landfill is seen as a way to understand the fate of xenobiotic organic compounds in leachate pollution plumes, with respect both to the degradation processes favourable at the various redox levels and to the hydraulic retention times available in the various zones allowing the processes to proceed.

The purpose of this paper is to identify the major redox zones downgradient from an actual abandoned, uncontrolled landfill and to correlate the concentration of organic matter to these redox zones.

The Site. Waste disposal at Vejen Landfill (Vejen, Denmark) started in 1962 and terminated in 1976. The landfill received municipal as well as industrial waste (demolition, pharmaceutical, and pesticides). The landfill has no engineered leachate collection system, and a substantial fraction of the leachate is leaking into a shallow, unconfined aerobic sandy glacioalluvial aquifer. The aquifer is confined on the bottom by a clay deposit found at approximately 20 m depth close to the landfill, but rising to only 10 m depth at 400 to 500 m downgradient from the landfill. Occasionally small clay lenses are found in the aquifer; one substantial lens is stretching out in the aquifer from below the landfill.

MATERIALS AND METHODS

Monitoring Concept. Based on the general criticism of performing direct redox potential measurements by electrode (Germanov *et al.* 1959; Lindberg & Runnels 1984) and some preliminary boring and field sampling at the site, it was decided to base the redox status assignment of the groundwater on measurements of oxygen, nitrate, nitrite, ammonium, manganese, iron, sulphide, and methane in groundwater samples. To minimize the disturbance of these redox sensitive parameters during sampling and to allow for a detailed mapping over depth, where the steepest redox gradients are expected, a combined boring, sampling, and field analysis system was developed.

Field Sampling and Analysis. All groundwater samples are taken from 2.5-cm (1-in.) iron pipes rammed into the ground by a motor-driven hammer. The iron pipe is fitted with a tip supplied with a 10-cm-long screen and a Teflon® check valve. Samples are taken after every 0.5 m of ramming through a Teflon tube lowered into the iron pipe by pressuring the pipe with nitrogen. The Teflon tube is connected to a multiboard device (Figure 1) to allow sampling of groundwater and direct analysis of oxygen, pH, redox potential, temperature, and chemical conductivity by flow cells. Sample filtering takes place in one or two containers that can be pressurised with nitrogen. For volatile components like methane and sulphide, sampling takes place in an unbroken water stream prior to filtering. All samples, except those for methane and sulphide analysis, are pressure-filtered through a 0.1-µm filter. Alkalinity is measured in the field by Grantitration with diluted sulphuric acid. All other analysis, except those performed in the flow cells, are performed in the laboratory.

Preservation of samples is done at the site in connection with the sampling. Samples for TOC and metal determination are preserved by adding sulphuric acid and nitric acid, respectively. Samples for ammonium are preserved with sulphuric acid, and those for nitrate and nitrite determination are preserved with mercury chloride. Samples for analysis of sulphate and chloride are not preserved, but kept cool. If sulphide is present, it is driven out of the sample with nitrogen gas at the site immediately after the sample has been taken. Water samples (6.5 mL) for methane analysis are transferred from the unbroken water stream to evacuated Venoject blood sample vials (13 mL) by use of a syringe. The samples for sulphide determination are preserved by adding a strong basic, antioxidating solution (Midgley & Torrance 1978). All samples except those for methane determination are kept in polyethylene bottles.

Laboratory Analysis of Collected Samples. Sulphide determination is done with a Radiometer selective sulphide electrode, F1212S. TOC analysis is performed on a Dohrmann TOC analyzer. Metals are quantified on a Perkin-Elmer 370 Atomic Absorption Spectrophotometer. All nitrogen, sulphate, and chloride analysis are performed by a standard autoanalyzer routine (Technicon Autoanalyzer II). Methane is quantified by injecting 100 to 250 µL of the air-phase from the Venoject vials to a GC equipped with an FI detector.

Field Transect. The groundwater quality was monitored in a 370-m-long transect containing nine borings downgradient of the landfill. In

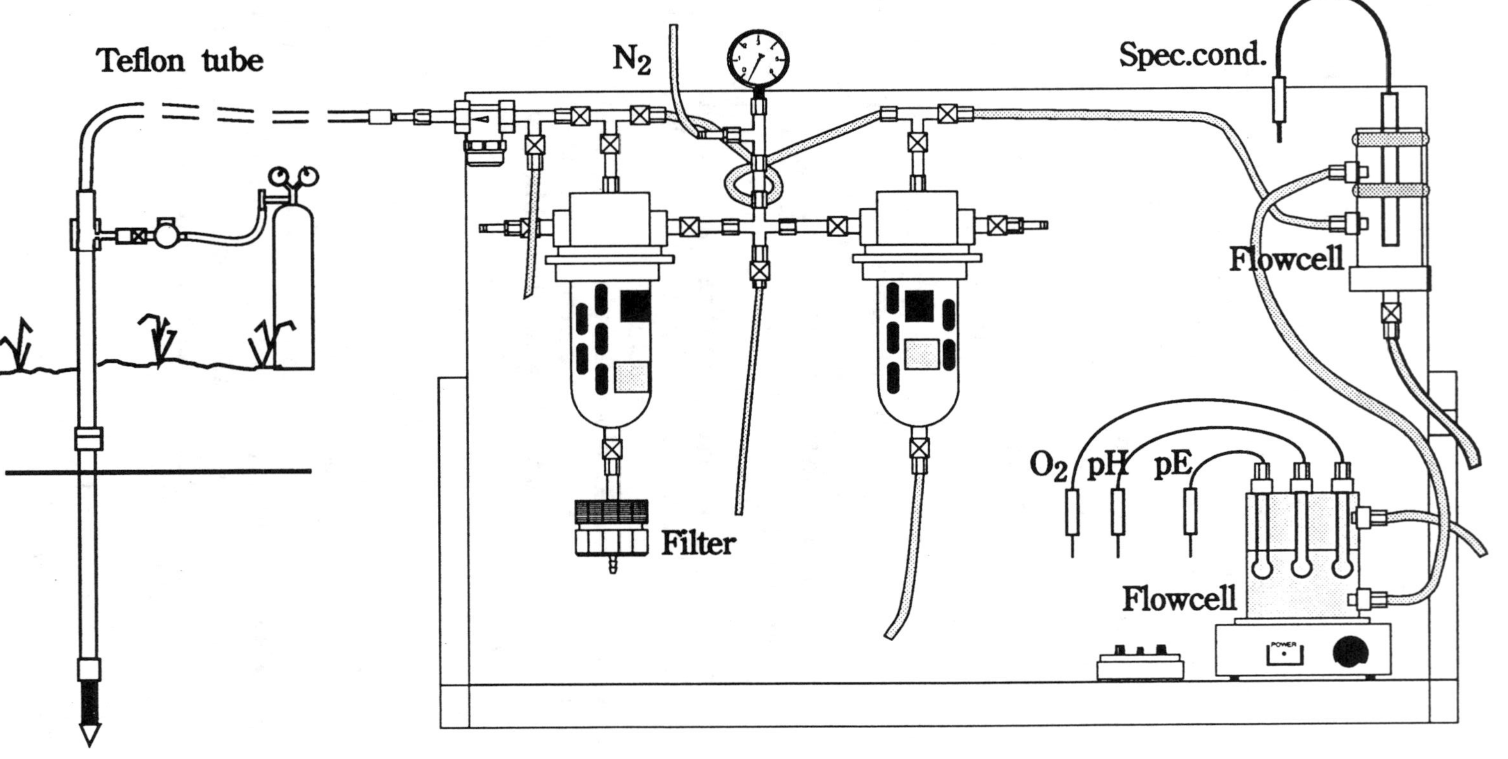

FIGURE 1. Multiboard for groundwater sampling and analysis. Board is depicted in large scale compared to pipe.

the borings groundwater samples were taken every 0.5 m, except for some few depths located outside any strongly concentrated gradients.

RESULTS AND DISCUSSION

Groundwater Quality

The vertical distribution of the measured parameters in borings representing three different distances from the landfill is shown in Figure 2 with Boring 1 (5 m) representing strongly polluted groundwater, Boring 19 (90 m) representing moderately polluted groundwater, and Boring 14 (365 m) representing slightly polluted groundwater.

The zone near the landfill contains strongly reduced species as methane, sulphide, and ammonium in concentrations up to 6.3 mg/L, 4.5 mg/L, and 262 mg N/L, respectively. High concentrations of sulphide are also found in Boring 2 (not shown) at a distance of 60 m from the landfill, but methane and ammonium concentrations are low in comparison with Boring 1. These strongly reduced species are all at insignificant concentration levels in Boring 19, as well as in all other borings located further downgradient. Presumably because of high amounts of sulphide and alkalinity, the concentrations of iron are low in the strongly reduced zone in Boring 1 compared to concentrations in more distant borings, including Boring 5. The groundwater in Boring 1 is well buffered over the entire profile. In all other borings pH increases with depth, as in Borings 19 and 14 (Figure 2).

The small concentrations of iron in Boring 14 (as well as other aerobic and denitrifying filters) also indicate that the analytical procedure of iron determination is functioning properly. A possible source of error in the analytical procedure is the migration through the filter of particles of iron(III) oxides subject to dissolution upon acidification.

Redox Zone Transect

Most (95%) characterized groundwater samples were assigned a redox status according to the scheme presented in Table 1. The criteria forming the redox scheme take into account the redox chemistry of the various species, the uncertainty of the characterization methods, and the fact that dissolved iron and manganese in strongly anaerobic environments are dominated by reduced species. The redox status assignment

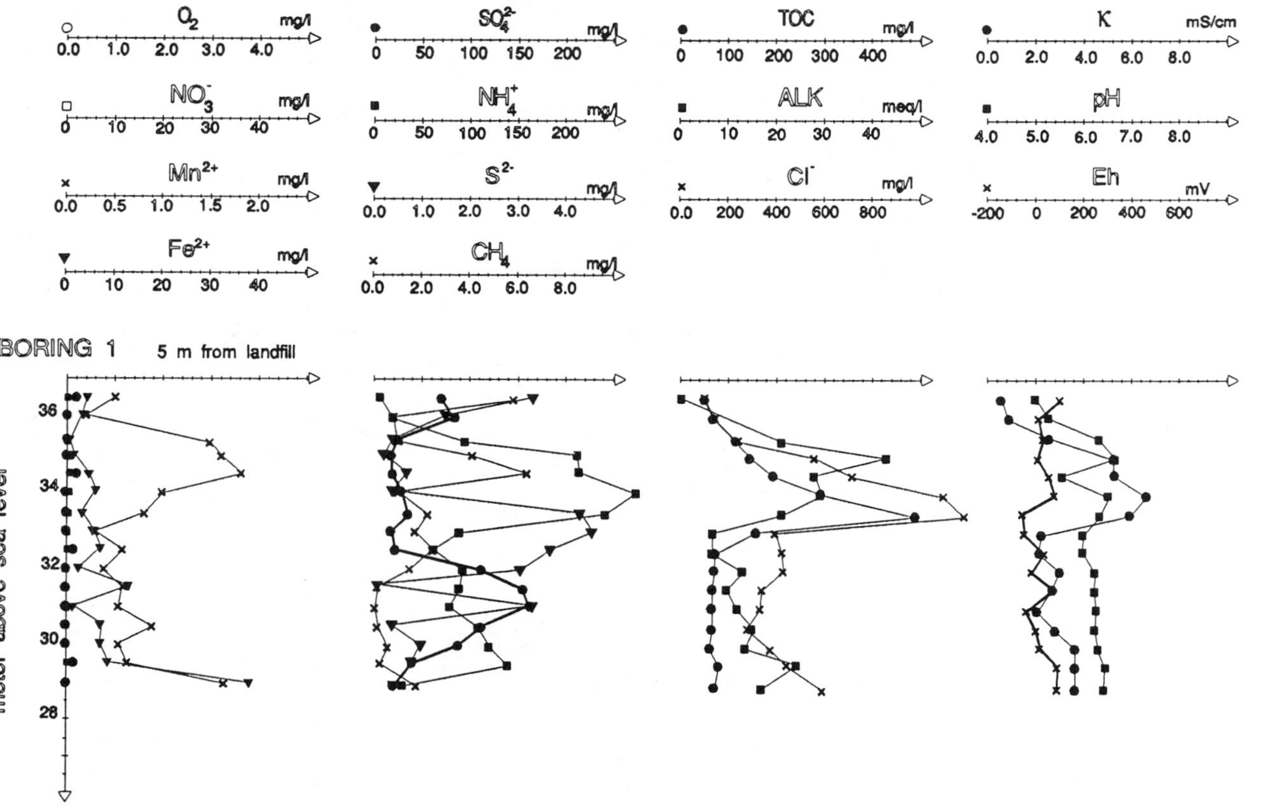

FIGURE 2. Selected profiles (Borings 1, 19, 14) for various groundwater parameters. All values are mg/L, except pH (–), Eh (mV), alkalinity (meq/L), and specific chemical conductivity (mS/cm).

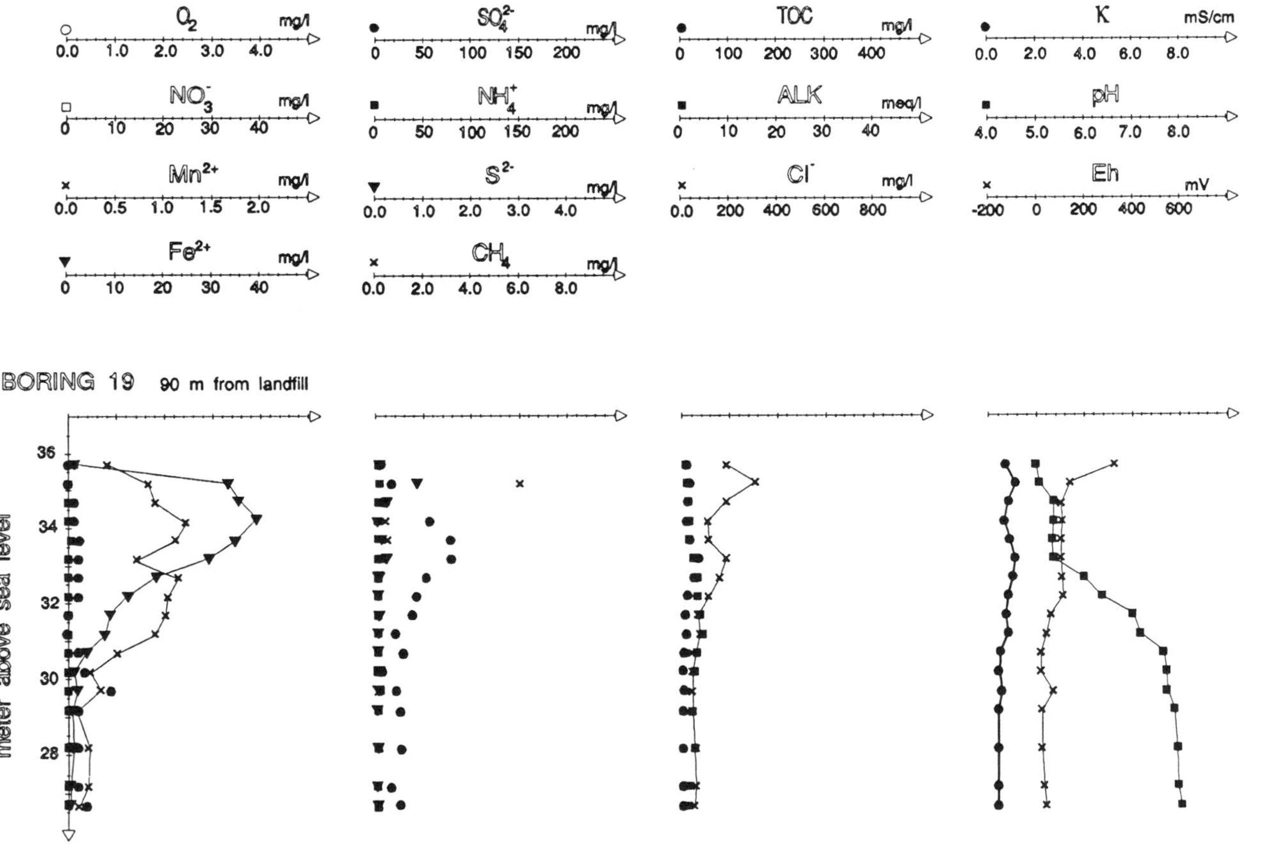

FIGURE 2. (Continued)

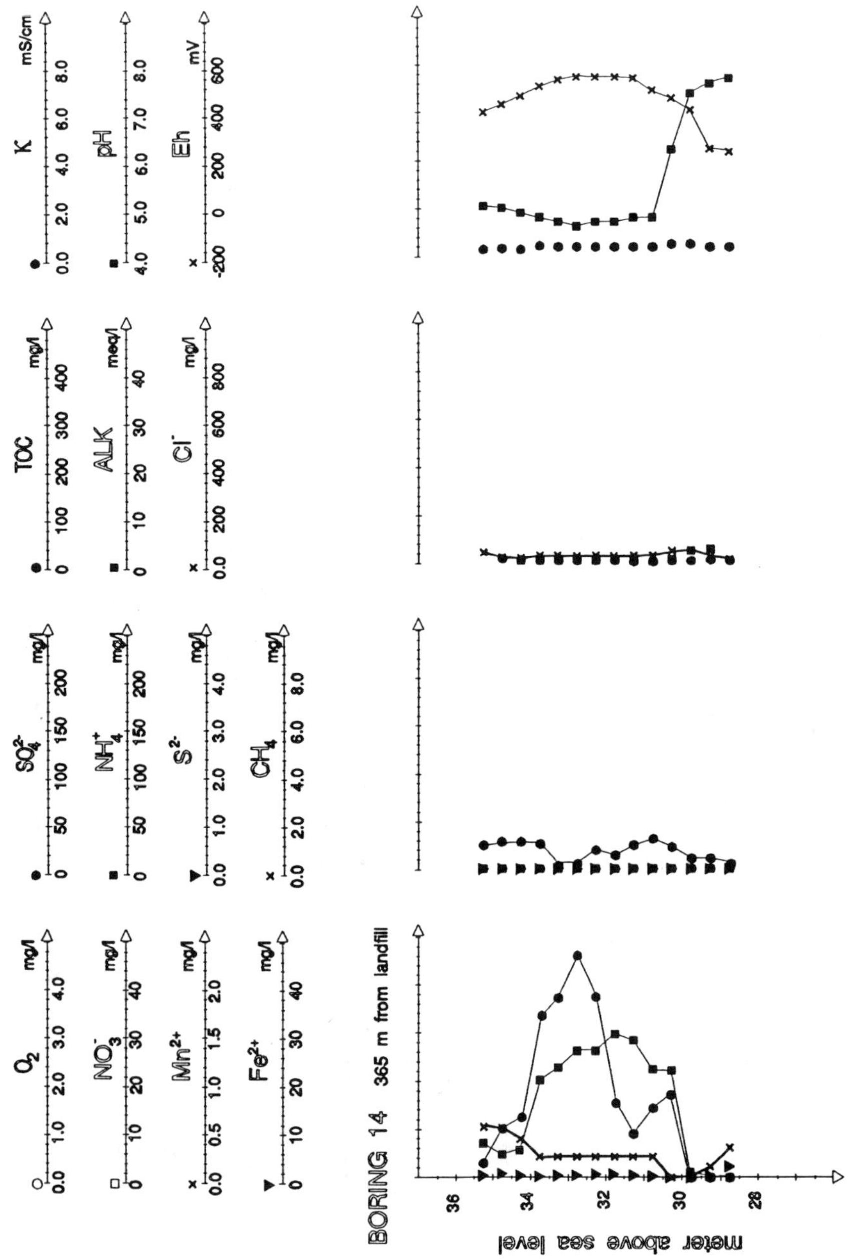

FIGURE 2. (Continued)

TABLE 1. Criteria for redox parameters used for assigning redox status to groundwater samples. All values in mg/L.

Parameter	Aerobic	Nitrate-reducing	Mn-reducing	Fe-reducing	Sulphate-reducing	Methanogenic
Oxygen	>1.0	<1.0	<1.0	<1.0	<1.0	<1.0
Nitrate	–	–	<0.2	<0.2	<0.2	<0.2
Nitrite	<0.1	–	<0.1	<0.1	<0.1	<0.1
Ammonium	<1.0	<1.0	–	–	–	–
Mn(II)	<0.2	<0.2	>0.2	–	–	–
Fe(II)	<1.5	<1.5	<1.5	>1.5	–	–
Sulphate	–	–	–	–	–	<40
Sulphide	<0.1	<0.1	<0.1	<0.1	>0.2	–
Methane	<1.0	<1.0	<1.0	<1.0	<1.0	>1.0

is based solely on the chemical characteristics of the groundwater, tacitly assuming that the chemical redox species are close to chemical equilibrium. Barcelona *et al.* (1989) refer to several investigations in which the assumption of chemical equilibrium is not valid. Approximately 5 percent of the samples do not comply with the criteria. This group contains groundwater samples violating the criteria with respect to a few of the parameters. These samples were evaluated individually in comparison with characteristics of adjacent samples.

The characteristics of the groundwater samples assigned to the same redox status are summarized in Table 2 in terms of parameter range and average. The large range observed for most of the parameters of each redox status shows the wide span of groundwater samples belonging to the same group. However, average values reveal that redox status is closely correlated to pollution level, showing high concentrations of chloride, alkalinity, TOC, methane, and ammonium in the methanogenic zone and levels approaching background values in the aerobic zone.

The assigned redox status of the groundwater samples creates the basis for a longitudinal transect showing the redox zones stretching from the landfill and 370 m downgradient (Figure 3). The redox zone map in Figure 3 shows continuous redox zones with only a few locally deviating redox zones. A methanogenic zone is found in close proximity to the landfill (<50 m), adjacent to a discontinuous, limited area of sulphate reduction. The groundwater samples assigned a manganese-reducing redox status do not constitute a continuous zone, but are

TABLE 2. Summary of groundwater characteristics (range, average, number of samples in parentheses) according to defined redox zones.

Parameter	Aerobic (43)	Nitrate-reducing (159)	Mn-reducing (14)	Fe-reducing (105)	Sulphate-reducing (31)	Methanogenic (14)
	1.0–7.0	0.0–0.9	0.0–0.5	0.0–0.9	0.0–0.3	0.0–0.2
Oxygen	3.01	0.15	0.13	0.07	0.06	0.05
	0.84–29.47	0–27.70	0–0.80	0–0.48	0–0.29	0–0.47
Nitrate	11.973	1.634	0.125	0.042	0.086	0.154
	0–0.48	0–0.20	0–0.14	0–0.03	0–0.02	0–0.13
Nitrite	0.032	0.014	0.017	0.002	0.002	0.015
	0–0.35	0–0.95	0–1.01	0–81.30	0–139.40	0–262.20
Ammonium	0.021	0.084	0.208	1.094	24.693	77.486
	0–0.5	0–0.5	0.3–0.9	0–1.2	0–1.6	0.1–1.8
Mn(II)	0.10	0.09	0.41	0.35	0.60	0.73
	0–5.6	0–7.1	0–1.4	1.5–39.0	0–76.0	0.5–38.0
Fe(II)	0.64	0.61	0.83	8.88	21.00	7.54
	2–301	6–95	6–70	6–155	0–156	1–38
Sulphate	42.3	35.3	34.6	57.5	51.7	14.5
	0	0	0	0–0.2	0–33.0	0–4.5
Sulphide	0.00	0.00	0.00	0.00	3.98	1.05
	0–0.14	0–1.60	0–0.51	0–4.48	0–5.87	1.03–6.30
Methane	0.033	0.064	0.078	0.391	0.550	2.524
	103–546	-91–546	20–319	-83–409	-29–169	-55–150
Redox pot.	385.0	155.1	123.0	113.0	85.3	63.1
	1.2–6.3	1.2–20.5	1.3–7.5	1.2–71.9	2.1–119.1	2.3–484.0
TOC	2.80	2.91	2.36	6.43	32.57	104.70
	0.0–1.6	0.0–5.3	0.3–5.2	0.0–10.1	0.3–23.9	0.7–42.9
Alkalinity	0.28	1.81	2.48	1.92	6.46	12.08
	16–71	14–108	16–188	15–341	13–764	13–1181
Chloride	30.1	36.3	48.4	75.5	225.3	384.7
	4.6–7.2	4.7–8.2	4.9–8.0	4.9–7.9	5.0–7.6	5.0–7.3
pH	5.27	7.00	7.18	6.38	6.27	6.51

All units are in mg/L, except alkalinity in meq/L, redox potential in mV, and pH (without unit). Nitrate, nitrite, and ammonium units are mg N/L. TOC unit is mg C/L.

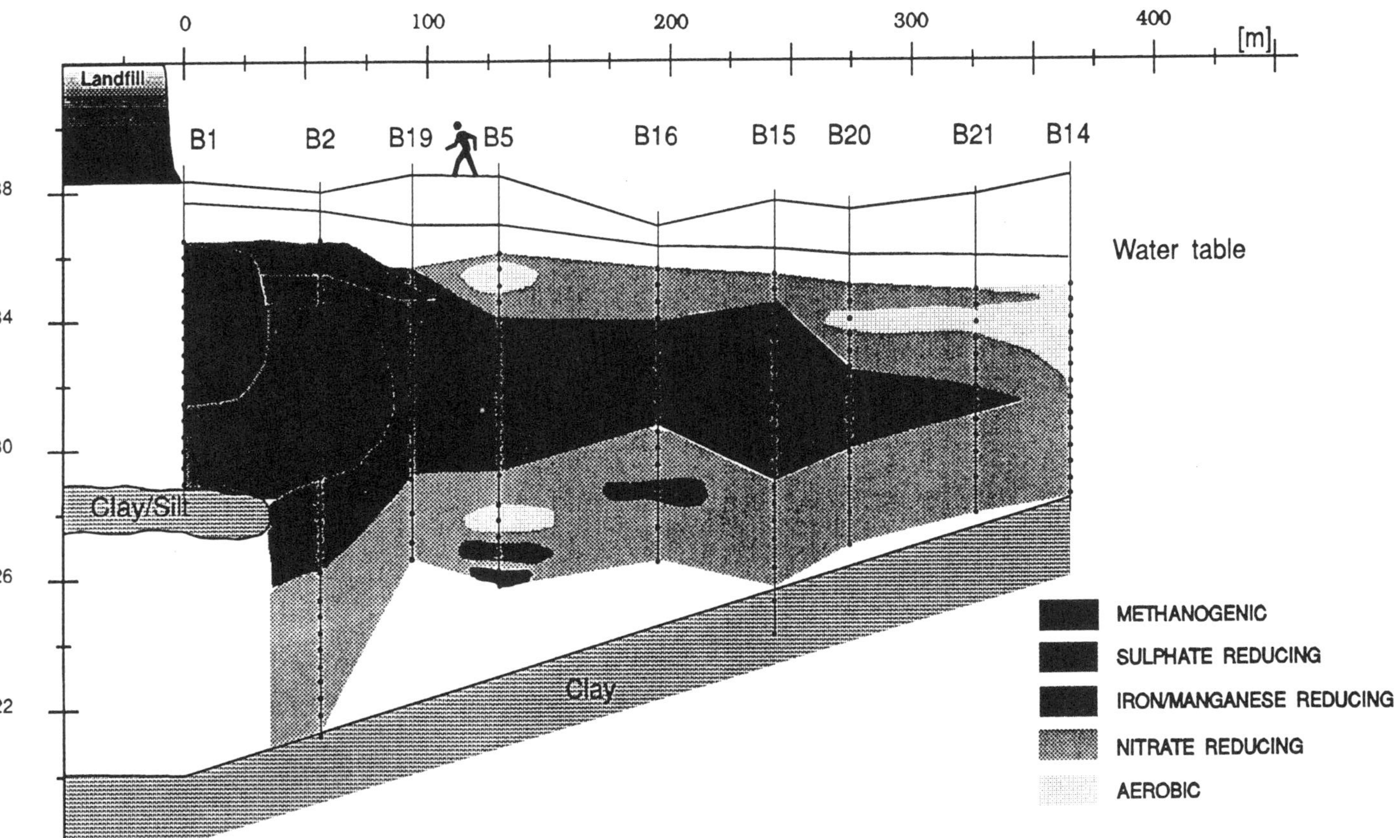

FIGURE 3. Longitudinal transect of redox zones at Vejen Landfill, Denmark.

usually found as small independent areas, often characterized by low TOC values. Only in a few cases do the manganese-reducing zones appear as boundaries between iron-reducing and nitrate-reducing zones. By far the most extensive zone is the iron-reducing zone stretching at least 350 m downgradient from the landfill. The iron-reducing zone is supposedly formed by reduction of iron oxyhydroxides naturally associated with the sediment, but reduced and dissolved due to the strongly anaerobic environment caused by the reducing organic matter of the leachate.

Close to the landfill the contaminated groundwater contains organic matter at a concentration level of 400 mg/L TOC. This concentration level drops rapidly downgradient of the landfill because of degradation and dilution. Degradation of organic matter is essential for developing strongly reduced zones and for maintaining a dense biomass potentially active in degrading xenobiotic organic trace compounds. To evaluate where the major part of the degradation of the organic matter takes place, Figure 4 was developed showing the concentrations of TOC, corrected for dilution, as a function of distance from the landfill. The dilution factor, F, accounts for the dilution based on chloride concentrations and is calculated $F = 1 + (C_0\text{-}C)/(C\text{-}C_B)$, where C is the chloride concentration of the groundwater sample, C_0 is the maximum chloride concentration in the groundwater 5 m from the landfill, and C_B is an average chloride concentration of the unpolluted groundwater beyond the pollution plume. Figure 4 shows that the major degradation of organic matter takes place within the first 100 m downgradient from the landfill. According to the redox zone map (Figure 3), this means that the major part is degraded in the methanogenic and sulphate-reducing zones. At the beginning of the iron-reducing zone, as illustrated by Boring 19, the actual TOC level is on the order of 25 mg/L TOC. However, this organic matter is apparently not completely recalcitrant since the redox processes proceeds in the iron-reducing zone.

The hydraulic properties of the aquifer (not presented here) suggest a groundwater flow velocity of 150 to 200 m/year, implying that the average hydraulic retention time is on the order of 3 to 4 months in both the methanogenic zone and the sulphate-reducing zone. In contrast to this, the iron-reducing zone may have a hydraulic retention time on the order of 12 to 18 months. These outlined aspects of retention times, redox status, and availability of degradable organic matter as a primary substrate may have implications for the fate of specific organic compounds present in the landfill leachate.

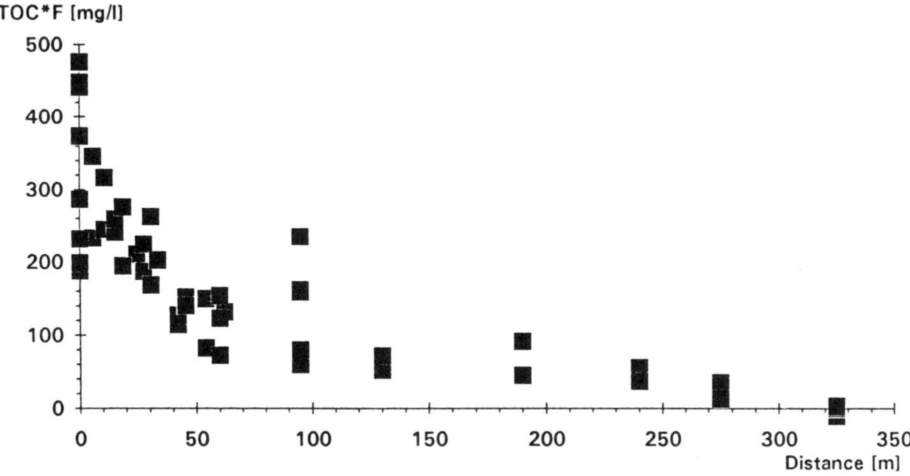

FIGURE 4. Concentration of organic matter (TOC) in groundwater downgradient from the Vejen Landfill as a function of distance. The concentrations have been corrected for dilution.

CONCLUSIONS

Based on groundwater sample characterization, a sequence of redox zones has been identified downgradient from the Vejen Landfill. Over a 350-m downgradient distance, the redox zones change from methanogenic, sulphate-reducing, over iron-reducing, to denitrifying and aerobic zones. Manganese-reducing zones were small and scattered. The iron-reducing zone is by far the largest zone, stretching for nearly 350 m. The major part of the organic matter in the leachate migrating into the aquifer is degraded within the methanogenic and sulphate-reducing zones. The hydraulic retention times of these zones are estimated to be 3 to 4 months each.

Information about redox status level, hydraulic retention times, and availability of organic matter as a primary substrate is seen as a key issue in understanding the fate of specific organic contaminants migrating with landfill leachate into aquifers.

Acknowledgments

This study is part of a major research programme focusing on the effects of waste disposal on groundwater. The programme is funded by the Danish Technical Research Council, the Technical University of Denmark, and the EEC. The field assistance of Tage V. Andersen and Thomas Larsen is gratefully acknowledged.

REFERENCES

Barcelona, M. J.; Holm, T. R.; Schock, M. R.; George, G. K. *Water Resources Research* **1989**, *25*(5), 991–1003.

Barker, J. F.; Tessman, J. S.; Plotz, P. E.; Reinhard, M. *J. Contam. Hydrol.* **1986**, *1*, 171–189.

Beckerath, K. von. *Müll und Abfall.* **1985**, *12*, 424–434.

Champ, D. R.; Gulens, J.; Jackson, R. E. *Can. J. Earth Sci.* **1979**, *16*, 12–23.

Först, C.; Stieglitz, L.; Roth, W.; Kuhnmünch, S. *Vom Wasser* **1989**, *72*, 295–305.

Germanov, A. I.; Volkov, G. A.; Lisitsin, A. K.; Serebrennikov, V. S. *Geochemistry* **1959**, *3*, 322–329.

Harkov, R.; Gianti, S. J., Jr.; Bozzelli, J. W.; LaRagina, J. E. *J. Environ. Sci. Health* **1985**, *A20*(5), 491–501.

Jackson, R. E.; Patterson, R. J. *Ground Water Monitoring Review* **1989**, *9*(3), 119–125.

Johnson, R. L.; Brillante, S. M.; Isabell, L. M.; Houck, J. E.; Pankow, J. F. *Ground Water* **1985**, *23*(5), 652–666.

Lesage, S.; Jackson, R. E.; Priddle, M. W.; Riemann, P. G. *Environ. Sci. Technol.* **1990**, *24*(4), 559–566.

Lindberg, R. D.; Runnels, D. D. *Science* **1984**, *225*, 925–927.

Midgley, D.; Torrance, K. *Potentiometric Water Analysis*; John Wiley & Sons: Chichester, 1978; p 356.

Reinhard, M.; Goodman, N. L.; Barker, J. F. *Environ. Sci. & Tech.* **1984**, *18*(12), 953–961.

Williams, G. M.; Ross, C.A.M.; Stuart, A.; Hitchman, S. P.; Alexander, L. S. *Q. J. Eng. Geol. London* **1984**, *17*, 39–55.

Disappearance of PAHs in a Contaminated Soil from Mascouche, Québec

R. N. Yong*, L. P. Tousignant, R. Leduc, E.C.S. Chan
McGill University

INTRODUCTION

Of the 365 sites classified by the Québec Ministry of the Environment in 1984 and 1985 as contaminated, 62 are considered hazardous. The site chosen for this study is one of these 62 sites and is located in the Mascouche municipality north of the city of Montreal. Site records indicate that it was used as a solid waste dump during the 1950s and 1960s. The northern side of the site was used as a dumping ground for semi-liquid wastes from the Montreal refineries.

In 1969, permission was received to operate an incinerator for liquid wastes from the petroleum refineries. However, it was discovered that after a year of operation the incinerator did not perform well and spilled liquid wastes into the surrounding soil. In 1974, the incinerator was shut down and all operations ceased. In addition to the incinerator, two open-air storage basins and numerous solid waste deposits occupy the present site. The two open-air concrete basins on the site originally contained a total of 660 m^3 of liquid petroleum wastes. These basins were fissured at several places and consequently leaked liquid petroleum wastes into the groundwater. In periods of heavy rain, the basins overflowed onto the surrounding soil terrain. In

* Geotechnical Research Centre, McGill University, 817 Sherbrooke St. W., Montreal, Québec H3A 2K6

1986, the liquid waste inside the basins was tranferred into large metal containers and left on site, and the basins and the incinerator were then covered with sand.

The site, which occupies an area of approximately 80 hectares, is mainly flat, with a few marshlike areas. The first 2 m of soil depth in this area consists of a sandy soil (Figure 1) underlain by 12 to 15 m of clay. The perched water table is located in the sandy soil layer and flows toward the southeast and southwest into the Mascouche River and the Milles-îles River. These rivers eventually join the des Prairies River and ultimately the St. Lawrence River. The "permanent" water table, situated in the rock underneath the clay, shows no trace of contamination yet.

Since contamination appears to be confined to the sandy soil layer, with traces in the top portions of the underlying clay, it appeared that *in situ* bioremediation using forced aeration to a nutrient-enhanced sandy soil system could yield significant benefit. However, before any field experiment could be conducted, it was necessary to undertake laboratory studies with the field soil to determine if forced aeration could produce effective results. In addition, the laboratory program was considered essential inasmuch as base values had to be established with respect to initial and boundary conditions. Accordingly, the disappearance of PAHs was monitored in the laboratory under forced

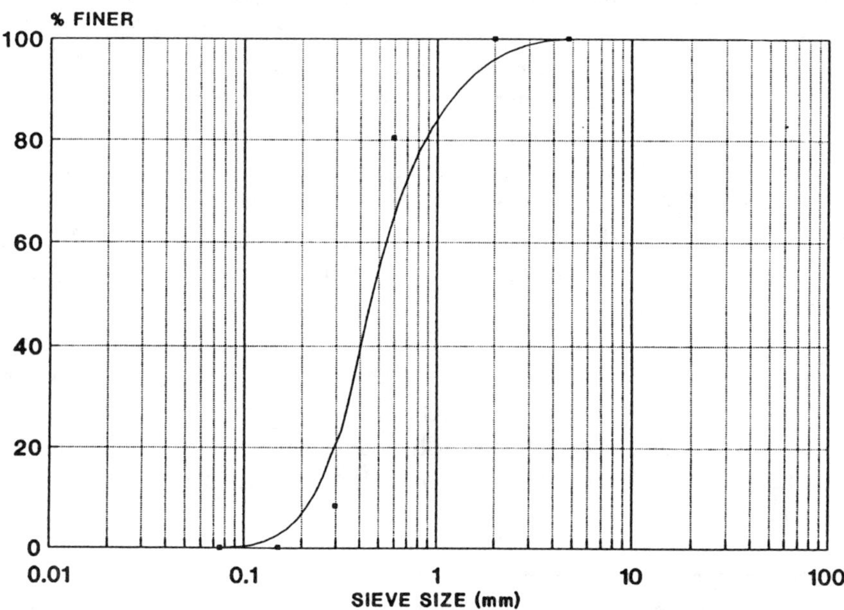

FIGURE 1. Grain size analysis of soil.

aeration conditions, and isolation and identification of PAH-degrading microorganisms were attempted. The results will be used to evaluate field implementation requirements, especially in regard to accountability for the parameters controlling the various aspects of the forced aeration technique.

MATERIALS AND METHODS

The sandy soil samples obtained from an area adjacent to the foundations of the old incinerator, at a depth of 1 m, were stored at 4 C. Analysis of contamination consisted of determining:

- Concentration of heavy metals (Zn, Cr, Pb, Cu, Ni, Cd) and potassium from atomic adsorption measurements on digested soil using the digestion technique described in HACH (1987)
- Concentration of nitrogen using the Total Kjieldahl Nitrogen test (TKN) and concentration of phosphorus using the ascorbic acid method (HACH 1987)
- Total organic carbon (TOC) using the wet oxidation test (U.S.D.A. 1954)
- pH of the samples according to the method of Dunn and Mitchell (1984)
- Grain size distribution of the sandy soil according to ASTM D421 and D422 (Bowles 1986)

The disappearance of PAHs and PCBs (polychlorinated biphenyls) was monitored at room temperature over two periods, 10 and 20 days, after initiation of forced aeration conditions. The contaminated soil was first placed in capped plexiglass cylinders (cells) provided with holes at both top and bottom caps. For the aeration procedure, compressed air at 0.0352 kg/cm^2 (0.5 psi) was introduced at the top (cap) and allowed to freely exit through the holes in the bottom cap. The (compressed) air from the compressor was forced through a gauze filter and humidified with sterile deionized water. Preliminary experiments designed to examine pressure effects indicated that pressures greater than 0.0352 kg/cm^2 caused excessive moisture loss in the test samples. Hence to avoid rapid desiccation of the test samples, it was decided to maintain an air pressure limit value of 0.0352 kg/cm^2. At this pressure, and for the type of soil tested, the average flow was approximately 0.79 L/min. Airflow was monitored every week for every cell using a rotometer.

Nine cells were used in the course of the laboratory study to assess the various parameters affecting the disappearance of PAHs. These cells were kept in the dark throughout the study period to avoid photodegradation of the samples. Cells A_1 and A_2 were filled with approximately 170 g of the sandy soil obtained from the site, with no added nutrients (subscripts 1 and 2 represent measurements after 10 and 20 days, respectively). The soil in each cell was subjected to forced air during the prescribed test period, and the concentration of PAHs was determined thereafter.

Cells AN_1 and AN_2 were identical to cells A_1 and A_2, but were supplemented with specificed amounts of nutrient salts (0.114 g NH_4NO_3 and 0.174 g KH_2PO_4 per cell), determined according to Parkinson (1974). Cells C_1 and C_2 were control cells containing the test soil samples, but without benefit of forced aeration and without nutrients. Cells AC_1 and AC_2 were control cells containing autoclaved test soil samples subjected to forced aeration. Autoclaving was conducted at 120 C and at standard pressure for 20 min. This procedure was adopted since it has been shown to have no effect on the concentration of the PAHs in the soil (Herbes & Schwall 1978).

Determination of PAH and PCB concentrations were performed by Analchem Laboratories, St. Hubert (Québec) according to the standards of the Québec Ministry of the Environment (1985). The only PAHs and PCBs measured were the ones listed on the PAH priority pollutants listed by the Québec Ministry of the Environment (1988).

Soil extract medium was used to isolate the microorganisms (Parkinson *et al.* 1971). After the specified test periods of 10 nd 20 days, 1 g of the test soil was taken from each cell and dissolved in 100 mL sterile deionized water. Dilutions ranging from 10^{-2} to 10^{-5} g/mL were made, and 0.1 mL of each dilution spread on soil extract agar (1.5%) (Parkinson *et al.* 1971). The plates were incubated at room temperature until distinct colonies formed. Isolated colonies were transferred to fresh soil extract agar plates and stored at 4 C.

The biodegrading activity of the isolated microorganisms in the presence of any contaminant was determined via optical density (O.D.) measurement (using a spectrophotometer at 420 nm) and population growth comparisons made with control samples. The isolated microorganisms were grown in soil extract broth, and 0.1 mL was transferred to 10 mL of minimal salt solution (Bailey & Coffey 1986). Duplicate tubes of each microorganism were then singly supplemented with a final concentration of 50 ppm of either phenanthrene, anthracene (American Chemical Co.), or fluorene (Eastman Kodak Co.). These hydrocarbons were first dissolved in acetone (Bumpus 1989). The controls contained only the minimal salt solution and the chemical. Assessment of growth of the microorganisms was by O.D. measurement. After a 28-day incubation period, the extent of growth was determined also by a plate count of microorganisms in the presence and absence of the above-described single contaminants.

RESULTS AND DISCUSSION

The grain size distribution for the sandy soil samples obtained from the Mascouche site and used for the study is given in Figure 1. The particle size analysis indicates that the soil is a silty sand. Figure 2 shows the concentration of heavy metals present in the test soil compared with the background concentration for a typical Québec soil and together with the level considered "contaminated" by the Québec Ministry of the Environment (1988). The results show that Pb and Cd far exceed the contamination levels established by the Québec government. The measured concentration of nutrients in the soil were: phosphorus 693 mg/kg, potassium 765 mg/kg, and nitrogen 220 mg/kg. These concentrations were increased to 903 mg/kg, 1,031 mg/kg, and 430 mg/kg, respectively, after the addition of nutrient salts in the AN cells.

Figure 3 shows the variation of the soil pH over the duration of the PAH disappearance experiments. All samples showed a linear decrease in pH with the exception of the control sample C. Sample AN demonstrated the greatest drop in pH. The sterile control sample C showed a slight increase in soil pH with time. The slightly acidic pH levels are not expected to affect the potential for biodegradability since the values are not below the limit conditions discussed by Leahy and Colwell (1990). Since most heterotrophic bacteria and fungi prefer near-neutral pH values, with fungi being more tolerant to acidic conditions (Leahy & Colwell 1990), fungi would be expected to be the main biodegrading agent.

The initial water content (27%) of the soil dropped significantly (to 3.1%) after 1 week of forced aeration, but only slightly thereafter. Since water content is essential for growth and metabolism of microorganisms (Leahy & Coldwell 1990), it was considered essential to

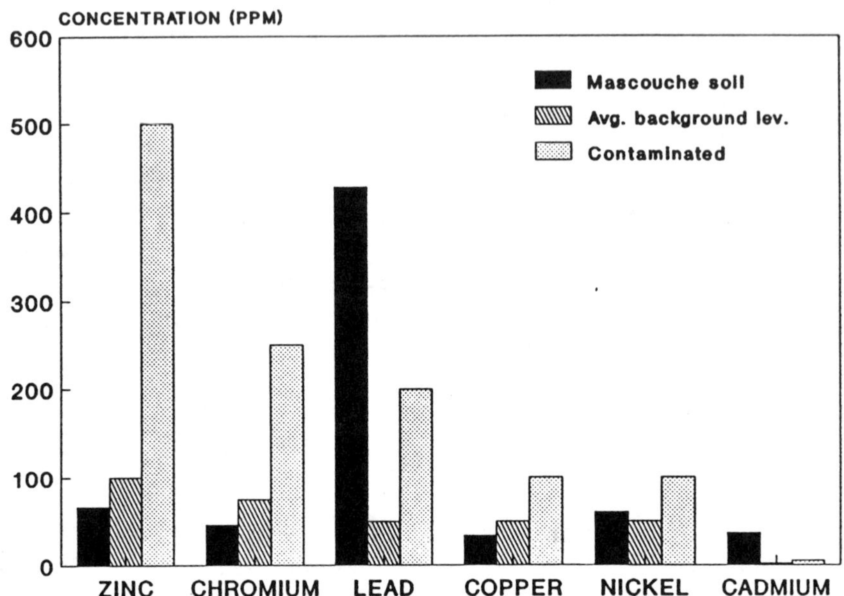

FIGURE 2. Heavy metals concentration.

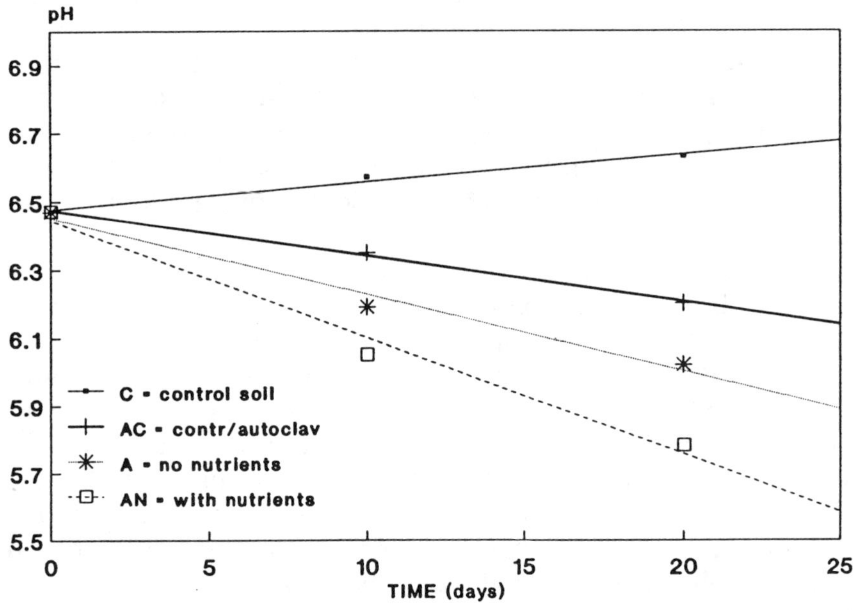

FIGURE 3. pH versus time.

maintain favourable water content conditions throughout the tests. Although pressure from forced aeration on the soil may have a slight inhibiting effect on the activity of the microorganisms (Leahy & Colwell 1990), this effect is considered minimal and insignificant for the pressure levels used. The total organic carbon (TOC) at the start of the experiment was approximately 2.76 percent. The small drop observed in the TOC for sample AN (2.66%) could mean that natural selection of microorganisms, as well as full development of the microbial population, may not have had time to occur (Kosson *et al.* 1987).

Disappearance of PAHs

The results shown in Figure 4 indicate biodegradation of PAHs in cell AN (the AN curve is lower than the AC curve) and the importance of added nutrients. A significant portion of the disappearance of PAHs occurred over the first 10 days. As shown in Figure 5, the AN curve is much lower than all the others, indicating that phenanthrene and/or anthracene was easier to biodegrade than the other contaminants. In Figure 6, the sample without added nutrients showed a much greater susceptibility to biodegradation of fluorene than the other samples.

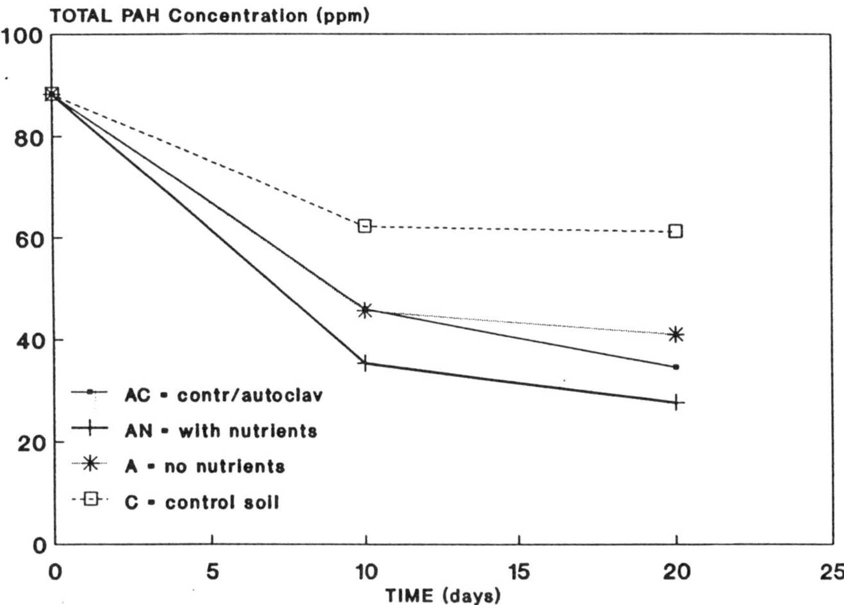

FIGURE 4. Disappearance of PAHs.

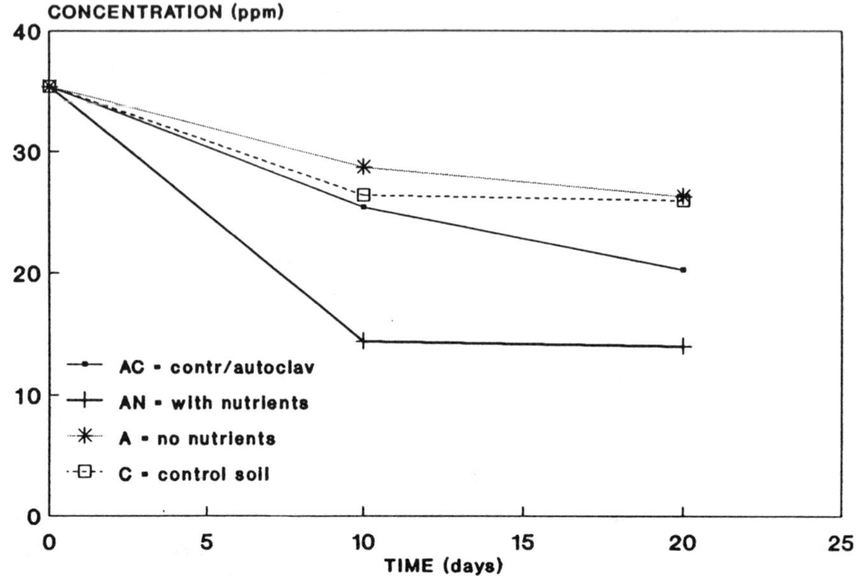

FIGURE 5. Degradation of anthracene and phenanthrene.

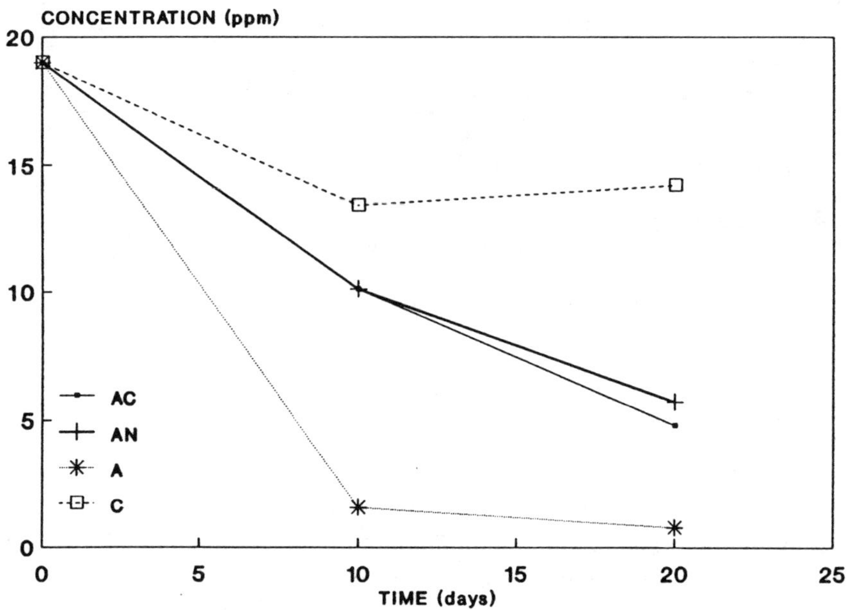

FIGURE 6. Degradation of fluorene.

These results indicate that the major factors in the disappearance of PAHs were nonbiological in nature. Approximately 80 percent of the PAH disappearance was due to physical or chemical parameters; only 20 percent was microbially transformed. The disappearance of anthracene and phenanthrene, however, demonstrates the effectiveness of bioremediation. The results in Figure 5 show that 52.4 percent of phenanthrene and anthracene was biodegraded. The results in Figure 6 indicate that microbial activity was responsible for the disappearance of fluorene. Since phenanthrene and anthracene show similar peaks in gas chromatograph measurements, they were added together in the results. However, phenanthrene is easier to degrade since it is approximately 20 times more soluble in water than anthracene (Leahy & Colwell 1990). Also, because of the bay area (Chakrabarty 1982) inherent in the biochemical structure of phenanthrene, it is exposed to enzymatic attack on three surfaces at the same time, whereas anthracene exposes only two of its surfaces. Fluorene demonstrated positive biodegradation as well. This is to be expected since it is more soluble in water than phenanthrene (*CRC Handbook* 1984), and mineralization rates of PAHs are related to aqueous solubilities rather than to total substrate concentration (Leahy & Colwell 1990).

Volatilization seems to be the major factor affecting the disappearance of PAHs in this experiment. Under natural conditions (i.e., no forced aeration), volatilization would have been negligible (Park *et al.* 1990). Forced aeration, combined with the decrease in water content and the low percentage of TOC (which increases the amount of contaminants exposed to air), stripped and volatilized most of the PAHs in the soil. PCBs such as Arochlor 1242, 1248, 1254, and 1260 were unaltered during the experiment and remained at 0.8, 0.2, 1.5, and 0.9 mg/kg, respectively, in the sandy soil.

Biological Tests

Of the 21 strains of microorganisms isolated from samples AN, A, and C, only 13 were actually tested for their degrading capacities. This is mainly because some of the strains were molds that could have been contaminants, and some strains could not be maintained. Only the results of strains that demonstrated positive degradation are included in this paper (Figures 7, 9, 11, 13, 15, 17, and 19). In Figures 7, 9, 13, 15, and 19, the plate count was higher for the minimal salt solution contaminated with phenanthrene (black bar graph) than for the control containing no contaminant (striped bar graph). This demonstrates that the microorganisms utilize phenanthrene as a source of carbon, and therefore, phenanthrene increases its population. In Figures 11, 13, and 18, the fluorene bar graph (black) is higher than the control bar graph (striped), showing that fluorene is utilized by those strains. In Figure 13, the bar graph for anthracene is higher than the control bar graph, indicating that this strain can use anthracene as its sole source of carbon. However, from Figures 9, 11, 15, and 19, the bar graphs indicate that anthracene has an inhibiting effect on growth of these strains.

Figures 8, 12, 14, 16, 18, and 20 show the variation in optical density over a period of 43 days. In the case of phenanthrene, the optical density decreased slightly before increasing. This was observed in most cases where phenanthrene was biodegraded. However, strains that showed no positive degradation, but some tolerance to phenanthrene, had the same shape of curve. In the cases where fluorene was biodegraded, the curve was mainly flat, so prediction of biodegrading capacity was not possible. In the case of anthracene, O.D. techniques proved impractical. O.D. curves must be compared to the control curve to avoid confusing a natural increase in population with growth of the microorganisms in the presence of a contaminant. Dean-Ross (1990) reported that growth of a phenanthrene-degrading, gram-negative bacteria could be measured by observing an increase in the optical density and in colony forming units in a liquid inoculum. However, Amador, Alexander, and Zika (1988) observed a decrease in optical density when monitoring the biodegradation of organic molecules in the presence of sunlight. Therefore, the initial drop in the O.D. curve for phenanthrene could indicate the emulsifying effect the bacterial or fungal enzymes may have on the chemicals (Leahy & Colwell 1990), as well as a certain degree of photodegradation and possibly adaptation (Leahy & Colwell 1990) during which many of the intolerant microorganisms die. The subsequent increase in optical density of the samples that demonstrates population increase could result from the growth of the microorganisms that increases the turbidity of the broth culture.

From the results of the biological tests performed on the individual strains (Figure 7, 9, 13, 15, and 19), phenanthrene seems to be the chemical that supported the most growth, thus is most susceptible to microbial attack. Anthracene showed the least susceptibility to microbial attack and had some inhibiting effects on certain strains (Figures 9, 11, and 15). This could explain the asymptotic return of curve AN in Figure 5, since most of the phenanthrene disappeared, while the more resistant anthracene remained in the soil. In Figure 15, no results are shown for fluorene because the plates were found to be

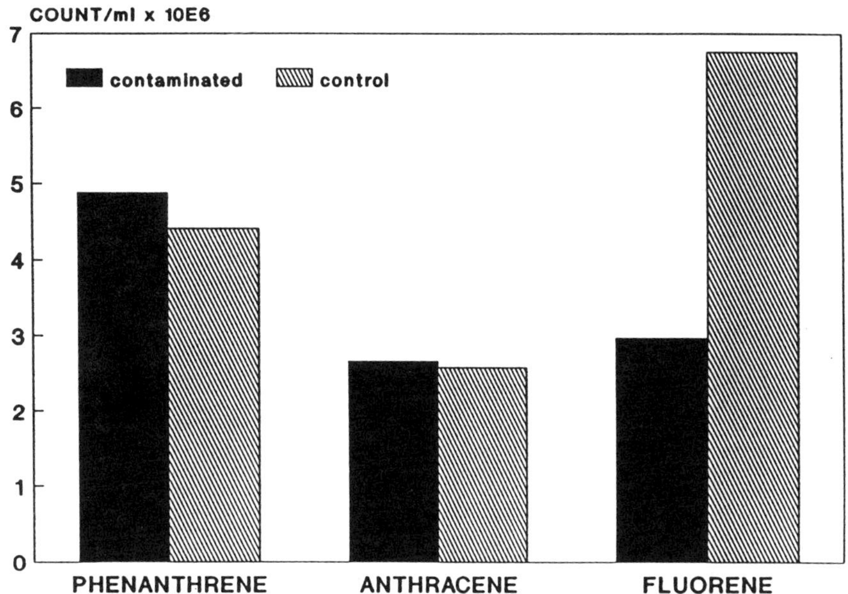

FIGURE 7. Plate count (strain #4).

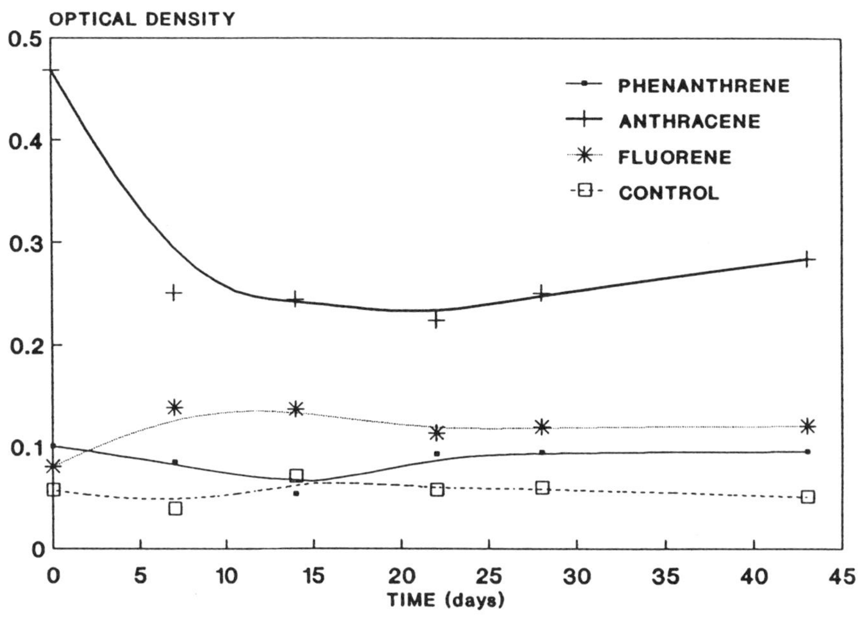

FIGURE 8. Optical density (strain #4).

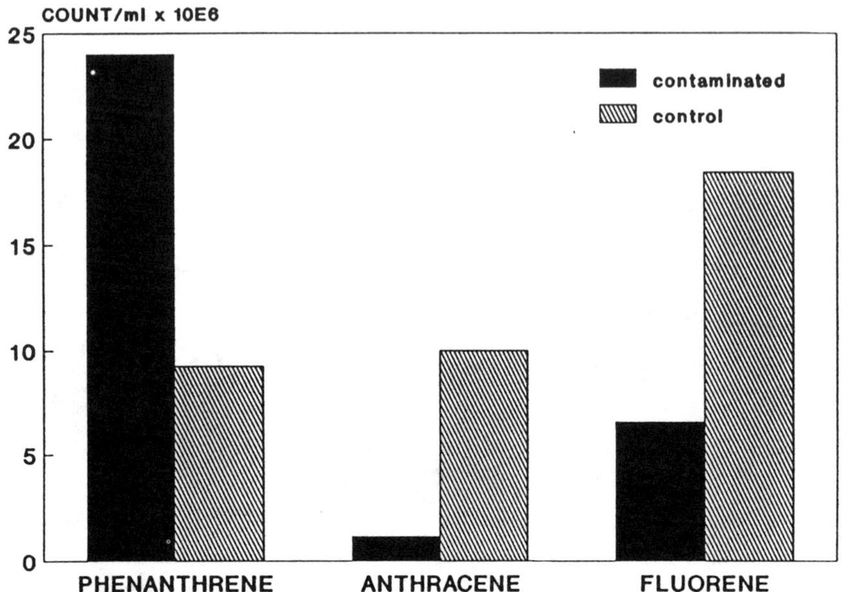

FIGURE 9. Plate count (strain #8).

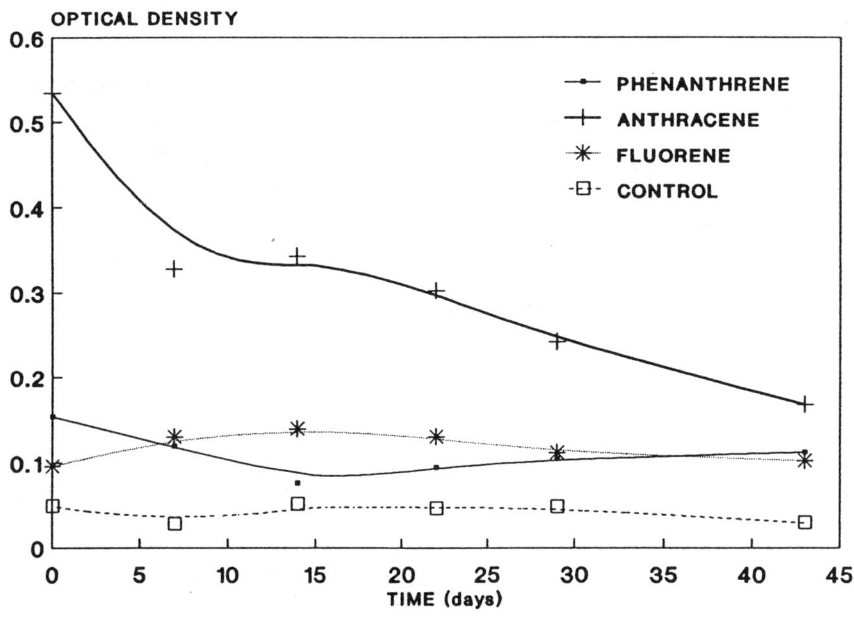

FIGURE 10. Optical density (strain #8).

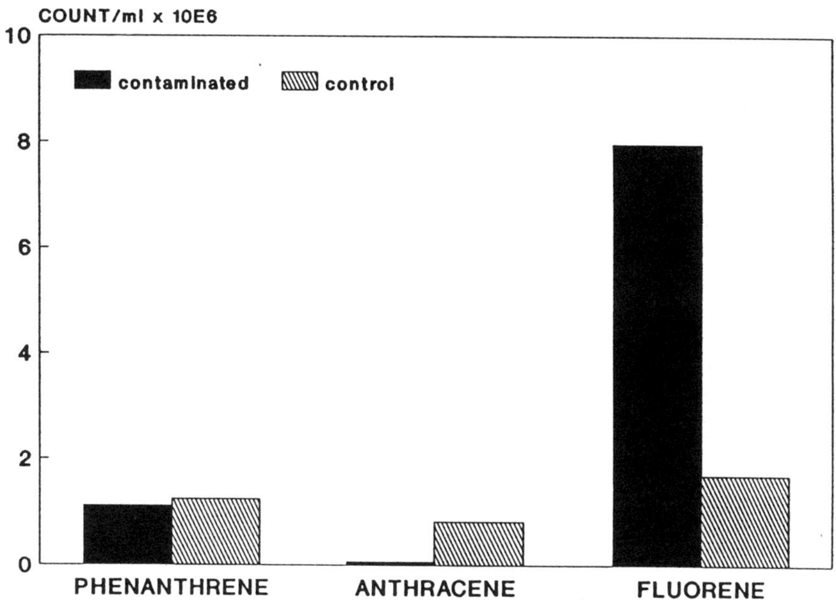

FIGURE 11. Plate count (strain #9).

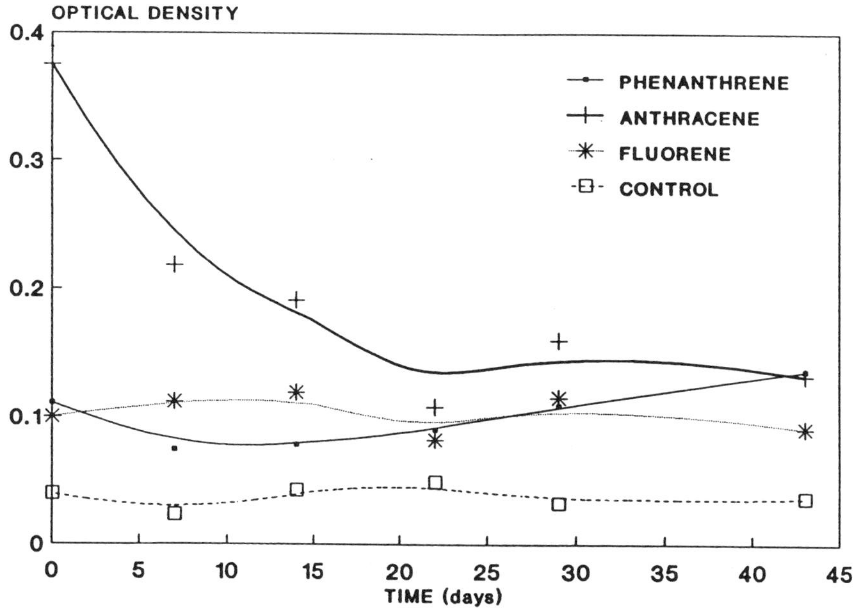

FIGURE 12. Optical density (strain #9).

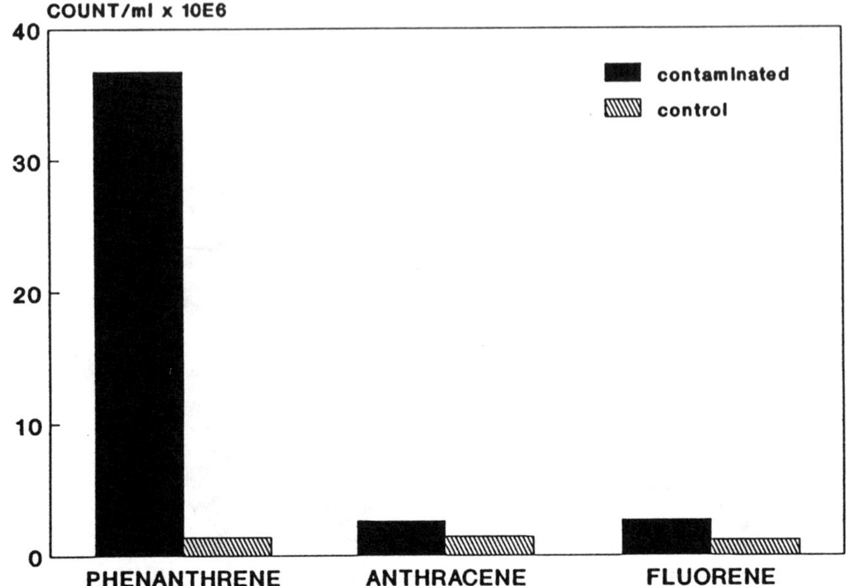

FIGURE 13. Plate count (strain #12).

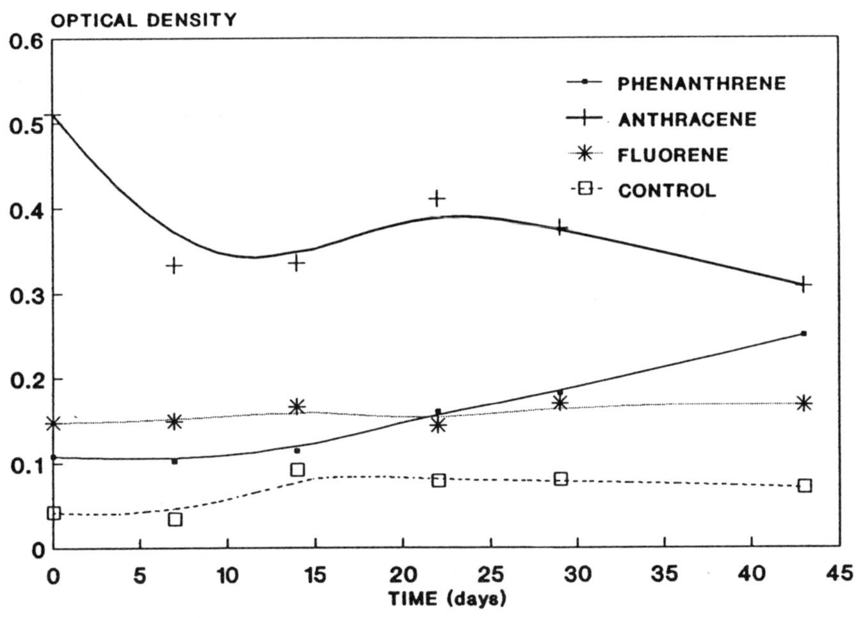

FIGURE 14. Optical density (strain #12).

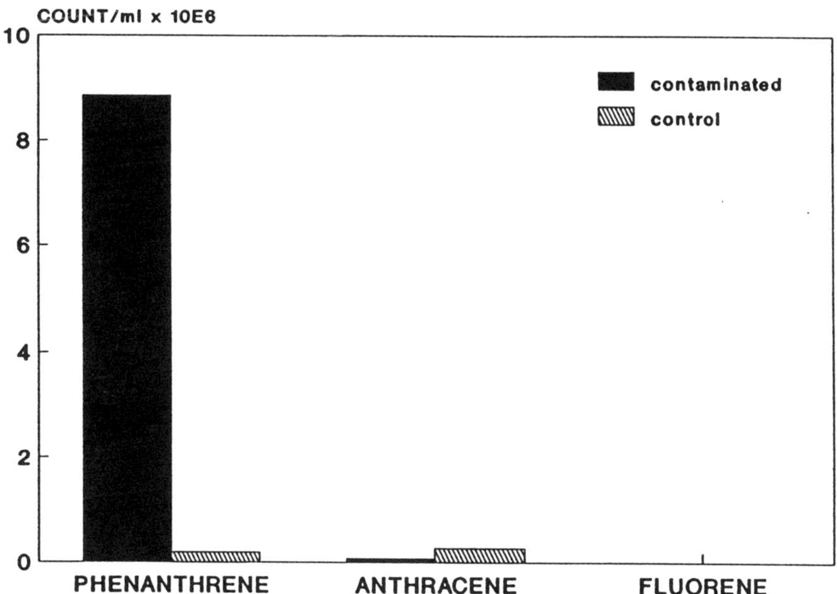

FIGURE 15. Plate count (strain #16).

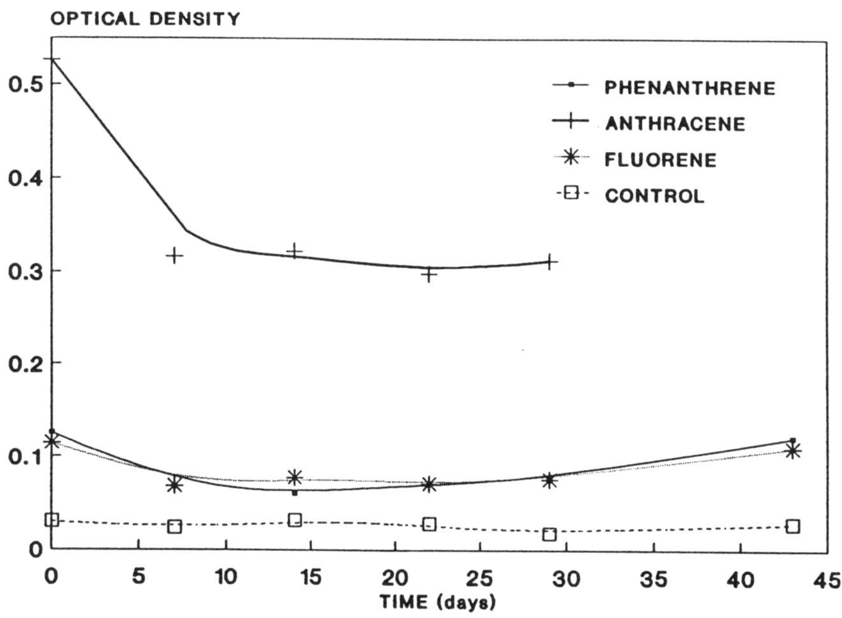

FIGURE 16. Optical density (strain #16).

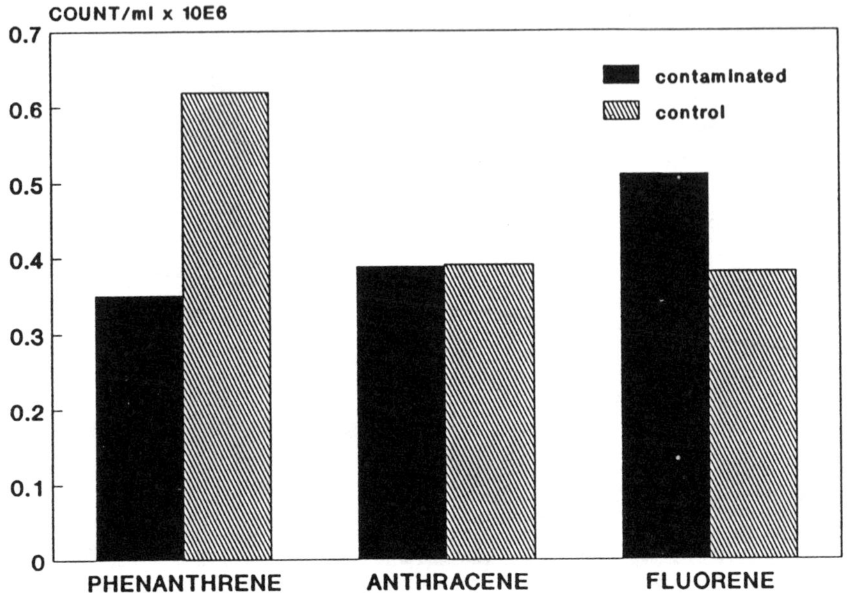

FIGURE 17. Plate count (strain #18).

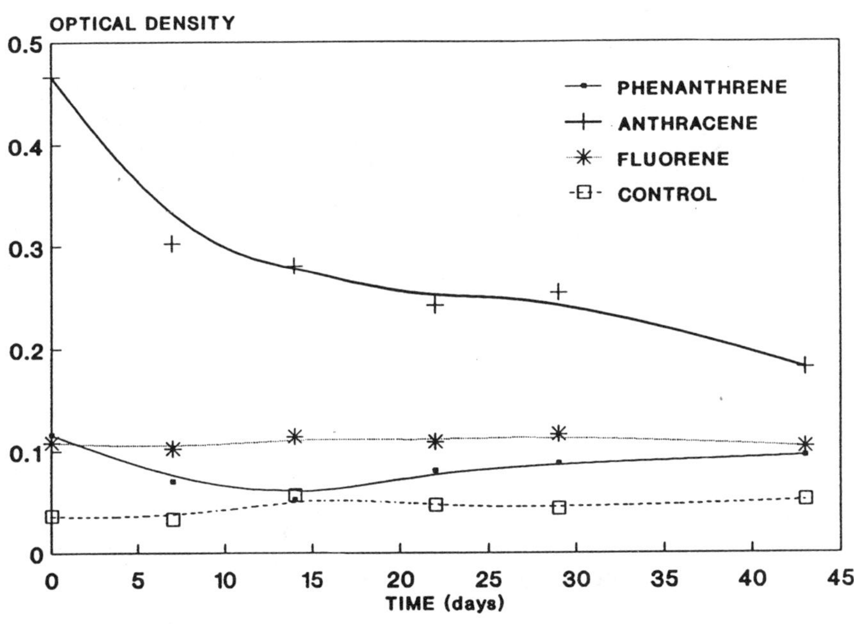

FIGURE 18. Optical density (strain #18).

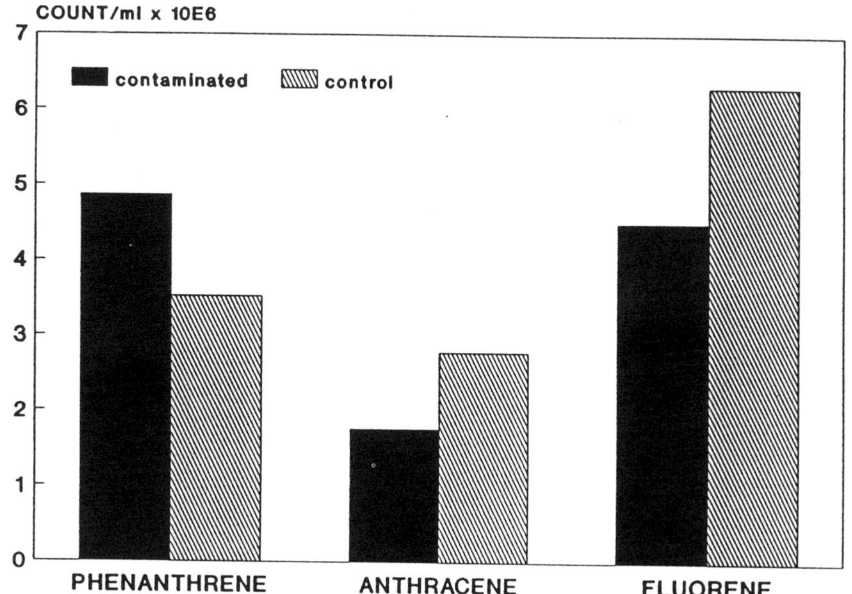

FIGURE 19. Plate count (strain #21).

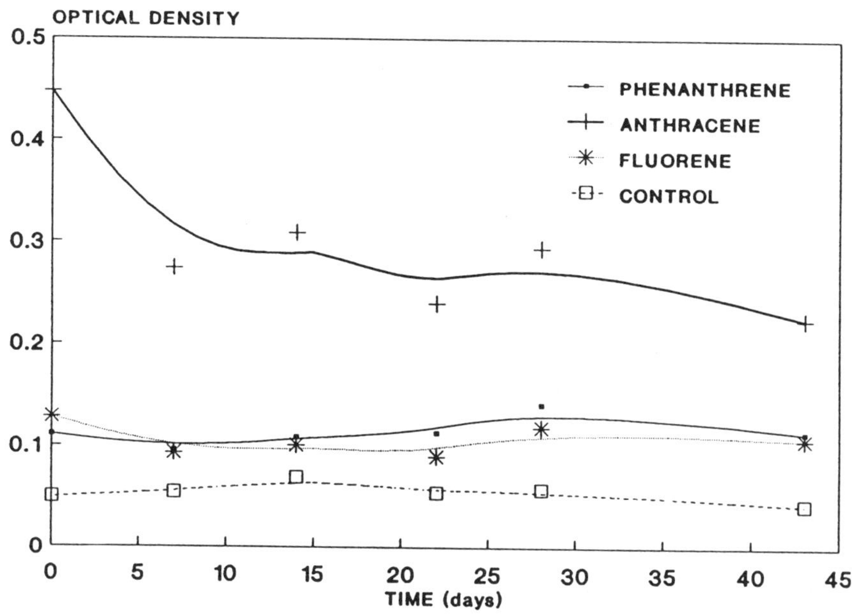

FIGURE 20. Optical density (strain #21).

contaminated during the count. Contamination also negated results for anthracene after 28 days.

Gram stains were performed on the strains that showed increased growth (positive degradation) in the presence of the three chemicals studied (Figures 7, 9, 11, 13, 15, 17, and 19). Strains 4, 8, and 21 were gram-negative rods, with strains 8 and 21 being pleormorphic. Strains 9 and 16 were yeasts. Strains 12 and 18 contained two microorganisms, which were difficult to separate at the start of the tests. Strain 12 consisted of a yeast and another fungus, and strain 18 consisted of a yeast and a gram-negative rod. All the yeast in these tests were of the same genera.

All the yeasts were identified as being *Candida parapsilosis* by the Royal Victoria Hospital Clinical Laboratories in Montreal. This type of yeast had been isolated from a soil in 1965 by Asahi Kaisha (Jong & Gantt 1987) and patented for its ability to degrade hydrocarbons. Equally, *Candida parapsilosis* has also been patented by Bioteknika International (Jong & Gantt 1987) for its ability to degrade petroleum. Parkinson (1974) stated that a yeast of the *Candida* genus could degrade petroleum in arctic soils. *Candida parapsilosis* can also produce citric acid (Jong & Gantt 1987), which could explain the decrease in soil pH in the sample that showed the most biodegradation, as well as in the other samples. The other strains of microorganisms are presently being identified.

CONCLUDING REMARKS

This study reinforces the view that base-value characterization of test technique—for assessment of effectiveness of treatment procedures and parametric evaluation—prior to scaled-up field experiments is necessary. The importance of nonbiological parameters in a forced aeration system has been demonstrated. During decontamination procedures involving this type of technique, a transfer of contaminants from the soil to the atmosphere could occur if no precautions are taken. If by-products and pathways of biodegradation are not fully documented, contaminants could be masked and may not be readily identified by gas chromatography. Nutrient salts appear to have made a difference in the biodegradation process in the sandy soil tested.

Acknowledgments

The authors wish to acknowledge the support of the Québec Ministry of Education via operating Grant FCAR.

REFERENCES

Amador, J. A.; Alexander, M.; and Zika, R. G. *Abstract of Papers*, Annual Meeting of the American Society for Microbiologists; American Society for Microbiologists: Washington, DC, 1988; Q-60.

Bailey, A.; Coffey, M. *Can. J. Micro.* **1986**, *32*, 562–569.

Bowles, J. E. *Engineering Properties of Soils and Their Measurement*, 3rd Ed.; McGraw-Hill Co.: New York, 1986.

Bumpus, J. *J. Appl. and Environ. Micro.* **1989**, *55*, 154–158.

Chakrabarty, A. M. *Biodegradation and Detoxification of Environmental Pollutants*; CRC Press Inc.: Cleveland, OH, 1982.

CRC Handbook of Chemistry and Physics, 64th Ed.; CRC Press Inc.: Cleveland, OH, 1984.

Dean-Ross, D. *Abstract of Papers*, Annual Meeting of the American Society for Microbiologists; American Society for Microbiologists: Washington, DC, 1990; Q-116.

Dunn, R.; Mitchell, J. J. of Geotech. Eng. **1984**, *110*(11), 1648–1665.

Gouvernement du Québec, Ministère de l'Environnement. *Procédure d'Evaluation des Caractéristiques des Déchets Solides et des Boues Pompables*; 1985.

Gouvernement du Québec, Ministère de l'Environnement. *Politique de Réhabilitation des Terrains Contaminés*; 1988.

HACH, Analysis of Wastewater, Liquids, Solids and Sludges: Selected Methods; HACH Co., 1987.

Herbes, S.; Schwall, L. *J. Appl. and Environ. Micro.* **1978**, *35*, 306–316.

Jong, S.; Gantt, M. *Catalogue of Fungi/Yeasts*, 17th Edition; American Type Culture Collection: Rockville, MD, 1987.

Kosson, D.; Agnihotri, G.; Ahlert, R, *J. Haz. Mat.* **1987**, *14*, 191–211.

Leahy, J.; Colwell, R. *Micro. Rev.* **1990**, 54, 305–315.

Park, K.; Sims, R.; Dupont, R. *J. Environ. Eng.* **1990**, *116*, 632–640.

Parkinson, D. Report No. 74-2, 1974; Environment Canada.

Parkinson, D.; Gray, T.; Williams, S. *Ecology of Soil Microorganisms*; IPB Handbook No. 19, 1971.

U.S.D.A. Agricultural Handbook No. 60; U.S. Department of Agriculture, 1954, p 105.

Laboratory and Field Tests for a Biological *In Situ* Remediation of a Coke Oven Plant

N.-Ch. Lund, J. Świniański, G. Gudehus, and D. Maier*
University of Karlsruhe

INTRODUCTION

For a biological *in situ* remediation of soils polluted with hydrocarbons, a geotechnical treatment has been developed. The treatment was examined in a large-scale sample test, where the biodegradation of hydrocarbons which are typical for coal tar processing products has been investigated. During this test an ozone-injection was made to pre-oxidize the hydrocarbons in order to increase their biological availability. At present, the treatment is tested *in situ* at a heavily contaminated part of an abandoned coke oven plant.

BIOLOGICAL *IN SITU* REMEDIATION WITH A COMBINED AIR-WATER FLUSHING

Certain microorganisms possess the ability to use hydrocarbons as a source of energy and carbon for their metabolism and to decompose them into carbon dioxide, water, and biomass (i.e., new microorganisms). Depending on the kind of hydrocarbons and the portion of the hydrocarbons that is decomposed into biomass, an aerobic

* Department of Soil Mechanics and Foundation Engineering, University of Karlsruhe, Postbox 6980, 7500 Karlsruhe, Federal Republic of Germany

biodegradation of 1.0 kg of hydrocarbons requires that approximately 1.5 to 3.5 kg of oxygen be provided to the microorganisms in the subsoil (Lund & Gudehus 1990a).

Current biological *in situ* remediation techniques solve this task by injecting water into the flushed contaminated subsoil. The water is circulated using infiltration and extraction wells and is saturated with air oxygen or pure oxygen, or is enriched with hydrogen peroxide or ozone. Sometimes nitrates are also added.

The success of this kind of water circulation system is questionable for technical, economic, and time reasons (Lund & Gudehus 1990a). Therefore, a new geotechnical treatment has been developed. The basic concept behind this treatment is that the oxygen needed for biodegradation is supplied to the microorganisms by creating a nearly horizontal airflow in the unsaturated subsoil with a network of positive (injection) and negative (suction) lances placed in the soil. Besides this aeration of the soil, a vertical flow of percolated water produced by a water sprinkling system installed on the surface is needed to activate the microorganisms in the contaminated soil. Depending on the solubility, the hydrocarbons attached to the soil dissolve into the water and are then accessible to the microorganisms. Together with the water, the needed inorganic nutrients, ammonia, and phosphate can be transported to the microorganisms.

Figure 1 shows the mode of operation between the airflow, the water flow, the hydrocarbons, and the microorganisms during a biological remediation of a contaminated soil. Depending on the water content of the soil, the airflow will occur in the bigger pores and the water flow in the smaller pores. If the hydrocarbons dissolve into the water and are then decomposed by the microorganisms, the consumed oxygen in the water is replenished by diffusion from the air. As a result, the concentration of oxygen in the water remains constant.

In addition, Figure 1 shows that the pollution in the subsoil is not only decreased through biological degradation, but also through the outflow of hydrocarbons with the air and water streams. After passing through the contaminated soil, the air contains volatile hydrocarbons and the water contains dissolved hydrocarbons.

LABORATORY TESTS WITH LARGE-SCALE SAMPLES

The method described above was tested first in two laboratory experiments (Test I and Test II). To imitate field conditions, undisturbed

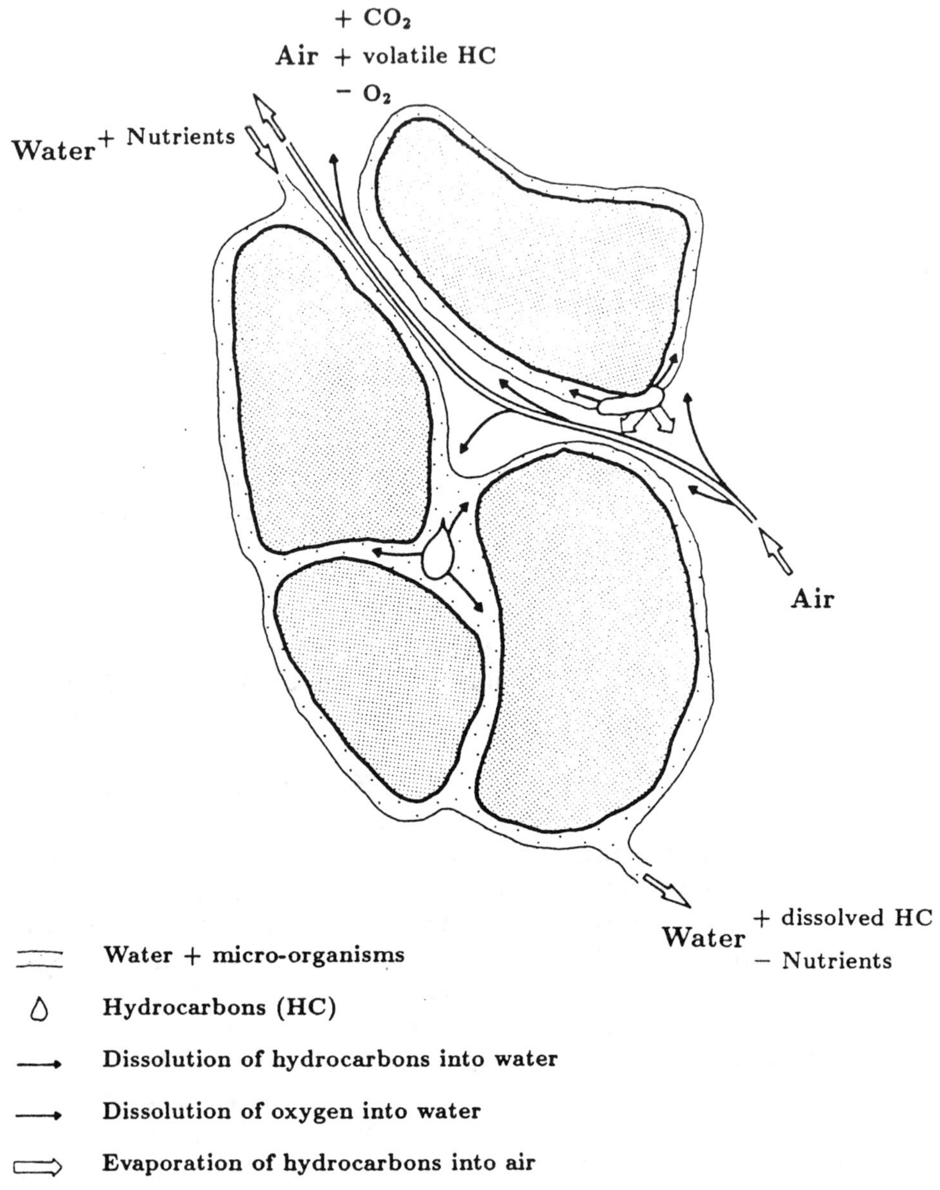

FIGURE 1. Mode of operation between air, water, hydrocarbons, and microorganisms in the unsaturated subsoil.

samples with a diameter of ϕ = 0.6 m and a height of h = 1.0 m were used, which were taken out from the ground using ground-freezing techniques (Figure 2). At the beginning of each test, the sandy, gravelly soil was artificially contaminated with a solution of hydrocarbons. The

FIGURE 2. Taking a large-scale sample using ground-freezing techniques.

solution was a mixture of the aliphatic hydrocarbons: *n*-decane (94.33 weight %), pristane and *n*-hexadecane (both 1.9 weight %), and the aromatic hydrocarbon naphthalene (1.9 weight %). The tests showed that the proposed procedure is suitable for activating microorganisms in a subsoil polluted with comparable hydrocarbons to decompose the contaminants. Thus, in Test II the initial concentration of hydrocarbons of 10,000 mg/(kg soil) was reduced to 1,500 mg/(kg soil) during 56 days. Detailed results of Tests I and II are described by Lund and Gudehus (1990b) and Lund (1991).

The next experiment (Test III), presented in the following section, was conducted as a preparation to a pilot biological *in situ* remedial action at a heavily contaminated part of an abandoned coke oven plant in Karlsruhe. The purpose of the test was to investigate the biological degradation of hydrocarbons that are typical for coal tar processing products. The degradation of these hydrocarbons is limited by their low biological availability for microorganisms, which depends mainly on the following factors:

- The solubility of the pollutants in water
- The chemical structure of the contaminants
- The ability of the microorganisms to take the organic compounds into their cells
- The ability of microorganisms to emulsify the hydrocarbons by secreting special substances before taking them into their cells
- The degree of contamination and the adsorbtion of the pollutants on the soil grains, which influences the local water and air permeabilities

Different methods for increasing the biological availability of hydrocarbons are discussed in the literature. For example, some solvent agents were used to improve their solubility in water (Battermann & Werner 1987), or acclimatized or mutant microorganisms were added into the treated soil (Bewley & Theile 1988). For biological wastewater treatment, the insertion of ozone is well known (Maier 1989). There, ozone is used for two purposes. First, the dissolved ozone in water is, as mentioned above, an oxygen donator for microorganisms. Second, as a strong oxidizing agent, ozone degrades the hazardous compounds directly or creates oxygenated compounds, which are more amenable for microorganisms.

Test Procedure

The large-scale sample for Test III (ϕ = 0.6 m, h = 0.63 m) was a mixture of sandy, gravelly polluted soil and contaminated broken concrete from the abandoned coke oven plant site in Karlsruhe, in proportion

≈ 2:1, respectively. To get a uniform, reference initial concentration of the contamination, 450 kg of the material were carefully homogenized and then installed in a pressure cell of the testing stand.

The test began with a hydraulic cleaning up of the sample which continued for a period of 10 days. During the soil flushing, a vertical water permeabilty of the sample of 1×10^{-6} m/sec was measured.

In spite of the low permeability, it was possible to inject, during the biological remediation (starting on the 11th test day), air and water containing ammonia and phosphate into the unsaturated sample. From the 48th test day onward, an alternating air and water supply was used to lower the degree of saturation of the sample. At the beginning of each cycle, the sample was saturated; during the next 24 h, water containing nutrients was injected. After the soil flushing, the sample was drained and aerated for several days.

After about 100 days of supplying the microorganisms with air oxygen and water containing nutrients, only a partial, rather slight reduction of contamination could be obtained (see Test Results section). Therefore, between the 132th and 137th test day an injection of ozone was applied. In this case the ozone was not, as in water treatment, dissolved in water, but in order to increase the concentration of the oxidizing agent, it was injected as a gas directly into the unsaturated sample. The ozone concentration in the supplying air was 30 g/m^3. During the injection, only residual concentrations of ozone were measured in the spent air. Finally, a net amount of 520 g/O_3 was delivered into the sample.

After the injection of ozone, the microorganisms were again supplied with air and water containing nutrients, which were injected into the sample simultaneously. The test was stopped on the 162th day. During the test, the temperature in the pressure cell was constantly controlled, averaging about 21 C.

Test Results

To control the concentration of the contamination in the different phases, five soil samples were taken from the large-scale sample during the test. The results of the chemical analyses of all the samples are presented in Table 1. Sample numbers 1, 2, and 4 were the mixed specimens taken from the whole soil material at the beginning and end of the test. Sample 3 was a local core specimen taken on the 104th test day. A sampling tube was driven into the cylindric sample at the distance of half of its radius. Sample L was also a local core specimen

TABLE 1. Results of analyses of the soil samples from Test III. All concentrations in mg/(kg soil < 2 mm); u.d. = under detection limit.

SAMPLE NUMBER:	1	2	3	4	L
SAMPLE TYPE:	mixed	mixed	local	mixed	local
TEST TIME [d]:	0	0	104	162	162
SUM PARAMETERS					
Organic compounds	11135	9875	11150	9100	7300
Aromatics (sum/non-polar)	4475 /126	3525 / 322	2856 / 48	3200 / < 10	2750 / < 10
Aliphatics (sum/non-polar)	900 / 70	720 / 220	832 / 296	900 / 175	750 / 315
POLYCYCLIC AROMATIC HYDROCARBONS (PAHs)					
Naphthalene	184	134	33	316	214
1-Methylnaphthalene	61	5	u.d.	40	26
2-Methylnaphthalene	37	28	u.d.	59	40
Acenaphthylene	95	78	17	36	25
Acenaphthene	36	33	u.d.	5	2
Fluorene	122	104	4	16	12
Anthracene	98	106	u.d.	u.d.	u.d.
Phenanthrene	326	276	29	u.d.	u.d.
Fluoranthene	224	196	133	122	88
Pyrene	182	156	145	113	85
Benzo(a)anthracene + Chrysene	172	165	153	50	56
Benzo(b)fluoranthene + Benzo(k)fluoranthene	109	90	110	u.d.	u.d.
Benzo(a)pyrene	65	57	131	u.d.	u.d.
Indeno(1,2,3-cd)pyrene + Dibenzo(a,h)anthracene	55	30	108	u.d.	u.d.
Benzo(g,h,i)perylene	50	11	85	u.d.	u.d.
Sum PAH	1816	1363	948	758	547
BENZENE, TOLUENE, XYLENE (BTX) AND PHENOLS					
Sum BTX	10.7	3.0	u.d.	u.d.	u.d.
Sum phenols	17.0	u.d.	u.d.	u.d.	u.d.

removed from the vicinity of the injection lance placed in the middle of the sample at the end of the test. Before the chemical analyses, each soil sample was sieved and dried by freeze-drying. To make all the analyses comparable, only soil material with grains smaller than 2 mm was used.

To estimate the sum of the organic compounds, each soil sample was extracted with two different solving agents (SOXHLETT-Extraction), and then the mass of the extracts was gravimetrically determined. Toluene and carbon disulfide (CS_2) were applied as solvents. The values presented in Table 1 are mean values of the two extracts. At the beginning of the test and on the 104th test day, the sum concentration of the organic compounds was about 11,000 mg/(kg soil < 2 mm). After the injection of ozone, the organic compounds in the vicinity of the lance were reduced by 30 percent.

The concentrations of the aliphatic and aromatic hydrocarbons were determined from the CS_2 extract by infrared spectroscopy (IS). The concentration of the hydrocarbons detected by the IS method was only about 45 to 50 percent of the total organic compounds. This can be attributed, in part, to the fact that the heterogroups of the organic compounds cannot be detected using IS. During the test only the non-polar aromatic hydrocarbons were reduced by 30 percent. In the case of the sum of the aliphatic hydrocarbons, virtually no changes could be found.

The concentration of the individual polycyclic aromatic hydrocarbons (PAHs) revealed various changes. The lower condensed PAHs (*e.g.*, naphthalene) were significantly reduced after the first 104 test days. During this time, however, the higher condensed PAHs did not change their concentrations (*e.g.*, pyrene) or even increased (*e.g.*, benzo(a)pyrene). But after the injection of ozone, some relevant higher condensed PAHs could not be detected at all (samples 4 and L), while some lighter PAHs, especially the naphthalene compounds, even increased their concentrations. Altogether the sum concentration of the analyzed PAHs could be reduced by 40 percent during the whole test. Benzene, toluene, xylene (BTX), and phenolic compounds were virtually nonexistent in the soil samples.*

During the test in water samples, the concentrations of the following and other parameters were determined: aromatic and aliphatic hydrocarbons, PAHs, BTX, phenols, dissolved organic carbon (DOC),

---

* The PAHs, BTX, and phenols were always determined in the toluene and CS_2 extract of the soil by gas chromatography. The values presented in Table 1 are the mean values of the results of both analyses. The scatter of results of the two analyses was negligible.

chemical oxygen demand (COD), ammonia, nitrate, nitrite, and phosphate.

During the hydraulic cleaning and only in the first days of the biological remediation, small concentrations of hydrocarbons (*e.g.*, sum of PAHs < 5 mg/L) could be found in the water samples. The values of the DOC were 220 mg C/L at the beginning and 30 mg C/L at the end of the biological remediation. A maximum of 650 mg/L and a minimum of 15 mg/L of the COD were also measured at the beginning and end of biological remediation, respectively.

The analyses of the nitrogen compounds showed that in the middle of the test duration a transformation of ammonia into nitrite and then into nitrate (nitrification) occurred. Already, in Test I and Test II nitrification processes appeared after the concentration of the pollutants decreased. Probably, when the concentration and/or biological availability of the hydrocarbons is low, the nitrifying bacteria grow better and hinder the biological activity of the heterotrophic bacteria degrading the organic compounds. Apart from the nitrification, a significant consumption of phosphates was observed in the sample.

In the spent air of the sample, the concentrations of CO_2 and O_2 were measured to observe the biological activity. The interpretation of these measurements is shown in the Figure 3, where the specific consumption of O_2 and the specific production of CO_2 are presented. After a take-off time at the beginning of the biological remediation, both curves reached their first maxima. Second maxima could be observed after the injection of ozone. The fluctuations that can be seen on the graphs result from the above-described cyclic procedure of the test. Totally, during the biological remediation of Test III about 1,540 g of oxygen were consumed and 1,330 g of carbon dioxide were produced.

Interpretation of the Results

Comparison of the analyses of the first three soil samples (1, 2, and 3) shows that by using a "natural" biological remediation, which means injecting air and water containing nutrients, only some of the hydrocarbons typical for gas-work contaminations (*e.g.*, the lower condensed PAHs) can be mineralized and transformation processes that format interim products predominated.

The above statement is also supported by a comparison of Test III with Tests I and II. In Test III, the maxima of O_2 consumption and CO_2 production were 4 to 6 times smaller than in Test II, in which the test conditions were similar except that the hydrocarbons had a

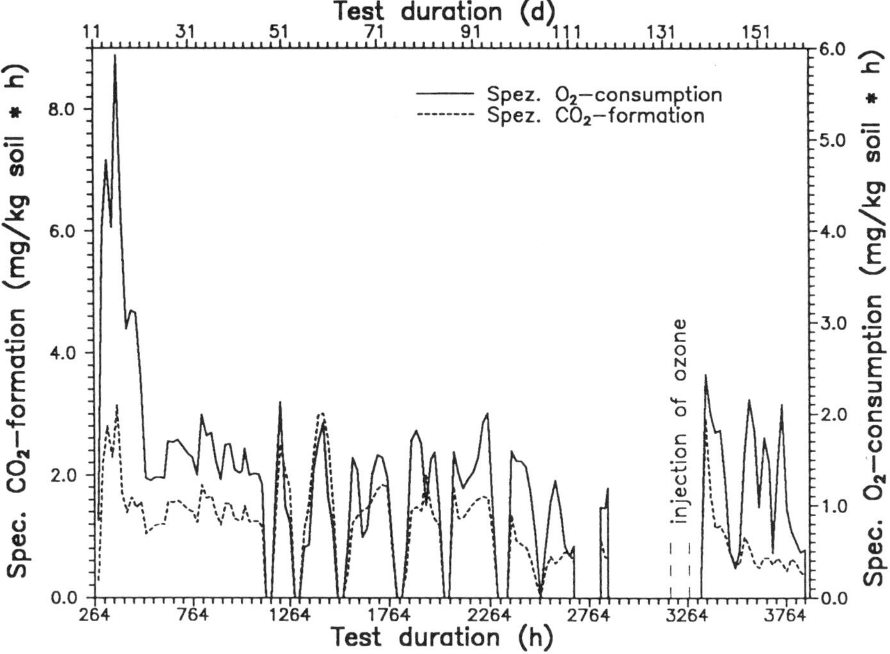

FIGURE 3. Specific oxygen demand and specific carbon dioxide formation during Test III.

higher biological availability. The CO_2 and O_2 measurements of Test II are shown in Figure 4.

It can be stated, as a hypothesis, that the transformation processes were the reason for the increase in concentrations of some higher condensed PAHs in the first 100 days of biological remediation. Thus, the biological activity in the sample could cause some compounds not detected previously to convert into the higher condensed PAHs found in sample 3.

The changes in concentration of the contamination after the injection of ozone are interesting and promising. Thus, a significant reduction of the sum of the organic compounds and aromatic hydrocarbons was detected after the ozonization. Different explanations are possible: One could be that the oxidation of the organic compounds could partially result in their degradation into CO_2 directly. Next, some compounds could have been cracked and/or changed into low-condensed PAHs, which could explain the increase of the concentrations of the naphthalene compounds. These products are better biologically degradable, which was confirmed by the reduction of the

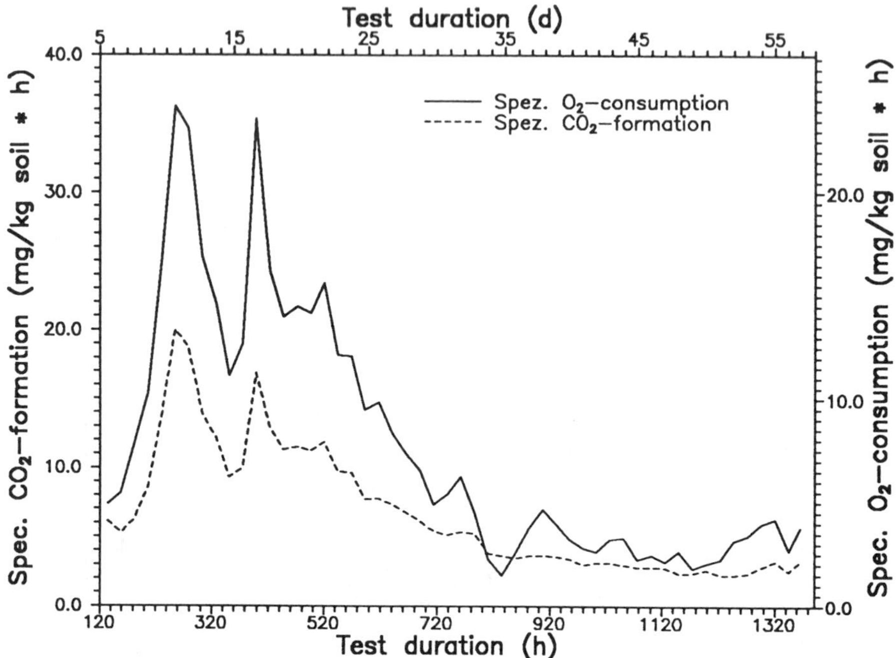

FIGURE 4. Specific oxygen demand and specific carbon dioxide formation during Test II.

naphthalene compounds in the first part of the biological remediation (sample 3) and the increase of CO_2 formation and O_2 consumption after the injection of ozone. Reduction of the organic compounds cannot be explained by washing out of the soil because higher concentrations of contaminants were not found in the water samples after the injection of ozone.

Taking into account the mean concentrations of the organic compounds (samples 1 and 2) and the whole grain distribution curve of the testing material, it can be estimated that the large-scale sample of Test III contained about 2.0 kg of organic compounds at the beginning of the test. At the end of the test (sample 4), about 1.7 kg of organics remained in the whole sample. The reduction by 0.3 kg lies in the range of biological degradation of hydrocarbons which can be approximately estimated from the consumption of oxygen and the production of carbon dioxide. The part of the contamination removed from the soil with the spent air (not measured) and water was not included in this simple balance.

Conclusions from Test III

The individual results of Test III showed various trends; therefore, the explicit interpretation of the test is not possible until now. For example, reducing the higher condensed PAHs below the detection limits at the end of the test is very promising, especially in comparison with other published test results (*e.g.*, Bewley & Theile 1988). On the other hand, the sum parameters (*e.g.*, the concentration of organic compounds or the aliphatics) revealed only a slight decrease. Therefore, it should be assumed in the remediation practice that in estimating degradation rates, not only individual concentrations of hydrocarbons but also sum parameters should be used.

As a final conclusion, the hypothesis can be proposed that the combination of a natural biological remediation with an ozonization can be an efficient treatment to clean up polluted soils at the sites of coke oven plants. To minimize the treatment costs, injection of ozone should occur once, before the biological remediation, and repeated later if necessary. A long-term, complete chemical oxidation (if possible at all) should be avoided because of high costs.

FIELD TEST

At present, the above-described natural biological remediation is being tested under field conditions at a contaminated part of an abandoned coke oven plant in Karlsruhe.

During the geological investigation at the gas-work site, the following general subsoil profile was determined: directly below the surface a fill layer 2 to 3 m thick was found. The fill material is contaminated and heterogenous and comprised remainings of concrete walls, brick walls, and foundations. Below, a sandy, gravelly, natural aquifer with lenses of different permeability extends about 17 m deep. To about the 10-m depth, this soil is heavily contaminated with gas-work substances. The aquifer is underlaid by impervious clayey soils.

The test site has the form of two rectangular fields (each 9 × 15 m) separated from the surrounding area by sealing walls (Figure 5). The walls end at a depth of 17 m in the clayey soil. In this closed test area, it is possible to lower the groundwater table in order to work in unsaturated conditions. Also, in this manner the groundwater outside of the test fields is protected against additional contamination during the remedial operation.

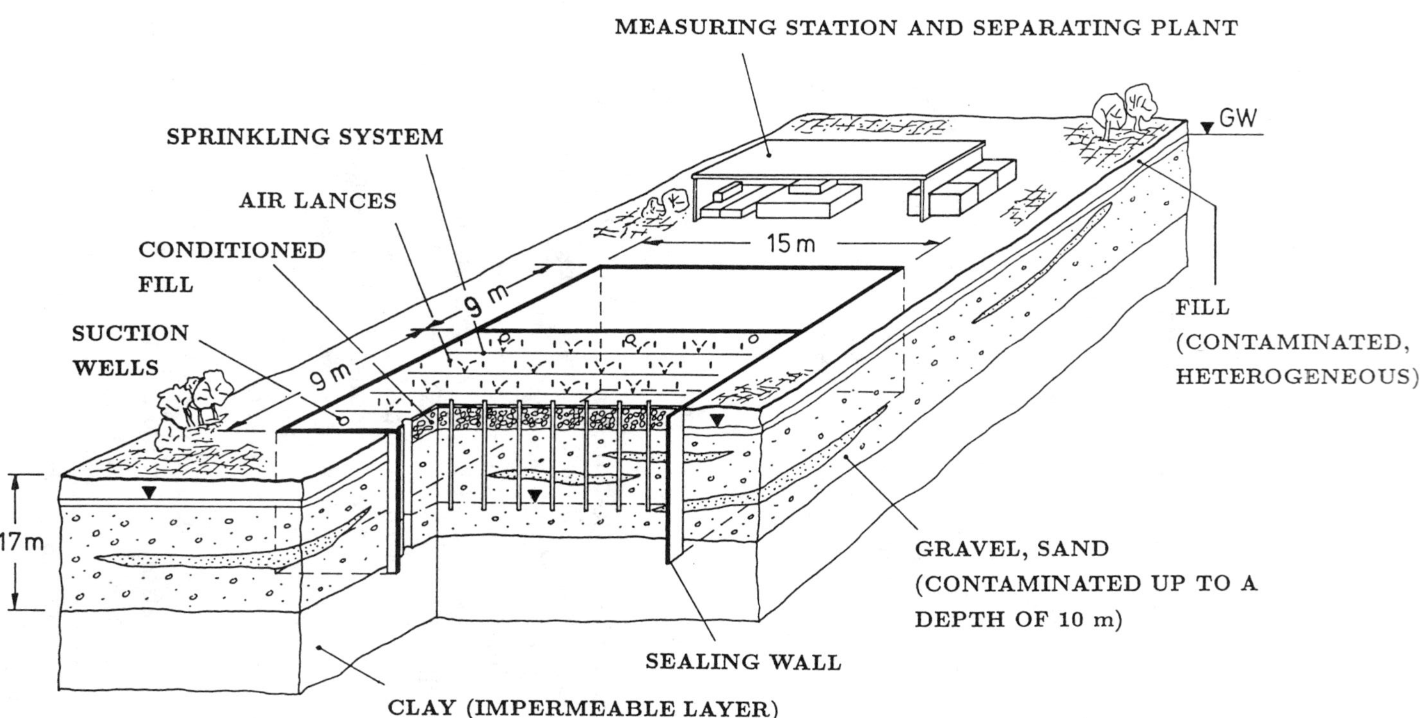

FIGURE 5. Schematic view of the field test equipment at the coke oven plant.

Before installing the walls, the fill material inside the test area was specially prepared (Figure 6): the soil was excavated and the old concrete, brick walls, and foundations were sorted out and broken. Then, this homogenized material was mixed with excavated soil and poured back into the excavation. Totally, 4,000 m^3 of soil material were removed and 350 m^3 of concrete were broken. An alternative to conditioning the fill was depositing it as waste, which would have cost about $100,000. The conditioned fill material will also be treated during the biological remediation.

The scheme of the surface and subsurface installations of field 1 of the testing plant are shown in the Figure 7. Among other measures, a net of air lances was lowered down into the soil, and a water sprinkling system was installed. By creating an air pressure gradient between individual lances or groups of lances, it is possible to impose an airflow in the unsaturated soil. For the aeration of less-permeable layers or lenses, a system of packers has been developed. The water sprinkling system produces a flow of percolated water in the treated soil.

In chemical analyses of soil samples from the field 1, removed from a depth of 7 m, a maximum concentration of the sum of the organic compounds of 35,000 mg/(kg soil < 2 mm) was detected. In total for field 1, approximately 9,000 kg of organic compounds were estimated from the analysis of the three core borings. The spectrum of the contaminants was the same as in samples 1 and 2 from Test III.

At the beginning of the test in field 1, the saturated subsoil was hydraulically cleaned to withdraw the movable contaminants and to enrich the soil with nutrients. During the soil flushing, it appeared that removing a significant amount of the pollutants was practically impossible. This result was unexpected considering the permeability of the soil and the high concentration of the contamination detected at some depths. One explanation is that during earlier natural groundwater flow the movable compounds were already washed out, or because of the high viscosity and high local concentration of the hydrocarbons, the local permeability in the subsoil is very low.

The biological remediation, which began in November 1990, will be realized in two phases. First, as in Test III, a so-called natural biological remediation is applied. Later, additional means are planned to increase the biological availability of persistent organic compounds. Currently, more laboratory tests with soil ozonization are under way to examine whether the potential capital expenditures for a field ozonization plant are fully justified.

FIGURE 6. Conditioning of the fill: breaking off the concrete walls and excavating.

SUMMARY

A method for biological remediation of soils contaminated with hydrocarbons was presented. To supply microorganisms with oxygen, ordinary air is injected into the unsaturated soil. The nutrients, ammonia and phosphate, which are also necessary for the microorganisms, are delivered with water percolated from the surface into the soil.

The presented method was used in a laboratory test with a large-scale sample contaminated with hydrocarbons that were typical for coal tar processing. The test results showed that because of the low biological availability, with a supply of oxygen and nutrients, only some of the gas-work hydrocarbons could be degraded. Therefore, additional means were applied. In this test ozone was injected. After the ozonization, the higher condensed polycylic aromatic hydrocarbons could not be detected anymore.

At present, the bioremedial *in situ* treatment is realized in a heavily contaminated part of an abandoned coke oven plant in

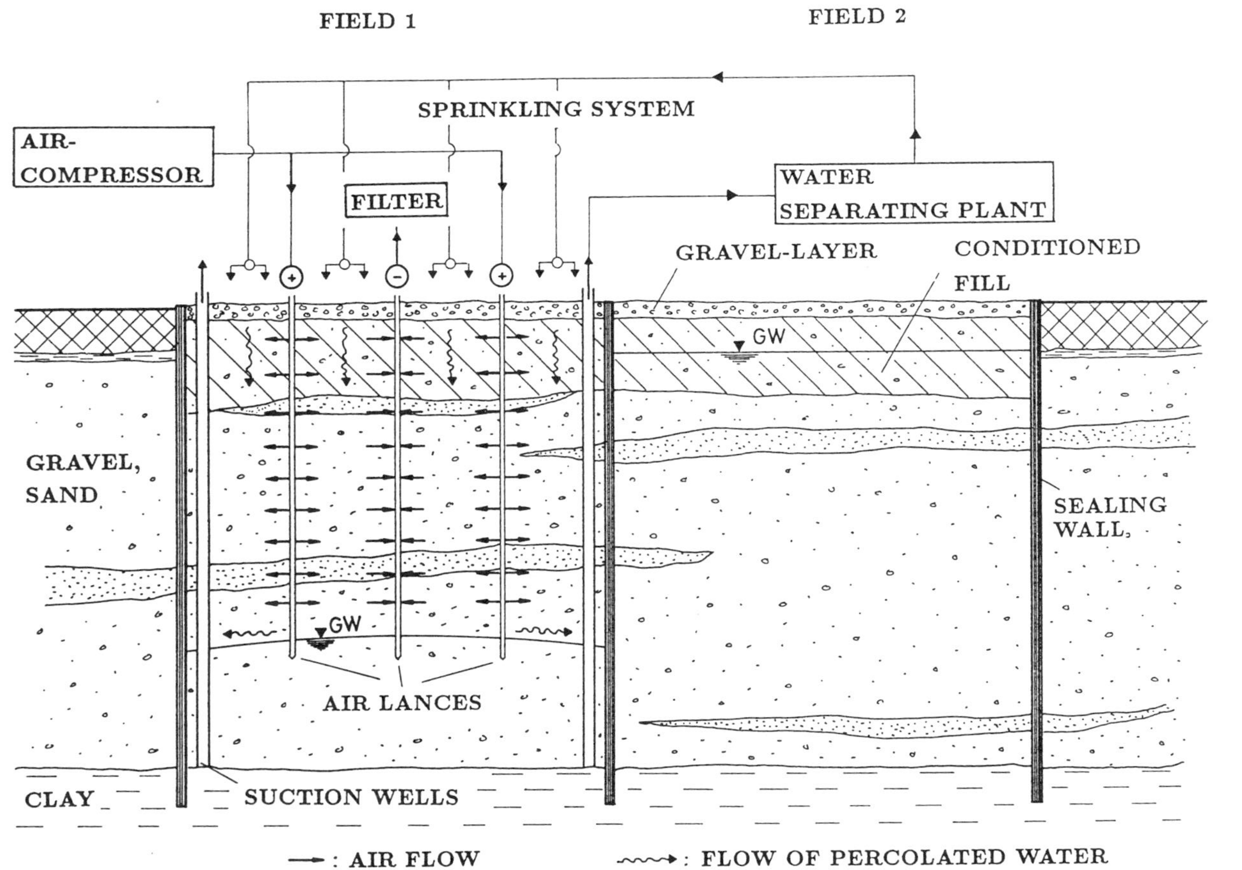

FIGURE 7. Principal draft of the installations in field 1.

Karlsruhe. Preparations to the field test and some initial remarks were described.

REFERENCES

Battermann, G.; Werner, P. "Feldexperimente zur mikrobiologischen Dekontamination." Presented at BIGTECH-Conference (Remediation of Contaminated Sites), Berlin, FRG, Nov. 1987.

Bewley, R.I.F.; Theile, P. "Dekontaminierung eines Gaswerkgeländes durch Einsatz von Mikroorganismen." Presented at 2. Int. TNO/BMFT-Conference on Contaminated Soil, Hamburg, FRG, Oct. 1988; Kluwer Academic Publishers.

Lund, N.-Ch. "Beitrag zur biologischen *in situ*-Reinigung kohlenwasserstoffbelasteter körniger Böden." Ph.D. Dissertation, Report-No. 119, Institute of Soil Mechanics and Rock Mechanics of the University of Karlsruhe, FRG, 1991.

Lund, N.-Ch.; Gudehus, G. "Biologische *in situ*-Sanierung kohlenwasserstoffbelasteter Böden." In *Vorträge zur Baugrundtagung in Karlsruhe, FRG*; Deutsche Gesellschaft für Erd- und Grundbau e.V., 1990a.

Lund, N.-Ch.; Gudehus, G. "Large-Scale Sample Tests for a Biological *In Situ* Remediation of Soils Polluted by Hydrocarbons." Presented at 3. Int. KFK/TNO-Conference on Contaminated Soil, Karlsruhe, FRG, Dec. 1990b; Kluwer Academic Publishers.

Maier, D. "Gesicherte Erkenntnisse zu den oxidativen Wasseraufbereitungsverfahren." In *Wasserchemie für Ingenieure*, DVGW-Schriftenreihe Wasser Nr. 205; Eschborn, FRG, 1989.

Aerobic Groundwater and Groundwater Sediment Degradation Potential for Xenobiotic Compounds Measured *In Situ*

*Peter E. Holm**, *Per H. Nielsen, Thomas H. Christensen*
Technical University of Denmark

Degradation of xenobiotic organic compounds in aquifers is a key issue in evaluation of the fate of pollutants migrating from landfills, abandoned industrial sites, and accidental spills, and in developing remedial actions involving *in situ* treatment methods. The microbial degradation of specific organic compounds in aquifers seems to be associated primarily with microbial biomass fixed to the sediment; for example Harvey *et al.* (1984) found that in a groundwater sediment the number of fixed bacteria per volumen was 26 to 2,000 times higher than the number of bacteria suspended in the groundwater. However, direct evidence of the significance of suspended versus fixed bacteria with respect to degradation of xenobiotic compounds is not available. This issue is important for two reasons. Firstly, water samples without sediment are very easy to obtain and therefore attractive for use in laboratory studies on degradation potentials, assuming that water samples reflect the degradation potential of the aquifer. Secondly, if a significant degradation potential is associated with the suspended biomass, the degradation potential is physically mobile by natural gradient flow or enforced by pumping.

This paper presents *in situ* measurements of the degradation potential associated with groundwater, and groundwater plus sediment,

* Department of Environmental Engineering, Groundwater Research Centre, Technical University of Denmark, Building 115, DK-2800 Lyngby, Denmark

for a mixture of 11 xenobiotic compounds in the dilute, aerobic part of a landfill leachate pollution plume.

Experimental Design and Methods

***In Situ* Tester.** The *in situ* tester concept was presented by Gillham *et al.* (1990), but for this study a modified tester has been developed (see Figure 1). The tester contains two separate compartments, one for groundwater plus sediment and one for groundwater only. Both compartments contain semi-confined, batch-like volumes to avoid uncontrolled dilution by displacing groundwater entering the tester during the experiment. The upper part for groundwater consists of a column with a coil (internal diameter 8 mm), while the lower part (for sediment plus groundwater) is a bottomless column. Both compartments are connected separately to the ground surface by tubes (see Figure 1). The tester is made of high-quality stainless steel. The tubing is made of Teflon®.

Installation. The testers were installed in cased boreholes made by hand auger and sand bucket. During placing of the tester, a slight N_2 flow out of the screen at the top of the lower compartment prevented clogging of the screen. The tester was driven down, filling the lower compartment with aquifer material and groundwater. Afterwards the well casing was withdrawn. A minimum of 5 L were pumped from the tester prior to sampling for groundwater characterization (500 mL).

Loading of Tester with Xenobiotic Compounds. A 5-L Teflon bag was filled with groundwater from the lower compartment by a peristaltic pump, and spiked with 200 mL of distilled water containing the 11 contaminants and tritium (hydraulic tracer). The resulting concentration levels for the 11 contaminants were approximately 100 µg/L. Four L of the spiked water were pumped to the lower compartment of the tester and 1 L to the coil of the upper compartment. The volume pumped to the lower compartment was 5 to 6 times the pore volume of the aquifer material (porosity 0.30) in the compartment to ensure an even distribution of the contaminants in the tester. The investigated contaminants exhibit retardation factors in the actual aquifer ranging from 1.01 to 2.96 (Larsen *et al.* 1989).

Sampling. Ten-milliliter samples were collected directly in 10-mL measuring flasks by means of a syringe connected to the tester tube (E,

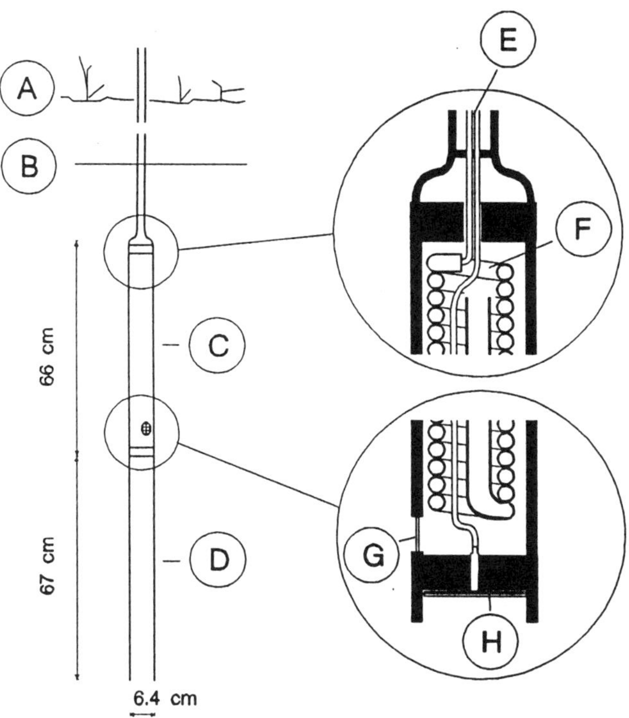

FIGURE 1. ***In situ*** **tester. (A) Ground surface. (B) Groundwater table. (C) Upper compartment containing coil for groundwater only. (D) Lower compartment for groundwater plus sediment. (E) Tubes for loading and sampling. (F) Coil with total volume of 350 mL. (G) Hydraulic contact with the aquifer. (H) Screen in top of the lower compartment.**

Figure 1). During the 94-day experimental period, samples were obtained 10 times.

Chemical Analysis. The contaminants (aromatics and chlorinated aliphatics) were analyzed by extracting the samples with pentane (isopropylbenzene was used as the internal standard for aromatics and trichlorobromomethane for chlorinated aliphatics) in a ratio of 100:1 (100 µL in 10-mL sample). Before extraction, 10 µL of 10 M NaOH were added for conservation. Aromatics and chlorinated aliphatics were detected on a Carlo Erba Mega 500 gas chromatograph with N_2 as the carrier gas (10 mL/min). The aromatic compounds were detected on a flame ionization detector (FID) and the chlorinated aliphatic compounds on an electron-capture detector (ECD).

pH, specific conductivity, and oxygen were measured in the field by electrodes. During the experimental period oxygen was measured by Winkler titration (modified for 12-mL volumes). Samples for total organic carbon (TOC), NO_3^-, and Cl^- were preserved and stored at 10 C for laboratory analysis (for details see Lyngkilde *et al.* 1990). Tritium activity was measured by scintillation counting.

Results and Discussion

Groundwater Characterization. Characterization of the groundwater pumped from the lower compartment of the tester showed that the groundwater zone was aerobic with sufficient oxygen (6 mg/L) to degrade all the added compounds to CO_2. To ensure that aerobic conditions were maintained during the experimental period (94 days), oxygen was measured during and at end of the experiment. These measurements showed at least 2 mg O_2/L in all compartments of the testers. pH, specific conductivity, and TOC were 4.6, 290 µS/cm, and 4.2 mg C/L. NO_3^- and Cl^- concentrations were measured at the levels 12 to 16 mg NO_3^--N/L and 31 to 37 mg/L. All measured concentrations are close to the background levels of the area (Pedersen *et al.*, in press), indicating that the testers were installed in the outskirt area of the landfill leachate pollution plume.

Contaminant Fate. All available data are summarized in Table 1 (two testers, two compartments each).

In general, the chlorinated aliphatics do not show any degradation, while all the other compounds are subject to degradation, ranging from complete degradation (less than 5% of initial concentrations) to a concentration decrease of only 20 percent during the 94-day experimental period. The lack of degradation of chlorinated aliphatic compounds is in accordance with the expectation for an aerobic, methane-devoid aquifer. The constant level of the chlorinated aliphatic compounds over time and the constant level of the tritium tracer (not shown) supported the fact that the *in-situ* tester system is not subject to unaccounted losses due to volatilization or dilution.

Degradation of aromatic compounds in the aerobic aquifer is in accordance with the general literature (*e.g.*, as summarized in Lyngkilde *et al.* 1989). Benzene and toluene degraded the fastest, followed by xylene, naphthalene, and biphenyl, while the chlorobenzenes degraded

TABLE 1. Degradation data for tester 1 and 2 for groundwater (w) and groundwater-plus-sediment (w+s) compartments, respectively.

	Tester Compartment			
	1		2	
	w	w+s	w	w+s
Benzene	70d[(a)]	28d	70%	38d
Toluene	70d	20d	75%	30d
o-Xylene	45%[(b)]	75%	60%	50%
Naphthalene	45%	49d	65%	65%
Biphenyl	40%	65%	55%	75%
1,4-Dichlorobenzene	35%	40%	60%	20%
1,2-Dichlorobenzene	35%	50%	55%	20%

(a) d refers to number of days to reach a concentration less than 5 µg/L (approximately 5%).

(b) % refers to the percentage degradation observed within the experimental period of 94 days.

the least during the experiment. For the groundwater plus sediment compartments, benzene and toluene were degraded completely within 20 to 40 days without any significant lag phases.

Comparison of the fate curves (see Table 1) for groundwater-plus-sediment and groundwater-only compartments shows that the degradation potential of the aquifer (as represented by the groundwater-plus-sediment compartments) also is reflected in the groundwater (as represented by the groundwater-only compartment). All the studied aromatic compounds were found to degrade in both compartments, and in both compartments benzene and toluene were identified as the most easily degraded compounds. In general, although not completely consistent, concentrations of compounds decreased initially most rapidly in the groundwater-plus-sediment compartment, but a direct comparison of the rates (or the percentages listed in Table 1) is not warranted. The two types of compartments do not offer comparable conditions: The biomass potentially active in degrading the xenobiotic compounds is much higher per volume in the lower compartment, the compound concentrations are lower in the lower compartment due to sorption to the sediment, and finally as degradation reduces the concentrations in the lower compartment, desorption from the sediment may occur. Thus, the results can be interpreted in terms of degradation potentials but do not provide meaningful rate data.

Comparing the results obtained in the two testers located only 0.8 m apart show that compound-ranking with respect to degradation is similar, while the rates differ (estimated as a factor of 2). The two testers are located as close as technically possible and approximate a duplicate experiment. The observed uncertainty should lead to caution in interpreting degradation rates determined in *in situ* testers. The variability of the fate curves from the groundwater-only compartments seems to be larger than in the groundwater-plus-sediment compartments. This could be related to the fact that the particle fraction associated with the groundwater is not well defined because of filtering effects possibly caused by the screen at the top of the lower compartment from which all the feed water is obtained. This filter may also filter suspended biomass and remove various larger-size fractions of the suspended bacteria. This would accordingly lead to less suspended biomass in the groundwater-only compartment. This again restricts the use of the double compartment *in situ* tester to field determination of degradation potentials, not measuring degradation rates.

Finally, no significant lag phases have been observed in any of the tester compartments. Lag phases are often found in laboratory groundwater batches (*e.g.*, Lyngkilde *et al.* 1989), but the results of these experiments may indicate that lag phases perhaps do not deserve much attention since they may be artifacts caused by the extensive manipulation of the biomass in laboratory batches.

Conclusions

Based on the duplicate field test on the degradation of 11 xenobiotic contaminants in an aerobic aquifer the following conclusions can be made:

- The potential of the aquifer to degrade aromatic contaminants (as measured in the groundwater-sediment compartment of the *in situ* tester) is also reflected by the groundwater that can be pumped from the aquifer (as measured by the groundwater-only compartment of the *in situ* tester).

- While data from the groundwater-only compartment may illustrate the degradation potential associated with the groundwater, the rates observed should not be applied directly to field conditions.

- The *in situ* determined degradation rates (as measured in the groundwater-plus-sediment compartment) vary substantially for the duplicate tests, indicating large variability in the aquifer and suggesting caution in ascribing single values for degradation rates in the aquifer.

- In the tests performed, no significant lag phases were observed in contrast to what often is found in laboratory test systems.

The field test performed indicates that the degradation potentials of an aquifer with respect to xenobiotic contaminants may be evaluated by laboratory batch test based on sampled groundwater, but apparently any observed lag phases should not be given any significance and the observed rates should not be applied to field conditions.

Acknowledgments

Anja Foverskov performed the main part of the inorganic analysis and Bent Skov helped install the testers. John Lyngkilde has been indispensable because of his outstanding knowledge of analytical chemistry. These contributions are gratefully acknowledged.

REFERENCES

Gillham, R. W.; Robin, M.J.L.; Ptacek, C. J. *Ground Water.* **1990,** *28,* 666–672.

Harvey, R. W.; Smith, R. L.; George, L. *Appl. Environ. Microbiol.* **1984,** *48,* 1197–1202.

Larsen, T.; Kjeldsen, P.; Christensen, T. H.; Skov B.; Refstrup, M. In *International Symposium on Contaminant Transport in Groundwater*: Stuttgart, 1989.

Lyngkilde, J.; Christensen, T. H.; Larsen, T.; Kjeldsen, P.; Skov, B.; Foverskov, A. *2nd International Landfill Symposium*; Sardinia, 1989.

Lyngkilde, J.; Christensen, T. H.; Skov, B.; Foverskov, A. Submitted for publication in *In Situ and On-site Bioreclamation Symposium*, 1990.

Pedersen, J. K.; Bjerg, P. L.; Christensen, T. H. *J. Hydrol.*, in press.

Alkane and Crude Oil Degrading Bacteria from the Petroliferous Soil of India

*Indrani Roy, Ajit K. Mishra, Anup Kumar Ray**
Bose Institute

It has been estimated (Morris 1971; Walkup 1971) that approximately 0.5 percent of transported crude oil finds its way into seawater, largely through accidental spills and discharge of ballast and wash water from oil tankers. The toxicity of crude and refined oil to marine ecology and even more directly to human life is well documented (Boesch *et al.* 1974; Nelson-Smith 1971).

Some microorganisms are well known for their ability to degrade a variety of hydrocarbons present in crude oil. Oil spills at sea or on land have demonstrated the hydrocarbon-degrading potential of these organisms. The use of microorganisms in the natural decomposition of oil spills as well as residues in oil tankers and storage drums has been reported (Atlas 1981). These hydrocarbon-oxidizing microbes multiply with availability of the proper substrate (Griffiths *et al.* 1982; Reisfield *et al.* 1972). In addition to the requirement of suitable cell substrate and specific metabolic and genetic potential, a number of general nutritional conditions are necessary for hydrocarbon utilization by microorganisms; nitrogen is one such essential element (Atlas & Bartha 1972; Reisfield *et al.* 1972).

Under laboratory conditions, nitrogen may be supplied in soluble form (inorganic salts of ammonia or nitrate of urea). Since most natural aquatic environments are deficient in utilizable forms of nitrogen, it is necessary to add the same exogeneously, but because of rapid dilution the added source of nitrogen does not remain effective. Many have reported on the character of hydrocarbon-utilizing bacteria under both laboratory and field conditions (Abbott & Gledhill 1971; Atlas & Bartha 1973; Gibson 1971; Treccani 1974; Zajic & Panchal 1976;

* Department of Organic Chemistry, Indian Association for the Cultivation of Science, Jadavpur, Calcutta 700 032, India

Zobell 1973;). The need for nitrogen supplements may be overcome by appropriate choice of microbes with the genetic capacity to fix molecular nitrogen.

Here we are reporting the isolation of a strain of *Pseudomonas stutzeri* from the petroliferous soil of India. This strain has the capacity to degrade alkane and crude oil and to fix nitrogen.

To isolate and screen alkane utilizing bacteria from petroliferous soils, diluted soil samples from established oil fields (Ankleswar) and exploratory areas (Jammu & Rajole) were prepared. A closed chamber was used, as described by Roy *et al.* (1989), which was saturated with *n*-hexane atmosphere as the only carbon and energy source; in this, cells plated in a mineral salt (MS) agar medium were incubated. The same condition was maintained where other gaseous hydrocarbons (*n*-heptane, *n*-octane) were applied as the carbon source for further experiments. The capability of alkane and crude oil degradation was tested qualitatively for further screening of isolated crude oil degrading bacteria.

Sterilized *n*-dodecane, *n*-hexadecane (v/v), and thick crude oil (w/v) were added separately at the final concentration of 1 percent to sterilized MS medium. Crude oil was sterilized by autoclaving at 8 lb/in.^{2} steam pressure for 1 h, and *n*-dodecane and *n*-hexadecane were sterilized using a millipore membrane filter (0.2 μ). Separate 100-mL flasks containing 30 mL medium were inoculated with the 10 identified bacterium. Incubation was done at 30 C (stationary condition) for 15 days. Closed chamber was used only for naturally volatile hydrocarbons.

The bacterial strains and their growth are shown in Table 1. The growth of different isolates in the presence of different hydrocarbon sources in the MS medium showed that strain numbers 16, 25, 27, 30, 32, and 49 were potent *n*-hexane utilizers. Strain 49 showed good growth in octane. It was remarkable that strain number 16 could utilize all the hydrocarbon sources used as well as crude oil. Strain 11 could utilize both short- and long-chain alkanes, but was unable to grow in crude oil. Strains 12, 25, 27, 29, 30, 32, 49, and Jm 2-2 were unable to utilize long-chain hydrocarbons.

Studies on aerobic degradation of thick crude oil by *P. stutzeri* ANKBI-16 (Table 1) were performed in two 100-mL flasks containing 30 mL MS medium supplemented with 0.05 percent yeast extract. Crude oil (sterile) was added to each of the flasks, and inoculum (4% of the 10^8 cell/mL) was added to only one flask (experimental). The flasks were incubated at 30 C on a shaker incubator. After 4 days, the oil adhered to the surface of the control flask, and a large lump of

TABLE 1. Growth of *n*-hexane utilizing strain on different hydrocarbons.

Strain No.*	Identified as	Growth (qualitatively)					
		n-hexane	n-heptane	n-octane	n-dodecane	n-hexadecane	Crude oil
11	*Azospirillum* sp. ANK BI-11	+++	+++	±	++	++	-
12	*Azotobacter chroococcum* ANK BI-22	+++	±	±	-	-	-
16	*Pseudomonas stutzeri* ANK BI-16	++++	+	++++	++++	++++	++++
25	*Pseudomonas stutzeri* ANK BI-25	++++	+	+++	-	-	-
27	*Pseudomonas stutzeri* ANK BI-27	++++	+	+	-	-	-
29	*Pseudomonas* sp. ANK BI-29	-	-	-	-	-	-
30	*Pseudomonas stutzeri* ANK BI-30	++++	±	+	-	-	-
32	*Pseudomonas stutzeri* ANK BI-32	++++	±	+	-	-	-
49	*Micrococcus luteus* ANK BI-49	++++	+	++++	±	±	-
J, 2-2	*Micrococcus agilis* Jm 2-2	+++	±	+	-	-	-

++++ excellent growth, +++ good growth, ++ moderate growth, + slight growth, ± questionable growth, - nongrowth.

* All strains except no. 11 were identified by the Czechoslovak Collection of Microorganisms.

crude oil was floating over the clear medium. The surface of the experimental flask was very clear, with only a thin layer of oil over the surface. Each flask was emptied into a 50-mL cylinder for clearer observation (Figure 1).

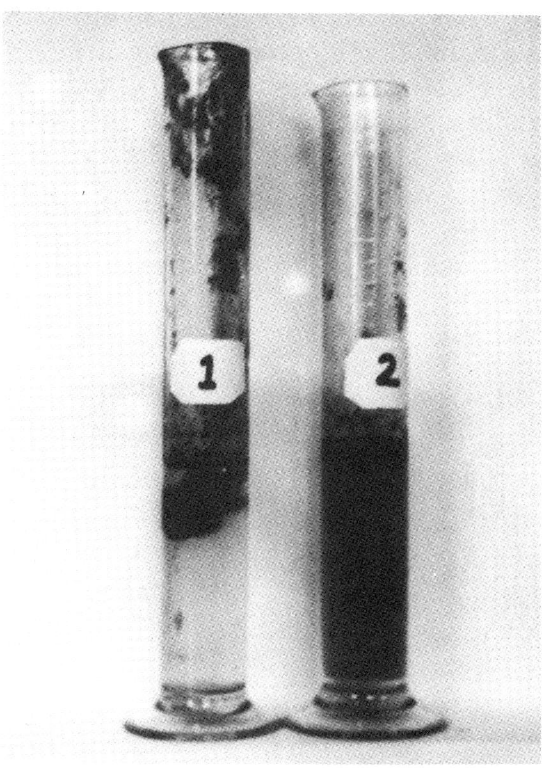

FIGURE 1. Oil slick dispersion by emulsification under the influence of *P. stutzeri* ANKBI-16 grown on crude oil. Cylinder no. 1 is the control flask without bacteria, whereas cylinder no. 2 shows the dispersed oil obtained after microbial treatment. Note that at the time of pouring, oil has stuck to the surface of cylinder no. 1 with a lump floating over the medium and cylinder no. 2 shows clear pouring as a result of emulsification by the bacteria.

Chemical analysis of the above two oil samples (microbially treated and untreated) was done by extracting the oil with chloroform. The chloroform extracts were collected in two flasks and evaporated at room temperature. The crude oil thus collected was passed through a millipore filter (pore size 0.45 µm) after dissolving in spectroscopic *n*-hexane. The hexane soluble portion was used for gas-liquid chromatographic (GLC) analysis for chemical investigation. These samples (1 µl) were injected into a Pye Unichem Chromatograph (model GSD with flame ionization detector) using an OV-101 glass column with nitrogen as the carrier gas at a flow rate of 30 mL/min. Column oven temperature was maintained by programming 50 to 150 C at 40 C/min. Injector and detector temperatures were 250 C and 300 C, respectively.

The results obtained through GLC analysis are shown in Figure 2A. Figure 2A is the GLC profile of the control portion, whereas

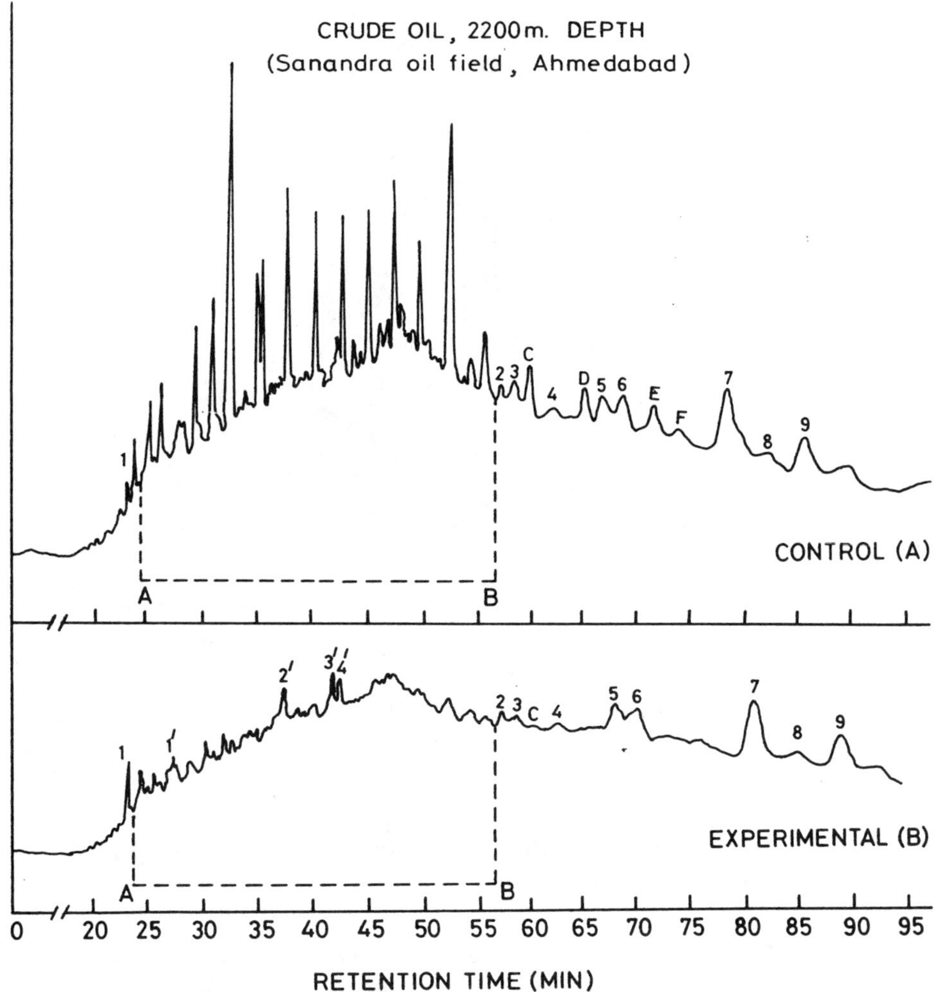

FIGURE 2. Gas-liquid chromatographic profile of the control (A) and microbially treated (B) crude oil.

Figure 2B is that of the experimental flask. Comparing the two figures shows that most of the crude oil fractions were consumed in the microbially treated sample. The oil fractions remaining with the AB region and peaks designated by C, D, E, and F of Figure 2A almost or completely disappeared as a result of microbe treatment, as shown in Figure 2B. Only the oil fraction designated as 1–9 and 1′–4′ within, the AB region remained unchanged in comparison with the consumed peaks.

Prolonged microbial treatment on crude oil is demonstrated in Figure 3, where it was observed that some black particles actually settled down at the bottom of the cylinder. The rate of this sedimentation had been found to increase with incubation time.

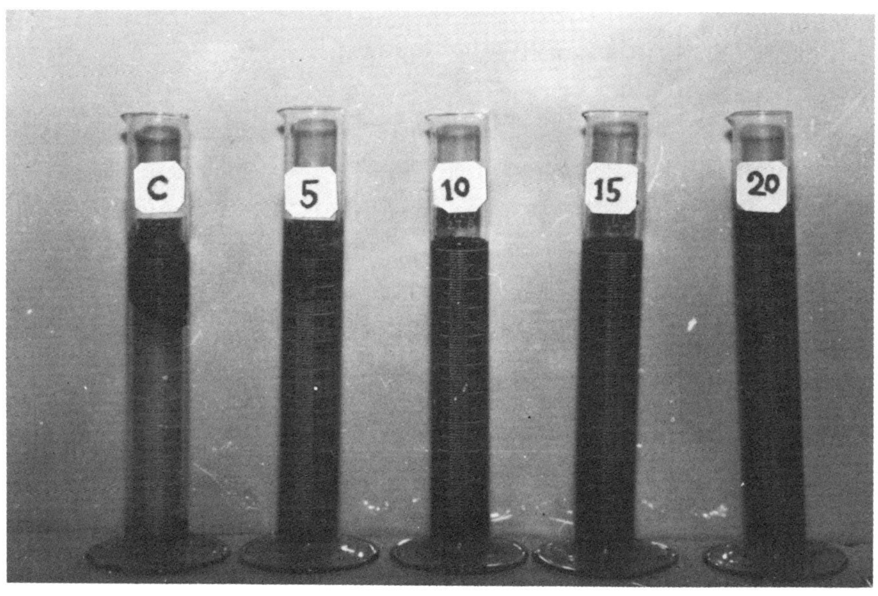

FIGURE 3. Separation of bitumen particles from the oil-bitumen association (5 to 20 days incubation with bacteria where C represents control).

Nitrogenase activity of the *P. stutzeri* ANKBI-16 was determined by performing the acetylene reduction (AR) test following the method described by Roy *et al.* (1988). Sodium succinate and *n*-hexadecane were tested separately both as electron donor and carbon sources for AR activity of the strain in nitrogen-free semisolid MS medium supplemented with 0.005 percent yeast extract. Figure 4 shows the amount of ethylene formed by the organism in the presence of sodium

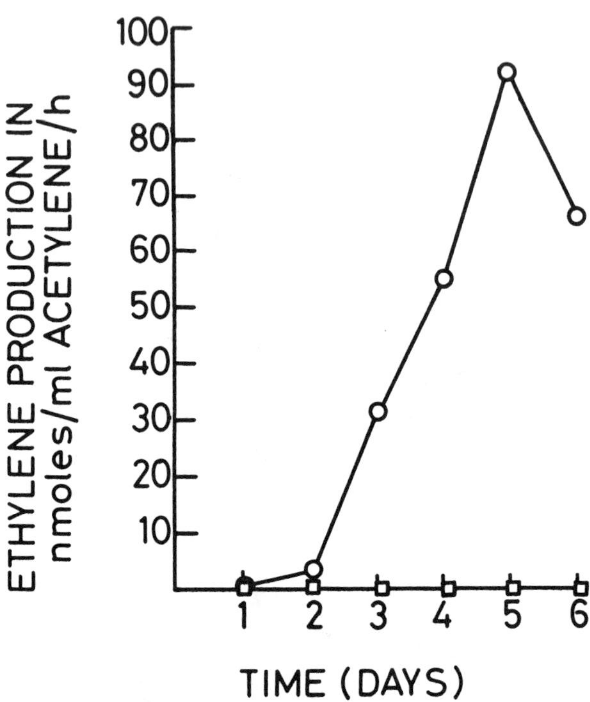

FIGURE 4. Acetylene reduction value for *P. stutzeri* ANKBI-16 in a semisolid N^2-free MS medium with different carbon source: ○, sodium succinate; □, *n*-hexadecane.

succinate with rich visible growth. No AR activity was observed in the presence of *n*-hexadecane, although slight visible growth was observed, which may have resulted from the presence of the trace amounts of yeast extract in the medium. It might become possible for the strain to fix atmospheric nitrogen in the presence of *n*-hexadecane by adjusting environmental conditions. Although several researchers have reported hydrocarbon utilization and related topics (Abbott & Hou 1973; Alikhan *et al.* 1963; Chakrabarty 1974; Chakrabarty 1986; Chakrabarty *et al.* 1978; Fennewald & Shapiro 1977, 1979; Griffin & Traxler 1981; Keeven & Decicco 1989; Minoda & Omori 1976; Omori *et al.* 1975; Shapiro *et al.* 1980; Stanier *et al.* 1966; Van der Linder & Huybregste 1967; Van Eyk & Bartels 1968), as well as nitrogen fixation capacity (Barraqui *et al.* 1986, 1988; Krotzky & Warner 1987) by the *Pseudomonas* spp, no report has yet found this species to have both properties. It was observed that the soil exposed to the natural gas pipeline leaks contained more nitrogen than normal soil (Harper 1939; Schöllenberger 1930). Utilization of natural gas and simultaneous nitrogen fixation by some unidentified soil microbes have been reported (Davis 1952; Davis *et al.* 1964).

An *Azospirillum* spp. reported from this laboratory (Roy *et al.* 1988) was able to fix nitrogen while growing on *n*-dodecane. Growth of this organism on long-chain hydrocarbons was sluggish, and in crude oil there was no growth (Table 1). Table 1 indicates that *P. stutzeri* ANKBI-16 is a good hydrocarbon as well as crude oil utilizer compared with other isolates. It is evident that *P. stutzeri* ANKBI-16 has a greater potential for crude oil emulsification (Figure 1) as well as degradation capability (Figure 2A and B). Prolonged microbiol treatment led to the settling of black carbonlike particles at the bottom of the cylinder (Figure 3). These are presumed to be dissociated bitumenlike particles that remain after microbes have consumed the oil portion of the oil droplets. From the studies cited previously, we can conclude that this aerobic bacteria is capable of both oil degradation and nitrogen fixation (Figure 4). It has also been observed (Goswami 1990) that the organism prefers hydrocarbons over conventional carbon sources like glucose, indicating immense genetic potential for hydrocarbon utilization. This dual capability has generated considerable interest in genetic manipulation of the bacteria to develop an improved strain that can fix nitrogen during hydrocarbon utilization. It might be possible to apply such a modified strain in open aquatic environments to remove oil spills.

Acknowledgments

This work was supported by Grant No. 2(157)/85 of the oil and Natural Gas Commission of India. We are thankful to Dr. M. Kocur of Czechoslovak Collection of Microorganisms for identification of the strains.

REFERENCES

Abott, B. J.; Gledhill, W. E. *Adv. Appl. Microbiol.* **1971**, *14*, 249–388.

Abott, B. J.; Hou, C. T. *Appl. Microbiol.*, **1973**, *26*, 86–91.

Alikhan, M. Y.; Hall, A. N.; Robinson, D. S. *Nature*. **1963**, *198*, 289.

Atlas, R. M. *Microbiol. Rev.* **1981**, *45*, 180–209.

Atlas, R. M.; Bartha, R. *Biotechnol. Bioeng.* **1972**, *14*, 309–318.

Atlas, R. M.; Bartha, R. *Residue Rev.* **1973**, *49*, 49–85.

Barraqui, W. L.; Padre, B. C., Jr.; Watanaba, I.; Knowles, R. *J. Gen. Microbiol.* **1986**, *132*, 237–241.

Barraqui, W. L.; Dumont, A.; Knowles, R. *Appl. Environ. Microbiol.* **1988**, *54*, 1313–1317.

Boesch, D. F.; Herschner, C. H.; Milgram, J. H. *Oil Spills and the Marine Environment;* Ballinger Publ: Cambridge, 1974; p 114.

Chakrabarty, A. M. U.S. Patent, 3 813 316, 1974.

Chakrabarty, A. M. In *Biotechnology;* Rehn, H. J. & Reed, G., Ed.; Verlag Chemie Publ.: Weinheim, Federal Republic of Germany, 1986, Vol. 8, pp 516–530.

Chakrabarty, A. M.; Friello, D. A.; Bopp, L. M. *Proc. Natl. Acad. Sci. USA.* **1978**, *75*, 3109–3112.

Davis, J. B. *Bull. Am. Assoc. Petrol. Geologists.* **1952**, *36*, 2186–2189.

Davis, J. B.; Coty, V. F.; Stanly, J. P. *J. Bacteriol.* **1964**, *88*, 468–472.

Fennewald, M. A.; Shapiro, J. A. *J. Bacteriol.* **1977**, *132*, 622–627.

Fennewald, M. A.; Shapiro, J. A. *J. Bacteriol.* **1979**, *136*, 264–269.

Gibson, D. T. *Crit. Rev. Microbiol.* **1971**, *1*, 199–223.

Goswami, I. Ph.D. Thesis, University of Calcutta, April 1990.

Griffin, W. M.; Traxler, R. W. *Develop. Indust. Microbiol.* **1981**, 22, 425–435.

Griffiths, R. P.; Caldwell, B. A.; Cline, J. D.; Broich, W. A.; Morita, R. Y. *Appl. Environ. Microbiol.* **1982**, *44*, 435–446.

Harper, H. I. *Soil. Sci.* **1939**, *48*, 461–466.

Keeven, J. K.; Decicco, B. T. *Appl. Environ. Microbiol.* **1989**, *55*, 3231–3233.

Krotzky, A.; Warner, D. *Arch Microbiol.* **1987**, *147*, 48–57.

Minoda, Y.; Omori, T. *Proc. Int. Ferment. Symp. 5th.* **1976**, p 423.

Morris, B. F. *Science* **1971**, *173*, 430–432.

Nelson-Smith, A. In *Water Pollution by Oil*, Hepple, P., Ed.; Elsevier: New York, 1971; pp 273–280.

Omori, T.; Jigami, T.; Minoda, Y. *Agric. Biol. Chem.* **1975**, *39*, 1775–1779.

Reisfield, A. ; Rosenberg, E. ; Gutnick, D. *Appl. Microbiol.* **1972**, *24*, 363–368.

Roy, I.; Sukla, S. K.; Mishra, A. K. *Current Microbiol.* **1988**, *16*, 303–309.

Roy, I.; Sukla, S. K.; Mishra, A. K. *FUEL* **1989**, *68*(3), 311–314.

Schöllenberger, C. J. *Soil Science* **1930**, *29*, 261–266.

Shapiro, J. A.; Benson, S.; Fennewald, M. A. *Proc. Annu. Symp. Sci. Basic. Med.* **1980**, *4*, 1.

Stanier, R. Y.; Palleroni, N. J.; Doudoroff, M. *J. Gen. Microbiol.* **1966**, *43*, 159–271.

Treccani, V. In *Industrial Aspects of Biochemistry;* Spencer, B., Ed; Fed. Eur. Bioch. Soc.: Berlin, 1974; pp 533–547.

Van der Linder, A. C.; Huybregste, R. *J. Microbiol. Serol.* **1967**, *33*, 381–385.

Van Eyk, J.; Bartels, T. J. *J. Bacteriol.* **1968**, *96*, 706–712.

Walkup, P. C. *J. Water Pollut. Control Fed.* **1971**, *43*, 1069–1072.

Zajic, J. E.; Panchal, C. J. *Crit. Rev. Microbiol.* **1976**, *5*, 39–66.

Zobell, C. E. *The Microbiol Degradation of Oil Pollutants*, Res. Pub. No. LSU-SG-730; Centre for Wetland: Baton Rouge, LA, 1973.

Case History of the Application of Hydrogen Peroxide as an Oxygen Source for *In Situ* Bioreclamation

*M. D. Lee**, *R. L. Raymond, Sr.*
Du Pont Environmental Remediation Services

The first applications of *in situ* bioreclamation to remediate hydrocarbon-contaminated aquifers used sparged air as the oxygen source (Raymond *et al.* 1976, 1978). However, the limited solubility of oxygen in groundwater (8 to 12 mg/L at typical groundwater temperatures) controls the rate at which the native microbial population can biodegrade the hydrocarbon contaminants (Lee *et al.* 1988). Liquid or compressed oxygen can supply up to 40 mg/L of dissolved oxygen and has been applied at some sites, but still limits the rate at which the bioremediation process can proceed. Hydrogen peroxide (H_2O_2), which is completely soluble in water, decomposes to form one molecule of water and one half molecule of oxygen. Decomposition of H_2O_2 may be catalyzed by iron, by fluctuations in solution pH, or by microbial enzymes such as catalases or peroxidases. Concentrations as low as 200 ppm H_2O_2 can be toxic to microorganisms. However, with exposure to gradually increased H_2O_2 concentrations, the microbial population appears to acclimate to hydrogen peroxide.

This case history focuses on the application of *in situ* bioreclamation for the treatment of an aquifer in Long Island, New York, which was contaminated with gasoline from a leaking underground storage tank. Hydrogen peroxide was used as the oxygen source for the *in situ* bioreclamation of the gasoline. Soil venting and air stripping of the groundwater were also used at the site.

* Du Pont Environmental Remediation Services, 500 W. Dutton's Mill Road, Suite 102, Aston PA 19014

Site Description

A gasoline station on Long Island in New York state lost an estimated 106,000 kg of gasoline to the subsurface (Litchfield *et al.* 1989). A plume of free product reached a residential subdivision, contaminating the groundwater and generating gasoline vapors in the basements of surrounding homes. The aquifer consisted of unconsolidated glacial outwash deposits, which were primarily coarse sands and gravels with localized clay lenses. The highly permeable aquifer could sustain pumping rates in excess of 1,300 L/min with groundwater flow in excess of 1.2 m/day based upon the movement of chloride tracers from the infiltration gallery toward a recovery well. The water table fluctuated between 4.42 and 4.88 m below land surface.

More than 82,000 kg of gasoline were skimmed from the groundwater table during a free product recovery program. The groundwater collected in conjunction with the free product recovery program was treated by passage through an air stripper. Free product was found over an area of approximately 95 by 51 m after the recovery program had ceased to be effective (Figure 1). An estimated 24,600 kg of gasoline remained adsorbed in the aquifer, distributed vertically above and below the water table for 2.3 m. The contaminated soils above the water table were treated by vacuum extraction.

Methods

The movement of dissolved oxygen, chloride, ammonia, and phosphate in the groundwater was followed in selected monitoring wells. Dissolved oxygen was determined using YSI dissolved oxygen meters. Chloride and phosphate concentrations were determined using Hach Co. reagents (Loveland CO). Ammonia was determined by the Nessler's Reagent method. The numbers of both heterotrophic bacteria and hydrocarbon-utilizing bacteria in the groundwater were determined. The heterotrophic bacterial counts were made on nutrient agar incubated for 7 days. The hydrocarbon-utilizing bacterial counts were made on a mineral medium containing inorganic nutrients and incubated under the gasoline vapors for 4 weeks. Soil samples were taken from the site over the course of the project to determine the concentrations of total gasoline in the soil and the numbers of both heterotrophic and hydrocarbon-utilizing bacteria. The soil samples were collected in the intervals of 3.05 to 3.66, 3.66 to 4.27, 4.27 to 4.88, and 4.88 to 5.49 m below land surface and immediately frozen on dry

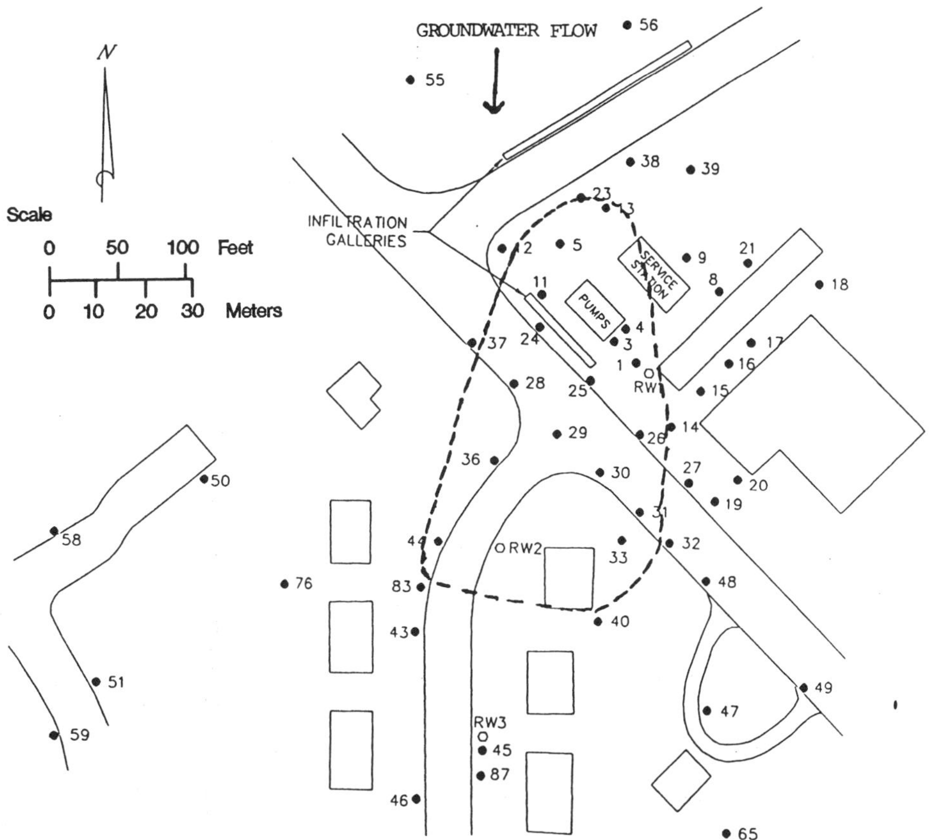

FIGURE 1. Site map showing extent of free product accumulation in April 1985 following free product recovery program.

ice. Total gasoline was determined by an in-house procedure that involved extraction of 10-g portions of the frozen sample in 20 mL iso-octane and quantitation on an ultraviolet spectrophotometer at 265 nm, a wavelength characteristic of gasoline. The detection limit for the total gasoline method is 10 mg/kg.

Bioreclamation Effectiveness

In situ bioreclamation was begun on the site with the addition of a nutrient mixture containing ammonium chloride, sodium phosphate, and trace inorganics (Litchfield *et al.* 1989). The concentrations of nutrients required for optimal biodegradation under the site-specific conditions were determined in a laboratory investigation. After 30 days of adding this inorganic nutrient solution, H_2O_2 additions were begun, with a gradual increase in the concentrations of the H_2O_2. The nutrient formulation changed during the bioreclamation process depending upon the results of the monitoring program and the rate of groundwater flow. The formulation of the nutrient solution is considered proprietary.

The effectiveness of the *in situ* bioremediation program can be evaluated for both groundwater and soil data from this site. Because of the tremendous quantity of information gathered during the monitoring portion of this project, the discussion will concentrate on well 26 (shown in Figure 1). The well 26 data are representative of the processes that occurred throughout the site. Well 26 was located 73 m away from the closest infiltration gallery. The wells and soil sampling locations changed during the project, and consequently all well and soil sampling locations were not sampled throughout the monitoring program. The microbial counts were started during month 14, but the nutrient analyses did not start until month 18.

Well 26 Groundwater. Groundwater samples taken prior to nutrient addition averaged 0.9 mg/L phosphate, 3 mg/L ammonia, 21 mg/L chloride, 2×10^5 colony forming units per milliliter (cfu/mL) heterotrophic bacteria, and 1×10^4 cfu/mL hydrocarbon-utilizers. Elevated average chloride and ammonia concentrations were found for month 18 for well 26, an indication that the nutrient solution had reached this area before the monitoring program began at the well (Figure 2). The maximum monthly average chloride concentration was 160 mg/L, and the maximum ammonia concentration was 25 mg/L. Phosphate levels were generally less than 1 mg/L, but increased to 3 mg/L by the end of the monitoring program (month 46). The phosphate added as part of the nutrient feed was probably adsorbed onto the aquifer material and utilized by the microorganisms. Dissolved oxygen appeared in detectable concentrations in well 26 by month 20 and peaked at greater than 20 mg/L from months 39 to 41. The heterotrophic microbial counts in the groundwater increased from 10^4 to 10^6 cfu/mL, and the hydrocarbon-utilizers increased from 10^3 to 10^6 cfu/mL when the dissolved oxygen concentrations were elevated as a result of the treatment program (Figure 3). The numbers of hydrocarbon-utilizing bacteria were greater than the heterotrophic population for much of the period after month 32. The higher counts of hydrocarbon-utilizing microorganisms may be a reflection of the different plating procedures and an indication that substantial numbers of the organisms could

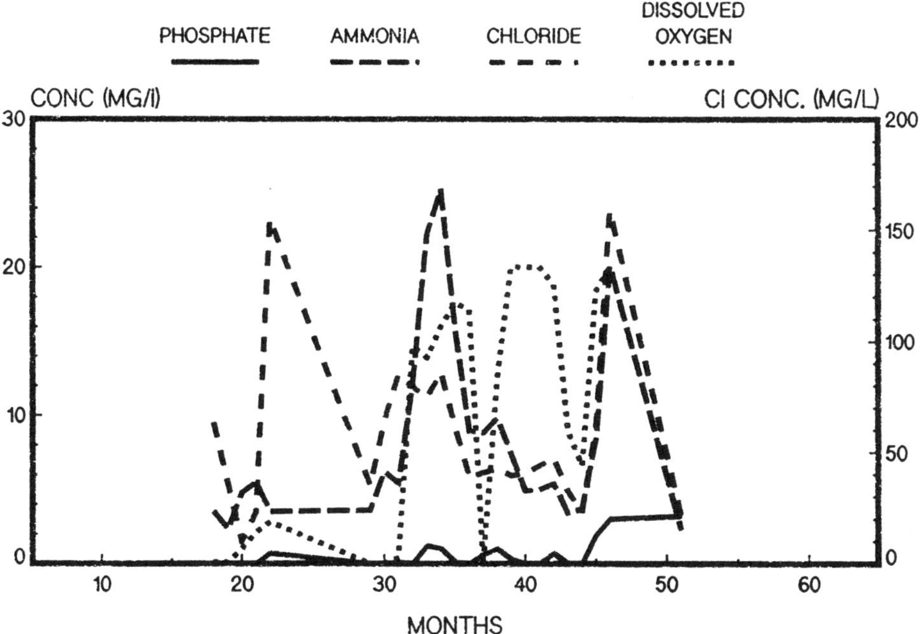

FIGURE 2. Monthly average of nutrient concentrations in well 26 over time.

utilize gasoline as their substrate. The microbial counts in the groundwater from well 26 declined after month 35 when the gasoline in the aquifer at this area was depleted.

Core 26 Data. The first set of core samples collected in the vicinity of well 26 during March 1984 contained an average of 5,400 mg/kg of total gasoline (Figure 4). There were averages of 6,000 cfu/g heterotrophic bacteria and less than 100 cfu/g hydrocarbon-utilizers. Over the next 45 months of treatment, the total gasoline concentrations in the soil and aquifer samples in the vicinity of well 26 declined to non-detectable concentrations (less than 10 ppm) in samples collected from 3.66 to 5.49 m and to 90 mg/kg in the interval from 3.05 to 3.66 m. The microbial numbers increased to in excess of 10^6 and 10^5 cfu/g of heterotrophs and hydrocarbon-utilizers, respectively.

Total Gasoline Removal. The reduction in the quantity of gasoline over time throughout the site is shown in Figure 5. In April 1985, following the free product recovery and the initiation of the *in situ*

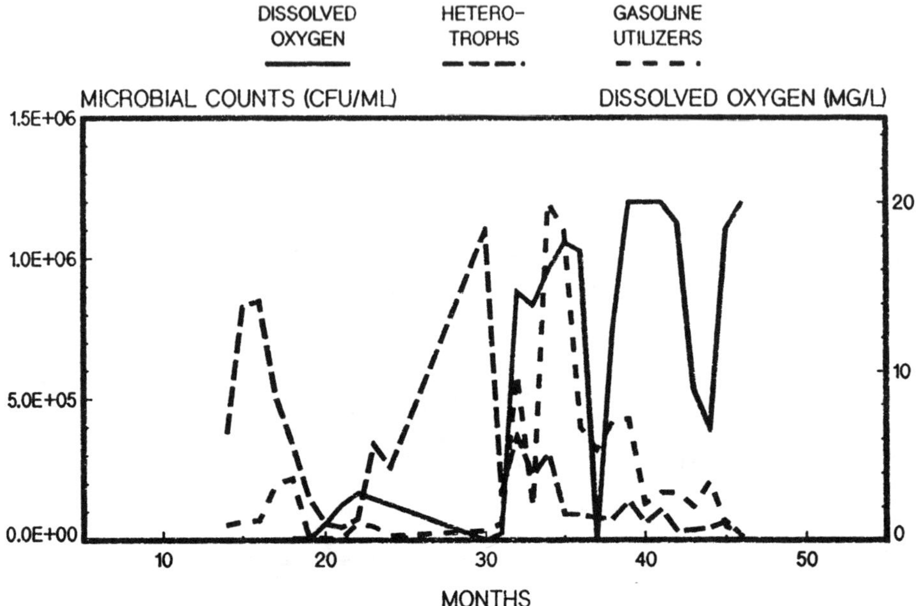

FIGURE 3. Monthly average of microbial counts and dissolved oxygen in well 26 over time.

bioreclamation program, there was an estimated 24,600 kg of gasoline within the 9,560 m^3 of contaminated soil based upon an average concentration of 1,605 mg/kg of gasoline and a soil density estimated to be 1,600 kg/m^3. The estimated quantity of gasoline decreased greatly between months 0 and 21 of the *in situ* bioreclamation treatment. There was an increase in the month 33 samples, possibly as a result of more intensive sampling in the contaminated locations. The soils with absorbed gasoline diminished to a 0.61-m zone in a small area surrounding wells 28 and 29 by February 1991, with an estimated total of 270 kg of gasoline remaining at the site. *In situ* bioreclamation activities were discontinued for months 49 to 64. The remainder of the site soil and groundwater has reached the treatment goal of nondetectable gasoline.

Relative Contribution of Treatment Technologies. The relative contribution of *in situ* bioreclamation to the removal of the 24,330 kg of gasoline can be estimated. Based upon an average flow rate of 1,320 L/min and average concentrations of 75 µg/L total benzene, toluene, and xylenes (BTX) for the groundwater going into the air stripper, as recorded in 1986, 52 kg of total BTX were removed per year

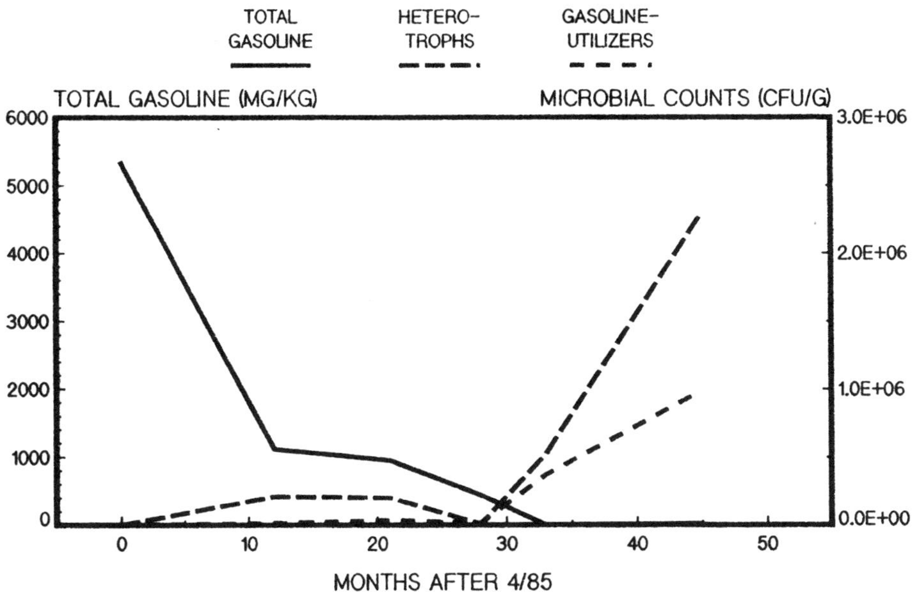

FIGURE 4. Total gasoline and microbial counts in core 26 with depth.

by the air stripper. These analyses showed that the total BTX made up 6.7 percent of the total dissolved volatile organics in gasoline. Over the 8-year treatment period, an estimated 6,240 kg of gasoline were removed by the air stripper. An estimated 450 kg of volatiles were removed by the soil venting system based upon measurements of the hydrocarbon in the air following the soil venting. These measurements were made using a organic vapor analyzer with a PID detector. The soil venting system could only treat the soil above the water table. The remaining 17,640 kg of gasoline removed can be attributed to *in situ* bioreclamation.

Conclusions

The *in situ* bioremediation program using hydrogen peroxide at this site, in combination with soil venting and air stripping of the withdrawn groundwater, has successfully treated the aquifer to less than 10 mg/kg of total gasoline except for one small area. *In situ* bioreclamation was estimated to have accounted for 72 percent of the 24,330 kg of gasoline removed. The microbial numbers in the

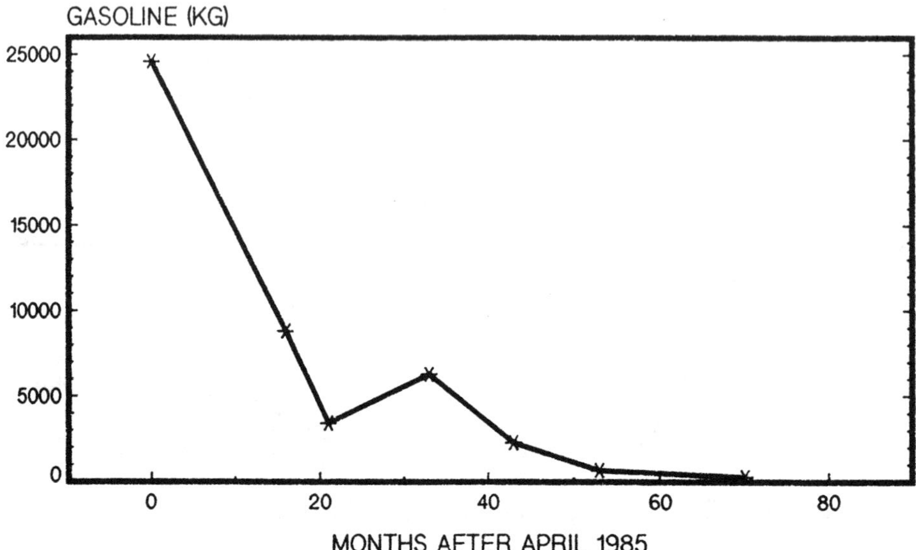

FIGURE 5. Reduction in estimated quantity of gasoline.

groundwater and soils increased when the dissolved oxygen levels in the groundwater increased as a result of the hydrogen peroxide additions. Continued treatment of the area with residual fuel should reach treatment goals in the near future.

REFERENCES

Lee, M. D.; Thomas, J. M.; Borden, R. C.; Bedient, P. B.; Ward, C. H.; Wilson, J. T. *CRC Critical Reviews in Environmental Control* **1988,** *18,* 28–89.

Litchfield, C. D.; Lee, M. D.; Raymond, R. L., Sr. 1989. *Proceedings,* 1989 Hazmat Central, 1989.

Raymond, R. L.; Jamison, V, W.; Hudson, J. O. *AIChE Symposium Series* **1976,** *73,* 390–404.

Raymond, R. L.; Jamison, V. W.; Hudson, J. O.; Mitchell, R. E.; Farmer, V. E. "Field Application of Subsurface Biodegradation in a Sand Formation"; final report to the American Petroleum Institute on Project No. 307-77; Washington, DC, 1978.

Gasoline Spill in Fractured Bedrock Addressed with *In Situ* Bioremediation

Richard A. Bell[*], *Adam H. Hoffman*
John Mathes & Associates, Inc.

This case study is a brief description of a project in which *in situ* bioremediation was successfully used to address a gasoline spill in fractured bedrock. The owner of a major Midwestern manufacturing facility in eastern Missouri sought the services of John Mathes & Associates, Inc. (Columbia IL) upon discovering a leak in an underground gasoline pipeline in late 1984. Gasoline odors and deterioration of an asphalt parking lot were the first indications of the gasoline leak, which proved to be extensive. Gasoline had migrated into the fractured limestone and contaminated the shallow bedrock aquifer.

A combination of air stripping, hydrocarbon and water separation, and venting systems was also utilized to remediate the vadose zone in addition to the biological system. This was the first application of *in situ* bioremediation approved by the Missouri Department of Natural Resources (MDNR). The system was shut down in April 1990 when remediation goals were achieved after 32 months of operation. The shutdown was performed under a plan approved by the MDNR. Gasoline contamination in the groundwater and soil had been significantly reduced (to nondetectable) in most groundwater monitoring wells across the facility.

Defining the Problem

Field investigations demonstrated that the geology at the facility consisted of 4.5 to 6.1 m of relatively impermeable till on top of 1.5 to 3.0 m of weathered limestone over a thick sequence of fractured and jointed limestone bedrock. The relatively impermeable till layer was breached at locations where the foundation of the building at the

[*] John Mathes & Associates, Inc., 210 West Sand Bank Road, Columbia IL 62236

facility penetrated to bedrock, which allowed gasoline to migrate down into the fractured limestone bedrock. The weathered zone and the fractured bedrock are also the water-bearing units at the facility.

Although the amount of gasoline actually lost is unknown, accounting figures indicated that the loss could be as high as 113,000 L. Initial actions focused on defining and containing the plume. A collection pump system was able to recover less than 3,800 L of free gasoline product from the area where the leak occurred. Groundwater monitoring wells and boreholes were then installed to define the extent of the contaminant plume. The plume was estimated to be approximately 120 by 240 m located directly under the spill near the center of the facility, far from the facility boundaries, which meant that off-site migration was not an immediate concern. After site investigation activities were completed, several different methods of clean-up were considered. The volume of the leakage, the difficult site geology, and the location of the spill in relation to essential operations at the manufacturing facility ultimately led to the recommendation of an *in situ* bioremediation solution.

Laboratory Treatability Study

A laboratory treatability study (TS) was conducted that modeled site conditions. Concerns about nutrient loading and feed rates, possible phosphate precipitation, decomposition of the hydrogen peroxide, and biological kinetics and degradation rates were addressed and allayed through laboratory tests and analysis.

The TS demonstrated that the *in situ* bioremediation solution was feasible. The existing microbial population in the groundwater was found to contain naturally occurring organisms capable of being stimulated to degrade gasoline by the addition of oxygen (via hydrogen peroxide) and nutrients (nitrogen and phosphorus). Furthermore, the clean-up could be accomplished in a reasonable time because the necessary bacterial population was shown to be indigenous to the facility. The laboratory TS demonstrated that hydrocarbons were degraded at a rate of 1 kg of gasoline per 4.64 kg of oxygen.

Bioremediation System Design

State and federal approval for this remedial action was required. Several departments in the MDNR (the lead agency) were involved in the approval process. Class V Underground Injection Control (UIC) permits were also required prior to construction of the system.

The surface treatment equipment associated with this remedial system included an oil separator, a surge tank, packed tower aeration (PTA), and injection feed pumps (Figure 1). A nutrient addition system and an automatic hydrogen peroxide system were also installed. A series of nine recovery wells at the plume perimeter and nine injection wells inside the contaminant plume were constructed to provide hydrodynamic control of groundwater. This configuration allowed a recirculation of recovered and injected groundwater to treat the contamination plume, while effectively controlling migration.

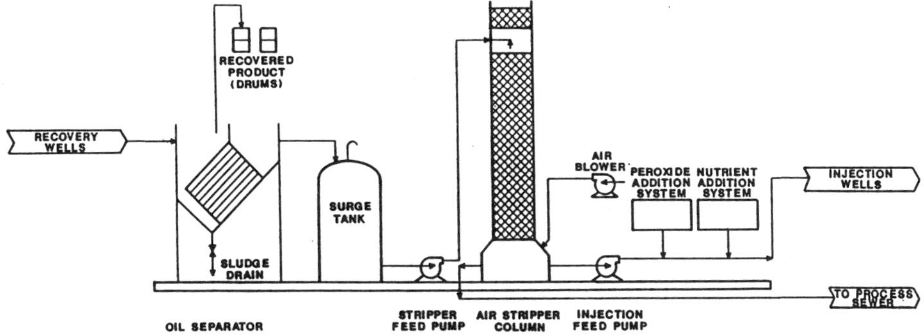

FIGURE 1. Remediation system schematic diagram.

The original *in situ* bioremediation system design was modified to include soil venting to address contaminants in the vadose zone (the unsaturated zone above the groundwater surface). Soil venting was estimated to cost one-seventh to one-eighth the cost of equivalent removal by soil bioremediation. This cost advantage occurred mainly because existing air pollution permits for the manufacturing facility allowed off-gases to be exhausted directly to the atmosphere without treatment.

Several venting systems were installed for removal of gasoline vapors from the vadose zone and to volatilize volatile organic compounds (VOC) adhered to the overlying till soil. A shallow horizontal venting system was buried 1.1 m deep to remove vapors from the backfill around the building's foundation, where a large portion of the gasoline had migrated. A deep vertical venting system was installed to reach gasoline vapors trapped in deeper areas of the vadose zone where the contaminant plume existed, approximately 6 to 8 m below the ground surface and immediately above the groundwater surface. Portable soil venting systems were also used to augment bioremediation at several recovery well locations where elevated hydrocarbon concentrations had been detected.

Startup and Operation

The *in situ* bioremediation system operations started in August 1987 with an initial hydrogen peroxide feed rate of 100 mg/L and then was slowly raised over a period of about 2 months to a concentration of 300 mg/L, which was the original target concentration based on laboratory results. Within several months of startup, the results of standard enumeration tests demonstrated that microbial populations measured in groundwater samples from the recovery wells increased from 10^3 to greater than 10^7 CFU/mL (colony forming units). The majority of these microbes were gasoline degraders. After 6 months of operation, the hydrogen peroxide concentration was raised to 700 mg/L, which was found not to be inhibitory to the biological system and allowed remediation to progress more quickly and cost effectively. Hydrogen peroxide was fed at a concentration of 700 mg/L throughout the operation of the *in situ* bioremediation system. Approximately 91 kg of nutrients were used each month.

A groundwater flow rate through the treatment system was maintained at an average of approximately 160 L/min during the life of the project. Over 220 million L of water were removed during the operation of the remediation system, of which approximately 87 percent was reinjected to the aquifer. The remaining water was pumped to the facility wastewater treatment plant to maintain a depression in the groundwater surface in the area of the plume, reducing the probability for migration of the contaminants. Periodic chemical analysis of groundwater samples for compounds such as benzene, toluene, and xylene, nutrient concentrations, and microbial enumeration indicated that the remediation was progressing well until late 1989, when microbial populations began to decline. By the spring of 1990, the microbial populations had declined to the estimated background levels of 10^2 to 10^4 CFU/mL. This decrease in microbial populations indicated a decrease in the carbon source (petroleum hydrocarbons) in the aquifer.

System Shutdown and Performance Summary

In April 1990, after 32 months of operation, this *in situ* bioremediation system reduced concentrations of gasoline contaminants in the groundwater to levels acceptable to the MDNR and was permitted to be shut down. The decrease in the bacterial population at the facility was the first indication that remediation of groundwater contamination at this facility was nearing completion. Also, nitrate levels in a number of the wells had increased several milligrams per liter, indicating that a breakthrough of the nutrients had begun to occur. Samples of the groundwater were taken in April and May to measure the remaining concentrations of contaminants. Another post-operation monitoring event occurred in mid-August that confirmed the contaminant concentrations had been reduced sufficiently to satisfy the MDNR. Because no statistically significant increase in groundwater concentrations was observed from the April/May samples to the August samples, the MDNR provided written approval to dismantle the treatment system.

Maximum concentrations in June 1987, prior to startup of the system, were in the range of 20 to 30 mg/L of total benzene, toluene, and xylene (Figure 2). Maximum total concentrations of these compounds in April 1990, at the time of system shutdown, were in the range of 0.05 to 0.10 mg/L, with concentrations below the analysis detection limit in samples from most of the wells (Figure 3).

Results of *in situ* bioremediation monitoring data indicate that an average of approximately 1,200 to 1,400 kg of gasoline per month were removed during the life of project. The estimated quantities of contaminants removed by the venting and PTA systems were calculated from analytical data of samples collected from the discharge of these systems. The estimated quantity of biological degradation that occurred was based on the kinetic data of the hydrocarbon removal rate measured during the treatability study.

Conclusion

This was the first *in situ* bioremediation system to be installed in the State of Missouri and was shut down with MDNR approval after successfully remediating the gasoline contamination at this site. During

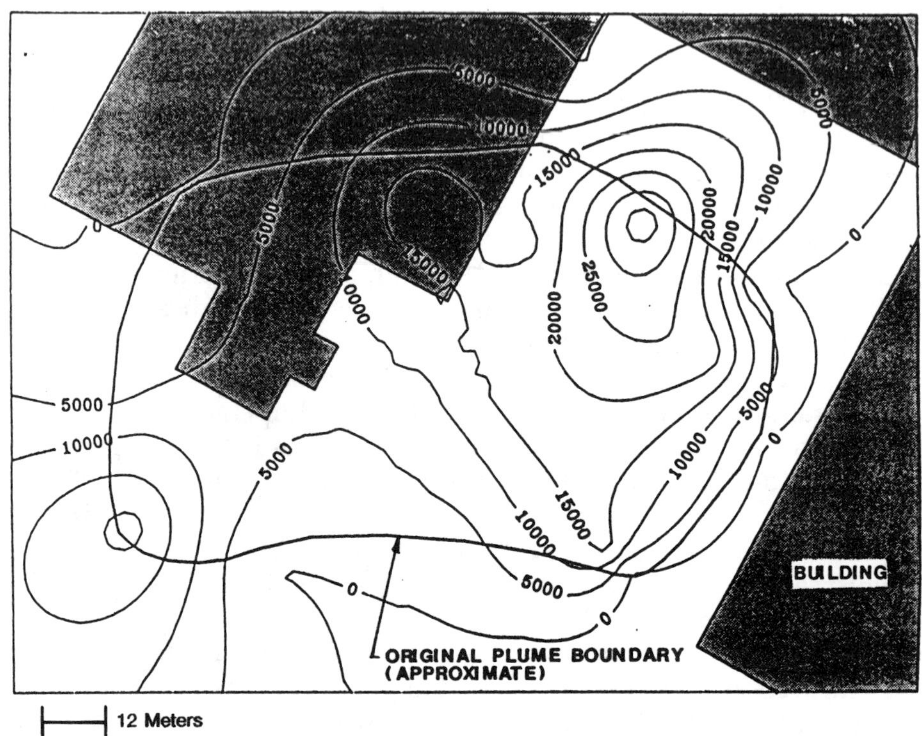

FIGURE 2. Initial benzene, toluene, and xylene groundwater concentration (μg/L)—June 22, 1987.

the 32-month operating life of this remediation system, biological degradation accounted for most of the removal (nearly 88 percent) of the total amount of more than 38,000 kg of gasoline that was removed from the subsurface. Additional VOCs were removed from the groundwater by the PTA system prior to the water being reinjected to the aquifer, and by venting of the vadose zone. Application of *in situ* bioremediation at this site met the facility owner's criteria for being a cost-effective and rapid remedial method.

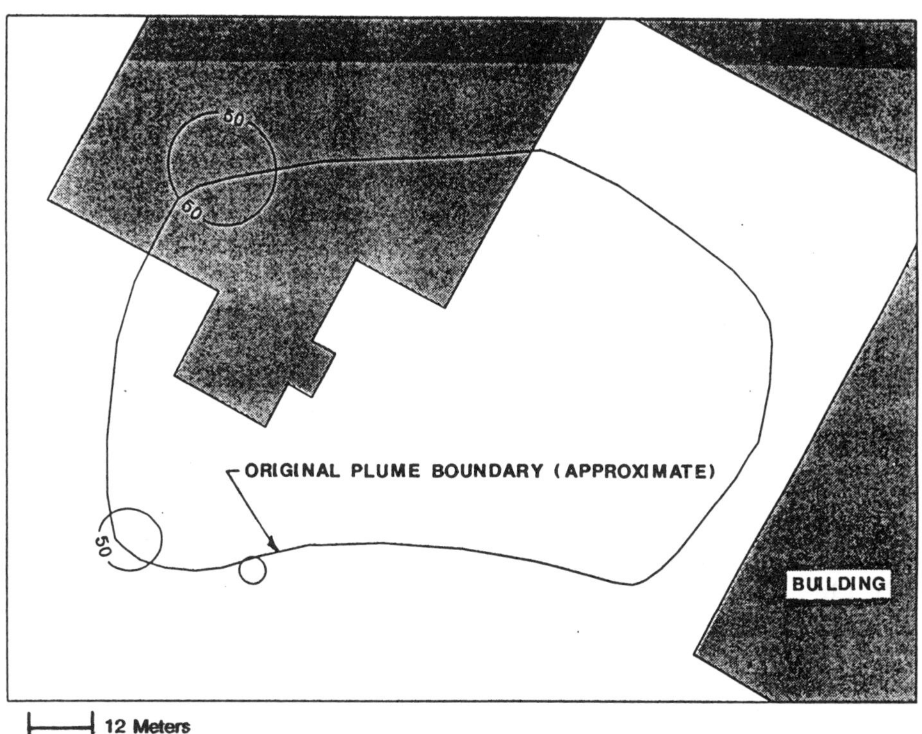

FIGURE 3. Final benzene, toluene, and xylene groundwater concentration (μg/L)—April/May 1990.

Integrated Site Remediation Combining Groundwater Treatment, Soil Vapor Extraction, and Bioremediation

*Richard A. Brown**, *Jeffrey C. Dey*
Groundwater Technology, Inc.
Wayne E. McFarland
Stearns & Wheeler

In October 1988, an apparent act of vandalism caused a large gasoline spill at a bulk petroleum storage plant in southern New Jersey, which is underlain by the Cohansy Aquifer, a sole source drinking water aquifer. Approximately 32,000 L of premium unleaded gasoline were spilled onto the ground around four underground storage tanks. A soil vapor survey was performed within two days of the loss to assess the extent of contamination. Vapor samples were drawn from many locations surrounding the spill site and plotted to determine areas of maximum contamination. The primary area of contamination was the immediate tank pit area where the gasoline had pooled. High concentrations of gasoline extended to the south and west of the tank pit.

In all, it is estimated that approximately 4,600 m^3 of soil were contaminated by the spill over an area of approximately 240 m by 300 m. Approximately 29,000 L of gasoline were adsorbed in the soil of the spill area. An additional 2,600 L were adsorbed in soil, which was excavated and stockpiled immediately following the spill. Because of the significant depth to groundwater (5.5 to 6.1 m) and the quick emergency response actions, it was estimated that less than 380 L of gasoline reached the groundwater under the spill area.

Groundwater samples were taken from the six monitoring wells and the recovery well. The results showed that the groundwater was contaminated, but that contamination was confined to the immediate area of the spill. Maximum contaminant levels in groundwater were 2,940 µg/L BTEX and 8,700 µg/L MTBE in the recovery well.

* Groundwater Technology, Inc., 100 Youngs Road, Mercerville NJ 08619

Soil contamination is found in the general area of the tank pit and extends from grade to the water table. Groundwater is contaminated, and the plume extends southeast of the spill area along the direction of groundwater flow. The spread of the groundwater plume was partially blocked by the existence of a trough in the water table elevations running west to east and located to the south-southwest of the spill area.

Remedial Alternatives

Two general alternatives were compared for dealing with the spill. The first involved excavation and replacement of all contaminated soil in the spill area, combined with continued groundwater treatment to remove dissolved contaminants and prevent off-site migration of any contaminated groundwater. Soil excavation would require removal of the storage tanks and loading rack. Estimated cost for this alternative, including groundwater treatment, was over $800,000. This option would have effectively put the facility out of the liquid fuel storage business.

The second alternative was to treat the contaminated soil on site using soil vapor extraction and bioremediation. Gasoline is relatively volatile, having an average vapor pressure of 50 to 100 mm Hg. It is biodegradable, having a BOD_5 (5-day biological oxygen demand) of 0.08 to 0.12 mg/L. It is, however, relatively insoluble, having a maximum solubility of ~150 mg/L. Based on these properties, remediation of significant quantities of gasoline is best accomplished by the use of volatilization and biodegradation. These were combined with a groundwater recovery system to ensure containment of the contamination during remediation. With the *in situ* system, the petroleum bulk plant could remain in operation during remediation. Off-site liability would not be an issue since no contaminants would leave the site. The estimated cost for on-site treatment was $500,000 to $600,000.

Figure 1 shows the integrated, *in situ* system. The groundwater system captures and treats contaminated groundwater. It also aids in the transport of nutrients and oxygen for the bioremediation system. The vent system removes gasoline from the vadose zone both by direct volatilization and by supplying oxygen to the bioremediation system. Finally, the bioremediation system degrades gasoline in vadose and saturated zone soils and in groundwater. It also increases the removal of adsorbed organics by solubilizing them so they can be captured and removed by the groundwater system. A description of the system elements follows.

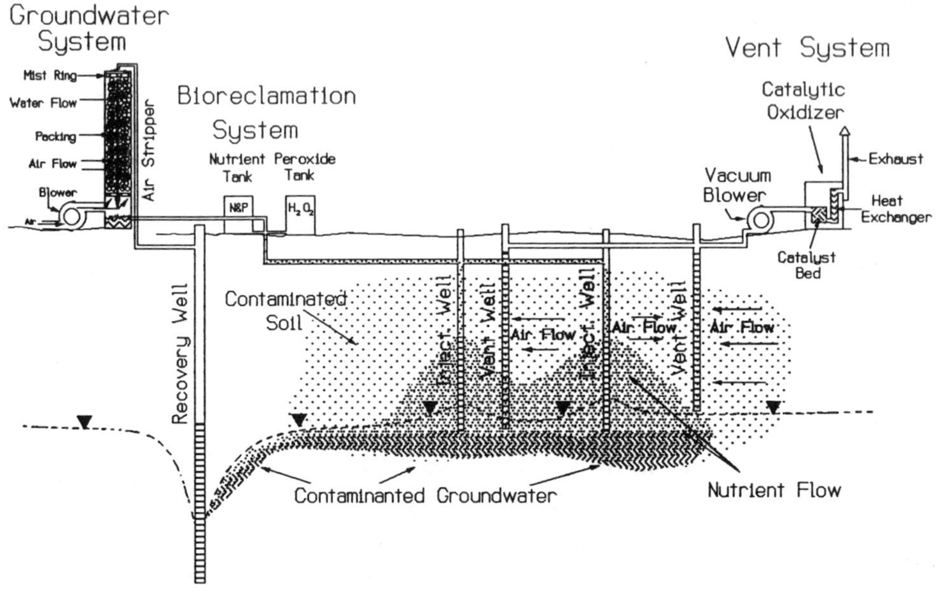

FIGURE 1. Conceptual view of the integrated remedial system.

Groundwater Extraction and Treatment

An initial recovery well, RW-1, was located and installed based on the results of the soil gas survey. Groundwater was pumped from this recovery well at approximately 76 L/min as an interim measure to establish hydraulic control of contaminated groundwater. The extracted groundwater was treated using air stripping. Treatment efficiencies ranged from 98 to 99.9 percent for BTEX to 60 to 85 percent for MTBE. The air stripper removes approximately 0.05 kg/h of hydrocarbon compounds, which represents an equivalent removal of about 260 L of gasoline per month.

After the site was hydraulically isolated and the remedial system installed, a 72-h pump test was then performed to determine the hydraulic characteristics of the water table aquifer. Based on the results of this test a second recovery well, RW-2, was installed 15 m SE of RW-1 to a total depth of 50 ft below grade. Optimal capture zones were obtained by pumping RW-1 at 20 L/min and RW-2 at 95 L/min. Hydraulic control was further enhanced by the upgradient injection of nutrient enriched air stripper effluent. This upgradient injection increased the gradient across the area of dissolved hydrocarbon compound occurrence, thereby increasing the rate at which contaminants were recovered and nutrients circulated.

Soil Vapor Extraction

Soil vapor extraction was applied in two parts: the first treated excavated soils from the spill area, and the second treated the remaining contamination in vadose zone soils. Five vapor extraction wells were installed in the spill area to allow soil vapor extraction of approximately 4,600 m^3 of contaminated soil in the unsaturated zone. Vapor wells were 10 cm in diameter and approximately 6.7 m deep. Separate laterals from each well were manifolded to a high-vacuum blower, and separate ball valves on each lateral allowed dedicated control of vapor flow from the different wells to optimize hydrocarbon removal in the vapor phase. Operating flow rate from the vapor wells is approximately 3.6 m^3/min.

Hydrocarbon removal from the vent system was approximately 38 kg of hydrocarbons per day, equivalent to approximately 53 L of gasoline per day. Vapor treatment was required to meet New Jersey air emission standards. Alternatives for vapor removal included carbon adsorption and catalytic oxidation. Because site remediation could ultimately result in the removal of almost 23,000 kg of hydrocarbons, it was determined that catalytic oxidation would be the most efficient means of vapor treatment.

Bioremediation

Bioremediation was used to treat adsorbed and dissolved contaminants in the saturated zone. Bioremediation was used in conjunction with soil vapor extraction to accelerate the removal of adsorbed hydrocarbons in the unsaturated zone. An added benefit of the process is that it removes adsorbed hydrocarbons in the saturated zone, which are not as effectively removed by soil vapor extraction.

To determine nutrient requirements, optimum conditions for biodegradation are determined by laboratory simulation of conditions using actual soil samples from the spill site. As can be seen, the site requires addition of ~100 mg nutrients/L gasoline to remove the gasoline.

Nutrient-amended groundwater is continuously injected through five independently valved injection wells and through a series of shallow infiltration lines. Groundwater from the recovery wells is air stripped and amended with nutrients at 100 mg/L and hydrogen peroxide at 1,000 mg/L. The water content in the vadose zone is kept at ~50 percent of residual saturation to maintain a balance between bioremediation and soil vapor extraction. The bioremediation system is fully balanced with the groundwater recovery system and integrated with the soil vapor extraction system.

Operating Results

The performance of the different system components is depicted in Figure 2. These results were calculated based on monitoring pump effluents (groundwater extraction, soil vapor extraction) and by the amounts of nutrients and peroxide used (bioremediation). The nutrient and peroxide consumption was adjusted for utilization efficiency as determined by laboratory experiments. Additionally, the results were calculated on a mass balance basis. As can be seen, the bulk of the removal has been by the vapor extraction system. The soil vapor extraction system provided a rapid and substantial response to the spill. Bioremediation also provides significant removal. As can be seen from the data, the percentage removal by the other processes has increased with time. This is to be expected as the vapor extraction system removes the highly volatile and mobile fractions, leaving a less mobile residue. This residue is more responsive to bioremediation than to soil vapor extraction. Hence, the increase in the relative importance of bioremediation. Using an integrated system to treat groundwater and soil contamination provided an effective and rapid response to large scale organic contamination. The combination of bioremediation, soil vapor extraction and groundwater recovery employed at this site has removed close to 15,500 kg of gasoline in under 18 months of operation, a >73 percent reduction of the original 21,000 kg loss. The system has effectively reduced groundwater contamination in the spill area by over 90 percent from the maximum levels. The integrated system has been effective because it has made best use of the properties of the contaminant and the nature of the site in remediation.

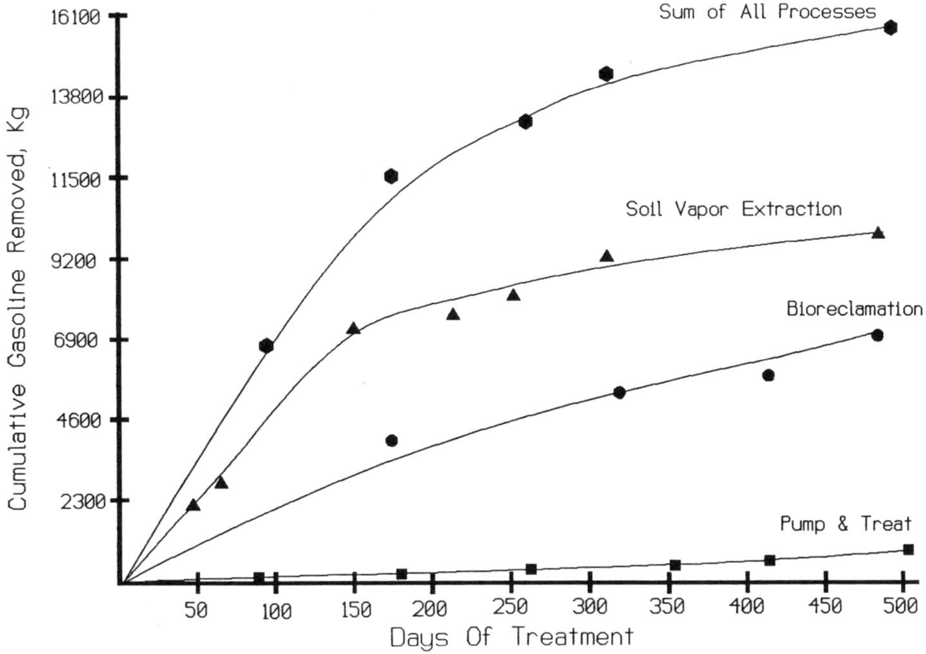

FIGURE 2. Treatment performance of the integrated system.

Application of *In Situ* Bioreclamation to a Low-Permeable Heterogeneous Formation: Evolution of a System in Response to Regulatory and Technical Issues

*Richard A. Brown**, *Cliff Harper, James Oppenheim*
Groundwater Technology, Inc.

A remedial solution is rarely composed of a single element, nor is it ever a static decision. Effective remediation requires blending different processes at the onset and changing the relative importance of each element with time, as goals, degree of contamination, or events change the complexity of the problem. Remediation is an integrated, iterative, and evolutionary process.

The site discussed in this paper is illustrative of this process of remedial management at a complex site. First, the geology of the site is extremely complex. Second, there have been (at least) two subsequent product losses since the start of remediation. Finally, the regulatory/social context for the site has continually changed. In response to these factors, the remedial system has evolved from an initial response of soil excavation and groundwater treatment to an integrated groundwater recovery, soil vent, and bioreclamation system.

The study site (Figure 1), is located in northwestern Vermont, in a town of approximately 6,000. The surficial geology encountered at the site consists of the littoral deposits of the Champlain Sea. These deposits consist predominately of marine sands overlying marine clays. Observations made during on-site monitoring well drilling activities have generally shown the top 5 to 7 m of the stratigraphic column to consist of fine sands interbedded with layers of medium sands, pebbles, silt, and clay.

* Groundwater Technology, Inc., 310 Horizon Center Drive, Trenton NJ 08691

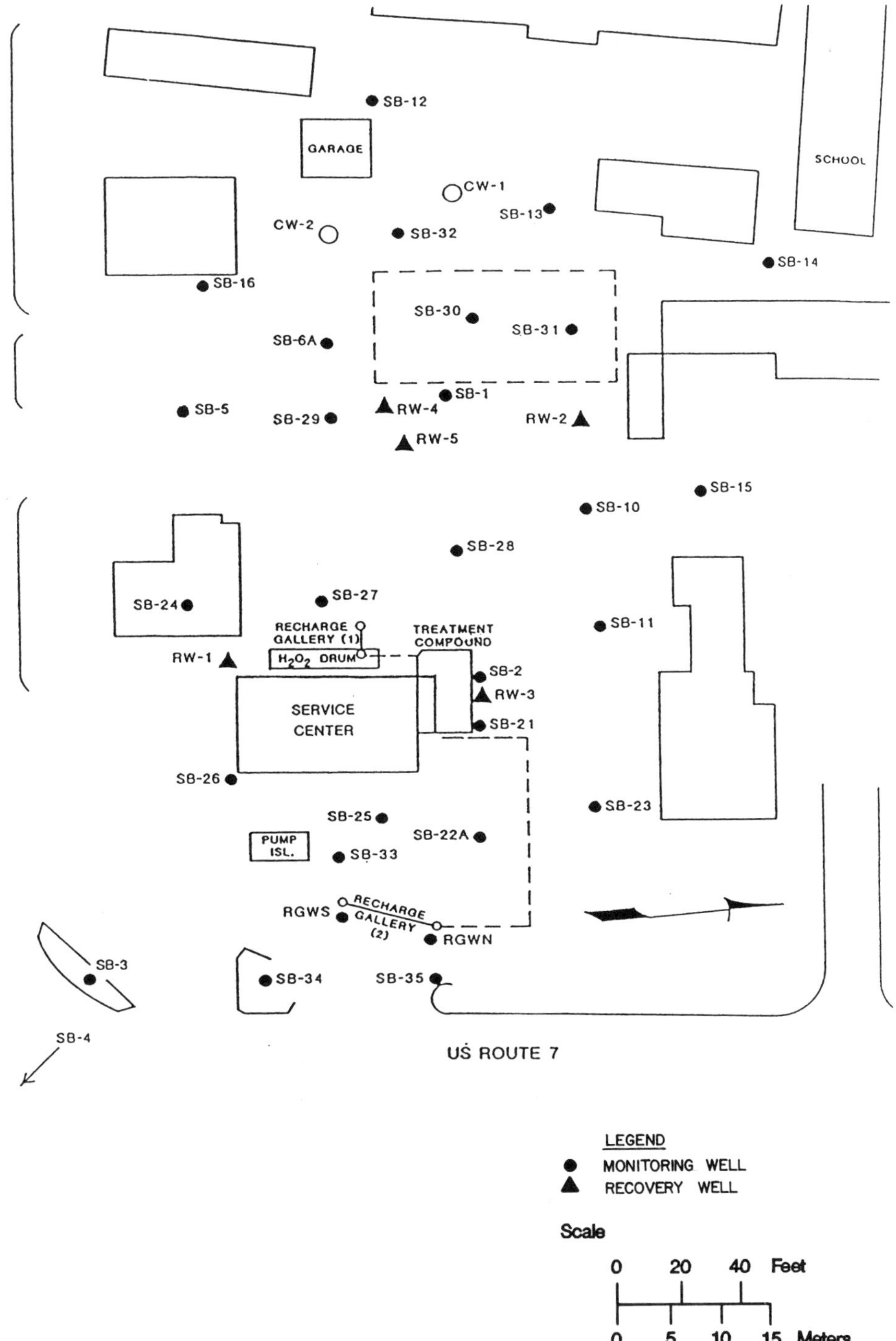

FIGURE 1. Study site.

Groundwater in the area occurs as a shallow water table system (1–2 m) within the marine sands; the clays act as a vertical barrier to groundwater movement. While general groundwater velocities and pumping rates are very low, the contamination does migrate rapidly throughout the site through the dispersed lenses of coarser grain material. These lenses are not large enough nor interconnected enough to be significant in terms of capture or transport of nutrients, but they are, none the less, the primary pathway of contaminant migration.

Contamination History

A loss of an undetermined amount of gasoline from a leaking underground storage tank impacted the shallow, nonpotable groundwater system. As early as April 1985, gasoline vapors were detected in residences. During high water, product was observed in the garden area (homeowner) near the station.

Subsequent to the implementation of this initial remedial system, a second product loss event was detected during site activities in the spring of 1988. This second loss was due to a distribution line leak. The loss may have occurred over a period of time. The remedial system was adjusted in response to the second identified product source.

In the summer of 1989, a third product loss source was detected during line testing. This product loss event appears to have been a short-term event and minor size. This was evident by the fact that no free-phase product appeared in any monitoring well as a result of this event.

In August 1985, after an initial investigation, the State of Vermont Agency of Environmental Conservation (AEC) issued an order to initiate remediation of the contamination consisting of tank removal and soil excavation. In February 1986, after implementation of this initial remedial response, additional contaminated soil was found during the digging of a test pit east of the service center. The Vermont AEC requested that these soils be excavated. In August of 1988, a site closure plan was developed and agreed to by the Vermont ANR. This plan presented the following criteria:

1. There shall be no phase-separated product above 0.01 feet [~0.3 cm] in any of the wells throughout the site for a period of one full year.

2. There shall no longer be any off-site migration of hydrocarbon vapors from the site to the surrounding dwellings.

3. The concentrations of dissolved hydrocarbons in the groundwater are shown to be stable or declining throughout the entire site.

Remedial Performance

There have been five remedial elements employed at the site. Most of these have been modified with time to respond to site conditions. These five elements have been soil excavation, free product recovery, groundwater recovery, soil vapor extraction, and bioreclamation. The first two of these five were significant only in the first year.

Approximately 270 m^3 of contaminated soil were removed from two areas of the site: the former tank pit area and the area east of the garden. The excavation did reduce contamination in the tank pit area. The success of the excavation was limited because of the shallow groundwater, less than a meter, during excavation in the garden area. The clean fill that was placed in the excavation was recontaminated by fluctuations in the water table that drove phase-separated product into the clean fill.

From 1986 to early 1988, approximately 325 L of phase-separated product was recovered from recovery wells RW-1, RW-2, and RW-3. The product recovery system consisted of a dual pump system. The lower pump was a groundwater depression pump which created a "collection" point to which product could drain. The upper pump was a product pump activated by a float switch.

The effectiveness of the product recovery system has been limited by the geology of the site. The general soil conditions are low permeable sands. The cone of depression of the recovery wells is quite limited. Pump tests have shown that the cone of depression for the recovery wells generally extends about 3 m or less. As a result, the capture of free phase material, which relies upon the gradient for drainage, has been quite limited.

The soil vapor extraction system consists of a number of shallow soil vent points, screened from 0.6 m below grade into the water table and exhaust points in the basements of the surrounding residences. The soil vent system consists of five separate legs (Figure 2). These legs have been installed over a period of several years in response to

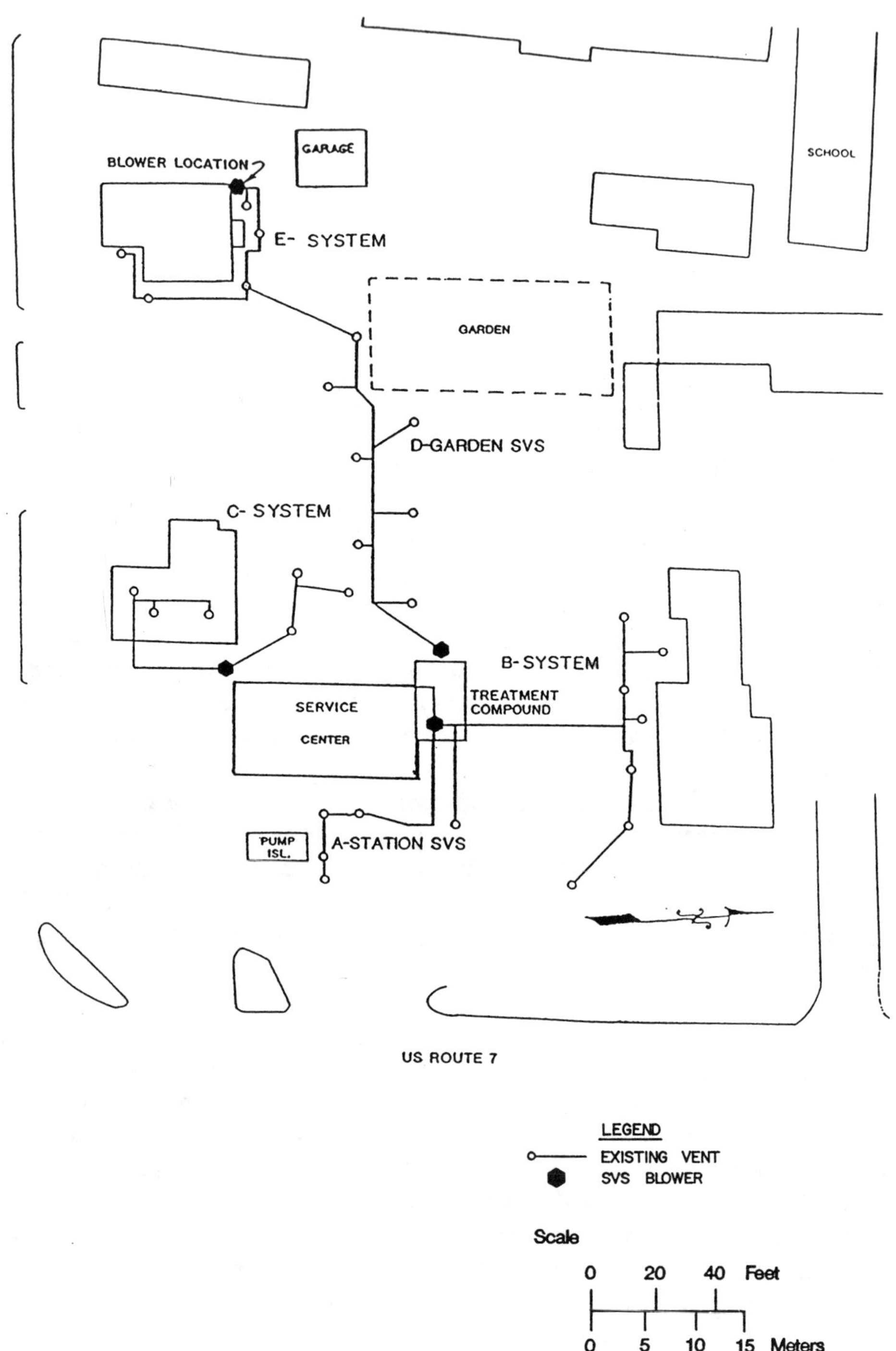

FIGURE 2. Soil vent system schematic.

contamination issues. This system has recovered a total of 700 kg since its start-up.

The third part of the SVS system was the "D" leg. This system was installed to address the known high levels of contamination in the garden area. Since its start-up, the "D" system has recovered 1,400 kg of hydrocarbons.

The "A" leg was installed in the front of the station. This was in response to the detection of high levels of contamination under the station blacktop during a soil survey. Since its start-up, the "A" system has recovered 1,420 kg of hydrocarbons.

The final leg, the "E" system, was installed in January 1989. The impetus for the system was "political." The resident in this area complained of vapors. Production from this system has borne out that there was no contamination, as only 1 kg has been recovered to date. The soil vapor extraction system has been one of the more successful parts of the site remediation plan.

The initial bioreclamation system is pictured in Figure 3. Recovered groundwater was air stripped and then amended with nutrients and hydrogen peroxide. The amended groundwater was then reinjected through two galleries, one in the front of the station, the other in the back. The nutrients consisted of a mixture of ammonium chloride and mono- and di-sodium phosphates. Hydrogen peroxide was added at a level of 1,000 to 2,000 mg/L.

The influence of nutrients was quite limited from this system. The tight clayey soils retarded and adsorbed nutrients limiting the influence to 12 to 18 m. The front gallery had a somewhat larger influence due to the fill that was replaced in the excavation north of the station.

In the fall of 1988, an additional injection system was installed. This system consisted of a number of shallow injection points which were air sparged as well as receiving nutrient/peroxide amended water. With the installation of this additional injection system, the influence of nutrients was significantly increased. The nutrient concentration was 500 mg/L and the peroxide concentration, 1,000 mg/L.

The bioreclamation system appeared to have an impact on the adsorbed phase contamination in the areas where nutrient transport was achieved. Based on nutrient utilization and decrease in dissolved concentration, the bioreclamation system was estimated to have removed approximately 1,540 kg of gasoline since its start-up. This estimate was based upon the amount of hydrogen peroxide used and an assumed hydrogen peroxide use efficiency of 15 percent. Laboratory

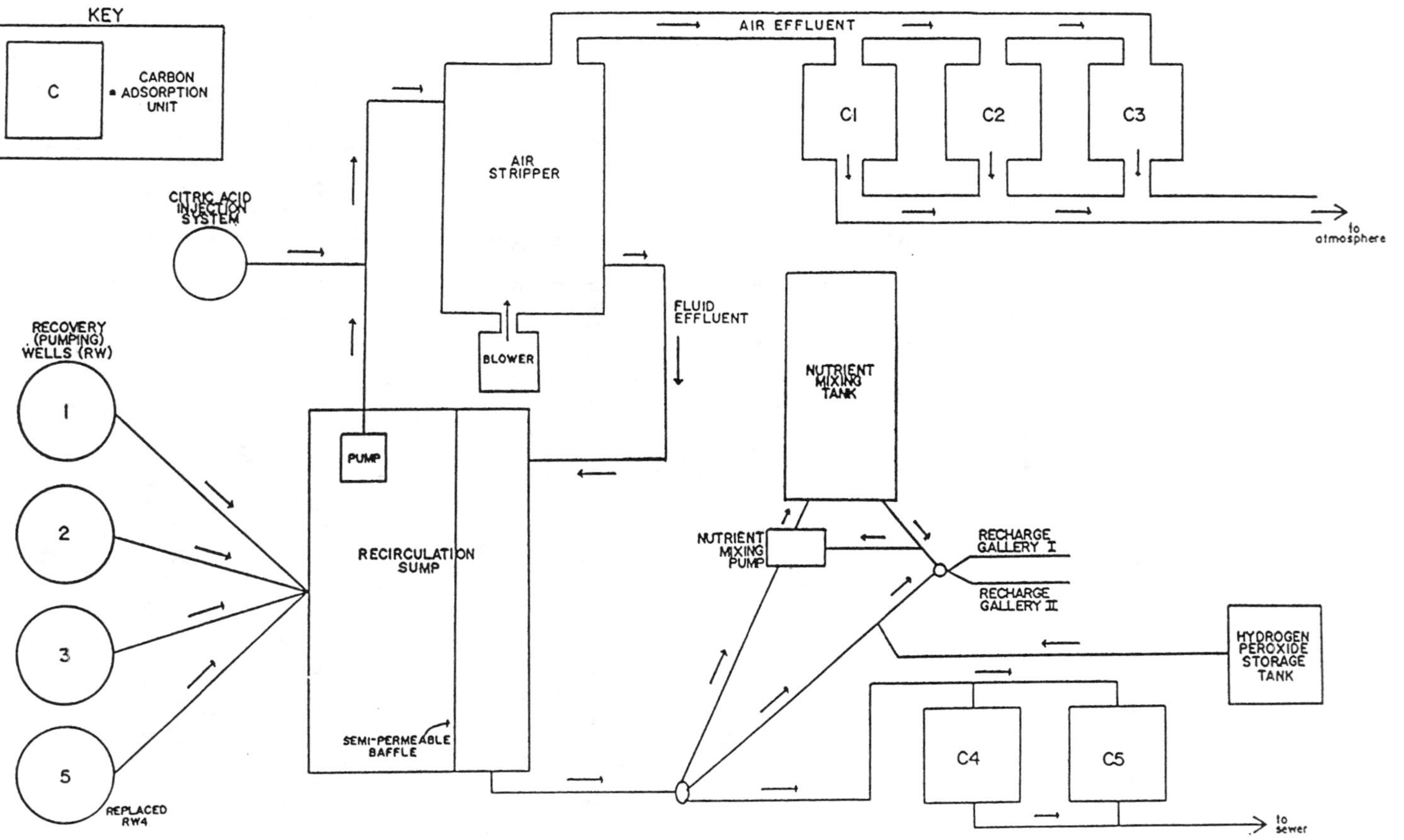

FIGURE 3. Groundwater treatment and bioreclamation system.

hydrogen peroxide stability tests were the basis for the hydrogen peroxide use efficiency.

The last element of the remedial system is the groundwater recovery system.

After installation of four pumping wells, the area of severe contamination was significantly reduced by the spring of 1987. However, the second loss event recontaminated much of the site based on 1988 dissolved BTEX contour lines. At this point, a recovery trench was installed. These combined elements once again led to a significant reduction in severe groundwater contamination.

The recovered groundwater has been air stripped prior to discharge. Based on the total hydrocarbon concentration in the stripper exhaust, 2,350 kg of hydrocarbons have been removed by the groundwater recovery system.

Summary of Performance

The remediation of the site has been a complex process. The geology is complex, which makes transport and recovery difficult and unpredictable. Additionally, there have been subsequent losses which have complicated the remedial performance. Lastly, there has been an evolving regulatory context which has changed goals and timing.

The remedial system has been effective in removing a considerable quantity of hydrocarbons. The system performance is summarized below.

Remedial Element	Hydrocarbon Removed (kg)
Soil excavation	1,860 (est.)
Free phase recovery	975
Soil vent system	5,200
Bioreclamation	1,540 (est.)
Groundwater recovery	2,350
TOTAL	11,925

The goals of no free phase, no vapor phase, and stable or declining groundwater concentrations have been achieved. The site is currently in closure monitoring.

Biotransformation of Monoaromatic Hydrocarbons under Anoxic Conditions

Harold A. Ball[*], *Martin Reinhard, Perry L. McCarty*
Stanford University

Aromatic hydrocarbons contained in gasoline are environmental pollutants of particular concern since they are relatively soluble in water, many are toxic, and some are confirmed carcinogens (*e.g.*, benzene). Although most gasoline constituents are readily degraded in aerobic surface water systems, the groundwater environment associated with hydrocarbon spills is typically anaerobic, thus precluding aerobic degradation pathways. Oxygen can be supplied to the aquifer either by aeration or by addition of oxidants such as peroxide. However, this substantially increases the cost of remediation. In the absence of oxygen, degradation of gasoline components can take place only with the utilization of alternate electron acceptors such as nitrate, sulfate, carbon dioxide, and possibly ferric iron or other metal oxides. Benzene, toluene, and xylene isomers were completely degraded by aquifer- or sewage sludge-derived microorganisms under denitrifying (Kuhn *et al.* 1985, 1988; Major *et al.* 1988; Zeyer *et al.* 1986) and methanogenic conditions (Grbić-Galić & Vogel 1987; Vogel & Grbić-Galić 1986; Wilson *et al.* 1986, 1987). Recently, a pure culture was found to degrade toluene and *m*-xylene using nitrate or nitrous oxide as an electron acceptor (Zeyer *et al.* 1990).

This paper presents initial results of our ongoing study to develop and characterize microbial consortia capable of transforming aromatic hydrocarbons under nitrate-reducing conditions, and understand the effect of environmental factors on the biotransformation processes.

[*] Western Region Hazardous Substance Research Center, Stanford University, Stanford CA 94305-4020

Experimental

Our first experiments focused on finding cultures with the ability to degrade aromatic compounds under denitrifying conditions. Microcosms were inoculated with material from various sources: aquifer solids from a contaminated aquifer (both from the contaminated zone and adjacent to the contaminated zone), sediments from an oil refinery treatment pond, and activated sludge from a refinery treatment facility (all previously exposed to hydrocarbons), as well as effluent from a secondary wastewater treatment facility. Batch bottle microcosms were the primary methodology used for developing these cultures. Denitrifying media were prepared from a modification of a methanogenic medium (Owen *et al.* 1979) supplemented with 6 mM sodium nitrate. Microcosms were prepared in an anaerobic glove box from the deoxygenated microbial medium and inoculated as indicated above. Sterile controls were prepared by autoclaving a replicate sample bottle. A mixture of monoaromatic compounds (benzene, toluene, chlorobenzene, ethylbenzene, *p*-xylene, *o*-xylene, phenol, and naphthalene) was added to the microcosms at 8 mg/L for each compound along with 25 mg/L acetate. Samples were extracted in ether and analyzed by a gas chromatograph equipped with a photoionization detector. Liquid samples were removed for inorganic analysis by ion chromatography.

In a second experiment, microcosms were prepared without added acetate, and aromatic compounds (benzene, toluene, ethylbenzene, *m*-xylene, *p*-xylene, *o*-xylene) were added alone and in mixtures at 8 mg/L each. Other microbial media were evaluated (Shelton & Tiedje 1984; Zeyer *et al.* 1986). In a third experiment, microbial consortia were enriched further for biotransformation of ethylbenzene. In these experiments, the microcosms were prepared in bottles containing 20 percent headspace from which a gaseous sample was taken and analyzed by direct injection onto the gas chromatograph. Liquid samples were removed for analysis of nitrate and nitrite by ion chromatography.

Results

In our first experiments, toluene and phenol were degraded after 222 days of incubation in the microcosms inoculated with aquifer solids, and toluene, ethylbenzene, and phenol were degraded in the microcosms inoculated with the sediment material from the refinery pond or activated sludge from the refinery treatment facility. No aromatics were degraded in microcosms inoculated with secondary wastewater. Acetate was completely degraded in all the microcosms, with commensurate loss of nitrate. Toluene was degraded without any adaptation period in all the samples that were previously exposed to petroleum products.

In the second experiment, toluene, ethylbenzene, and *m*-xylene were biotransformed under anoxic conditions with nitrate as the electron acceptor by a microbial consortium that was enriched from the oil refinery treatment pond sediment. The transforming consortia have been maintained for over 1 year with no loss of activity. Benzene, *o*-xylene, and *p*-xylene were not transformed. Microcosms were set up in the modified Owen medium and the Zeyer medium and maintained at 35 C. Some microcosms contained a mixture of aromatic compounds to ascertain multicomponent effects on adaptation. Biodegradation, while selective, was enhanced in the modified Owen medium containing no organic cosubstrates other than the monoaromatic compounds of interest, but was fortified with essential vitamins. Toluene was readily degradable under denitrifying conditions and appears to be a preferential substrate as it was degraded both alone and in combination with benzene, ethylbenzene, or *p*-xylene. However, toluene was not degraded in combination with *p*-xylene when the consortium was prepared in the Zeyer medium, which contains a phosphate buffer but no vitamin additions. Ethylbenzene was degraded by the consortium alone and in combination with toluene, and *m*-xylene was degraded by the consortium only as a single substrate. After adaptation, kinetics of transformation by nonenriched cultures were typically fast, with complete removal of the aromatic within 3 to 4 days.

Consortia that degrade toluene, ethylbenzene, and *m*-xylene are readily obtained from the samples of oil refinery treatment pond sediment. However, enrichments grew in the modified Owen medium only if it had not been sterilized first by autoclaving. What changes in the medium were caused by autoclaving and resulted in this effect are unknown. Since transformation occurred in medium amended with the original sediment, it is possible that autoclaving removed (through precipitation, decomposition, or other chemistry) some essential nutrients that were present in and supplied by the sediment inocula.

In our third experiment, enrichments in the modified RAMM medium (Shelton & Tiedje 1984) were successful even when the medium had been sterilized by autoclaving. At 35 C and 20 C, biotransformation rates were higher in this medium than in the other medium tested. Figure 1 illustrates the degradation of ethylbenzene by a previously enriched consortium in the modified RAMM medium at

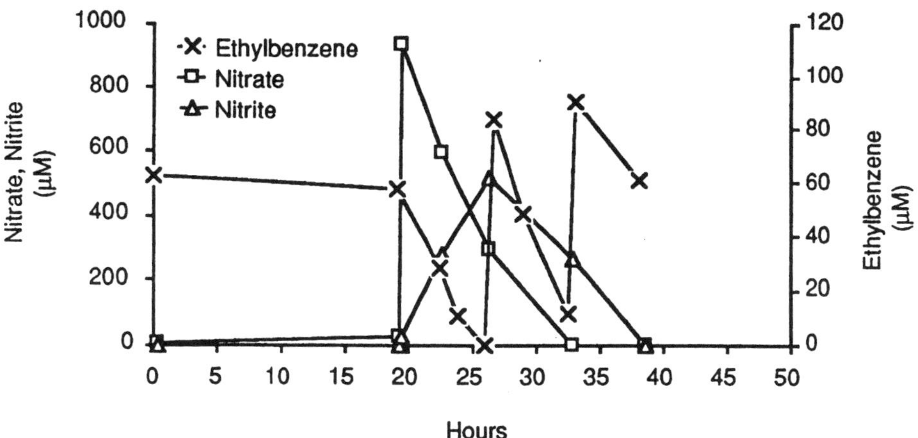

FIGURE 1. Ethylbenzene degradation at 20 C in the presence of nitrate.

20 C. The microcosm actively degraded ethylbenzene for 4 days until nitrate and nitrite were depleted, at which point ethylbenzene degradation stopped (this point is given as time zero in the plot). Nitrate was added to the microcosm after 19 h, and ethylbenzene degradation resumed. Ethylbenzene was added to the microcosm after 26 and 33 h, and degradation continued while nitrate or nitrite were present. As shown in Figure 1, nitrite appeared as a transient in the overall denitrification that occurred. During this period, there was neither transformation of nitrate and ethylbenzene nor production of nitrite in the sterile control. Present was 80 mg/L total suspended solids, which represents the upper limit for the microbial population present.

Theoretical stoichiometry would require 8.4 M nitrate per mole ethylbenzene if the nitrate were reduced completely to nitrogen gas and the ethylbenzene were oxidized to carbon dioxide (i.e., no intermediates or cells produced). When potential cell growth is considered (McCarty 1975), this ratio decreases to 5.3 M nitrate per mole ethylbenzene, assuming again that the nitrate is reduced to nitrogen gas and the ethylbenzene is oxidized to carbon dioxide or cells (no intermediates produced). In this experiment, 190 µM nitrate were consumed in the degradation of 35 µM ethylbenzene for a molar ratio of 5.4, very close to the prediction when cell growth is included.

Conclusions

It appears that microorganisms with the ability to degrade aromatic compounds are not ubiquitous. Although sewage seed contains a very diverse population of microorganisms, this mixed culture did not adapt to the aromatic compounds tested under denitrifying conditions. It is also clear that the composition of the microbial medium significantly affects both the potential for biotransformation of aromatics and transformation rate. Toluene was a preferential substrate for biotransformation under the different conditions and in the various media tested. Further investigation is needed to evaluate why toluene is so readily biotransformed. Finally, the nitrate utilization results give evidence that biodegradation of ethylbenzene takes place under denitrifying conditions where nitrate is available as an electron acceptor.

Acknowledgments

This project was supported by a grant from the Department of the Navy, NCEL, Port Hueneme (IAG-RW-97934004-0), and by the Orange County Water District through the Western Region Hazardous Substance Research Center sponsored by the Environmental Protection Agency (R-815738).

REFERENCES

Grbić-Galić, D.; Vogel, T. M. "Transformation of Toluene and Benzene by Mixed Methanogenic Cultures." *Applied and Environmental Microbiology* **1987,** *53,* 254–260.

Kuhn, E. P.; Colberg, P. J.; Schnoor, J. L.; Wanner, O.; Zehnder, A.J.B.; Schwarzenbach, R. P. "Microbial Transformations of Substituted Benzenes During Infiltration of River Water to Ground Water: Laboratory Column Studies." *Environmental Science and Technology* **1985,** *19,* 961–968.

Kuhn, E. P.; Zeyer, J.; Eicher, P.; Schwarzenbach, R. P. "Anaerobic Degradation of Alkylated Benzenes in Denitrifying Laboratory Aquifer Columns." *Applied and Environmental Microbiology* **1988,** *54,* 490–496.

Major, D. W.; Mayfield, C. I.; Barker, J. F. "Biotransformation of Benzene by Denitrification in Aquifer Sand." *Ground Water* **1988,** *26,* 8–14.

McCarty, P. L. "Stoichiometry of Biological Reactions." *Progress in Water Technology* **1975,** *7,* 157–172.

Owen, W. F.; Stuckey, D. C.; Healy, J. B.; Young, L. Y.; McCarty, P. L. "Bioassay for Monitoring Biochemical Methane Potential and Anaerobic Toxicity." *Water Research* **1979,** *13,* 48–492.

Shelton, D. R.; Tiedje, J. M. "General Method for Determining Anaerobic Biodegradation Potential." *Applied and Environmental Microbiology* **1984**, *47*, 850–857.

Vogel, T. M.; Grbić-Galić, D. "Incorporation of Oxygen from Water into Toluene and Benzene during Anaerobic Fermentative Transformation." *Applied and Environmental Microbiology* **1986**, *52*, 200–202.

Wilson, B. H.; Smith, G. B.; Rees, J. F. "Biotransformation of Selected Alkylbenzenes and Halogenated Aliphatic Hydrocarbon Compounds in Methanogenic Aquifer Material: A Microcosm Study." *Environmental Science and Technology* **1986**, *20*, 997–1002.

Wilson, B. H.; Bledsoe, B.; Kampbell, D. "Biological Processes Occurring at an Aviation Gasoline Spill Site." In *Chemical Quality of Water and the Hydrologic Cycle*; Averett, R. C.; McKnight, D. M., Eds.; Lewis: Chelsea, MI, 1987; pp 125–137.

Zeyer, J.; Kuhn, E. P.; Schwarzenbach, R. P. "Rapid Microbial Mineralization of Toluene and 1,3-Dimethylbenzene in the Absence of Molecular Oxygen." *Applied and Environmental Microbiology* **1986**, *52*, 944–947.

Zeyer, J.; Eicher, P.; Dolfing, J.; Schwarzenbach, R. P. "Anaerobic Degradation of Aromatic Hydrocarbons." In *Biotechnology and Biodegradation*, Kamely, D., *et al.*, Eds.; Advances in Applied Biotechnology Series, Vol. 4; Gulf: Houston, 1990; Chapter 3.

Anaerobic Degradation of Toluene and Xylene—Evidence for Sulphate as the Terminal Electron Acceptor

*E. A. Edwards**, *L. E. Wills, D. Grbić-Galić, M. Reinhard*
Stanford University

Petroleum products frequently contaminate soil, sediment, and groundwater from leaks in underground storage tanks, improper disposal techniques, and inadvertent spills. Of the many constituents of petroleum, the nonoxygenated, homocyclic aromatic compounds that include benzene, toluene, ethylbenzene, and xylenes (BTEXs) are of particular concern because they are confirmed or suspected carcinogens, even at very low concentrations (Dean 1985). BTEXs are relatively water

* Western Region Hazardous Substance Research Center, Department of Civil Engineering, Terman Engineering Building, Stanford University, Stanford CA 94305-4020

soluble and thus are frequently detected in groundwater used for municipal drinking water supplies.

Anaerobic conditions often develop in natural ecosystems and in leachate plumes emanating from contaminated sites because oxygen is depleted by aerobic microorganisms consuming easily degradable substrates (Wilson *et al.* 1986a, b). A better understanding of anaerobic processes is required to predict the fate of pollutants in the subsurface, to design effective remediation schemes, and to best utilize these biological reactions in the treatment of water both *in situ* in the subsurface and in aboveground reactors.

The anaerobic transformation of BTEXs has been observed in contaminated leachate plumes and aquifers (Kuhn *et al.* 1985; Reinhard *et al.* 1984). In the laboratory, the transformation of certain BTEXs has been demonstrated under denitrifying conditions (Kuhn *et al.* 1985; Kuhn *et al.*1988; Major *et al.* 1988; Zeyer *et al.* 1986), under methanogenic conditions in microcosms (Grbić-Galić & Vogel 1987; Vogel & Grbić-Galić 1986; Wilson *et al.* 1986a, b; Wilson *et al.* 1987), by iron-reducing organisms (Lovley *et al.* 1989), and recently under sulphate-reducing conditions (Haag *et al.* 1990).

The objective of this research is to study anaerobic transformations of BTEX compounds in microcosms containing aquifer solids from a gasoline-contaminated aquifer with conditions conducive to the growth and activity of sulphate-reducing bacteria.

Materials and Methods

Aquifer Material. Samples from a gasoline-contaminated sandy silt (*i.e.*, Seal Beach CA) were taken on May 12, 1987, using a steam-cleaned, 0.76-m-diameter bucket auger at a depth of 0.25 m above the water table. The sediment samples were stored in 20-L plastic containers with airtight snap lids at 4 C.

Medium. A medium designed to support sulphate-reducing bacteria was prepared that had the following constituents per liter of deionized water: 10 mL of phosphate buffer (27.2 g/L KH_2PO_4, 34.8 g/L K_2HPO_4), 10 mL of salt solution (53.5 g/L NH_4Cl, 7.0 g/L $CaCl_2 \bullet 6H_2O$, 2.0 g/L $FeCl_2 \bullet 4H_2O$), 2 mL of trace mineral solution [0.3 g/L H_3BO_3, 0.1 g/L $ZnCl_2$, 0.75 g/L $NiCl_2 \bullet 6H_2O$, 1.0 g/L $MnCl_2 \bullet 4H_2O$, 0.1 g/L $CuCl_2 \bullet 2H_2O$, 1.5 g/L $CoCl_2 \bullet 6H_2O$, 0.02 g/L Na_2SeO_3, 0.1 g/L $Al_2(SO_4)_3 \bullet 16H_2O$, 1 mL/L H_2SO_4], 2 mL of $MgSO_4 \bullet 7H_2O$ solution (62.5 g/L)], 1 mL of redox indicator stock solution (1 g/L resazurin),

10 mL of saturated bicarbonate solution (260 g/L $NaHCO_3$), 10 mL of vitamin stock solution (0.02 g/L biotin, 0.02 g/L folic acid, 0.1 g/L pyridoxine hydrochloride, 0.05 g/L riboflavin, 0.05 g/L thiamine, 0.05 g/L nicotinic acid, 0.05 g/L pantothenic acid, 0.05 g/L PABA, 0.05 g/L cyanocobalamin, 0.05 g/L thioctic acid), and 10 mL of an amorphous ferrous sulphide solution [39.2 g/L $(NH_4)_2Fe(SO_4)_2 \bullet 6H_2O$, 24.0 g/L $Na_2S \bullet 9H_2O$] that had been washed three times using deionized water to remove free sulphide. Sodium lactate was added initially at a concentration of 37 mg/L. The initial sulphate concentration was 25 mM (Na_2SO_4); additional sulphate was added as degradation occurred.

Chemicals. Chemicals were purchased from Sigma (St. Louis MO) or Aldrich (Milwaukee WI) and were greater than 99.9 percent pure. Radio-labelled [^{14}C]toluene (ring- and methyl-labelled) and [^{14}C]*o*-xylene (methyl-labelled) were also purchased from Sigma.

Microcosms. Microcosms were prepared in 250-mL (8 oz) screw-cap bottles and sealed with Mininert™ valves (Alltech cat. #95326). The bottles and caps were acid washed and sterilized before use. One hundred grams of aquifer material was transferred into each 250-mL bottle in an anaerobic glove box (Coy Laboratory Products, Ann Arbor MI). Control bottles were sealed with Mininert valves and were autoclaved for 20 minutes at 121 C on 3 consecutive days. Finally, 100 mL of medium were added to each bottle, which was then sealed with a Mininert valve. 2-Bromoethanesulphonate (BESA) was added at a concentration of 1 mM to inhibit methanogens. In total, eight microcosms were established: two autoclaved controls, two amended with BESA, and four experimental. All bottles were incubated anaerobically at 20 C in the dark.

The bottles were spiked with substrates in the glove box. A stock substrate solution was prepared containing benzene, toluene, ethylbenzene, *p*-xylene, and *o*-xylene; the initial concentration of each compound was approximately 5 mg/L. As degradation occurred, additional substrate was added by injecting 1.5 µL of the pure compound with a 10-µL syringe.

Enrichment. Enrichment cultures were prepared by transferring sediment (10 g) and liquid (30 mL) from active microcosms to clean, autoclaved 250-mL bottles. These bottles were then filled with 170 mL of sulphate-reducing medium and spiked with the appropriate substrates.

In some instances, these bottles were spiked with [^{14}C]-labelled substrates.

Analytical Procedures. To monitor the concentration of substrates, 300 µL of headspace were sampled from each bottle approximately every 10 days using a 500-µL gastight syringe. The gas samples were injected onto a Carlo Erba Fractovap 2900 Series Gas Chromatograph (GC) equipped with a Photoionization Detector (PID, HNU Systems Inc., Model PI-52-02; 10 eV lamp). Liquid samples were analyzed monthly on a Dionex Series 4000i Ion Chromatograph (IC) using an electrochemical conductivity detector to monitor sulphate concentrations. The data from the GC/PID and IC were collected and processed with the Nelson Analytical Inc. 3000 Series Chromatography Data System.

The operating conditions for the GC/PID column (DB-625 fused silica megabore capillary column, 30 × 0.53 mm interior diameter) were an injection port temperature of 240 C, a detector temperature of 250 C, helium carrier gas at a column head pressure of 0.7 kg/cm^2, helium make-up gas at a flow of 30 mL/min, an isothermal temperature of 75 C, and a splitless injection (split closed for 30 sec). The operating conditions for the IC were as follows. The eluant was 0.75 mM/2.2 mM sodium bicarbonate/sodium carbonate (2 mL/min); the regenerant was 0.025 N sulfuric acid; and the output range was 100 µS.

^{14}C activity was determined on a Tricarb Model 4530 scintillation spectrometer (Packard Instrument Co., Downers Grove IL). Three separate 1-mL liquid samples were counted for each analysis to determine the ^{14}C activity in the volatile, nonvolatile, and CO_2 fractions as described by Grbić-Galic and Vogel (1987).

Results

Compounds Degraded and Rates of Degradation. In all of the microcosms that were initially fed a mixture of benzene, toluene, ethylbenzene, *o*-xylene, and *p*-xylene (initial concentration of each compound approximately 5 mg/L), toluene was the first compound to be degraded. Toluene was degraded to less than 0.15 mg/L in about 40 days, followed by *p*-xylene in 72 days, and finally *o*-xylene after 104 days. Neither benzene nor ethylbenzene were degraded after 270 days of incubation. None of the compounds disappeared in autoclaved controls. Toluene, *p*-xylene and *o*-xylene were refed periodically as needed. Degradation of all three compounds began immediately after refeeding and proceeded rapidly initially. Microcosms acclimated to toluene, *p*-, and *o*-xylene also degraded *m*-xylene without a lag. After several feedings, the rates of degradation appeared to slow down. When the microcosms were replenished with fresh, sterile mineral medium, the rates of degradation increased again (see Table 1). Sulphate was not limiting in this series of microcosms (initial SO_4^{2-} concentration = 25 mM; final SO_4^{2-} concentration = 5.5 mM).

TABLE 1. Initial rates of degradation in Seal Beach sediment microcosms.

Feeding Day	Rates of Degradation (mg/L/day) Toluene	*p*-Xylene	*o*-Xylene
30	0.34 ± 0.01[(a)]		
95	0.38 ± 0.09	0.56 ± 0.07	
115			0.52 ± 0.2
135	0.23 ± 0.13	0.63 ± 0.04	0.54 ± 0.13
189	0.09 ± 0.05	0.24 ± 0.14	0.12 ± 0.02
237[(b)]	0.43 ± 0.06	1.58 ± 0.84	0.77 ± 0.39
258	0.46 ± 0.01	1.14 ± 0.05	N/A

(a) Mean and range of rates from three or four replicate microcosms.
(b) Microcosms replenished with fresh sterile defined mineral medium on day 237.
N/A: not available.

Microcosms with or without 1 mM 2-bromoethanesulphonate amendment performed identically. No methane was detected in the headspace of any microcosm (data not shown). The resazurin redox indicator remained colorless throughout the experiments.

Dependence on Sulphate. Figure 1 demonstrates the dependence of toluene degradation on the presence of sulphate in the microcosm. The same effects were observed for *p*- and *o*-xylene degradation (data not shown). Degradation proceeds as long as sulphate is present, ceases when sulphate becomes depleted, and resumes upon addition of sulphate.

Enrichment Cultures. Enrichments inoculated with both liquid and a small amount of solid material from active microcosms into defined mineral medium retain activity toward toluene, *p*-, *o*-, and *m*-xylene.

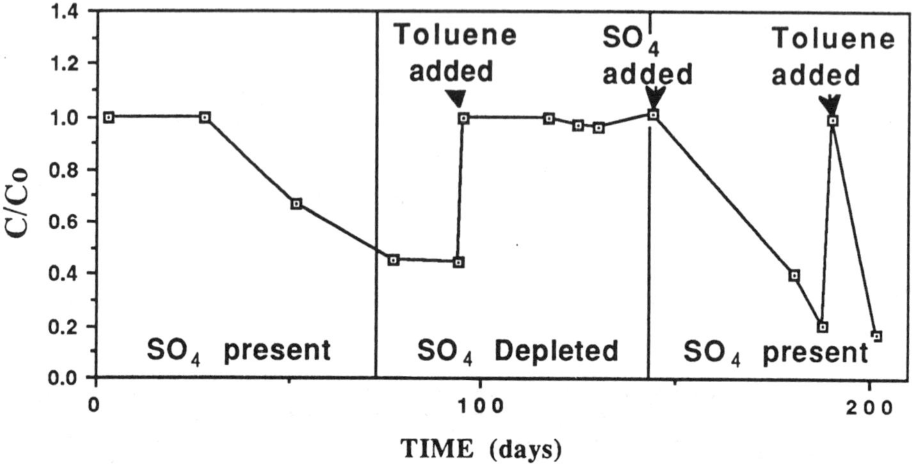

FIGURE 1. Effect of sulphate on toluene degradation. Normalized toluene concentration is plotted against time. The initial sulphate concentration was 5 mM. As degradation proceeds, the sulphate concentration drops to below the detection limit during the "depleted" phase. Degradation resumes upon addition of 20 mM sulphate.

Enrichments containing little or no sediment are needed for mass balance estimations to minimize complications due to sorption and unknown carbon sources and electron acceptors present in the sediment. Table 2 compares the calculated (theoretical) amount of sulphate required to degrade the known amount of aromatics added to the enrichment to the measured decrease in sulphate concentration. The theoretical amount of sulphate required was calculated for sulphate reduction to sulphide assuming 4.14 M of sulphate required per mole of toluene and 4.83 M of sulphate per mole of xylene. In addition, it was assumed that all 10 percent of the hydrogen gas in the headspace initially was used as an energy source requiring 0.23 M of sulphate per mole of H_2. These stoichiometric coefficients were calculated following the method of McCarty (1971), assuming that the fraction of substrate utilized for energy is 0.92. In these enrichment cultures, sulphate consumption consistently equalled or exceeded theoretical sulphate required, further demonstrating the role of sulphate-reducers in the metabolism of toluene and xylenes.

Certain enrichments were spiked with [^{14}C]-labelled substrates (either ring-labelled toluene, methyl-labelled toluene, or methyl-labelled *o*-xylene). All of the initial volatile activity was recovered as

TABLE 2. Comparison of theoretical and measured sulphate consumed in enrichment cultures.

	Total Substrates Consumed[(a)]	Theoretical SO_4^{2-} Required (mM)	Measured SO_4^{2-} Consumed (mM)
Enrichment A	7 µL toluene 3 µL *p*-xylene	2.25	3.3
Enrichment B	3 µL *p*-xylene	0.87	0.88
Enrichment C	5 µL toluene 5 µL *p*-xylene 2 µL *o*-xylene	2.71	5.00

(a) Enrichments consisted of 200 mL culture in 250-mL bottles which were periodically refed 1 or 2 µL of each substrate as needed.

[^{14}C]-labelled CO_2 in all cases, indicating complete mineralization of the substrates to CO_2.

Summary and Conclusions

We have provided preliminary evidence for the complete mineralization of toluene and xylenes under sulphate-reducing conditions. The Seal Beach aquifer material used in this study was exposed to gasoline and contained relatively high concentrations of sulphate because of its proximity to an intertidal marsh. The relatively short adaptation times observed in this study may be due to preexposure of the sediment *in situ* to these same compounds with sulphate present. The order in which the compounds were degraded (toluene, *p*-xylene, *o*-xylene) is the same as previously found in column studies using the same sediment (Haag *et al.* 1990). Benzene and ethylbenzene were recalcitrant under the experimental conditions used. The complete mineralization of toluene and *o*-xylene to carbon dioxide was confirmed using [^{14}C]-labelled substrates.

The degradation of toluene and xylenes proceeds simultaneously in microcosms fed substrate mixtures. The rates of degradation are more rapid in fresh medium, flushed with N_2/CO_2 and in the presence of sediments. The rates of degradation increased upon addition of fresh medium, implying that either some nutrient became limiting or that the concentration of an inhibitor was increasing. Iron is to be tested as a possible limiting nutrient because of the copious amount of iron sulphide precipitate formed concomitantly with the production of sulphide from sulphate reduction. The effects of other environmental conditions (such as pH, macro- and micro-nutrients) on the rate of degradation and on the rate of cell growth need to be determined.

As shown in Figure 1, the degradation of aromatic hydrocarbons in the microcosms depends on SO_4^{2-} availability. More sulphate was consumed than could be attributed to substrate degradation (Table 2) perhaps because sulphate was also being used as an electron acceptor for the degradation of unknown organics present on the aquifer solids. Absolute proof that sulphate is the terminal electron acceptor for toluene or xylene degradation is difficult to obtain because there is no mass transfer between the electron donor (substrate) and the electron acceptor (sulphate); only electrons are transferred. Accurate mass balances for substrate, CO_2, SO_4^{2-}, and sulphide are needed. To further confirm sulphate as the terminal electron acceptor, the behaviour of the cultures in the presence of molybdate, a specific inhibitor of sulphate reducers, and in the presence of alternate electron acceptors such as oxygen and nitrate needs to be investigated. ^{14}C-labelled substrates will be used to achieve accurate carbon mass balances.

Acknowledgments

This project was supported by a grant from the U.S. Department of the Navy, NCEL, Port Hueneme (IAG-RW-97934004-0) through the U.S. Environmental Protection Agency supported Western Region Hazardous Substance Research Center at Stanford University, the Orange County Water District, and a grant from the U.S. Air Force (AFOSR 88-0351) awarded to D. Grbić-Galić We appreciate the assistance of Dr. Harry F. Ridgway and D. Phipps in obtaining sediments from the Seal Beach site.

REFERENCES

Dean, B. J. *Mutation Research* **1985**, *145*, 153–181.

Grbić-Galić, D.; Vogel, T. M. *Environ. Microbiol.* **1987**, *53*(2), 254–260.

Haag, F.; Reinhard, M.; McCarty, P. L. *Environ, Toxicol. Chem.* Submitted, June 1990.

Kuhn, E. P.; Colberg, P. J.; Schnoor, J. L.; Wanner, O.; Zehnder, J. B.; Schwarzenbach, R. P. *Environ. Sci. Technol.* **1985**, *19*(10), 961–968.

Kuhn, E. P.; Zeyer, J.; Eicher, P.; Schwarzenbach, R. P. *Appl. Environ. Microbiol.* **1988**, *54*(2), 490–496.
Lovley, D. R.; Baedecker, M. J.; Lonergan, D. J.; Cozzarelli, I. M.; Phillips, E.J.P.; Siegel, D. I. *Nature* **1989**, *339*, 297–300.
Major, D. W.; Mayfield, C. I.; Barker J. F. *Groundwater* **1988**, *26*(1), 8–14.
McCarty, P. L. In *Organic Compounds in Aquatic Environments*; Faust, S. D.; Hunter, J. V., Eds.; Marcell Dekker, 1971; pp 495–531.
Reinhard, M.; Goodman, N. L; Barker, J. F. *Environ. Sci. Technol.* **1984**, *18*(12), 953–961.
Vogel, T. M.; Grbić-Galić, D. *Appl. Environ. Microbiol.* **1986**, *49*(5), 1080–1083.
Wilson, B. H.; Smith, G. B.; Rees, J. F. *Environ. Sci. Technol.* **1986a**, *20*, 997–1002.
Wilson, J. T.; Leach, L. E., Henson, M.; Jones, J. N. *GWMR* Fall **1986b**, 56–64.
Wilson, B. H.; Bledsoe, B.; Campbell, D. In *Chemical Quality of Water and the Hydrologic Cycle*; Averett, R. C.; McKnight; D. M., Eds.; Lewis Publishers: Chelsea, MI, 1987; pp 125–137.
Zeyer, J.; Kuhn, E. P.; Eicher, P.; Schwarzenbach, R. P. *Appl. Environ. Microbiol.* **1986**, *52*(4), 944–947.

Biodegradation of Oil- and Creosote-Related Aromatic Compounds under Nitrate-Reducing Conditions

*John Flyvbjerg**, *Erik Arvin, Bjørn K. Jensen, Susan K. Olsen*
Technical University of Denmark

Oil- and creosote-contaminated groundwater typically contains a complex mixture of phenolic compounds, aromatic hydrocarbons with one to three rings, and nitrogen, sulphur, and oxygen-containing heterocyclic compounds (Arvin *et al.* 1988; Ehrlich *et al.* 1982; Turney & Goerlitz 1990). It is well established that most of these chemicals are easily biodegraded in the presence of oxygen, but comparatively little is known about their biodegradability under anaerobic conditions. However, the past 10 years have seen an increasing interest in the potential of nitrate- reducing bacteria for pollutant destruction. This is

* Department of Environmental Engineering, Building 115, Technical University of Denmark, DK-2800 Lyngby, Denmark

because nitrate-reducing redox conditions often exist between the aerobic and strictly anaerobic zones in polluted aquifers, and because the addition of nitrate to contaminated sites would be a feasible *in situ* technique due to the low cost and high solubility of this electron acceptor. Degradation of phenol and the cresols during nitrate-reducing conditions was first demonstrated by Bakker (1977) and later confirmed in other studies (Bossert & Young 1986; Ehrlich *et al.* 1983; Hu & Shieh 1986; Tschech & Fuchs 1987). To our knowledge, there are no reports in the literature about the fate of the dimethyl-phenols (DMP) and other higher substituted methyl phenols during nitrate-reducing conditions. Concerning the aromatic hydrocarbons, the traditional point of view has been that these compounds were resistant to biodegradation under nitrate-reducing conditions. However, recent studies have shown that this is not necessarily the case for toluene and the xylenes (Dolfing *et al.* 1990; Jensen *et al.* 1988; Kuhn *et al.* 1985, 1988; Major *et al.* 1988; Zeyer *et al.* 1986) nor apparently for naphthalene and acenaphthene (Mihelcic & Luthy 1988).

The purpose of this study was to investigate the potential for biodegradation of phenols and aromatic hydrocarbons in creosote-contaminated groundwater during nitrate-reducing conditions. The biodegradation experiments were performed as simple batch microcosm tests.

MATERIALS AND METHODS

Description of Batch Experiments

The experiments were performed with groundwater from a creosote-contaminated aquifer in Fredensborg, Denmark. The site is polluted from a public gasworks operated between 1906 and 1958. The groundwater was sampled in autoclaved, stainless steel containers from an existing observation well using a small submergible centrifugal pump. Prior to sampling, the well water was disinfected with hydrogen peroxide to avoid contamination of the groundwater samples with surface and well bacteria.

The batch experiments were conducted in 5.5-L stirred glass containers equipped with supply and sampling valves (Figure 1). The groundwater was transferred from the steel containers to the glass batches aseptically. Oxygen absorbed by the water during sampling was removed by flushing with nitrogen gas for 30 minutes through the sterile filter at valve D followed by adding a stoichiometrically balanced

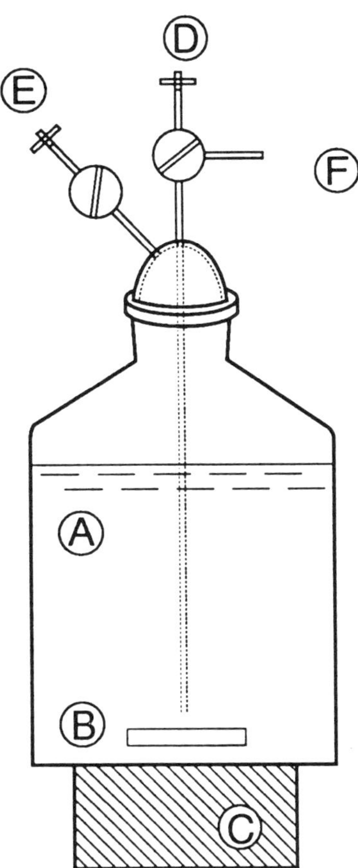

FIGURE 1. Batch reactor. (A) Glass container containing groundwater. (B) Rod for stirring. (C) Magnetic stirrer. (D) Valve for addition of substrates. (E) Valve for addition of gas. (F) Sampling valve.

amount of sulphite (Na_2SO_3) to reduce the residual O_2. Finally, nitrate (6 to 16 mg NO_3-N/L) and phosphate (1 mg P/L) were added to the batches. All chemicals were injected as aqueous solutions with a syringe through valve D. Duplicate batch reactors were incubated in the dark at 10 C (the approximate groundwater temperature) and at 20 C. A bottle with acidified groundwater (pH = 1.2) was used as a control for evaporation losses and abiotic processes. A batch with well water treated with hydrogen peroxide was incubated as a control for the efficiency of the disinfection procedure used prior to sampling the groundwater. The control batches were incubated at 20 C. Samples from the batch reactors were obtained by introducing a slight nitrogen pressure at valve E and then releasing the pressure through the sampling valve, F. During the experiments, organic substrates and nitrate were injected into the batch reactors through valve D.

Chemical Analysis

Phenols and Aromatic Hydrocarbons. Aromatic hydrocarbons and derivatized phenols (see below) were extracted from the groundwater with pentane. The compounds were identified on a Carlo Erba QMD 1000 GC/MS. Upon identification, analysis was performed with a Dani 8500 gas chromatograph equipped with a flame ionization detector.

The derivatization and extraction procedure was as follows. Borax (disodium-tetraborat-decahydrate, $Na_2B_4O_7 \bullet 10\ H_2O$) (200 mg) was added to a 10-mL volumetric flask. Ten milliliters of sample was added and the flask was shaken until the salt dissolved. The resulting pH was around 9.8. Then 500 µL of pentane with undecane and 2-bromophenol as internal standards was added, followed by 50 µL acetic acid anhydride. The solution was immediately shaken vigorously for 1 min. One microliter of the organic phase was injected into the GC. The detection limits were approximately 0.01 mg/L for the aromatic hydrocarbons and 0.02 mg/L for the phenols.

Nitrate and Nitrite. NO_3-N and NO_2-N were measured on a Technicion Autoanalyzer II by the hydrazine sulphate reduction procedure (Kamphake *et al.* 1967).

Oxygen. O_2 was measured by the azide modification of the Winkler method each time a batch was sampled. O_2 was never present above the detection limit (0.05 mg/L) in any of the experiments.

RESULTS

Groundwater Chemistry. Concentrations of the phenols and aromatic hydrocarbons found in the groundwater, and data on some inorganic species, are shown in Table 1. Neither oxygen nor nitrate and nitrite were detected, whereas the concentration of ammonia was high. These data indicate a strict anaerobic environment in the groundwater around the sampling well.

TABLE 1. Data for groundwater.

Organic Compounds		Inorganic Compounds	
Aromatic Hydrocarbons			
Benzene	0.90 mg/L	O_2	BD[(a)]
Toluene	0.15 –	NO_2-N	BD
o-Xylene	0.03 –	NO_3-N	BD
m- and *p*-Xylene	0.08 –	NH_4-N	100 mg/L
Napthalene	0.50 mg/L	pH	7.1
		Temp.	8.5 C
		Smell	Coal tar
		Colour	Yellow-brown
Phenols			
Phenol	2.00 mg/L		
o-Cresol	0.65 –		
m-Cresol	0.78 –		
p-Cresol	BD		
2,3-DMP	0.05 mg/L		
2,4-DMP	0.15 –		
2,5-DMP	0.15 –		
2,6-DMP	BD		
3,4-DMP	BD		
3,5-DMP	0.22 –		

(a) Below detection limit.

Biodegradation Experiments. The results of the biodegradation experiments are shown in Table 2. Toluene, phenol, *o*-cresol, *m*-cresol, and 2,4-dimethylphenol (2,4-DMP) were removed from the groundwater, whereas benzene, the xylenes, naphthalene, 2,3-DMP, 2,5-DMP, and 3,5-DMP were resistant to biodegradation. The compounds were removed in the same sequence at both 10 C and 20 C, with *o*-cresol being the compound requiring the longest lag phase prior to its degradation. The total time for complete removal of the compounds and the duration of the lag phases were about 2 to 4 times faster at 20 C than at 10 C.

Simultaneous with biodegradation, nitrate was consumed and nitrite produced. Forty to 80 percent of the nitrate was reduced to nitrite. Nitrate consumption ranged from 1.6 to 2.0 mg NO_3-N per

TABLE 2. Biodegradation of phenols and aromatic hydrocarbons in creosote-contaminated groundwater under nitrate-reducing conditions. For initial concentrations of compounds, see Table 1.

	10 C[a]		20 C[b]	
Compound	Degradation Time (days)	Lag Phase (days)	Degradation Time (days)	Lag Phase (days)
Benzene	R[c]		R	
Toluene	60	30	30	7
o-Xylene	R		R	
m- and *p*-Xylene	R		R	
Naphthalene	R		R	
Phenol	30	20	10	5
o-Cresol	95	70	39	30
m-Cresol	37	25	12	8
2,3-DMP	R		R	
2,4-DMP	20	10	12	5
2,5-DMP	R		R	
3,5-DMP	R		R	

(a) Total incubation time: 200 days.
(b) Total incubation time: 60 days.
(c) Resistant to biodegradation.

milligram of aromatic degraded, with the largest nitrate consumption in the batch with the highest nitrite production.

After 26 days of incubation, a mixture of toluene, 2,4-DMP, *p*-cresol, and 3,4-DMP were added to a final concentration of 1.0 to 1.5 mg/L of each compound in the 20 C batches. 3,4-DMP and *p*-cresol were not initially present in the groundwater. After a short lag phase, all the phenols in the mixture were degraded during 4 days (Figure 2). Toluene was degraded after 1 week (Figure 3). In the second addition of substrates, the concentrations of the compounds were slightly increased. Degradation now started without a lag phase (Figures 2 and 3). Simultaneous with the degradation of toluene and the phenols, nitrate and nitrite were exhausted (Figure 2). The third addition was a mixture of *p*-cresol, 2,4-DMP, and 3,4-DMP to a final concentration of

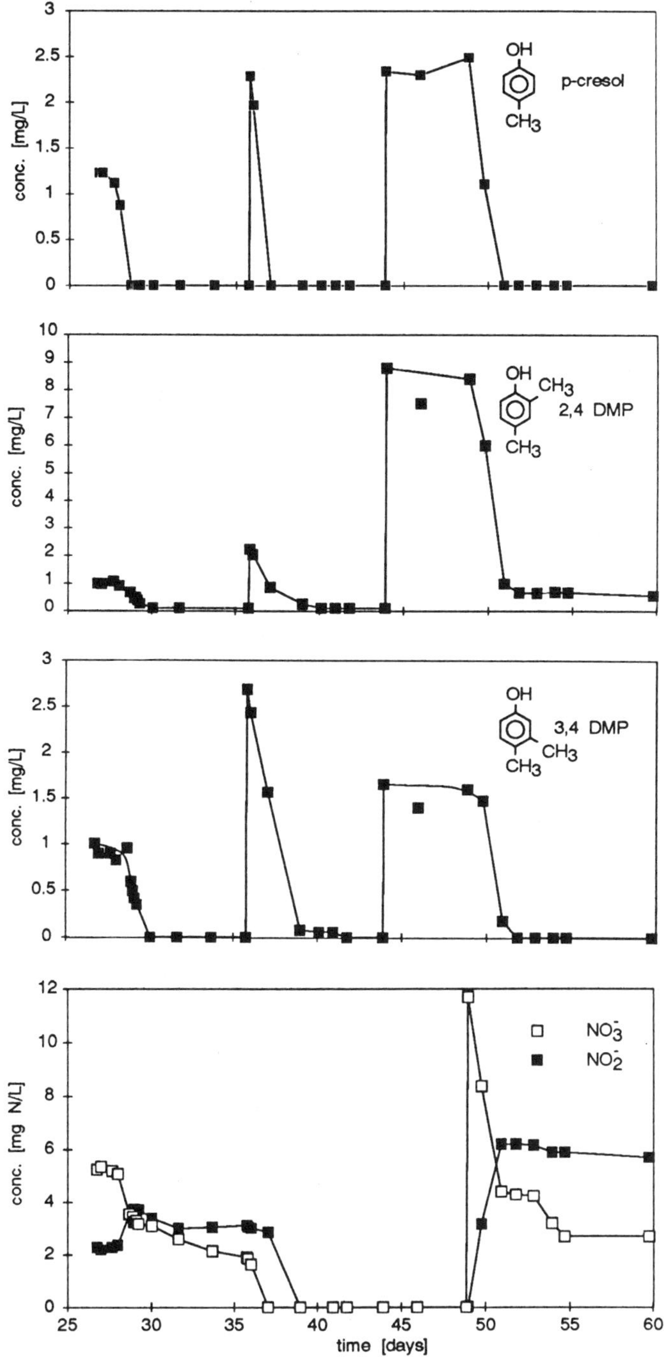

FIGURE 2. Biodegradation of phenols added to the groundwater culture at 20 C.

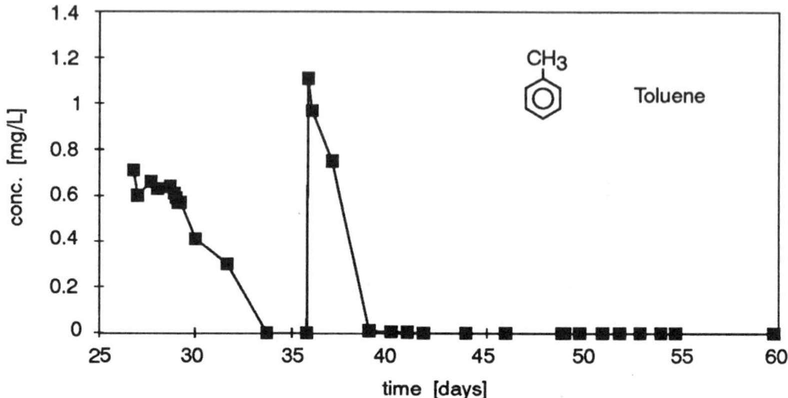

FIGURE 3. Biodegradation of toluene added to the groundwater culture at 20 C.

2.5 mg *p*-cresol/L, 8.5 mg 2,4-DMP/L, and 1.7 mg 3,4-DMP/L. During 6 days, the compounds were not degraded. Upon readdition of nitrate, biodegradation started immediately (Figure 2). The results show that nitrate was the terminal electron acceptor for degradation of these phenols.

Control Batches. No compounds disappeared from the acidified control batch. The batch containing disinfected well water was oversaturated with oxygen due to decomposition of hydrogen peroxide. There was no significant disappearance of phenol and toluene in this batch, but the cresols, benzene, and naphthalene were 30 to 50 percent reduced after 100 days incubation. However, no microbial activity (measured as ATP) was detected, so the observed decrease can probably be explained by chemical oxidation due to the high initial concentration of hydrogen peroxide (4,000 mg/L) in this batch.

CONCLUSION

Toluene, phenol, *o*-cresol, *m*-cresol, *p*-cresol, and the dimethylphenols 2,4-DMP and 3,4-DMP were degraded by bacteria in creosote-contaminated groundwater under nitrate-reducing conditions. Benzene, the xylenes, naphthalene, 2,3-MP, 2,5-DMP, and 3,5-DMP were not degraded after 200 days of incubation at 10 C and 60 days at 20 C. Simultaneous with degradation of the creosote compounds, nitrate was consumed and nitrite produced. If nitrate was exhausted, the groundwater bacteria instead utilized nitrite as electron acceptor. It was furthermore shown that the degradation of *p*-cresol, 2,4-DMP, and 3,4-DMP stopped when there was no nitrate or nitrite in the groundwater.

These results demonstrate the potential for *in situ* remediation of oil- and creosote-contaminated groundwater by nitrate injection. However, it is not possible to clean the groundwater for the whole spectrum of pollutants by nitrate reduction. Subsequent aerobic treatment seems to be necessary.

Acknowledgments

We thank J. Lyngkilde for assistance with the GC/MS analysis.

REFERENCES

Arvin, E.; Godsy, E. M.; Grbić-Galić, D.; Jensen, B. Presented at the International Conference on Physiochemical and Biological Detoxification of Hazardous Wastes, Atlantic City, NJ, May 3–5, 1988.

Bakker, G. *FEMS Letters* **1977**, *1*, 103–108.

Bossert, D. I.; Young, L. Y. *Appl. Environ. Microbiol.* **1986**, *52*, 1117–1122.

Dolfing, J.; Zeyer, J.; Binder-Eicher, P.; Schwarzenbach, R. P. *Arch. Microbiol.* **1990**, *154*, 336–341.

Ehrlich, G. G.; Goerlitz, D. F.; Godsy, E. M.; Hult, M. F. *Ground Water* **1982**, *20*, 703–710.

Ehrlich, G. G.; Godsy, E. M.; Goerlitz, D. F.; Hult, M. F. *Developments in Industrial Microbiology* **1983**, *24*, 235–245.

Hu, L. Z.; Shieh, W. K. *Biotechnol. Bioeng.* **1986**, *30*, 1077–1083.

Jensen, B. K.; Arvin, E.; Gundersen, A. T. Presented at Cost 691/681 Workshop on Organic Contaminants in Wastewater, Sludge & Sediments: Occurrence, Fate and Disposal. Brussels, Oct. 26–27, 1988.

Kamphake, L. J.; Hannah, S. A.; Cohen, J. M. *Water Resour. Res.* **1967**, *1*, 205–216.

Kuhn, E. P.; Colberg, P. J.; Schnoor, J. L.; Wanner, O.; Zehnder, J. B.; Schwarzenbach, R. P. *Environ. Sci. Technol.* **1985**, *19*, 961–968.

Kuhn, E. P.; Zeyer, J.; Eicher, P.; Schwarzenbach, R. P. *Appl. Environ. Microbiol.* **1988**, *54*, 490–496.

Major, D. W.; Mayfield, C. I.; Barker, J. F. *Ground Water* **1988**, *26*, 8–14.

Mihelcic, J. R.; Luthy, R. G. *Appl. Environ. Microbiol.* **1988**, *54*, 1182–1187 and 1188–1198.

Tschech, A.; Fuchs, G. *Arch. Microbiol.* **1987**, *148*, 213–217.

Turney, G. L.; Goerlitz, D. F. *GWMR*, **1990**, Summer, 187–198.

Zeyer, J.; Kuhn, E. P.; Schwarzenbach, R. P. *Appl. Environ. Microbiol.* **1986**, *52*, 944–947.

Biodegradation of Toluene by a Denitrifying Enrichment Culture

*Claus Jørgensen**, *Erik Mortensen, Bjørn K. Jensen, Erik Arvin*
Technical University of Denmark

INTRODUCTION

In-situ bioreclamation of gasoline-polluted water and soil is often limited by the amount of available electron acceptors. Its low water solubility makes oxygen unsuitable for bioreclamation of heavily polluted environments; therefore, the use of alternative electron acceptors is necessary. Until the middle of the 1980s, it was believed that molecular oxygen was required as a reactant for the initial oxidation of the aromatic ring in aromatic hydrocarbons, offering no alternatives to oxygen. Since then several works on anaerobic biodegradation of aromatic hydrocarbons have been published. Kuhn *et al.* (1985) found that the three isomers of xylene were degraded with NO_3^- as the electron acceptor in column experiments. Jensen *et al.* (1988) found that *o*-xylene was degraded only when toluene was degraded, and suggested a cometabolic relationship. Zeyer *et al.* (1986) showed that toluene and *m*-xylene were degraded with NO_3^- as the electron acceptor. Major *et al.* (1988) reported on the degradation of benzene under nitrate-reducing conditions, but the experiments could not be repeated (J. F. Barker, pers. com.). Hutchins *et al.* (1991) showed that toluene, ethylbenzene, the xylenes, and 1,2,3-trimethylbenzene could be degraded under nitrate-reducing conditions, while benzene was undegradable. Michelcic and Luthy (1988) reported on the degradation of naphthalene and acenaphthene. Of the aromatic hydrocarbons, toluene appears to be the easiest degraded under nitrate-reducing conditions.

Today, very little information on the physiology of denitrifying aromatic hydrocarbon degrading bacteria is available. The purpose of the work described in this paper was to determine the toxicity of

* Department of Environmental Engineering, Technical University of Denmark, Building 115, DK-2800 Lyngby, Denmark

toluene to a mixed culture of denitrifying bacteria enriched from denitrifying sewage sludge.

MATERIALS AND METHODS

Batch experiments were carried out in 1.25-L glass bottles equipped with glass stoppers and glass valves for sampling (Figure 1). During the experiments a slightly positive pressure was maintained in the bottles to avoid leakage of oxygen into the bottles and to make sampling possible. An 11th-generation enrichment culture originating from denitrifying sewage sludge was used as inoculum. The culture assimilated 44 percent ± 4 percent of the metabolized carbon (^{14}C-toluene study, unpublished data). When NO_3^- was present in excess, the maximum specific growth rate and the Monod saturation constant of the culture were determined to be 0.2 h^{-1} and 0.15 mg toluene/L, respectively, and the maximum specific utilization rate to be 0.7 mg toluene/(mg protein • h) (Jørgensen *et al.* 1991). The medium was modified from Platen and Schink (1989) by addition of a 10 mM phosphate buffer; 0.9 L of mineral medium, buffers, and $NaNO_3$ were added to the bottles before they were autoclaved. Vitamins and trace metals were added and the pH was adjusted to 7.2 ± 0.1. The headspace was purged 4 times for 4 minutes with oxygen-free N_2/CO_2 gas (O_2 < 0.5 ppm, CO_2 = 300 ppm); the bottles were shaken vigorously between purgings. Toluene (Merck, analytical grade) was then added, and the bottles were sealed and allowed to equilibrate. The remaining oxygen (< 0.5 mg O_2/L) was reduced by addition of Na_2SO_3 in stoichiometrically balanced amounts. Finally, 0.1 L of inoculum was added to a final volume of one liter, and the bottles were incubated in the dark at 20 C on magnetic stirrers. The initial bacterial concentration corresponded to 0.8 ± 0.7 mg protein/L.

Six bottles were prepared with increasing concentrations of toluene (6 to 240 mg/L) and NO_3^- (13 to 520 mg NO_3^--N/L) to assure the availability of the electron acceptor. In addition, two bottles were prepared with either the high concentration of toluene and the low concentration nitrate or vice versa. 3.7 mM NH_4^+ was added to all bottles as N-source. An uninoculated bottle was used as control.

Single samples of toluene and NO_3^-/NO_2^- were measured as follows: after extraction of a 10-mL sample in one milliliter of pentane with 4 mg heptane/L as the internal standard, toluene was measured isothermically at 50 C on a DANI 8520 gas chromatograph equipped with a 30 m J&W DB5 capillary column and FID detector. Peak areas

FIGURE 1. Glass bottle for batch experiment: A positive pressure is maintained in the bottle during the experiment to avoid oxygen influx and to make sampling possible.

were calculated on a Shimatzu C-R3A Chromatopac integrator. NO_2^- was measured on a Technicon autoanalyzer^{TMII} at 545 nm after reaction with sulphanilamide and N-(1-naphthyl)-ethylendiamine at pH 2 (Kamphake *et al.* 1967). NO_3^- was measured as NO_2^- after reduction with hydrazinsulphate. Protein was measured at 595 nm on a Perkin Elmer Lamda 2 spectrophotometer after reaction at low pH with Coomassie Brilliant Blue G-250 (Sigma) as described by Gälli and McCarty (1989). Bovin serum albumin (Sigma) was used as the standard sample. Oxygen was measured by the azide modification of the Winkler method. Oxygen concentrations were below detection limit (0.05 mg O_2/L) at all times during the experiment.

RESULTS AND DISCUSSION

Degradation of toluene at six concentrations and the corresponding accumulation of NO_2^- are shown in Figure 2. It is seen that toluene is rapidly degraded at concentrations up to 22 mg/L, followed by reduction of NO_3^- to NO_2^- (Figure 2a). At an initial concentration of 63 mg toluene/L, degradation proceeded rapidly until day 3, after which the rate of degradation slowed (Figure 2b). At this point, NO_3^- was low (data not shown). The slower rate may have resulted from the low NO_3^- concentration and a switch to use of NO_2^- as the terminal electron acceptor, or to an inhibitory effect of NO_2^-. At a toluene concentration of 117 mg/L, the rates of toluene degradation and NO_3^- reduction were lower than at 63 mg toluene/L. At 240 mg toluene/L, no degradation was observed and reduction of NO_3^- was absent. The data suggest that toluene partly inhibits the mixed culture at 117 mg/L and completely inhibits it at 240 mg/L. Toluene degradation was inhibited at 240 mg toluene/L and 13 mg NO_3-N/L, while no inhibition was observed at 6 mg toluene/L and 520 mg NO_3-N/L (data not shown), which proves that toluene and not nitrate was the inhibiting factor. The control showed a decrease in toluene concentration of approximately 10 percent, due to sampling.

At an initial concentration of 117 mg/L, the degradation of toluene ceased after 15 to 20 days. At this point, the concentration of NO_2^- was approximately 90 mg N/L. This suggests that NO_2^-, as the product of nitrate reduction, inhibits the toluene degradation. NO_2^- is known to inhibit denitrifying bacteria at 28 to 56 mg N/L (Tiedje 1988). When the initial toluene concentration was 63 mg toluene/L, NO_2^- did not inhibit degradation completely at 120 mg N/L. This inconsistency is probably due to a combined toxic effect of NO_2^- and toluene, making the culture more sensitive to NO_2^- at high toluene concentrations.

From a series of earlier experiments similar to the ones described above, NO_3^- consumptions have been calculated. NH_4^+ was added as N-source. In cases where NO_3^-/toluene ratios were above 2.1 mg N/mg toluene (≈14 mM NO_3/mM toluene), NO_3^- consumption was 1.4 ± 0.23 mg N/mg toluene (≈9.2 ± 1.5 mM NO_3/mM toluene). Toluene concentrations were below 22 mg/L and the number of experiments [n] = 12. The degradation was followed by accumulation of nitrite in stoichiometric amounts. The data presented in Figure 2 indicate that NO_3^- consumption at high toluene concentration is somewhat higher. In cases where NO_3^-/toluene ratios were below 1 mg N/mg toluene (≈7 mM NO_3/mM toluene), NO_3^- consumption was 0.69 ± 0.21 mg N/mg toluene (≈4.6 mM NO_3/mM toluene, [n] = 7), and no

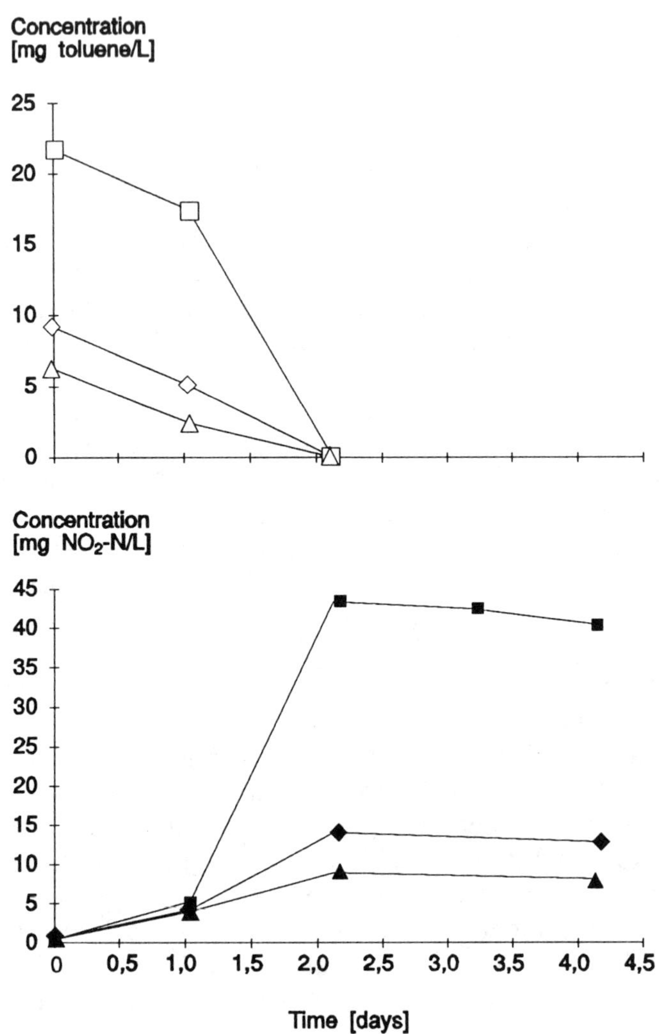

FIGURE 2a. Degradation of toluene at initial concentrations of 6 mg/L (Δ), 9 mg/L (◊), 22 mg/L (□), and the corresponding nitrite production (filled markers).

accumulation of NO_2^- was observed. When the NO_3^- concentration is low relative to the toluene concentration, the availability of the electron acceptor becomes limiting to degradation, and the bacteria switch to nitrite as the electron acceptor, which is then reduced to N_2.

As discussed in the introduction, toluene appears to be the more easily degradable of the monoaromatic hydrocarbons under nitrate-reducing conditions. This study confirms that toluene is easily

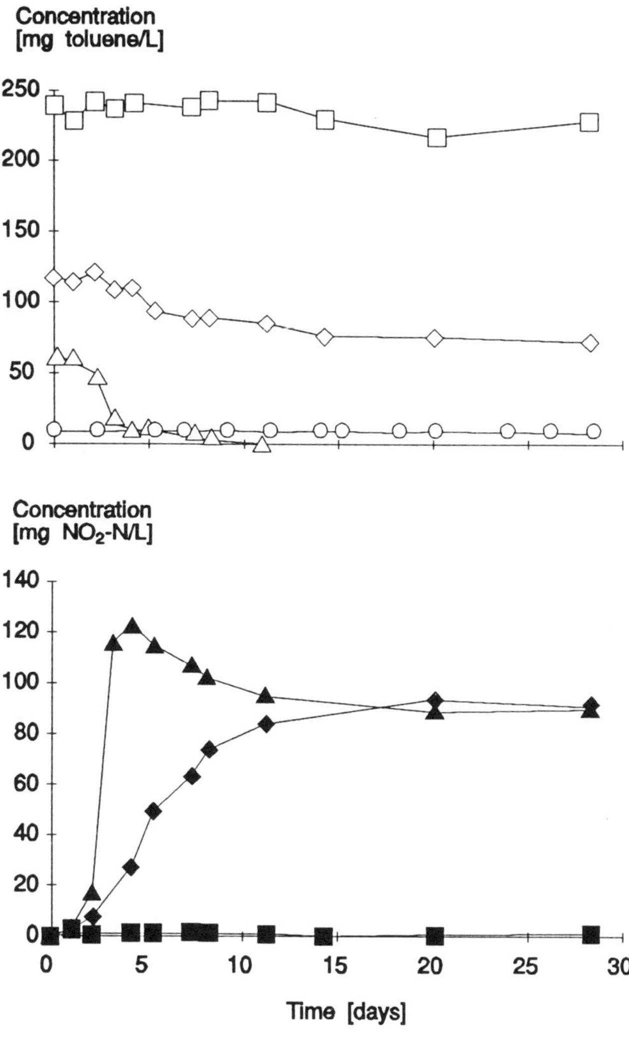

FIGURE 2b. Degradation of toluene at initial concentrations of 63 mg/L (Δ), 117 mg/L (◊), 240 mg/L (□), uninoculated control (○), and the corresponding nitrite production (filled markers).

degraded under nitrate-reducing conditions. It follows that NO_3^- can be used as an electron acceptor in bioreclamation of gasoline-polluted aquifers, enhancing at least the degradation of toluene. NO_3^- consumption is in the range of 0.5 to 1.6 mg N/mg toluene (≈3 to 11 mM NO_3/mM toluene), provided that NH_4^+ is available as N-source. However, several problems may be encountered. First, toluene and presumably other aromatic hydrocarbons may be present in concentrations that are toxic to nitrate-reducing bacteria and thereby inhibit the microbial degradation. These data suggest that the toxic level of toluene is approximately 120 mg toluene/L. However, extrapolation of quantitative data from controlled batch experiments to natural environments may lead to incorrect conclusions. Second, the reduction of NO_3^- may lead to accumulation of NO_2^- to toxic concentrations. This is likely to occur if the NO_3^-/toluene ratio in the aquifer is above the threshold at which nitrite reduction does not take place. This threshold is probably in the range of 1 to 2 mg NO_3^--N/mg toluene (≈7 to 13 mM NO_3/mM toluene). In cases of bioremediation where NO_3^- is used as the electron acceptor, concentrations of NO_2^- that are toxic to nitrate-reducing bacteria as well as aerobic to bacteria may be reached.

CONCLUSION

Bioreclamation of gasoline-polluted aquifers may be enhanced by addition of NO_3^-. In cases of aquifers heavily polluted with gasoline, the levels of aromatic hydrocarbons may be high enough to inhibit the nitrate-reducing bacteria. Injection of too high concentrations of NO_3^- may lead to accumulation of nitrite, toxic to nitrate-reducing bacteria.

Acknowledgments

This work was supported by the Center for Environmental Biotechnology, Denmark.

REFERENCES

Gälli, R.; McCarty, P. L. *Appl. Environ. Microbiol.* **1989**, *55*, 837–844.

Hutchins, S. R.; Sewell, G. W.; Kovacs, D. A.; Smith, G. A. *Environ. Sci. Technol.* **1991**, *25*, 68–76.

Jensen, B. K.; Arvin, E.; Gundersen, A. T. In proceedings from *COST 691/681 Workshop on Organic Contaminants in Wastewater, Sludge & Sediments: Occurrence, Fate and Disposal*, Brussels, Oct. 26–27, 1988.

Jørgensen, C.; Flyvbjerg, J.; Jensen, B. K.; Arvin, E.; Mortensen, E; Olsen, S. K. In proceedings from *COST 641 Workshop on Anaerobic Biodegradation of Xenobiotic Compounds*, Copenhagen, Nov. 22–23, 1991 (in press).

Kamphake, L. J.; Hannah, S. A.; Cohen, J. M. *Water Resour. Res.* **1967**, *1*, 205–216.

Kuhn, E. P.; Colberg, P. J.; Schnoor, J. L.; Wanner, O.; Zehnder, A. J. B.; Schwarzenbach, R. P. *Environ. Sci. Technol.* **1985**, *19*, 961–968.

Major, D. W.; Mayfield, C. I.; Barker, J. F. *Ground Water* **1988,** *26,* 8–14.
Mihelcic, J. R.; Luthy, R. G. *Appl. Environ. Microbiol.* **1988,** *54,* 1188–1198.
Platen, H.; Schink, B. J. *Gen. Microbiol.* **1989** *135,* 883–891.
Tiedje, J. M. In *Biology of Anaerobic Microorganisms;* Zehnder, A. J. B., Ed; Wiley Interscience: New York, 1988; Chapter 4.
Zeyer, J.; Kuhn, E. P.; Schwarzenbach, R. P. *Appl. Environ. Microbiol.* **1986,** *52,* 944–947.

A Field Experiment for the Anaerobic Biotransformation of Aromatic Hydrocarbon Compounds at Seal Beach, California

M. Reinhard[*], *L. E. Wills, H. A. Ball, T. Harmon*
Stanford University
D. W. Phipps, H. F. Ridgway
Orange County Water District
M. P. Eisman
Naval Civil Engineering Laboratory

Biotransformation of aromatic hydrocarbons under anaerobic conditions is of interest because dissolved oxygen is rapidly consumed in groundwater contaminant plumes of hydrocarbon fuel (for a general review of the topic, see Grbić-Galić 1990). Since mass transfer of oxygen is generally slow under groundwater conditions, anoxia may develop and remain over extended distances and long periods of time.

Anaerobic biotransformation of aromatic hydrocarbons has been demonstrated under different redox regimes including nitrate-reducing (Ball *et al.*, this volume; Hutchins *et al.* 1991; Zeyer *et al.* 1986), iron-reducing (Lovley *et al.* 1989), and fermentative-methanogenic conditions (Grbić-Galić & Vogel 1987). Recently, laboratory evidence has been

[*] Western Region Hazardous Substance Research Center, Stanford University, Stanford CA 94305-4020

obtained for the degradation of alkylbenzenes including toluene under sulfate-reducing conditions (Beller *et al.* 1991; Edwards *et al.*, this volume; Haag *et al.* 1990). Although the potential of anaerobic bioremediation processes is very significant, the fundamental environmental factors that govern transformation rates are poorly understood.

The long-term objective of this study is first to determine transformation rates under the conditions of the Seal Beach site, and second to explore the feasibility of inducing nitrate- and sulfate-reducing conditions and fermentative-methanogenic conditions in field bioreactors. Both laboratory studies and field studies in bioreactors are being conducted. This paper reports on the experimental design of the bioreactors and initial results.

Experiments are being conducted in reactors filled with aquifer material taken from the contaminated zone. The reactors were designed for operation in either a flow-through mode or a recirculation mode. During this phase of the study, defined initial conditions were established by pumping anaerobic water containing aromatic contaminants from the aquifer through the reactors and determining compound behavior without added nutrients and electron acceptors. In future experiments, the effects of nutrients will be investigated. To study slow reactions, the reactors may be operated in a recirculation mode and the concentrations of substrates and nutrients in the recirculating fluid determined.

Site Description. The Seal Beach study site is located approximately 1 km from the coast and adjacent to an intertidal marsh and wildlife refuge. A leaking storage tank discharged 20,000 to 30,000 L of unleaded gasoline into a shallow sandy aquifer consisting of silty/sandy alluvial deposits. Schroeder (in press) provided detailed descriptions of the hydrogeological conditions at the site and the extent of the plume. The gasoline front has been migrating radially approximately 100 m away from the source. Free product has been found 10 to 20 m from the tank.

The water quality of the site has been characterized by Ridgway *et al.* (1988). The site groundwater had a relatively high concentration of total dissolved solids (1,000 mg/L). Within the core of the anaerobic gasoline plume, nitrate was nearly depleted (<1 mg/L). Sulfate was present throughout the plume at concentrations of approximately 100 mg/L. Peripheral to the plume, nitrate and sulfate concentrations were approximately 2 and 150 mg/L, respectively.

To obtain evidence for the biochemical conditions of the site, methane and carbon dioxide were measured in the monitoring well headspace. The methane concentration ranged from 9,000 ppm in the well closest to the tank to undetectable in wells outside the plume. All monitoring wells within the contaminated area contained measurable quantities of methane, but concentrations tended to be higher in wells closer to the tanks. High carbon dioxide concentrations (>2%) coincided with high methane concentrations, but the reverse was not always the case. Elevated concentrations of carbon dioxide (>0.1%) were found in several wells with undetectable or low concentrations of methane. The observed high methane concentrations suggest that strongly methanogenic conditions had developed close to the tanks. This is in agreement with laboratory observations (Haag *et al.* 1990) in which methane production in column studies began to accelerate rapidly after more than a year.

Bioreactor Design and Operation

The reactor shown in Figures 1 and 2 was fabricated by Orange County Water District personnel. Each 200-L stainless steel tank is composed of a 122-cm-long tube (30 cm i.d.) equipped with a flange at the bottom and a dome-shaped cover at the top to allow collection of gases. The volume of each empty reactor is 90 L. The influent line feeds into a sand pad placed at the bottom of the reactor. Six side ports enable sampling from within the reactor along the flow path. The cover is equipped with a gas sampling valve and a pressure gauge. The bioreactors were packed with native aquifer material and charged with contaminated goundwater.

The reactors are fed by a peristaltic feed pump (range 12 to 400 mL/min) (ColePalmer Masterflex) with a neoprene tube insert and a recirculation piston pump (range 12 to 300 mL/h) (Fluid Metering Inc.) made of stainless steel, ceramic, and Teflon®. All lines are of stainless steel or Teflon.

The source water (well SB2) contained toluene and other monoaromatic hydrocarbon compounds at a concentration of approximately 1 mg/L. The aromatics were measured using a purge-and-trap technique, and the electron acceptors were measured using ion chromatography.

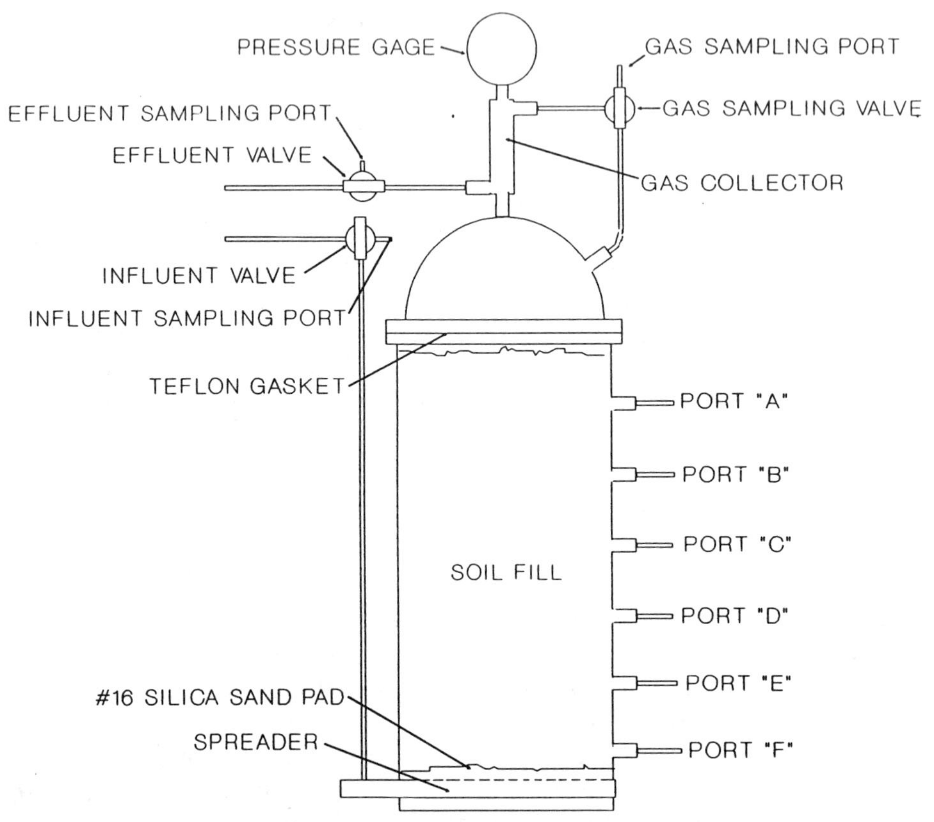

FIGURE 1. Schematic of anaerobic bioreactor.

Results and Discussion

Hydrodynamic Characterization of the Bioreactors. The hydrodynamic dispersion coefficients (D_{HD}) of the reactors were determined by injecting a spike of bromide as the conservative tracer and by measuring the breakthrough curves. The breakthrough curves were modeled using the advection/dispersion equation and an analytical solution after Hunt (1978). The reactors were assumed to be one-dimensional, semi-infinite columns packed with solids.

Table 1 summarizes the parameters used for the three reactors and the resulting dispersion coefficients found. The average pore water velocity was calculated from the effluent flow rates and the effective cross-sectional area. The porosity estimate for SBII was also used for reactors SBIII and SBIV.

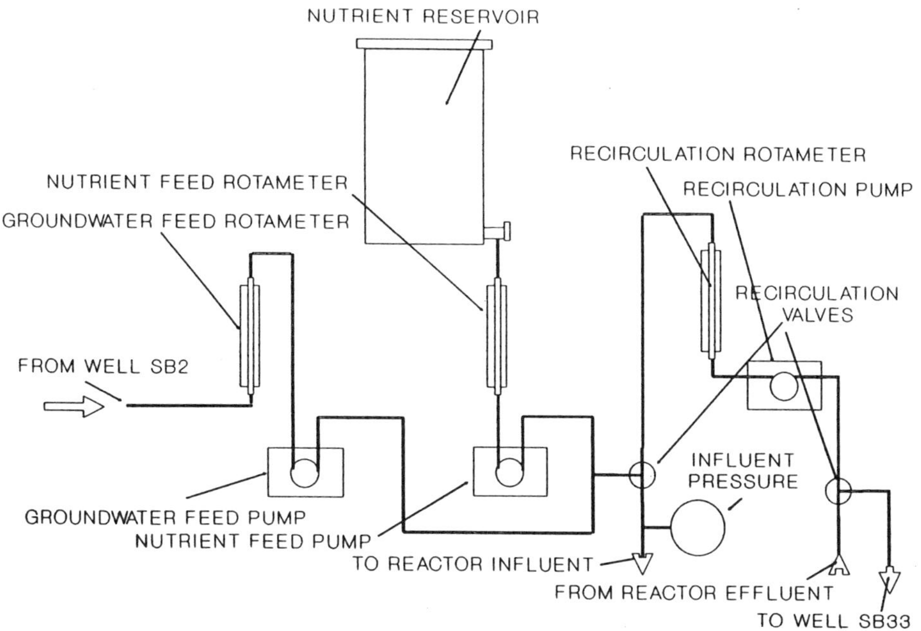

FIGURE 2. Schematic of pumping system.

TABLE 1. Hydrodynamic characteristics of reactors SBII, SBIII, and SBIV.

Reactor	Effluent Flow Rate (L/h)	Average Pore Water Velocity (m/h)	Hydrodynamic Dispersion Coefficient (m^2/h)
SBII	3.28	0.14	0.071
SBIII	2.96	0.13	0.017
SBIV	1.92	0.16	0.054

The values of the dispersion coefficents correspond to visual best fits of the data achieved with the model solution. The fit for the SBII tracer study is shown in Figure 3. The values of the dispersion coefficients differ significantly, by a factor of 4 in the case of reactors SBII and SBIII. The dispersion coefficients are of the same order of magnitude as those obtained by Chrysikopoulos *et al.* (1990) in a field tracer study of similar scale. The aquifer of that study was known to be very

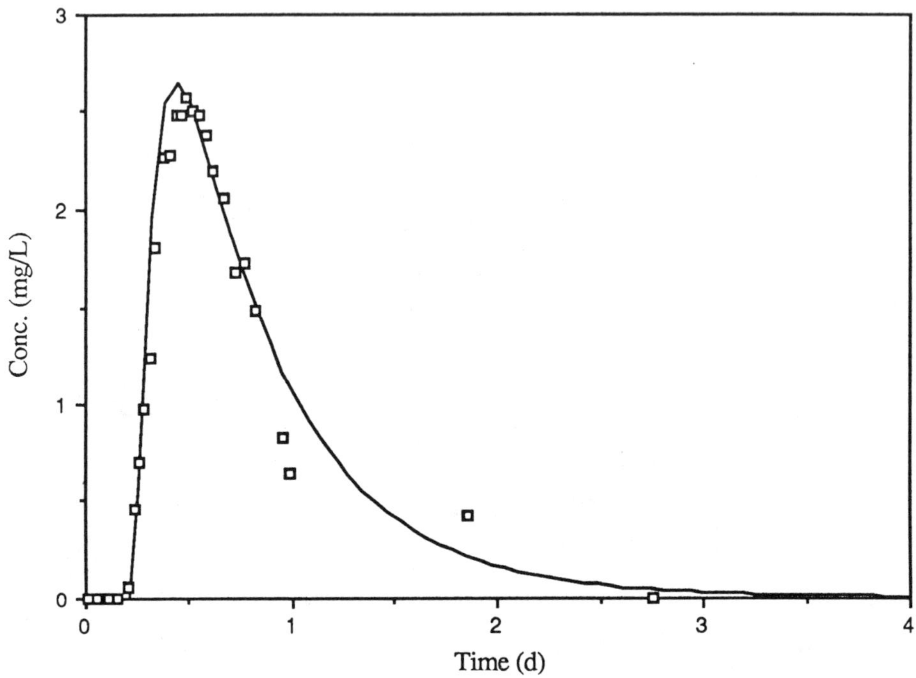

FIGURE 3. Breakthrough curve of bromide tracer and model fit of SBII.

heterogeneous. The high dispersion coefficients are probably due to clay lumps. Clay lumps formed when the auger broke up the clay lenses present in the area of sampling.

Values of the equilibrium distribution coefficients (Kd) for the various compounds and the Seal Beach solids were determined by Haag *et al.* (1990) based on breakthrough curves. The best estimate Kd values were used to determine retardation factors for the compounds. The retardation factor indicates the degree of slowing of the center of mass of a sorbing contaminant breakthrough curve relative to a nonsorbing tracer, such bromide. Figure 4 shows the breakthrough simulations for an infinite pulse input of benzene, toluene, and *o*-xylene with the bromide simulation from Figure 3. Of the three compounds shown, benzene sorbs the least, and is least retarded. As the shape of the curves in Figure 4 indicates, the degree of sorption significantly influences the spreading of the breakthrough response.

Nonequilibrium, or rate-limited, sorption may limit the rate of biodegradation in an aquifer (Semprini *et al.* 1990). The limitation is due to the slow desorption of contaminants from intraparticle pores, or

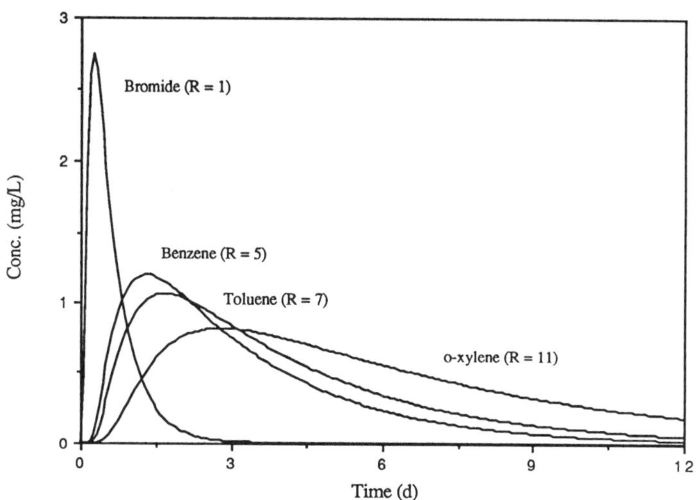

FIGURE 4. Simulated breakthrough curves for bromide, benzene, toluene, and *o*-xylene.

from layers of relatively low transmissivity. The bioreactors, composed of alluvial sands interspersed with clay lumps and silt, are probably influenced significantly by nonequilibrium sorption phenomena.

Compound Removal in the Flow-Through Mode. After filling the reactors with aquifer material in June 1990, the reactors were filled with contaminated water from well SB33. The reactors were sealed and left without movement of water for 4 months to let the reactors become anaerobic and to reestablish field biochemical conditions. On October 9, 1990, the reactors were charged with contaminated water from well SB33 by pumping at a rate of 50 mL/min. The intake was later changed to well SB2, where contaminant concentrations were higher. In the influent, approximately 0.5 mg/L O_2 were measured (using an oxygen probe) which could account for up to 0.2 mg/L hydrocarbon removal, assuming that no other O_2 consuming species were present.

During the following 6 weeks, influent and effluent samples were analyzed for benzene, toluene, ethylbenzene (EtBenz), *m,p*-xylene (*m,p*-Xyl), and *o*-xylene (*o*-Xyl), along with nitrate, phosphate, and sulfate (Table 2). The influent for the three reactors was sampled and analyzed separately.

After the 4-month resting period ending on October 9, 1990, nitrate in all three reactors was nearly depleted, but sulfate was significantly lowered relative to the influent only in SBIII. It is not clear, however, if the latter resulted from aromatic degradation. The aromatic

TABLE 2. Influent/effluent data from Seal Beach bioreactors in the flow-through mode.

		Concentration[a] (mg/L)					Concentration[b] (mg/L)		
Date	Sample	Benzene	Toluene	EtBenz	*m,p*-Xyl	*o*-Xyl	NO_3	PO_4	SO_4
10/9/90	Influent	1.600	0.090	0.017	0.120	0.130	0.470	0.340	32.0
	SBII Effluent	2.000	0.056	0.044	0.002	0.041	0.250	0.102	56.0
	SBIII Effluent	1.900	0.012	0.006	0.340	0.230	0.290	1.200	2.5
	SBIV Effluent	1.600	0.070	0.001	0.043	0.230	0.190	0.410	36.0
10/22/90	Influent	0.760	1.000	0.040	0.370	0.170	7.400	1.800	7.5
	SBII Effluent	1.800	0.011	0.035	0.002	nd	0.180	nd	30.0
	SBIII Effluent	1.500	0.015	0.004	0.032	0.031	0.060	nd	17.0
	SBIV Effluent	0.530	0.120	nd	0.003	0.046	2.900	nd	57.0
10/25/90	Influent	1.100	1.100	0.011	0.470	0.220	7.400	0.090	73.0
	SBII Effluent	1.200	0.320	0.009	0.210	0.150	0.400	0.120	73.0
	SBIII Effluent	1.400	0.130	nd	0.420	0.170	1.100	0.150	76.0
	SBIV Effluent	1.200	0.180	nd	0.430	0.180	2.200	0.130	74.0
10/29/90	Influent	0.590	0.400	0.023	0.230	0.100	7.300	0.870	76.0
	SBII Effluent	0.830	0.048	nd	0.290	0.150	0.280	nd	77.0
	SBIII Effluent	0.770	0.027	nd	0.340	0.120	0.680	0.140	79.0
	SBIV Effluent	0.790	0.035	nd	0.340	0.120	1.800	0.260	77.0
11/5/90	Influent	0.560	0.800	0.028	0.440	0.200			
	SBII Effluent	0.590	0.038	nd	0.120	0.120			
	SBIII Effluent	0.680	0.041	nd	0.120	0.120			
	SBIV Effluent	0.790	0.074	nd	0.130	0.140			
11/8/90	Influent	0.410	0.380	0.001	0.470	0.220			
	SBII Effluent	0.370	0.024	nd	0.083	0.100			
	SBIII Effluent	0.230	0.009	nd	0.077	0.092			
	SBIV Effluent	0.320	0.011	nd	0.064	0.085			
11/19/90	Influent	0.930	1.200	0.009	0.610	0.280	6.000	2.900	69.0
	SBII Effluent	0.098	0.015	nd	0.030	0.062	0.720	3.000	67.0
	SBIII Effluent	0.230	0.008	nd	0.018	0.022	0.500	1.100	7.0
	SBIV Effluent	0.230	nd	nd	nd	0.009	2.400	1.100	68.0

nd = Not detected.
(a) Organics concentrations typically are averages of between 2 and 6 data points.
(b) Inorganics concentrations are obtained from single analyses.

concentrations in SBIII were not significantly lower at that time than in SBII and SBIV, and the aromatic concentrations were not high enough to cause a significant SO_4^{2-} reduction. Benzene and *o*-xylene appear to have been refractory in all three reactors.

Breakthrough appears to have occurred by October 22, and compound behavior under steady-state conditions can be studied by comparing influent and effluent concentrations over the period from October 22 to November 19. During this time, influent concentrations of the organics varied approximately ±50 percent. It is evident that by October 25, benzene and the xylene isomers had broken through without significant attenuation. By contrast, toluene and ethylbenzene were significantly decreased (by more than 90 percent in most cases). Ethylbenzene was present in the influent only at low levels (0.001 to 0.040 mg/L), but was not detected in any of the effluent samples after October 29. Generally, removal of all compounds appeared to increase with time. The downward trend of benzene and the xylenes cannot be attributed to biotransformation with certainty, although the observed concentration decrease in SBIII and SBIV are greater than the elution curves predicted by the model (assuming that the influent concentration decreased to zero at time zero). More data and better estimates of the retardation factor are needed to test this hypothesis. The observed sequence of removal (toluene > *p*-xylene > *o*-xylene > benzene) was also observed in laboratory data obtained using anaerobic columns packed with Seal Beach sediment (Haag *et al.* 1990).

Of the two electron acceptors present, nitrate and sulfate, only nitrate showed a clear removal. This was most pronounced in SBII and SBIII where more than 90 percent was removed. In SBIV, the NO_3^- removal was less, ranging from 60 to 76 percent. Disappearance of sulfate was not detected. Removal of 6 mg of nitrate would support oxidation of 1.3 mg of toluene, assuming that nitrate is reduced to N_2 and 100 percent is used for energy. The observed removals appear consistent with the proposition that nitrate is the electron acceptor, although an exact mass balance is difficult to establish because of the variability of the data.

Acknowledgments

This project was supported by a grant from the Department of the Navy, NCEL, Port Hueneme (IAG-RW-97934004-0) and the Orange County Water District through the U.S. Environmental Protection Agency sponsored Western Region Hazardous Substance Research Center (R-815738).

REFERENCES

Ball, H. A.; Reinhard, M.; McCarty, P. L. [This volume.]

Beller, H. R.; Edwards, E. A.; Grbić-Galić, D.; Reinhard, M. "Degradation of Monoaromatic Hydrocarbon Compound by Aquifer Derived Microorganisms under Anaerobic Conditions"; project report, R. S. Kerr Environmental Research Laboratory, EPA, Ada, OK, 1991.

Chrysikopoulos, C. V.; Robert, P. V.; Kitanidis, P. K. *Water Resourc. Res.* **1990**, *26*(6), 1189–1195.
Edwards, E. A.; Wills, L. E.; Grbić-Galić, D.; Reinhard, M. [This volume.]
Grbić-Galić, D. In *Soil Biochemistry*; Bollag, J.-M; Stotzky, G., Eds.; Marcell Dekker, Inc.: New York, 1990; Chapter 3.
Grbić-Galić, D.; Vogel, T. M. *Appl. Environ. Microbiol.* **1987**, *53*(2), 254–260.
Haag, F.; Reinhard, M.; McCarty, P. L. *Environ. Toxicol. Chem.* Submitted 1990.
Hunt, B. *J. Hydraulic Division (ASCE)* **1978**, *104*(HYI), 75–85.
Hutchins, S. R.; Sewell, G. W., Kovacs, D. A.; Smith G. A. *Environ. Sci. Technol.* **1991**, *25*(1), 68–67.
Lovley, D. R.; Baedecker, M. J.; Lonergan, D. J.; Cozzarelli, I. M.; Phillips, E.J.P.; Siegel, D. I. *Nature* **1989**, *339*, 297–300.
Ridgway, H. F.; Phipps, D. W.; Safarik, J.; Haag, F.; Reinhard, M.; Ball, H. A.; McCarty, P. L. "Investigation of the Transport and Fate of Gasoline Hydrocarbon Pollutants in Groundwater"; final report to USGS; Reston, VA, 1988.
Schroeder, R. "Delineation of a Hydrocarbon (Weathered Gasoline) Plume in Shallow Deposits at the U.S. Naval Weapons Station, Seal Beach CA"; U.S. Geological Survey Water Resources Investigations Report 89-4203, in press.
Semprini, L.; McCarty, P. L. *Ground Water.* Submitted 1990.
Zeyer, J.; Kuhn, E. P.; Eicher, P.; Schwarzenbach, R. P. *Appl. Environ. Microbiol.* **1986**, *52*(4), 944–947.

German Experiences in the Biodegradation of Creosote and Gaswork-Specific Substances

*Peter Werner**
Universität Karlsruhe

In the late 1970s in Germany, public interest in contaminated sites increased rapidly. First-aid measures included replacing the polluted soil with nonpolluted material. The more people investigated soil and groundwater, the more contaminated areas were found. Gradually, people became more and more aware of the problems. An increasing

* DVGW-Forschungsstelle am Engler-Bunte-Institute der Universität Karlsruhe, Richard-Willstätter-Alle 5, Federal Republic of Germany

feeling of responsibility helped spawn a large industry of remediation measures during the 1980s.

Besides landfill, encapsulation, extraction, and incineration, there is a strong interest in biological methods for the remediation of contaminated sites. The object of methods using microbial influences is a complete mineralization of the pollutants (Franzius *et al.* 1988).

The technical aspects of biological remediation using on-site and *in situ* methods are described in a great number of publications (*e.g.*, Annokkée 1990; Klein & Beyer 1990; Lotter *et al.* 1990; Riss & Ripper 1990; van den Berg *et al.* 1990; Weidner *et al.* 1990).

Abandoned gaswork sites have been scrutinized more recently. Since the introduction of natural gas about 20 years ago, the gasworks that had been used for about 100 years disappeared from nearly all cities. Before many of the gasworks were closed, gas production was converted from coal to mineral oil products. This paper focuses on soil and groundwater contamination from the time of coal degasifying because the 10- to 15-year interval during which mineral oil was used is much shorter and the decay of these products released considerably fewer pollutants.

In addition to gaswork sites, many coal gasification plants at the rivers Ruhr and Saar were closed during the constitutional change. In most cases, because of contamination, the sites cannot be sold and used for another purpose, since only completely remediated sites are easy to sell.

After stopping production, equipment and buildings were usually dismantled and the site was leveled. So these areas give a harmless impression at first sight. The condition of the subsoil, however, often depends on how carefully the closure of the factory and the waste disposal were carried out. In many cases, soil contamination was caused during war time.

In Germany more and more refineries have been closed for lack of profitability. Here, this soil usually is also highly contaminated with various hydrocarbons, mainly creosotes.

Since reunification, a virtually incalculable number of contaminated sites has been added.

Analysis of the Contaminants and Sampling Technique. Soil and seepage water samples are often highly contaminated. Table 1 shows detailed data that approximately correspond with maximum concentrations measured in abandoned gasworks. Pollution by the aromatic hydrocarbons benzene, toluene, and xylene (BTX), as well as polynuclear aromatic compounds (PAH), is obvious.

TABLE 1. Concentrations of pollutants in soil and seepage water of abandoned coke-oven and gaswork plants (maximum concentrations found so far).

Soil:	Benzene	*ca.* 5000 mg/kg
	Toluene	*ca.* 5000 mg/kg
	Xylene	*ca.* 5000 mg/kg
	Naphtalene	*ca.* 5000 mg/kg
	Phenanthrene	*ca.* 5000 mg/kg
	PAH (sum)	*ca.* 1000 mg/kg
	Cyanides (complexed)	*ca.* 1000 mg/kg
	Phenols	*ca.* 1000 mg/kg
Water:	Aromatics According to Maximum Solubility	
	Sulphate	*ca.* 3000 mg/L
	Nitrate	0 mg/L
	Oxygen	0 mg/L
	Ammonia	*ca.* 20 mg/L
	Iron	*ca.* 20 mg/L
	Manganene	*ca.* 10 mg/L

Furthermore, the samples showed high amounts of cyanides and ammonia, as well as partly high contaminations of heavy metals and sulphates in the water. The contaminants are distributed very inhomogeneously in the soil, both vertically and horizontally. For that reason the sampling technique is very important. The sample should be taken in a way that it is representative for a certain area. Thus, the analytical result can be assigned unmistakably to a certain amount of soil and give information about a possible remediation. Data taken from measurements of certain samples, such as blue stained clods stemming from complex cyanide compounds, are unrepresentative and irrelevant for evaluation. Such inappropriate sampling techniques allow the measuring of practically all "desired" values.

Procedure in the Application of Bioremediation. The procedure to check the applicability of biological methods for the remediation of contaminated sites using *in situ* or on-site measures is shown in Figure 1.

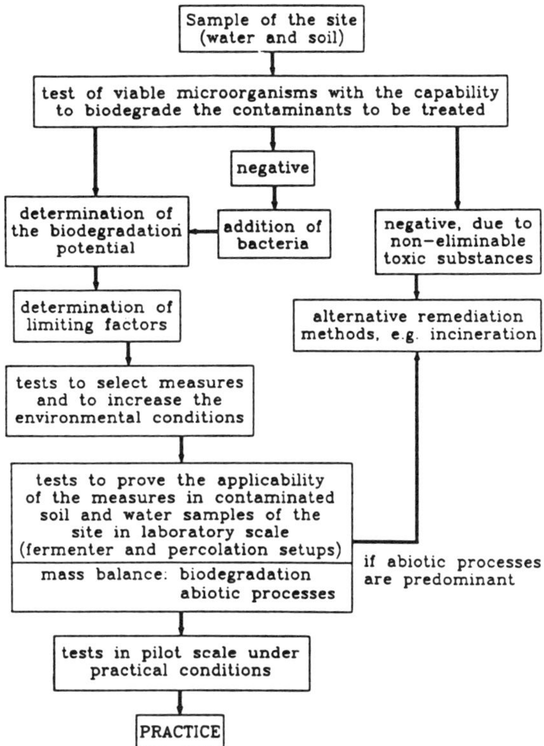

FIGURE 1. Procedure in the application of bioremediation.

The intricate scheme indicates that deciding whether these methods can be applied is time consuming. Although some tests can be carried out in parallel, the procedure takes a long time due to the complexity of the contaminations (compare Table 1). Experience shows that in the case of hydrocarbons at least 4 to 6 months are necessary.

Biodegradability of Aromatic and Polyaromatic Hydrocarbons. In principle, the pollutants occurring in the contaminated sites mentioned above are biodegradable. As a rule, a microflora, which is adapted to the hydrocarbons, can be found in the polluted soil. The biochemical pathways are not stressed here since they are discussed in the literature specialized in this subject (Thole & Werner 1986).

The experimental setup described below demonstrates the mineralization of aromatic hydrocarbons found in an abandoned refinery site. The contaminated soil was shaken in water, and the supernatant was treated microbially under aerobic conditions. A carbon-free nutrient salt solution was added to optimize conditions for the indigenous bacteria. Figure 2 shows the biodegradation of the aromatics during 2 weeks. No reduction could be found in a simultaneous sterile experimental setup to which 100 mg/L mercury chloride were added. This proves the mineralization of the contaminants.

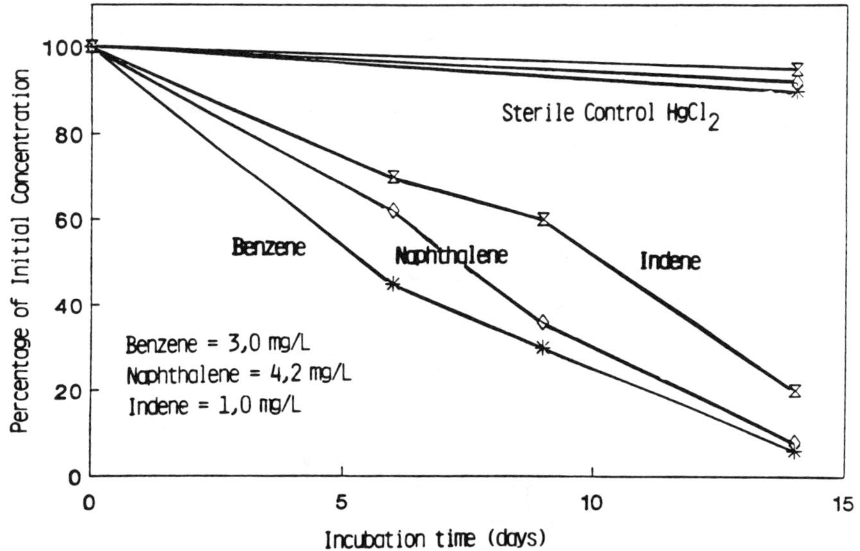

FIGURE 2. Degradation of aromatic hydrocarbons.

The described experiment was carried out under optimum environmental conditions, and the results must not be transferred to on-site remediation.

To demonstrate the behaviour of different soils contaminated by similar hydrocarbons, investigations in percolators were necessary.

The following example demonstrates the difficulties of biodegrading PAH-contaminated soils of a different geological structure.

- Model conditions: sand was artificially contaminated with a PAH-combination of 10 compounds (Thole & Werner 1986).

- Real conditions: pebbly sand of a PAH-contaminated gaswork plant.

Experimental Conditions. Of each type of contaminated soil, 2 kg were put in laboratory percolators (Böckle *et al.* 1991) and subjected to circular flushing under different operating conditions:

- Soil flooded completely; oxygen supply of the microorganisms through water.

- In the beginning, flooded soil was flushed from top to bottom (flow rate 1.3 m/h) or in case where oxygen was limited by too little soil permeability, from bottom to the top (flow rate 5 m/h).

The samples with artificially contaminated sand were inoculated with mixed cultures of PAH-adapted microorganisms. The soil of the gaswork plant contained sufficient microorganisms that were typical for the site and adapted to the pollutants (approx. 10^3/mL).

Nutrient compounds necessary for anabolic and energy metabolism, such as ammonium, nitrate, and phosphate, were added to the flushing water when their concentration reached a certain limit (ammonium $\geq$ 1 mg/L, phosphate $\geq$ 0.1 mg/L, nitrate $\geq$ 10 mg/L). At the beginning of the experiment, a nutrient salt solution was added to optimize the living conditions of the bacteria.

Microbiological Investigations. The examined microbiological parameters are the following:

- Determination of total cell counts (microscopically), colony counts (nutrient agar), and number of hydrocarbon-degrading bacteria (carbon source naphthalene)

- Determination of toxicity with bioluminescence test

Physico-Chemical Investigations. The physico-chemical parameters are listed below:

- Temperature, pH-value, water flow, oxygen content

- Dissolved organic carbon (DOC)

- Spectral absorption coefficient (SAC) at 254 nm
- PAH in water and soil (determination with GC/FID after extraction)
- Carbon dioxide of the outgoing air by precipitation as barium carbonate

Results of Experimental Setup. The results of two representative experiments with artificially contaminated sand and polluted soil from a gaswork plant are presented and discussed below.

In the first 3 days in which the percolator was operated with artificially contaminated soil, the total concentration of PAHs in the flushing water decreased from approximately 26 to 10 mg/L (Figure 3). This was mainly due to the fast reduction of the naphthalene concentration. The drop in concentrations of the higher molecular compounds was apparently slower. From the 35th day on, only traces of pollutants could be detected in the flushing water. The addition of an inorganic solubilizer to the flushing water (sodium pyrophosphate, 0.1% final concentration) resulted in a slight, short-time increase of the PAH concentration in the water. A similar course of concentration was

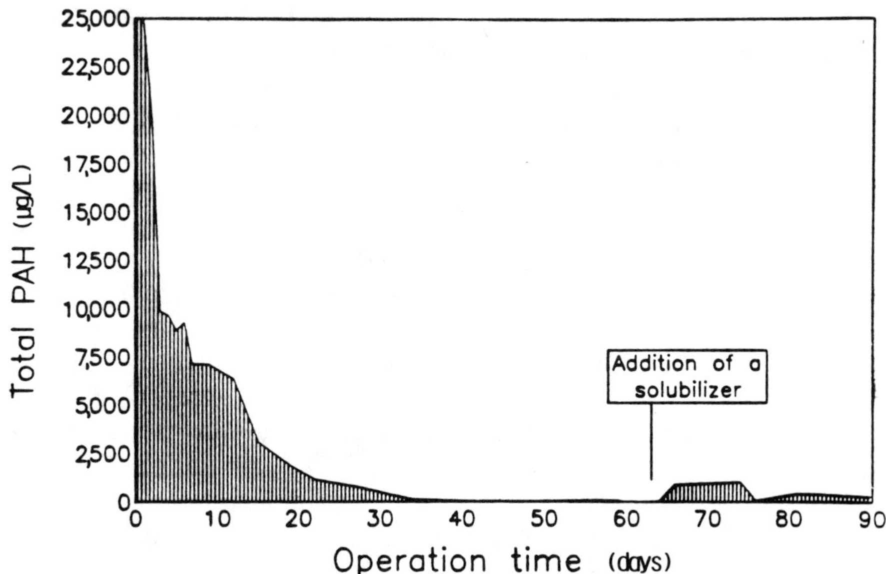

FIGURE 3. Course of PAH concentrations in the flushing water (artificially contaminated sand).

noticed in the experiment with gaswork soil. In spite of the higher contaminations in the soil (Figures 5 and 6), the concentration in the flushing water was clearly lower (Figure 4). After 4 days, naphthalene could no longer be detected. From the 11th day on, the water contained no more PAH. The addition of solubilizer had no effect on this soil.

The enormous differences in the PAH concentrations of both systems at the beginning of the experiment were mainly attributable to the different structure and composition of the two contaminated soils. Coarse clay and clay fractions of soils are known to have a high capacity of adsorption to PAH (Werner 1989), i.e., the amount of PAH that can be flushed by water is influenced not only by the solubility of the compounds, but also by the pedological structure of the soil. Since the gaswork soil showed a relatively high portion of coarse clay, the amount of PAHs flushed into water was clearly smaller in spite of a higher PAH contamination. One of the limiting factors of degradation mentioned in the introduction is the bioavailability of pollutants, which from these results, was expected to have a stronger effect on the PAH degradation in the gaswork soil than in the sand. The results of soil analyses, shown in Figures 5 and 6, confirm this expectation.

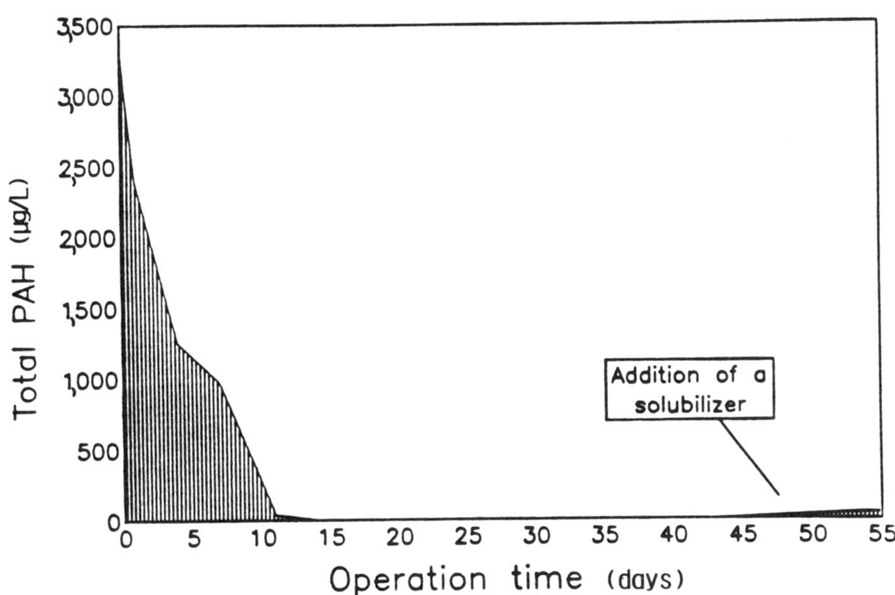

FIGURE 4. Course of PAH concentrations in the flushing water (gaswork soil).

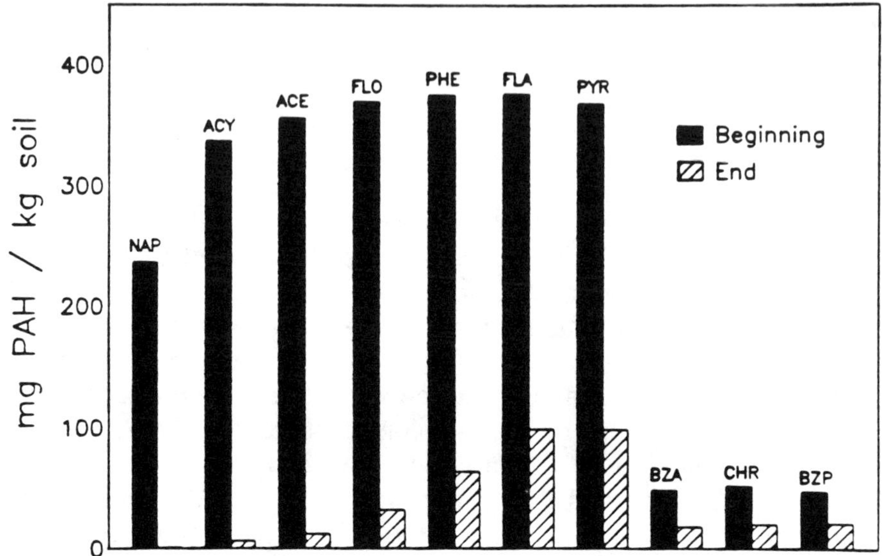

FIGURE 5. Initial and final PAH concentrations (artificially contaminated sand, testing time 90 days; NAP = naphthalene, ACY = acenaphthylene, ACE = acenaphthene, FLO = fluorene, PHE = phenanthrene, FLA = fluoranthene, PYR = pyrene, BZA = benz(a)anthracene, CHR = chrysene, and BZP = benzo(a)pyrene).

PAH analyses of the sand showed that treatment in laboratory percolators led to an apparent reduction of pollutants in the soil, totalling 85 percent in 90 days. The results of gaswork soil treatment were a great deal less successful, with 33 percent of pollutants removed in 55 days. A continuation of the test until the 90th day was not expected to bring about a further reduction of pollutants because of the strong adsorption of PAHs to the gaswork soil (Figure 4) and the greater number of higher molecular compounds. In both systems, the degradation rate decreased with the increasing molecular weight of the PAHs.

Under the applied experimental conditions, compounds like benzo(a)pyrene, benzo(g,h,i)pyrene, and indeno(1,2,3-cd)pyrene persisted completely in the gaswork soil.

Development of Hydrocarbon-Degrading Bacteria. In both systems, with artificially contaminated sand as well as with gaswork soil, the number of hydrocarbon-degrading bacteria increased within the first few days (Figures 7 and 8), but was reduced again in the further course of the experiment. The reason for the considerably higher numbers (up to a

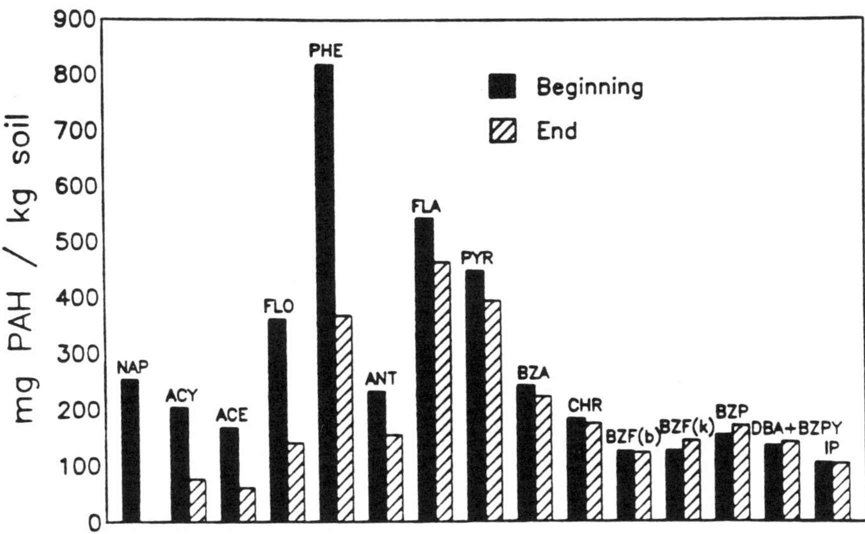

FIGURE 6. Initial and final PAH concentrations (gaswork soil, testing time 55 days; NAP = naphthalene, ACY = acenaphthylene, ACE = acenaphthene, FLO = fluorene, PHE = phenanthrene, ANT = anthracene, FLA = fluoranthene, PYR = pyrene, BZA = benz(a)anthracene, CHR = chrysene, BZF(b) = benzo(b)fluoranthene, BZF(k) = benzo(k)-fluoranthene, BZP = benzo(a)pyrene, DBA = dibenz(a,h)-anthracene, BZPY = benzo(ghi)perylene, and IP = indeno(1,2,3-cd)pyrene).

factor of 1,000) of hydrocarbon-degrading bacteria in the sand probably was the higher PAH concentration in the water and thus the better availability of PAHs. In both experiments, the curves of hydrocarbon-degrading bacteria correlated with the course of the PAH concentrations in the flushing water (compare with Figures 3 and 4).

Dissolved Organic Carbon (DOC), Spectral Absorption Coefficient at 254 nm (SAC), and Toxicity. In both experiments, the DOC values increased considerably during the first few days and sank again in the further course of the experiment (Figures 9 and 10). Due to their similarity, the SAC values are not presented graphically. The UV absorption at the beginning of the experiment was approximately 10 m^{-1} for artificially contaminated sand and 50 m^{-1} for gaswork soil. The rapid DOC decrease in the flushing water of the gaswork soil was

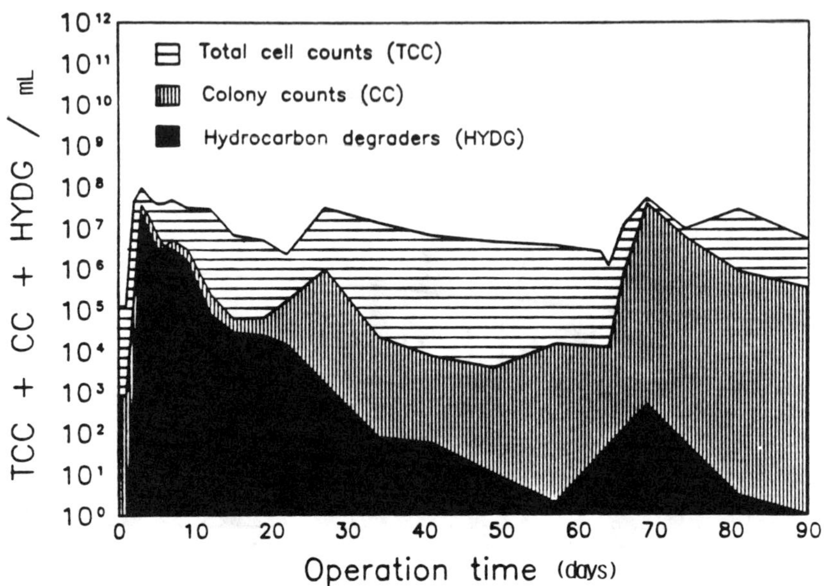

FIGURE 7. Microbiological data in the course of the experiment (artificially contaminated sand).

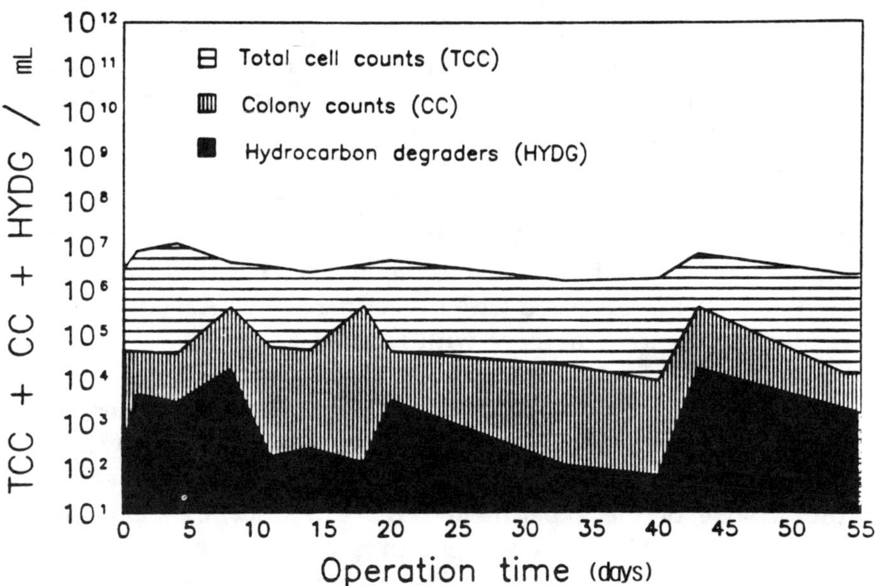

FIGURE 8. Microbiological data in the course of the experiment (gaswork soil).

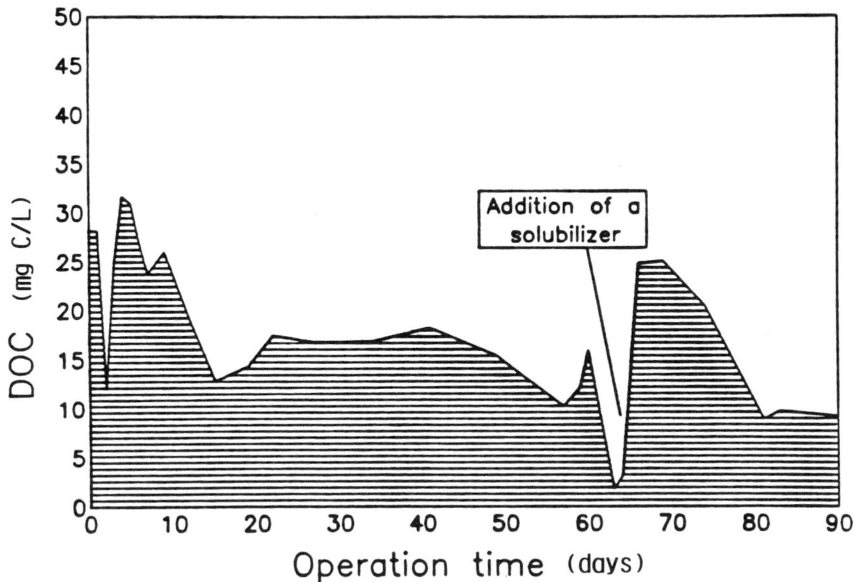

FIGURE 9. DOC concentrations in the course of the experiment (artificially contaminated sand).

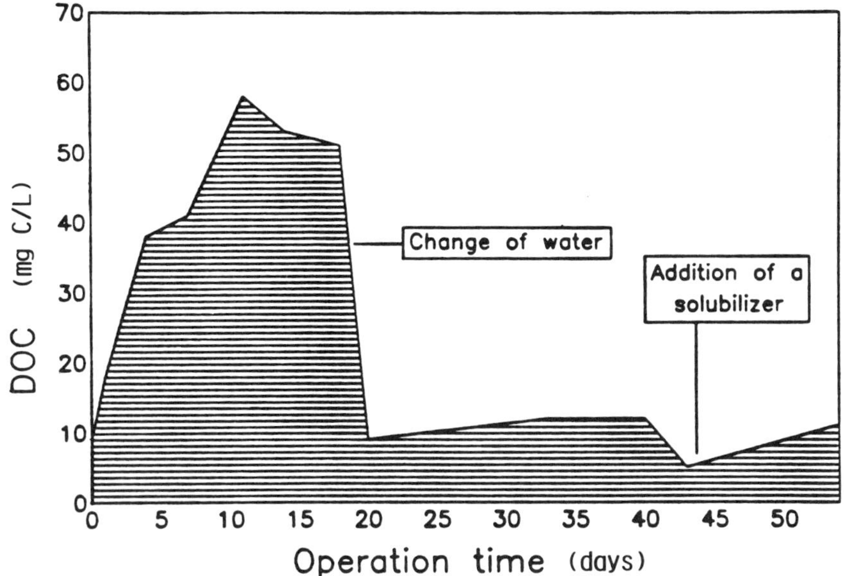

FIGURE 10. DOC concentrations in the course of the experiment (gaswork soil).

caused by a renewal of the water for technical reasons. This, however, had no consequence for the whole system since the PAH concentration in the flushing water already was below detection limit by this time.

The initial increase of dissolved organic compounds in water can only be explained by an enrichment of intermediate products of the microbial PAH degradation, i.e., mainly aromatic compounds, as the SAC analyses prove. The higher values of DOC and SAC caused by the addition of solubilizers indicate an enrichment of intermediate products in the soil. Toward the end of both experiments, the DOC and SAC values decreased again (DOC <10 mg/L, SAC <50 m^{-1}, which means that the intermediate products have been further degraded. These results imply the fact of metabolite formation that must be considered in any remediation measures. At the moment, however, very little is known about the importance and composition of intermediate products.

By measuring the CO_2 in the outgoing air of the system, it was possible to calculate the actual mineralization rate in artificially contaminated soil: 75 percent of the PAHs that could not be detected any more had been mineralized.

Toxicity measurements (inhibition of light emission of luminescent bacteria) brought similar curves for both experiments (Figures 11 and 12).

In the beginning, the high PAH concentrations in the flushing water led to an inhibition of light emission of 80 to 90 percent. According to the decrease in PAH concentration, the percentage of inhibition was reduced in the sand experiment (Figure 11). The concentration profile of fluoranthene or pyrene corresponded very well with toxicity values. In the percolator system with gaswork soil, the inhibition was still 70 percent when no PAHs were detectable in the water (Figure 12). The good agreement of the toxicity curves with the DOC values shows that unidentifiable compounds dissolved in water had also been responsible for the toxic effect of the samples.

Nitrate as an Oxygen Donor

As a rule, the amount of contaminants in the soil is so enormous that the dissolved oxygen in the flushing water is not sufficient for a complete mineralization. Therefore, oxygen is the limiting factor in most cases. To decrease the time of remediation, the system must be provided with an additional oxygen source.

Some experiences were made with the application of nitrates (Battermann & Werner 1984; Riss *et al.* 1988) and hydrogen peroxide

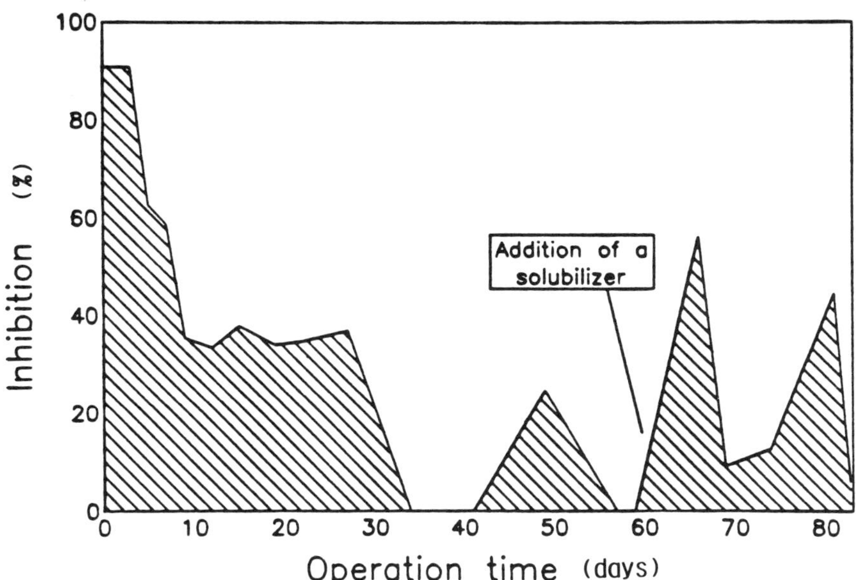

FIGURE 11. Toxicity (bioluminescence test) in the course of the experiment (artificially contaminated sand).

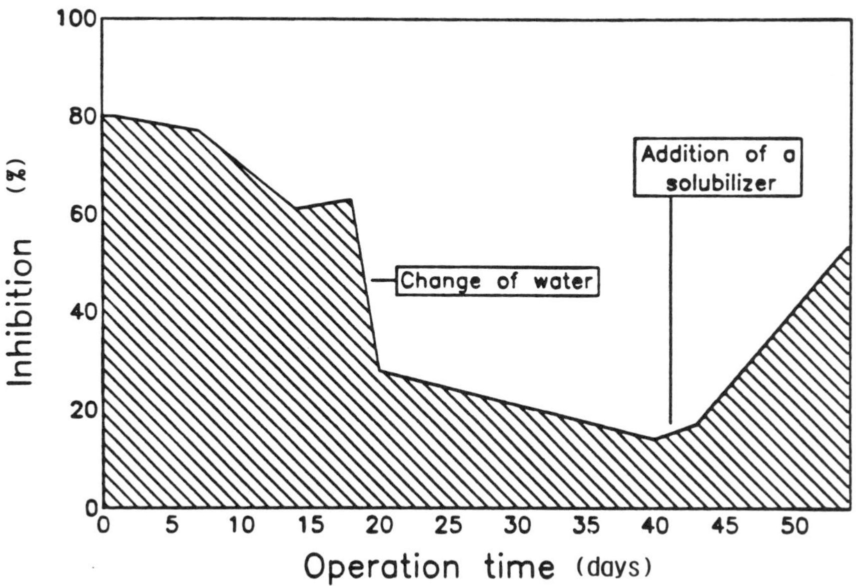

FIGURE 12. Toxicity (bioluminescence test) in the course of the experiment (gaswork soil).

(Barenschee *et al.* 1990). The examples mentioned cannot be transferred to the general practice of remediation.

Figure 13 shows the use of nitrate in the biodegradation pathway of an aliphatic hydrocarbon. According to the data in the literature, the initial oxidation only occurs with molecular oxygen (Höpner *et al.* 1989). Neither alcohol nor aldehyde can be mineralized in the presence of nitrate alone. Only the fatty acid can be oxidized during denitrification.

In an experimental setup similar to the above-described percolation, contaminated soil from a refinery was treated microbially. The hydrocarbon concentration was about 7 g/kg soil. The column was flooded and flushed during the experiment. The oxygen supplied by aerating the water was insufficient, as evidenced by dark reduced zones in the soil. Bacteria were not added because of the presence of oil-degrading microorganisms in the range of 10^4/mL. This value increased up to 10^6/mL after starting the experiment. It is not necessary here to stress all the parameters measured; of great interest, however, is the behaviour after adding nitrate to the system in final concentrations of 100 mg/L. The curves for nitrate and nitrite are shown in Figure 14. With the addition of nitrate, the number of denitrifiers also increased.

The dark areas disappeared after adding nitrate, which means that the biodegradation was optimized.

This run shows the beneficial effect of nitrate as an additional oxygen source for bioremediation.

Toxicity and Mutagenicity

Not only toxicity and mutagenicity of the initial substances are decisive for the risk assessment with relation to the environment, but also the development of these parameters during remediation. Experiments in our own laboratory, as well as those of other research groups, have shown that the toxicity of metabolic products is often higher than of the initial pollutants.

An example of this are the results of a degradation experiment, given in Figures 15 and 16. The problem arising in this context is the fact that the metabolites in general are soluble in water and can, therefore, without adequate safety measures drain off into the groundwater. If these substances are more toxic or mutagenic than the less mobile initial products, the risk of a negative impact on the environment

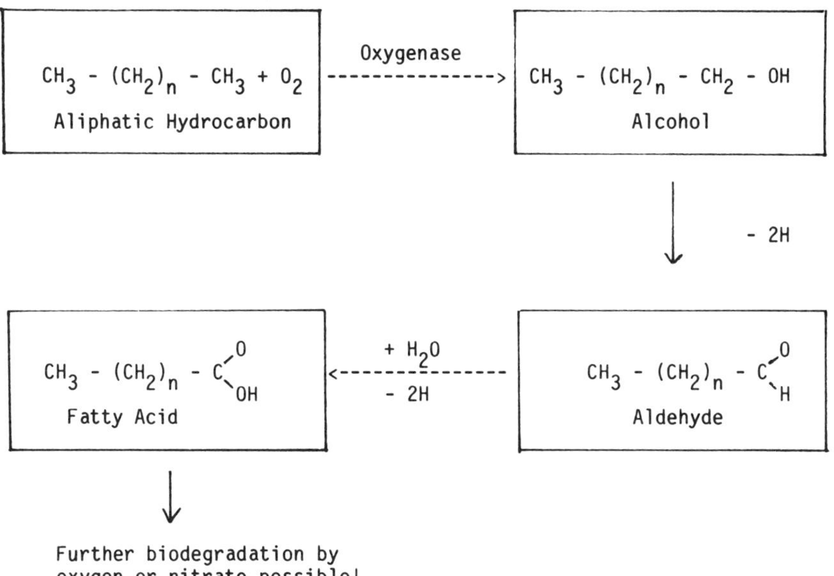

FIGURE 13. Biodegradation pathways of aliphatic hydrocarbons with respect to the use of nitrates.

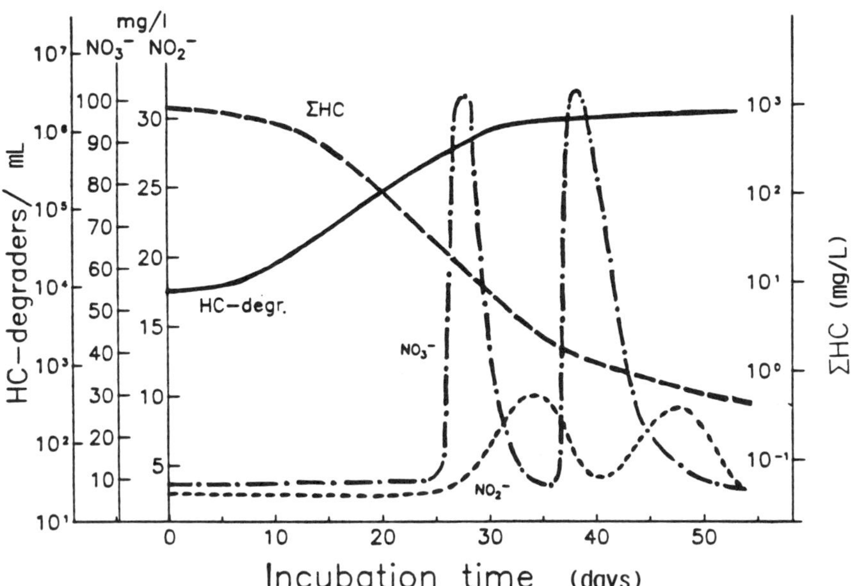

FIGURE 14. The effect of nitrate as additional electron acceptor.

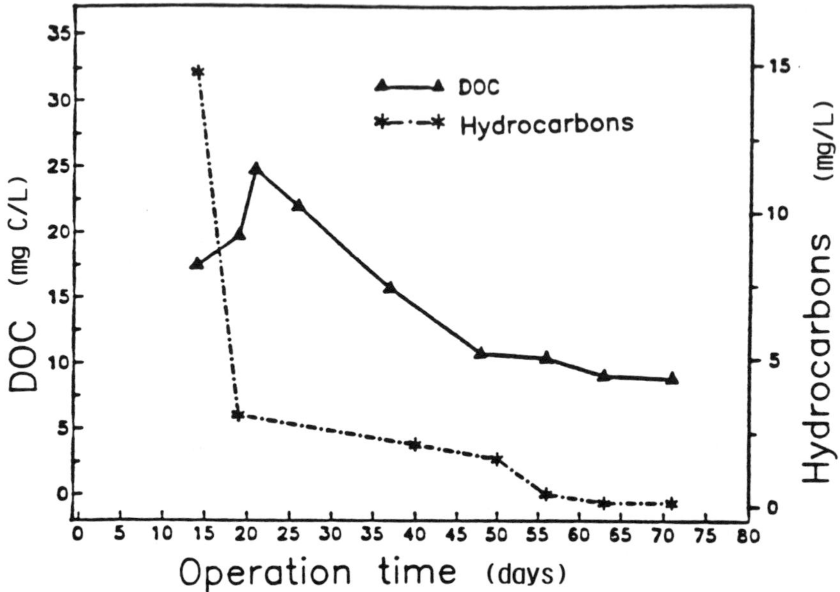

FIGURE 15. Degradation of a mixture of gaswork-specific contaminants (decane, hexadecane, pristane, naphthalene)—development of DOC and hydrocarbons.

through remediation is rather high. Few data exist about the behaviour of metabolic products, which are generally difficult to detect.

In the procedure to test the applicability of biological measures, the registration and identification of the metabolites and their toxicological importance should be taken into consideration.

Mixed Contaminations

Typical examples of mixed contamination can be found in abandoned gasworks. Analyses of soil and seepage water samples often show high amounts of pollutants, as the maximum concentrations given in Table 1 prove. The high contamination with aromatic hydrocarbons such as benzene, toluene, xylene, as well as PAHs, is well demonstrated. Furthermore, high values of cyanide and ammonium can be found. The water partly shows high concentrations of heavy metals and sulphates.

Substances such as ammonium (which is not a pollutant) can impede the mineralization of hydrocarbons. Nitrification processes, i.e.,

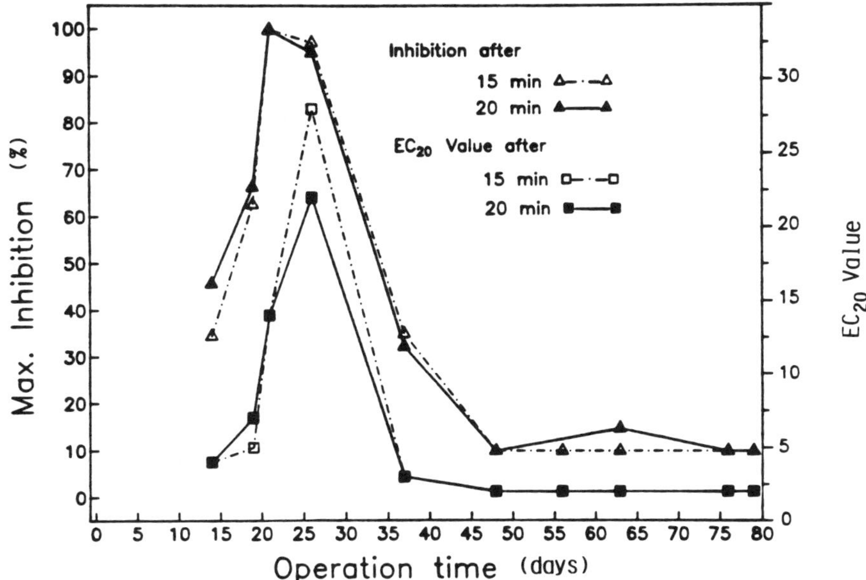

FIGURE 16. Degradation of a mixture of gaswork-specific contaminants (decane, hexadecane, pristane, naphthalene)—development of toxicity (bioluminescence test).

the oxidation of ammonium to nitrite or nitrate, consume oxygen, which is then no longer available for the oxidation of the pollutants themselves. This is also true for the readily degradable organic substances that compete with contaminants for oxygen.

Hydrocarbons often show a competitive degradation. BTX aromatics, mineral oils, and PAHs probably do not mineralize at the same degradation rate. Compounds of higher solubility are in general more readily degraded. According to laboratory and field experiences, *m*- and *p*-xylene are usually biodegraded more slowly than other BTX aromatics. In some cases, biodegradation of *m*- and *p*-xylene does not start until the concentration of the other contaminants has decreased to a minimum (Battermann & Werner 1987).

High concentrations of heavy metals can have a negative effect on bacterial growth. Contaminations with lead and/or mercury, which often occur in abandoned gasworks, are of special importance. Microbiological remediation of such soil is out of the question. Even if bacterial strains that resist heavy metals and mineralize hydrocarbons are increased, soil and groundwater cannot be considered as remediated because the heavy metals will remain. Presently, there is no biological procedure for the elimination of heavy metals.

Another problem are the cyanides, which are found in complexed forms in gasworks, mostly in the shape of the nontoxic "Prussian blue." Although they do not impede the degradation of pollutants, they cannot be mineralized. In contrast, the degradation of free cyanides is known and applied in sewage plants of coke oven works. Chemical analysis for the detection of complexed cyanides is only possible after special pretreatment, so that often no difference can be noticed between free and complexed cyanide.

Noncomplexed cyanides are easily biodegradable in concentrations up to about 15 mg/L, although they are some of the most toxic inorganic substances known. Experience shows that cyanide in abandoned industrial areas is complexed and therefore not or only less toxic and less soluble. On the other hand, it is almost nonbiodegradable in this form. Table 2 shows the occurrence of cyanides at the site and their toxic properties.

All contaminants are individually composed and therefore require individual treatment. The type and size of contaminations in connection with geohydrological conditions can only hint at the different possibilities of remediation.

General Conclusions

Although biological treatment of contaminated soil and groundwater is already in widespread use, many questions remain to be answered about the success and further optimization of these processes. The facts and data of the different measures applied in special cases are published in detail elsewhere (Franzius *et al.* 1988). To improve the systems, we must learn more about the problems that occur during biodegradation remediation.

One of the main problems is the bioavailability of the contaminants for the bacteria. The limiting factors are both solubility of the pollutants (biodegradation only occurs in the aqueous phase) and spatial separation, mainly due to geological conditions. Oxygen, however, is normally available in excess in the flushing water.

Furthermore, the question of additional and/or alternative electron acceptors has to be taken into consideration. Research is needed to decide which contaminants can be biodegraded with different oxygen sources.

TABLE 2. Occurrence and behaviour of cyanides.

Form	Solubility	Toxicity	Biodegrad-ability
Noncomplexed			
KCN, NaCn, NH_4CN Cyanides	high	high	+
KOCN, NaOCN, NH_4OCN Cyanates	high	not toxic	+
KSCN, NaSCN, NH_4SCN Thiocyanates	high	low	+
$Zn(CN)_2$	low	fairly toxic	?
Complexed			
$K3[Fe(CN)_6]$ red	high	low	–
$K_4[Fe(CN)_6]$ yellow	high	not toxic	–
$KFe[Fe(CN)_6]$ soluble Prussian blue	low	not toxic	+
$Fe_4[Fe(CN)_6]_3$ nonsoluble Prussian blue[(a)]	not soluble	not toxic	–

(a) Predominant in abandoned coal gasification plants.

Another problem is based on the mixture of several different pollutants contaminating a site. Work is needed to learn how to mineralize all pollutants in an acceptable time and with an acceptable level of effort. A site is not remediated if only one substance out of a whole consortium is eliminated. Therefore, the application of different methods, of which microbial degradation is one, is advisable.

Lastly, there is the problem of metabolites. If metabolites are not avoidable, their risk must be assessed and considered, when bioremediation is applied.

From this point of view, future research should mainly be focused on the questions mentioned above, which have to be considered with respect to the application of microbial methods.

REFERENCES

Annokkée, G. J. "MT-TNO Research into the Biodegradation of Soils and Sediments Contaminated with Oils and Polycyclic Aromatic Hydrocarbons (PAHs)." In *Contaminated Soil 1990*; Arendt, F., *et al.*, Eds.; Kluwer Academic Publishers: London, 1990; pp 941–946.

Barenschee, E. R.; Helmling, O.; Dahmer, S.; Del Grosso, B.; Ludwig, C. "Kinetic Studies on the Hydrogen Peroxide-Enhanced *in-Situ* Biodegradation of Hydrocarbons in Water Saturated Ground Zone." In *Contaminated Soil 1990*; Arendt, F., *et al.*, Eds.; Kluwer Academic Publishers: London, 1990; pp 1011–1018.

Battermann, G.; Werner, P. "Beseitigung einer Untergrundkontamination mit Kohlenwasserstoffen durch mikrobiellen Abbau." *Grundwasserforschung-Wasser/Abwasser* **1984**, *125*, 366–373.

Battermann, G.; Werner, P. "Feldexperimente zur mikrobiologischen Dekontamination." In *Abfallwirtschaft in Forschung und Praxis*; Franzius, V., Ed.; E. Schmid-Verlag, **1987**; Vol. 22, pp 167–185.

Böckle, K.; Stieber, M.; Werner, P.; Frimmel, F. H. "Biologischer Abbau von polycyclischen aromatischen Kohlenwasserstoffen in kontaminierten Böden." *Vom Wasser*, **1991**, in press.

Franzius, V.; Stegmann, R.; Wolf, K. *Handbuch der Altlastensanierung*; R. v. Decker's Verlag, G. Schenk: Heidelberg, 1988.

Höpner, T.; Harder, H.; Kiesewetter, K.; Dalyan, U.; Kutsche-Schmietenknop, I.; Teigelkamp, B. "Biochemical Aspects on the Hydrocarbon Biodegradation in Sediments and Soils." In *Toxic Organic Chemicals in Porous Media*; Ecological studies 73; Gerstl *et al.*, Eds.; Springer Verlag: Berlin, 1989; pp 251–272.

Klein, J.; Beyer, M. "Development of Microbiological/Adsorptive Methods for Remediation of PAH-Contaminated Soils." In *Contaminated Soil 1990*; Arendt, F., *et al.*, Eds.; Kluwer Academic Publishers: London, 1990; pp 1031–1032.

Lotter, S.; Stegmann, R.; Heerenklage, J. "Basic Investigations on the Optimization of Biological Treatment of Oil-Contaminated Soil." In *Contaminated Soil 1990*; Arendt, F., *et al.*, Eds.; Kluwer Academic Publishers: London, 1990; pp 967–974.

Riss, A.; Gerber, I.; Kreßler-Schmitt, M.; Meisch, H.-U.; Schweisfurth, R. "Altlastensanierung mittels Nitratdosierung: Laborversuche zum mikrobiellen Abbau von Heizöl." *Grundwasserforschung-Wasser/Abwasser* **1988**, *129*, 32–40.

Riss, A.; Ripper, P. "Soil and Groundwater Sanitation on the Area of the Waste-oil Refinery Pintsch-Öl, Hanau—Large Scale Tests for Biorestoration." In *Contaminated Soil 1990*; Arendt, F., *et al.*, Eds.; Kluwer Academic Publishers: London, 1990; pp 1033–1035.

Thole, S.; Werner, P. *Bodenkontamination mit polycyclischen aromatischen Kohlenwasserstoffen (PAK) und Möglichkeiten der Sanierung durch mikrobiellen Abbau*; Literaturstudie, Engler-Bunte-Institut der Universität Karlsruhe, 1986.

van den Berg, R.; Verheul, J.H.A.M.; Eikelboom, D. H. "*In Situ* Biorestoration of a Subsoil, Contaminated with Gasoline." *In Contaminated Soil 1990*; Arendt, F., *et al.*, Eds.; Kluwer Academic Publishers: London, 1990; pp 1025–1026.

Weidner, J.; Wichmann, K; Czekalla, C. "Hydraulic Remediation in Combination with Microbial Decontamination *In-Situ* and On-Site." In *Contaminated Soil 1990*; Arendt, F., *et al.*, Eds.; Kluwer Academic Publishers: London, 1990; pp 1021–1022.

Werner, P. "Experiences in the Use of Microorganisms in Soil and Aquifer Decontamination." In *Contaminant Transport in Groundwater*; Kobus, H. E.; Kinzelbach, W., Eds.; Balkema: Rotterdam, 1989; pp 59–63.

Engineering Design Aspects of an *In Situ* Soil Vapor Remediation System (Sparging)

Christopher J. Griffin, John M. Armstrong, Robert H. Douglass*
The Traverse Group, Inc.

INTRODUCTION

In 1969 an aviation gas (Avgas) spill occurred at the United States Coast Guard Traverse City (Michigan) Air Station due to a failure of a transfer pipe within the underground storage tank farm. Subsequent work at this site suggests that the total release from the underground tank farm was approximately 94,600 L. The Avgas has moved downwards and laterally to form a plume approximately 370 m long and 80 m wide. Subsurface conditions at this site consist of a fairly uniform, unconsolidated beach sand approximately 15 m thick, underlain by a gray to grayish blue silty clay deposit that is relatively impermeable (Twenter *et al.* 1985). The groundwater table at this site is located at a depth of 5 m, but over the past 6 years, a 1.5 to 2 m fluctuation in the water table has been observed which has caused the formation of a smeared zone of Avgas contaminated soil that is 0.4 to 0.5 m thick. The majority of the spill persists but is confined to the Air Station property by an interdiction field. The oily phase bound in the capillary fringe is a continuous source of groundwater contamination (Ostendorf *et al.* 1989). Different remediation designs flexible enough to handle the groundwater fluctuation were examined and one such option we considered we called "sparging."

* The Traverse Group, Inc., 2525 Aero Park Drive, Traverse City, MI 49684, (616) 947-2033

Sparging utilizes a compressor to inject air at some depth below the water table. As the air travels through the fuel-contaminated groundwater and soil, the oxygen in the air is used by the indigenous bacteria to biodegrade the contamination. The airflow will also enhance volatilization and stripping of remaining contamination. Sparging is a method for remediation of contamination contained within the capillary fringe, on the groundwater surface, and dissolved in groundwater.

The mass transfer capability of sparging should be flow rate dependent. At high flow rates volatilization, similar to soil venting, is expected to predominate; at low flows biodegradation is expected to predominate. The physical processes that occur related to volatilization are similar to those of soil venting and consist of pumping air through or across the contaminated material. Theoretically, sparging should be a more efficient process than soil venting when the contaminant is located within the capillary fringe (such as at the Traverse City Air Station). Sparging is expected to increase the rate of mass transfer by increasing contact between the contaminated surface area and air. At the Traverse City Air Station, the product-contaminated interval is very thin, and product removal by conventional soil venting is limited by vapor diffusion because the airflow is predominantly across rather than through the interval. To increase the removal rate, the airstream would have to travel through rather than across the surface of the water or fuel-saturated soil. This is not practical in a conventional soil venting system because the low pneumatic conductivity in the saturated and capillary zones restricts lateral airflow.

PRELIMINARY SPARGING TEST

Design and Procedure

The sparging test system consisted of a 5.1-cm outside diameter PVC well with a 0.61-m-long, 10-slot well screen. The top of the well screen was initially installed at a depth of 16.6 m below the ground surface. Air was injected into the well at flow rates of 11.9 and 27 cubic meters per hour (m^3/h). The airflow for the sparging tests was supplied by an electric-powered compressor for the 11.9-m^3/h flow rate and a gasoline engine driven compressor for the 27-m^3/h flow rate.

While injecting air, pressure readings were obtained at cluster well points located at selected depths and horizontal distances from the sparging well. The cluster wells consisted of 3.2-mm-diameter copper tubing with a 60-mesh stainless steel welded wire screen on the tip.

The cluster wells, located at horizontal distances of 0.6, 1.5, 3.0, 4.6, and 9.1 m from the sparging well, each had sampling points located at depths of 3.4, 4.6, and 4.9 m below ground surface.

These cluster points were located such that at the time of installation, the deepest point was in the capillary fringe, the next deepest was just above the capillary fringe, and the shallowest was located at least 1.2 m above the capillary fringe. The pressure readings were obtained by attaching a Magnehelic gauge to the cluster point using quick disconnect fittings. The Magnehelic pressure gauges measured in units of inches of water, accurate to 0.01 in. of water.

After the sparging tests were performed at both flow rates, the sparge well screen was raised approximately 3 m and the sparging repeated. Figure 1 presents a cross section of the sparging test. The tests were run for periods ranging from 12 to 48 h. Test runs were performed intermittently from June 6, 1990, through July 30, 1990, at flow rates of 11.9 and 27 m^3/h. A summary of the test dates, flow rates, injection depth, and depth to water is included on Figure 1.

The purpose of the study to date has been only to attempt to determine a radius of influence for sparging as a function of sparging flow rate and spraging depths. No evaluation of biodegradation enhancement due to sparging was performed during this study. The effectiveness of aerobic and anaerobic biodegradation at this site has been demonstrated by others (Hutchins *et al.* 1989; Ward *et al.* 1988).

The results of the test run at both flow rates, injected at a depth of 11.6 m below the water table, are shown as pressure versus time in Figures 2 and 3. As shown on these figures, variation in pressure can be seen in all of the cluster wells in the cluster points located within the capillary fringe (depth of 4.9 m). The data from this study indicate that a minimum radius of influence of 4.6 m is observed at both flow rates and at all depths of sparging.

Plotting pressure at lateral distance from the sparge well versus depth of the sparge well below the groundwater surface for the 27-m^3/h flow rate illustrates the relationship between sparging depth and radius of influence (Figure 4). Review of this plot indicates that sparging at depths of 8 m or more below the water table does not significantly increase pressure. In essence, air at a flow rate of 27 m^3/h should be injected at depths less than 8 m. However, a similar relationship was not observed for the flow rates of 11.0 m^3/h, possibly because of the effects of a rising water table.

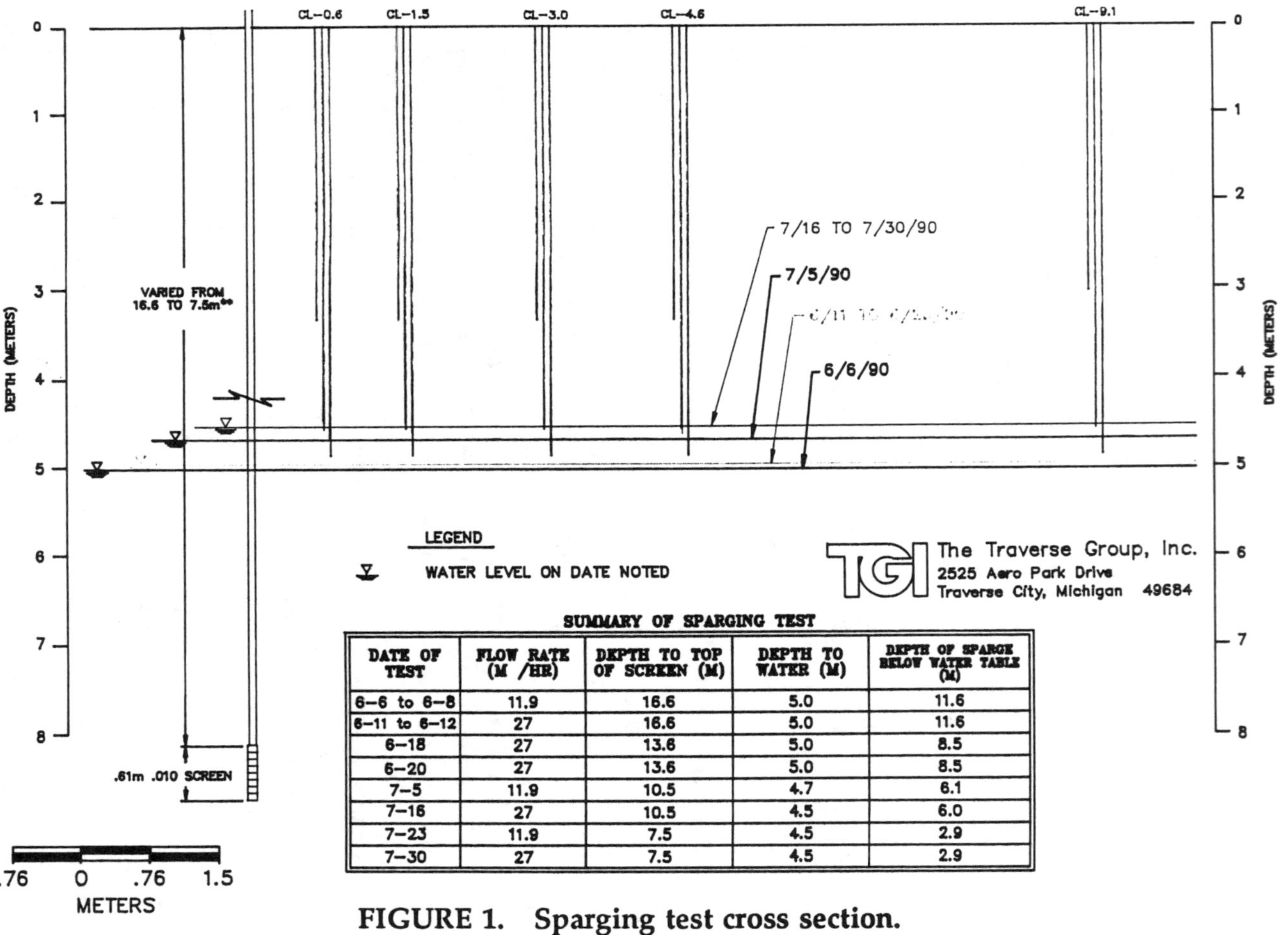

SUMMARY OF SPARGING TEST

DATE OF TEST	FLOW RATE (M /HR)	DEPTH TO TOP OF SCREEN (M)	DEPTH TO WATER (M)	DEPTH OF SPARGE BELOW WATER TABLE (M)
6-6 to 6-8	11.9	16.6	5.0	11.6
6-11 to 6-12	27	16.6	5.0	11.6
6-18	27	13.6	5.0	8.5
6-20	27	13.6	5.0	8.5
7-5	11.9	10.5	4.7	6.1
7-16	27	10.5	4.5	6.0
7-23	11.9	7.5	4.5	2.9
7-30	27	7.5	4.5	2.9

FIGURE 1. Sparging test cross section.

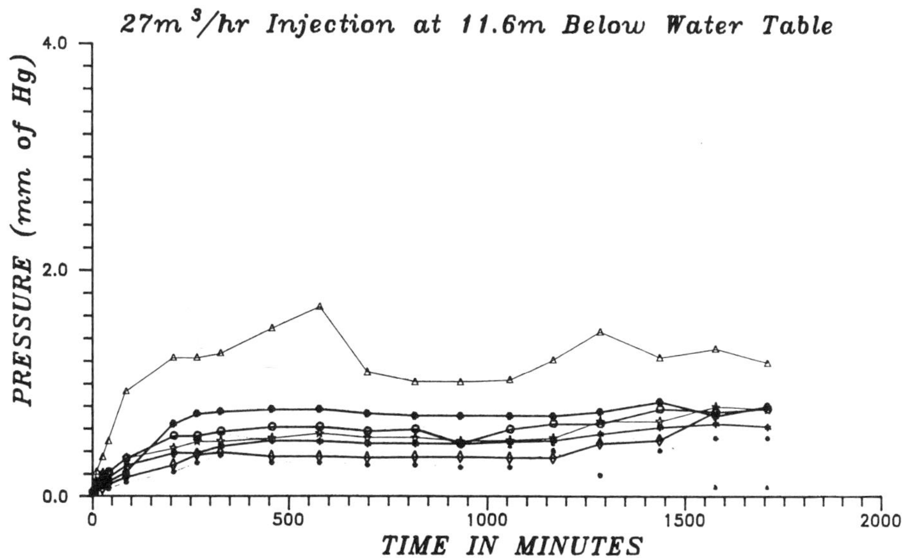

FIGURE 2. 27-m³/h injection at 11.6 m below water table.

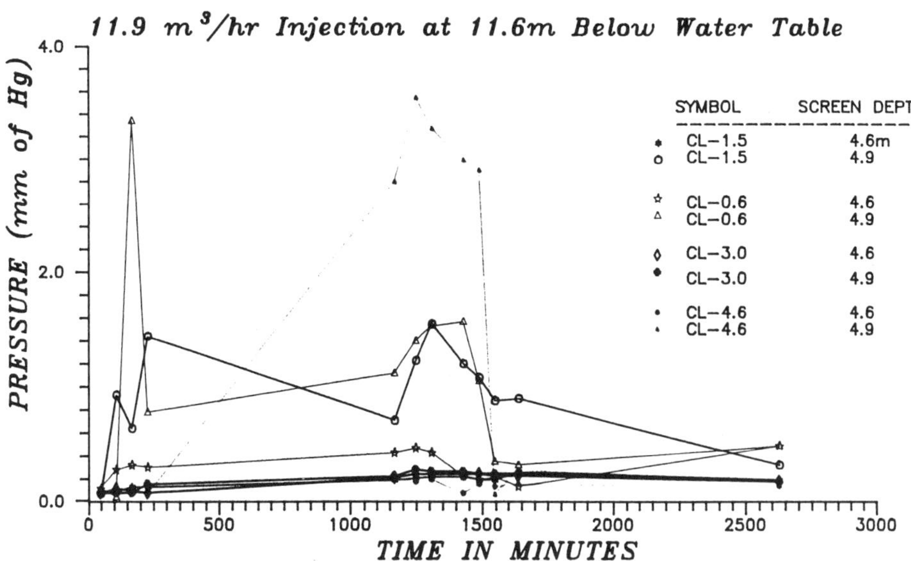

FIGURE 3. 11.9-m³/h injection at 11.6 m below water table.

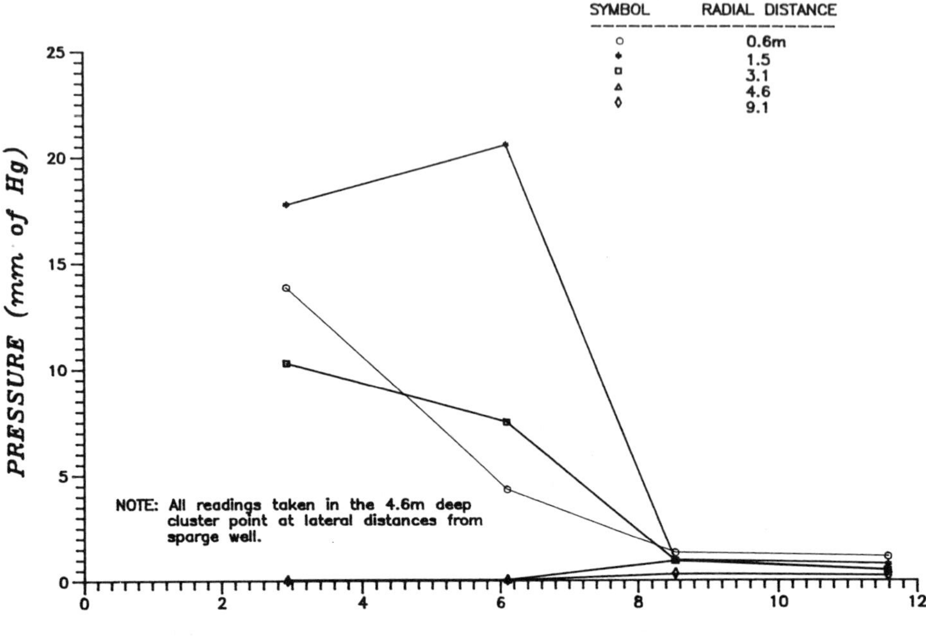

FIGURE 4. Plot of pressure vs. depth of sparging below the water table at 27-m^3/h flow rate.

SUMMARY

Unfortunately, test results to date are inconclusive for various reasons. First and foremost is the rise in the water table of approximately 0.3 to 0.5 m since May, 1990, when the test wells were installed. As shown on Figure 2, the monitoring points were located at depths of 3.4, 4.6, and 4.9 m below ground surface. The water table was at a depth of 5.0 m when these points were installed. However, since then the water level, as shown in Figure 1, has risen to a depth of 4.5 m below the ground surface, submerging two out of every three well screens and therefore preventing their use. In addition to the rising water table, variations in the pressure reading resulting from wind, rain, and changes in the barometric pressure were observed that could not be quantified using the present system layout and instrumentation. Finally, to evaluate more fully the effects of flow rates on the radius of influence, the sparging tests should have been conducted using a greater range of flow rate.

REFERENCES

Hutchins, S. R.; Downs, W. C.; Kampbell, D. H.; Smith, G. B.; Wilson, J. T.; Kovacs, D. A.; Douglass, R. H.; Hendrix, D. J. *NWWA/API Petroleum Hydrocarbons and Organic Chemicals in Ground Water* **1989**, 219–233.

Ostendorf, D. W.; Kampbell, D. H.; Wilson, J. T.; Sammons, J. H. *Research Journal Water Pollution Control Federation* **1989**, *61*, 1684–1690.

Twenter, F. R.; Cumming, T. R.; Grannemann, N. G. *Ground Water Contamination in East Bay Township, Michigan,* U.S. Geological Survey Water Resource Investigation Report 85-4064, 1985.

Ward, C. H.; Thomas, J. M.; Fiorenza, S.; Rifai, H. S.; Bedient, P. B; Wilson, J. T.; Raymond, R. L. In Situ *Bioremediation of Subsurface Material and Ground Water Contaminated with Aviation Fuel: Traverse City, Michigan, Proceedings,* A & WMA/EPA International Symposium, February 21–22, 1988; Cincinnati, OH, 1989; pp. 83–96.

Design and Preliminary Performance Results of a Full-Scale Bioremediation System Utilizing an On-Site Oxygen Generator System

Barbara J. Prosen[*], *William M. Korreck, John M. Armstrong*
The Traverse Group, Inc.

Enhanced aerobic biodegradation has been successfully used to remediate groundwater contaminated with benzene, toluene, ethylbenzene, and xylenes (Ward *et al.* 1989; J. Wilson *et al.* 1986). This process typically involves introducing oxygen-amended water into a contaminated groundwater plume to enhance the natural biodegradation occurring within the water phase. Sources of concentrated oxygen for this process may include hydrogen peroxide, liquid oxygen, or gaseous oxygen, which may be purchased in gas cylinders or generated on site. After careful consideration of each alternative, on-site oxygen

[*] The Traverse Group, Inc., 2525 Aero Park Drive, Traverse City MI 49684

generation was chosen as an effective and inexpensive source of oxygen for enhancing biodegradation at a contaminated site in northern Michigan.

Remediation Design

Oxygen is being generated on site to help remediate a groundwater plume at an oil and gas well site which has been contaminated with benzene, toluene, ethyl-benzene, and xylenes (BTEX). Free product and contaminated soil have been completely removed from the source area, leaving only the groundwater plume for further remediation. Total BTEX concentrations average over 2 mg/L in the plume hot spot. A one-well purge and treat system has been installed to prevent groundwater contamination from entering a nearby river. State regulations require the purge well to continue operation as a plume containment device until contamination within the aquifer has reached an "irreducible level." Since cleanup to an "irreducible level" may take many years with purge and treat technology (Hall 1989), enhanced bioremediation was integrated into the existing system to accelerate and enhance groundwater cleanup.

Natural biodegradation processes have been shown to reduce BTEX levels within some aquifers through contaminant consumption by indigenous microorganisms (B. Wilson *et al.* 1986). Natural biodegradation may be enhanced through the introduction of concentrated oxygen into the contaminated groundwater (Ward *et al.* 1989; J. Wilson *et al.* 1986). At this site, oxygen was introduced into treated groundwater from the carbon treatment system, and the oxygenated water was then distributed over the upgradient half of the plume through a buried drainfield.

Since one of the objectives of oxygenation was to decrease remediation time, fresh air introduction was not considered as an oxygen source because oxygen delivery rates would be only one-fifth of those achieved with concentrated oxygen. Hydrogen peroxide was discounted because of its hazardous nature and high costs ($3.20 to $4.63/kg, $1.45 to $2.10/lb oxygen). Although liquid oxygen is much less expensive than hydrogen peroxide ($0.70/kg, $0.32/lb oxygen), it would require more maintenance than was possible at this remote site. Gas cylinders were also discounted because of high maintenance requirements. On-site oxygen generation was chosen for its low cost (approximately $6,500 initial cost and $0.22 to $0.42/kg, $0.10 to

\$0.19/lb oxygen), low maintenance requirements, and high level of safety. Oxygen generated from this system is 90 percent pure.

Operation of the oxygen generation system required an air compressor, an oxygen generator, and a power source. All oxygenation equipment was placed inside an insulated building. Compressed air was passed through two air filters to remove solid particles and oily material and then fed into an oxygen generator at delivery pressures ranging from 5.1 to 7.5 atmospheres (75 to 110 psi). Inside the oxygen generator, the air was passed through additional filters before entering one of two columns packed with crystalline zeolite (Figure 1). The zeolite column acted as a molecular sieve and trapped nitrogen present in the air within its structure. The zeolite-packed column not in use was purged to remove the accumulated nitrogen, and the two columns were cycled regularly.

Oxygen was generated at a rate of 9.3 m^3/h (32 ft^3/h) and was collected in a pressurized holding tank to assure steady delivery into the effluent water stream. Oxygen was introduced into the 13.6 m^3/h (60 gpm) water stream through a two-micron metal porous element (Figure 1). The element was 1.91 cm (0.75 in.) in diameter and 15.24 cm (6 in.) long. Sizing of the element was based on water flow rate, water pipe diameter, and pressure differential between the air and water. After oxygen was sparged into the treated groundwater, the amended water was discharged into a drainfield for below grade distribution. Figure 2 illustrates the entire remediation process.

A five-arm drainfield was constructed to distribute the oxygen-amended water over the upper half of the plume. Each drainfield arm was constructed of polyvinyl chloride (PVC) drainfield pipe 10.2 cm (4 in.) in diameter. Two rows of holes offset by 120° ran down the length of the pipe. Each arm was connected to a separate 5.1-cm (2-in.) line that ran back to the treatment building. The drain pipe was installed with the holes facing upward to assure full flow throughout each arm. Drainfield lines were connected to a manifold and controlled separately inside the treatment building so that water could be run through any combination of drainlines. The drainfield configuration covered the upgradient half of the plume, including the plume hot spot. Since groundwater is only 1.5 m (5 ft) below grade at the site, the drainfield was placed immediately above the groundwater table. Remediation of the lower half of the plume will be accomplished through a combination of enhanced biodegradation as oxygenated water moves downgradient and the existing purge and treat system.

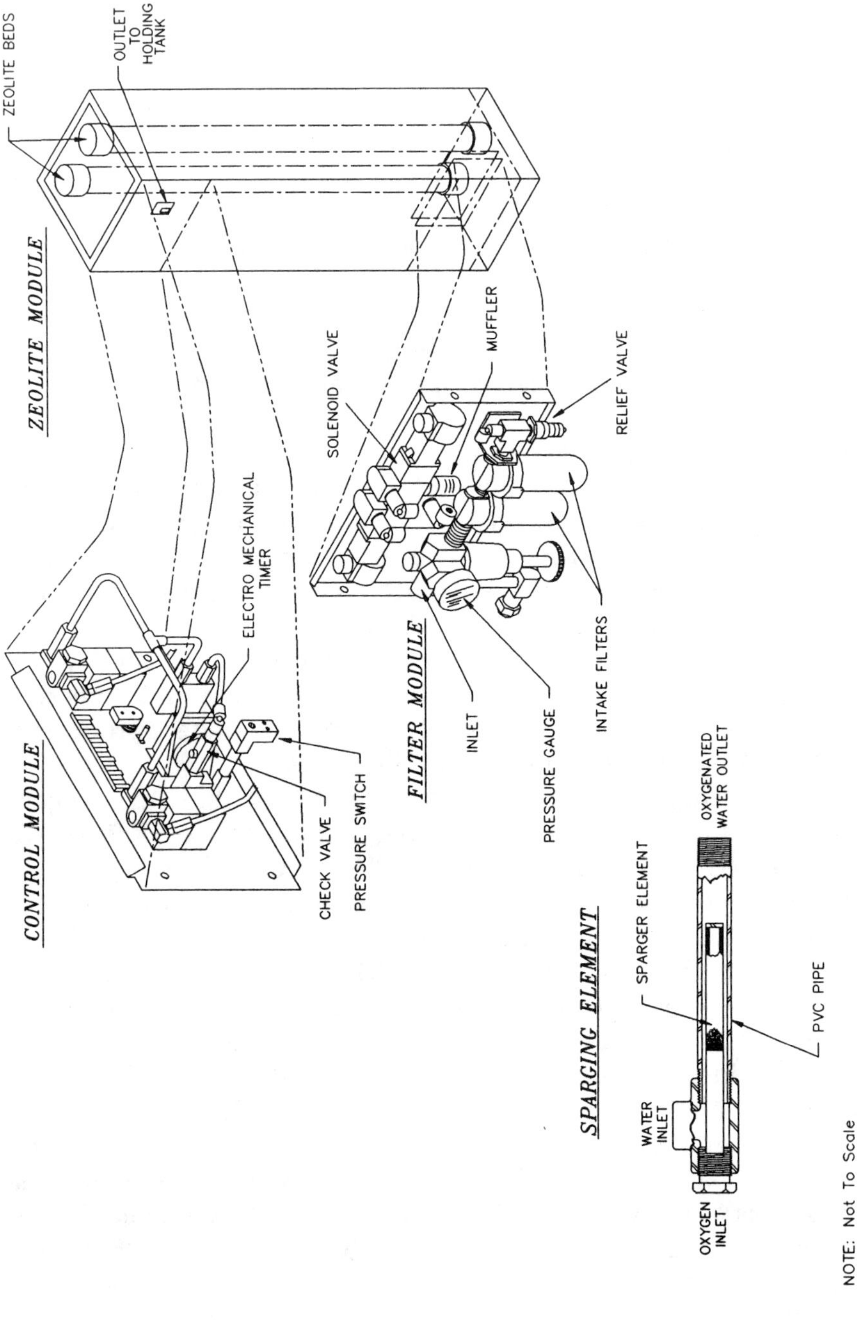

FIGURE 1. Oxygenation system.

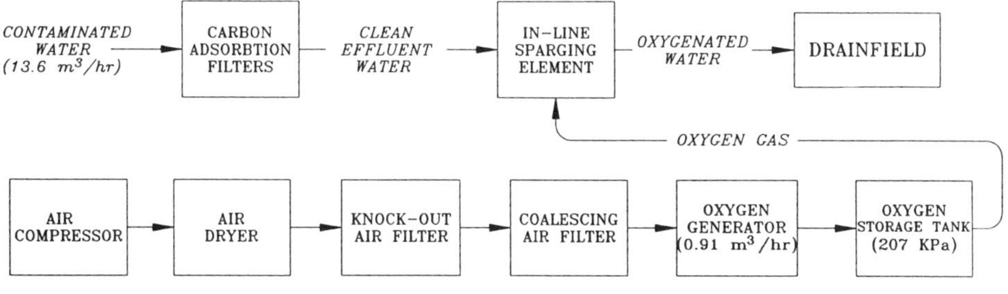

FIGURE 2. Process diagram.

Results

Operational data indicate that oxygen delivery pressures (pressure readings taken at the holding tank) have ranged from 1.02 to 3.40 atmospheres (15 to 50 psi) in the 5 months since system start-up. Iron bacteria have been found to accumulate on the sparging element, thus increasing the pressure differential between air and water flow. The highest dissolved oxygen concentrations in the water have been obtained immediately after cleaning the sparging element. Currently, the element is cleaned twice per month.

Dissolved oxygen concentrations in the effluent water have ranged from nondetectable to 59.0 mg/L. At this site, oxygen concentration has been found to be directly related to the amount of iron bacteria plugging the sparging element. Concentrations of 33.2 and 59.0 mg/L of dissolved oxygen have been obtained immediately after cleaning the element. Average dissolved oxygen concentration in the discharge water is 13.4 mg/L.

Dissolved oxygen concentrations measured in monitoring wells located near the drainfield have not followed any trends. Although an increase of 4.0 mg/L of dissolved oxygen concentration was observed approximately 3 months after start-up in a well located immediately adjacent to the drainfield, this type of increase does not appear to be consistent, either between monitoring wells or over time. Most monitoring wells have exhibited stable or decreasing dissolved oxygen concentrations since start-up of the oxygenation system.

Conclusions

Because of the relatively short time since implementation of the oxygenation system, only preliminary conclusions may be drawn. The in-line sparging element has proven to be an effective mechanism for oxygen transfer. Very high concentrations of dissolved oxygen have been observed in the effluent water when the element is not plugged with iron bacteria. Maintenance requirements on the system are minimal. Although periodic cleaning of the sparging stone is required, the oxygen generator needs no routine maintenance, and the compressor needs routine oil changes only. All air filters are inspected routinely, and none have been replaced thus far. Safety hazards have been minimized through pressurized tanks and automatic shutdowns.

Since the oxygenation system has been operating, trends have not been observed in dissolved oxygen concentrations within the monitoring wells. More time may be required before these trends develop. The fact that dissolved oxygen concentrations have been stable or have been reduced in many of the wells may indicate a microbial acclimation period. Nutrients may be added to the discharge water in the future to further stimulate microbial growth. While oxygen delivery rates through the sparging stone indicate that oxygenated water is reaching contaminated groundwater, continued monitoring is needed to determine the effects of oxygenation in the groundwater plume.

REFERENCES

Hall, C. W. "Practical Limits to Pump and Treat Technology for Aquifer Remediation." *Proceedings*, Prevention and Treatment of Groundwater and Soil Contamination in Petroleum Exploration and Production, May 9–11, Calgary Convention Centre, 1989; pp 29–29.7.

Ward, C. H.; Thomas, J. M.; Fiorenza, S.; Rifai, H. S.; Bedient, P. B.; Wilson, J. T.; Raymond, R. L. "*In Situ* Bioremediation of Subsurface Material and Ground Water Contaminated with Aviation Fuel: Traverse City, Michigan." *Proceedings*, A&WMA/EPA International Symposium, February 21–22, Cincinnati, OH, 1989; pp 83–96.

Wilson, B. H.; Bledsoe, B. E.; Kampbell, D. H.; Wilson, J. T.; Armstrong, J. M.; Sammons, J. H. "Biological Fate of Hydrocarbons at an Aviation Gasoline Spill Site." *Proceedings*, Petroleum Hydrocarbons and Organic Chemicals in Ground Water, NWWA/API, Houston, TX, 1986; pp 78–90.

Wilson, J. T.; Leach, L. E.; Henson, M.; Jones, J. N. "*In Situ* Biorestoration as a Ground Water Remediation Technique." *Ground Water Monitoring Review* **1986**, 7(4), pp 56–64.

Simulation of Enhanced *In-Situ* Biorestoration of Petroleum Hydrocarbons

*Robert C. Borden**
North Carolina State University

A general mathematical model is being developed to aid in the design and analysis of projects for the enhanced aerobic bioremediation of petroleum-contaminated aquifers. Development of the enhanced biotransformation model is proceeding in three steps: (1) development of an abiotic hydrocarbon dissolution model (Borden & Kao, in review); (2) coupling the dissolution model with existing equations for simulating aerobic biodegradation (Borden & Bedient, 1986); and (3) comparison with laboratory data.

Hydrocarbon Dissolution Model

A mathematical model of hydrocarbon dissolution has been developed based on classic liquid-liquid equilibrium theory. The model assumes that the residual hydrocarbon is distributed between two fractions, a fast fraction in equilibrium with the aqueous phase and a slow fraction in which mass transfer is limited. Overall, the model provides an excellent fit to the experimental data and requires a minimum of input parameters. A detailed description of the equation development and experimental validation is provided by Borden and Kao (in review).

Biotransformation Model

A general mathematical model is being developed for simulating the dissolution of trapped liquid hydrocarbon (oil), transport of the

* Civil Engineering Department, North Carolina State University, Box 7908, Raleigh, NC 27695

dissolved contaminants with the moving groundwater, and biodegradation of the dissolved contaminants by attached- and solution-phase microorganisms. The current version of the model was developed by modifying the hydrocarbon dissolution model to simulate enhanced aerobic biorestoration by addition of microbial kinetic terms describing the growth and decay of microorganisms and consumption of dissolved hydrocarbons and oxygen as described by Borden and Bedient (1986). The model allows for *n* individual hydrocarbon components which may be present in four phases: (1) an aqueous phase, (2) a fast oil phase in chemical equilibrium with the aqueous phase, (3) a slow oil phase in which mass transfer between the aqueous and oil phases is relatively slow and can be modeled as a first order reversible reaction, and (4) a sorbed phase in equilibrium with the aqueous phase. The rate and extent of biodegradation may be limited by the available oxygen supply. Microbial growth, organic substrate, and oxygen depletion are simulated using Monod kinetics modified to include inhibition at low oxygen concentrations.

Governing equations are developed by writing material balances for each phase and compound. The aqueous, fast oil, slow oil, and sediment phases will be identified by the superscripts *a, of, os,* and *s.* The supercript *o* indicates both fast and slow oil phases. In one dimension, material balance equations for compound *i* in the aqueous and slow oil phases may be written

$$\left(\theta^a + \rho^s K_i^s + \frac{\theta^{of}}{K_i^o}\right)\frac{\partial C_i^a}{\partial t} = \left(\frac{\partial}{\partial z}\cdot\frac{\theta a\, D\, \partial C_i^a}{\partial z}\right) - \frac{\partial\, q\, C_i^a}{\partial z}$$

$$- \left(\frac{M\,\mu_i}{Y_i}\cdot\frac{C_i^a}{K_i + C_i^a}\cdot\frac{O}{K_O + O}\right)$$

$$+ \,Km_i^{os}\left(K_i^o\, C_i^{os} - C_i^a\right)$$

and

$$\frac{\partial C_i^{os}}{\partial t} = Km_i^{os}\,\frac{\left(K_i^o\, C_i^{os} - C_i^a\right)}{\theta^{os}}$$

where

$$K_i^o = \frac{\gamma_i \, S_i \, \overline{MW}^o \, \rho^a}{\overline{MW}^a \, \rho^o}$$

Material balance equations are also needed to describe the variation in fast and slow oil saturation over time. The governing equations for the slow oil and fast oil saturations are

$$\frac{\partial \theta^{os}}{\partial t} = \frac{\overline{MW}^o}{\rho^o} \cdot \sum \; Km_i^{os} \left(C_i^a - K_i^o \, C_i^{os} \right)$$

and

$$\frac{\partial \theta^{of}}{\partial t} = \frac{\theta^{of} \, \overline{MW}^o}{\rho^o} \cdot \sum \left(\frac{1}{K_i^o} \cdot \frac{\partial C_i^a}{\partial t} \right)$$

Oxygen is assumed to be present only in the aqueous phase.

$$\frac{\partial O}{\partial t} = \left(\frac{\partial}{\partial z} \cdot \frac{D \, \partial O}{\partial z} \right) - \frac{\partial q O}{\theta^a \, \partial z} - \sum \left(\frac{M \, \mu_i}{F} \cdot \frac{C_i^a}{K_i + C_i^a} \cdot \frac{O}{K_O + O} \right)$$

Equation variables are defined below.

$C\beta i$	=	concentration of hydrocarbon i in the β phase
O	=	concentration of oxygen in the aqueous phase
M	=	concentration of aerobic hydrocarbon degrading organisms in all phases
D	=	longitudinal dispersion coefficient $= \alpha q/\theta^a$
α	=	longitudinal dispersivity
q	=	groundwater specific discharge
t	=	time
z	=	longitudinal coordinate
$\theta\beta$	=	β phase saturation
$\rho\beta$	=	β phase bulk density
$\overline{MW}\beta$	=	β phase average molecular weight
$\gamma\beta$	=	β phase activity coefficient

S = pure compound aqueous solubility
Km^{os} = slow oil–aqueous phase mass transfer rate
K^s = solid-aqueous phase distribution coefficient
μ_i = maximum specific growth rate of microorganisms on compound i
K_i = half saturation constant for compound i
Y = ratio of cells produced to hydrocarbon consumed
F = ratio of cells produced to electron acceptor consumed

The governing equations are solved using a fully upwinded, block-centered, finite difference solution. Specific discharge and longitudinal dispersivity were assumed constant throughout. Numerical dispersion effects were minimized by reducing the longitudinal dispersivity by the factor $(\Delta x - q\Delta t/2\theta^a)$. Simulation results for the nonreactive case ($K_i^s = \theta^{os} = \theta^{of} = 0.0$) were compared with an existing analytical solution and agreed within 1 percent.

Enhanced Biorestoration Column Experiment

The dissolution, transport, and biodegradation of gasoline constituents was measured in an 18 cm long by 2.6 cm diameter all glass and Teflon® column packed with a medium to fine aquifer sand. Prior to packing, the sand was mixed with microorganisms grown in the chemostats using toluene as the sole carbon source. This was done so that the metabolic characteristics of the community in the column would be uniform and well defined. Regular grade unleaded gasoline (10 mL) was introduced into the previously saturated column, followed by 250 mL of distilled water. The distilled water was pumped through at a high flow rate (10 L/d) to displace a portion of the gasoline. All water and gasoline produced during this initial period was collected and extracted to determine the mass of each hydrocarbon constituent remaining in the column. After this initial displacement, 7.4 mL of gasoline remained trapped within the column (relative oil saturation = 0.17). Oxygenated water amended with nitrogen, phosphorus, and other trace nutrients was then pumped into the column at a lower flow rate (0.5 L/d) for 6 weeks. Effluent samples were collected and analyzed for hydrocarbon components and dissolved oxygen.

Model Calibration and Testing

The enhanced biodegradation column experiment was modeled assuming the gasoline is composed of four major groups of compounds: benzene, toluene, total xylenes (m,p,o-xylene plus ethylbenzene), and low-solubility aliphatic hydrocarbons. Biological kinetic coefficients for benzene, total xylenes, and aliphatic hydrocarbons were assumed to be equal to the coefficients for toluene. Biological reaction coefficients for toluene were obtained from a series chemostat studies conducted by Lee (1989) using the same consortia used to inoculate the column. Partion coefficients were obtained from prior abiotic column experiments (Borden & Kao, in review).

Numerical simulation results are compared to experimental data from the biological column experiments in Figure 1. Overall, the model provides a fairly good prediction of the experimental data. Effluent benzene, toluene, and total xylene concentrations were accurately simulated during the early portion of the biorestoration column experiment. In order to match the xylene concentrations in the effluent, it was necessary to reduce the maximum specific growth rate for xylene to one-fifth that of toluene. Biological rate data were obtained for only one compound, toluene. Apparently, the population that developed in the chemostats (and added to the column) did not have the capablility to effectively degrade m-xylene (data not shown).

During the later portion of the experiment, the effluent benzene and toluene concentrations leveled out. This is probably due to a slow rate of mass transfer between the residual hydrocarbon and the aqueous phase. During this later portion of the experiment, the model slightly overpredicted benzene and underpredicted toluene concentrations in the effluent. A somewhat better fit to the experimental data could be obtained by adjusting the half saturation constants for benzene and toluene.

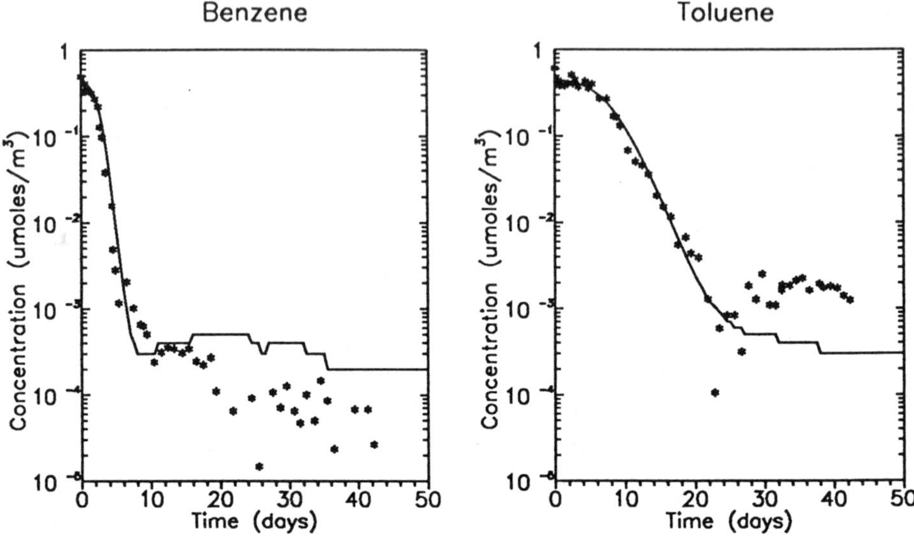

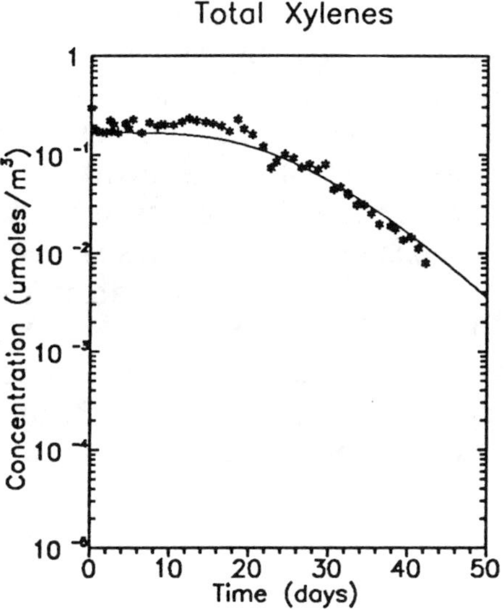

FIGURE 1. Comparison of numerical model simulations and experimental results from enhanced biorestoration column experiment.

REFERENCES

Borden, R. C.; Bedient, P. B. *Water Resourc. Res.* **1986**, *22*, 1973–1982.
Borden, R. C.; Kao, C. M. Submitted for review to *Res. J. Water Poll. Contr. Fed.*
Lee, W. E. Master's Thesis, North Carolina State University, 1989.

Modeling Bioremediation: Theory and Field Application

Hanadi S. Rifai, Gregory P. Long, Philip B. Bedient*
Rice University

Site Description

The U.S. Coast Guard Air Station at Traverse City, Michigan, is the site of a 37,850 to 75,700 liter aviation fuel spill which occurred in 1969. A plume of contamination migrated into the shallow drinking water aquifer and underneath a residential area, with alkylbenzene (BTEX) concentrations ranging from 0.03 to 30 mg/L (Rifai *et al.* 1989; Wilson *et al.* 1986).

An interdiction well field (Figure 1) was installed in April of 1985 to limit the contamination to the site. Within about 24 months, rapid dissipation of the downgradient BTEX plume was observed at a number of wells and is the subject of this modeling study.

Site Hydrogeology

The study area is underlain by lacustrine glacial deposits consisting of an upper sand and gravel layer (9 to 36 m thick) and underlying clay layer (up to 102 m thick). Hydraulic conductivity (K) ranges from 31

* Rice University, Department of Environmental Science and Engineering, P.O. Box 1892, Houston TX 77251

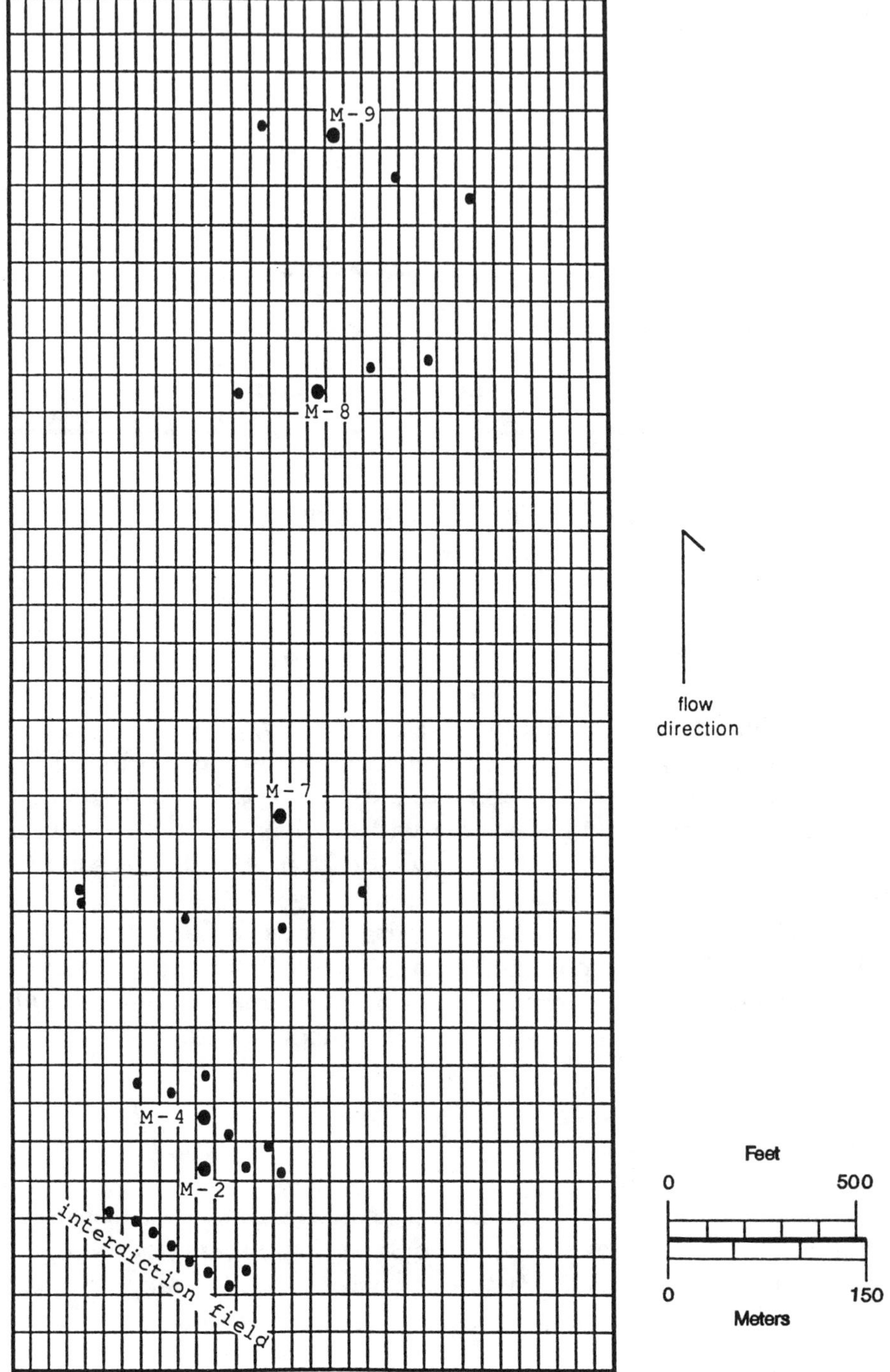

FIGURE 1. Base map and model grid.

to 46 m/day based on pump test data acquired several years ago. However, a lower dense sand has been identified and has a hydraulic conductivity of about 6 m/day. The hydraulic gradient averages 0.005, and effective porosity is in the range of 0.30.

Model Application

The modeling approach taken here starts with a well-defined initial source concentration in the aquifer downgradient of the pump-and-treat wells and attempts to match observed dissipation at a number of well-monitored points downstream. Both hydraulic conductivity and rate of biodegradation were used to match the observed dissipation in the aquifer over a 2-year period.

Modeling of the plume downgradient of the interdiction field was performed using Bioplume II, which is a modified version of the USGS Method of Characteristics (MOC) Solute Transport Model. This study was performed under steady-state conditions using an effective porosity of 30 percent, a longitudinal dispersivity of 1.5 m, a ratio of transverse to longitudinal dispersivity of 0.3, and a 32 by 36 grid (Figure 1). Saturated thickness ranges from 9 to 30 m, and the hydraulic gradient averages 0.005. These parameters are in line with previous model studies of the site (Smythe *et al.* 1989).

Initial model runs were made with no biodegradation taking place to evaluate the sensitivity of the model to observed total BTEX concentrations, based solely on the hydraulics of the system. Wells M-2, M-4, M-7, M-8, and M-9 (Figure 1) were selected to allow for the comparison of calculated to observed total BTEX concentrations. Also, the seven interdiction wells that were pumping as of April of 1985 were included in the model and are IN-1 through IN-7 (Figure 1). These wells were modeled with an average pumping rate of 110 m^3/day/well for the period of the simulations.

Hydraulic conductivities of 46 and 29 m/day were used corresponding to the high and low end values, respectively, as reported by Wilson *et al.* (1986). Based on data reported by Smythe *et al.* (1989) as well as the result of analysis of peak travel time between wells M-2 and M-4, K values of 15 and 22 m/day were also modeled. Model runs and brief descriptions are summarized in Table 1.

TABLE 1. Comparison of Bioplume II model runs.

Run	Hydraulic Conductivity (m/day)	Run Time (yrs)	Biodegradation	Comments
1	46	1	none	Used high end of reported hydraulic conductivities, calculated contaminant concentrations declined more rapidly than observed concentrations
2	29	1	none	Decreased hydraulic conductivity to be in line with values used in previous studies, calculated concentrations compared favorably with those observed
3	26	1	none	Further reduced hydraulic conductivity to evaluate sensitivity of the model to this parameter, calculated concentrations shifted somewhat with run two giving the best match
4	29	3	none	Went to a three-year run to evaluate the match between calculated and observed concentrations over a longer period, found a favorable match for the first two years with the calculated concentrations going to zero and not matching the leveling off seen in the observed data
5	15	3	none	Used a decreased hydraulic conductivity as reported by Smythe (1988), found calculated concentrations were high compared to observed
6	15	3	yes	From run five it was observed that decreasing the calculated concentrations might be accomplished using biodegradation, run six repeated run five with biodegradation turned on in Bioplume II, the calculated concentrations declined more like the observed but too rapidly, further work is on-going
7	22	3	yes	Initial oxygen concentrations change from 8 mg/L to 4 mg/L and run five repeated, similar results seen but with closer match to observed in the first year

Results without Biodegradation

The comparison of the 29 m/day curve to the observed data for wells M-2, M-4, and M-9 shows a good match of observed to modeled concentrations, suggesting that if the hydraulic conductivity of 29 m/day is a valid measure of the properties at the site, then the observed declines in BTEX concentrations can be explained largely by hydraulic flushing. Since the possibility that a lower K might be more appropriate for this site, especially for the lower part of the sand, values of 15 and 22 m/day for K were modeled. The resulting modeled concentrations at the lower K values shown in Figure 2 suggest that hydraulic flushing alone cannot account for the observed concentration declines. Given our knowledge of the biodegradation processes ongoing at Traverse City, further tests were performed to evaluate the relative effects of biodegradation.

Results with Biodegradation

Modeling of the dissipation at the site using a hydraulic conductivity of 22 m/day coupled with naturally occurring biodegradation was completed using an instantaneous reaction between the contaminant and the available oxygen. The instantaneous reaction decreases the concentration of the contaminant by an amount that is proportional to the available oxygen in the aquifer (it is assumed that about 3 units of oxygen are required to completely biodegrade 1 unit of BTEX). Dissolved oxygen concentrations are assumed to be 4 mg/L. These levels of oxygen are in line with data reported in Rifai *et al.* (1989). The results of these model runs are graphed in Figure 2, indicated by the open circles, and show that the calculated concentrations move closer to the observed concentrations. Well M-9 is completely degraded to zero in the first time step and will require additional modeling efforts.

Conclusions

The modeling of the dissipation of BTEX concentrations was greatly simplified by the cutoff of the variable source 460 m upgradient. The success of the modeling effort was largely a function of the accuracy of the initial downgradient plume and the hydraulics of the system. The modeling results are preliminary at this time, but show a distinct and

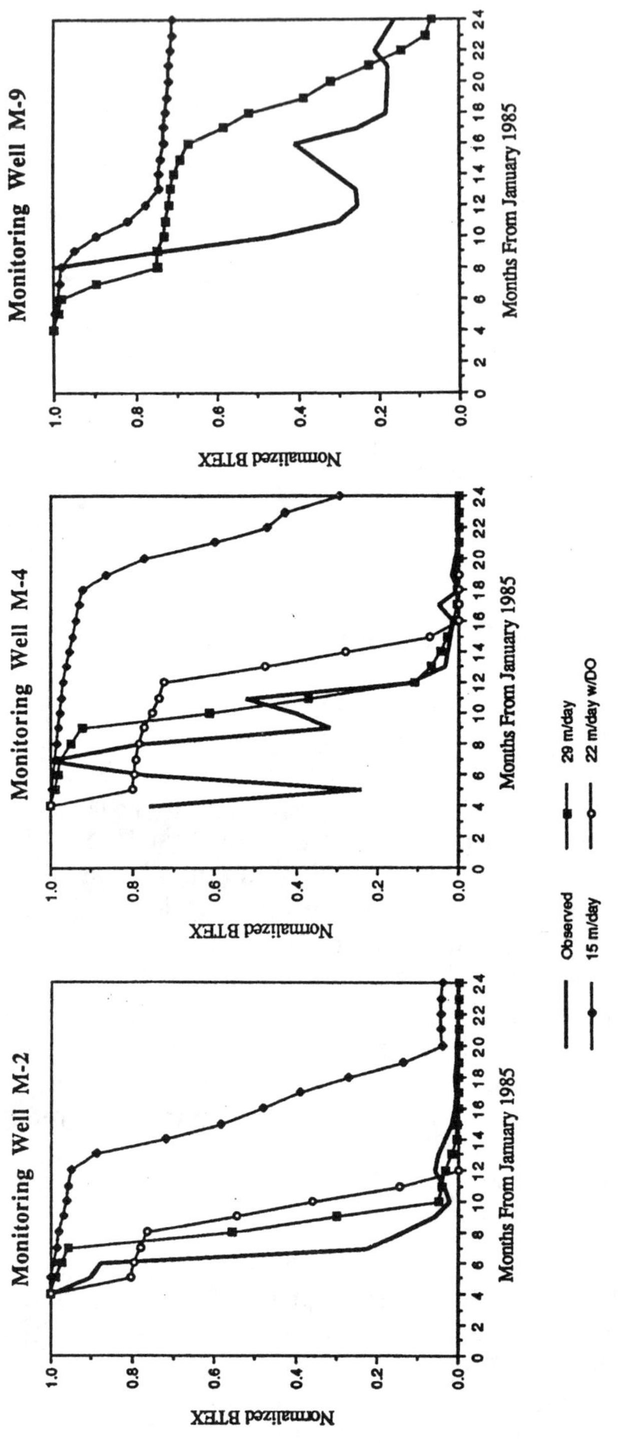

FIGURE 2. Observed BTEX declines at selected wells.

reasonable comparison to observed dissipation in five wells downgradient. The higher the K value used, the faster the flush out and decline, and the lower the K value used, the more biodegradation was required to match the decline.

REFERENCES

Rifai, H. S.; Bedient, P. B.; Wilson, J. T.; Miller, K. M.; Armstrong, J. M. *ASCE J. Environmental Engineering Division* **1989,** *114,* 1007–1019.

Smythe, J. M.; Bedient, P. B.; Klopp, R. A. *Proceedings,* Second National Outdoor Action Conference on Aquifer Restoration, Ground Water Monitoring and Geophysical Methods, Association of Ground Water Scientists and Engineers, Las Vegas, NV, 1989; pp 71–94.

Wilson, B. H.; Bledsoe, B. E.; Kampbell, D. H.; Wilson, J. T.; Armstrong, J. M.; Sammons, J. H. *Proceedings,* National Water Well Association/American Petroleum Institute Conference on Petroleum Hydrocarbons and Organic Chemicals in Ground Water: Prevention, Detection and Restoration, Houston, TX, 1986; pp 78–89.

In Situ Respirometry for Determining Aerobic Degradation Rates

*Say Kee Ong**, *Robert Hinchee*
Battelle
Ron Hoeppel
Naval Civil Engineering Laboratory
Rick Scholze
U.S. Army Corps of Engineers

Soil venting is an effective means of physically removing volatile hydrocarbons by volatilization while providing a source of oxygen for biological mineralization of the volatile and nonvolatile hydrocarbons in the

* Battelle, 505 King Avenue, Columbus OH 43201

soil. "Bioventing," as this technology is sometimes called, is gaining wide acceptance as a viable treatment alternative for hydrocarbon-contaminated soils. However, site-specific soil properties and the heterogenous nature of contaminated sites usually make short-term treatability tests necessary to assess its feasibility for a given site.

The objective of this paper is to outline the development of a field treatability test that can be applied *in situ* to contaminated sites to measure the aerobic biodegradation rates of organic hydrocarbons. This test is straightforward, simple and inexpensive to apply, and does not involve major analytical instrumentation.

In Situ Treatability Test

The *in situ* treatability test described here is a short-term modification of the *in situ* respiration test described by Hinchee *et al.* (in press) and Miller and Hinchee (1990). Biodegradation rates for this method are estimated by measuring the change in oxygen and carbon dioxide concentrations in the soil gas of contaminated soil and uncontaminated soil after it has been vented with air. One to three soil gas sampling probes, called *test probes*, are placed in contaminated soil, and air is injected for 20 to 24 h. Nominal air flow rates are 1 to 2 m^3/h. Initial modeling calculations showed that this time period and air flow rate are sufficient to aerate an adequate volume of contaminated soil to sustain the test for about 5 days with minimum interference by carbon dioxide and oxygen diffusion from either the ground surface or gas pores outside the vented zone. The test probes are usually located 3 to 6 m apart.

At the same time, two other probes, inert gas control and background control probes, are used as controls. The inert gas control probe is placed in contaminated soil typically 30 m or more away from the test probes. Inert gas such as argon or nitrogen is injected at the same rate as air for the test probes. Measurement of oxygen in this argon- or nitrogen-vented zone provides data on the diffusion of oxygen from the ground surface and the surrounding soil. The background control probe is placed in an uncontaminated site, and air is injected to monitor natural background soil respiration.

The gas sampling probes used are 1 cm (3/8 in.) i.d. stainless steel rods with a slotted well point assembly at one end. For good results, the slotted well assembly of the test probes must be placed in the contaminated soil. A shut-off valve is attached to the ground level end of the probe.

Before any air or argon gas injection, soil gas samples are withdrawn and carbon dioxide and oxygen concentrations of the soil gas are measured. After air and argon are turned off, carbon dioxide and oxygen concentrations of the soil gas are monitored periodically over time. Initial readings of oxygen and carbon dioxide in the soil gas are taken every 2 h and then progressively at 4- to 6-h intervals. The test is terminated when the oxygen concentration in the soil gas has depleted to about 5 percent. Measurements of carbon dioxide and oxygen concentrations are made using a Gastechtor Model 32520X. This instrument measures carbon dioxide by an infrared absorption method with a measurement range of 0 to 5 percent and oxygen with an electrochemical cell with a range 0 to 25 percent.

Typical results of an *in situ* treatability test performed at a site on the east coast of the United States are shown in Figures 1 and 2. This site is contaminated with JP-4 jet fuel. Oxygen depletion is of zero order (see Figure 1). The low consumption of oxygen at an uncontaminated site (background control probe) provided data for natural background soil respiration. For the inert gas control probe, the negligible increase in oxygen indicates that very little oxygen has diffused into the test zone during the test. To compare the respiration rates obtained

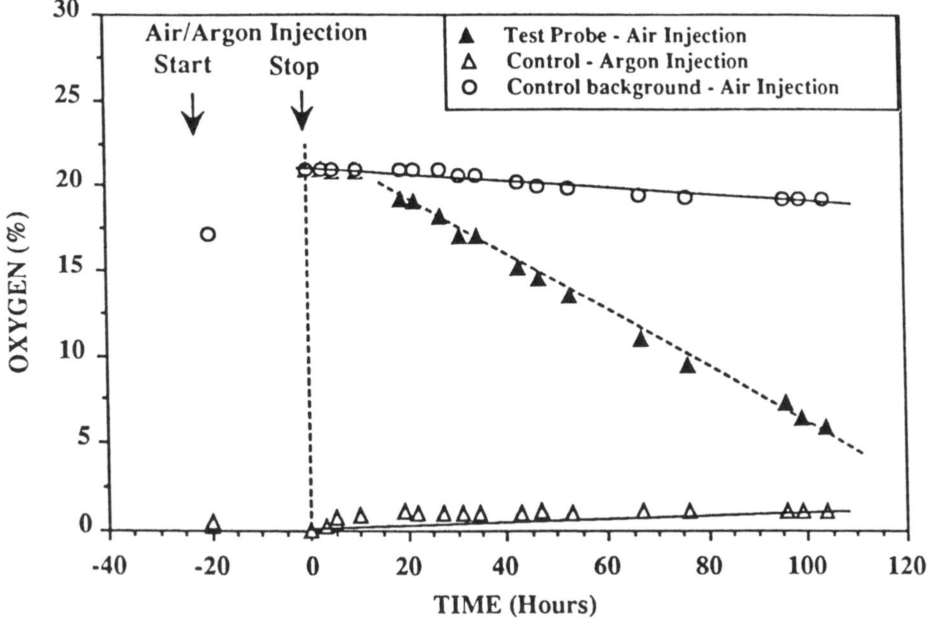

FIGURE 1. Change of carbon dioxide concentrations over time.

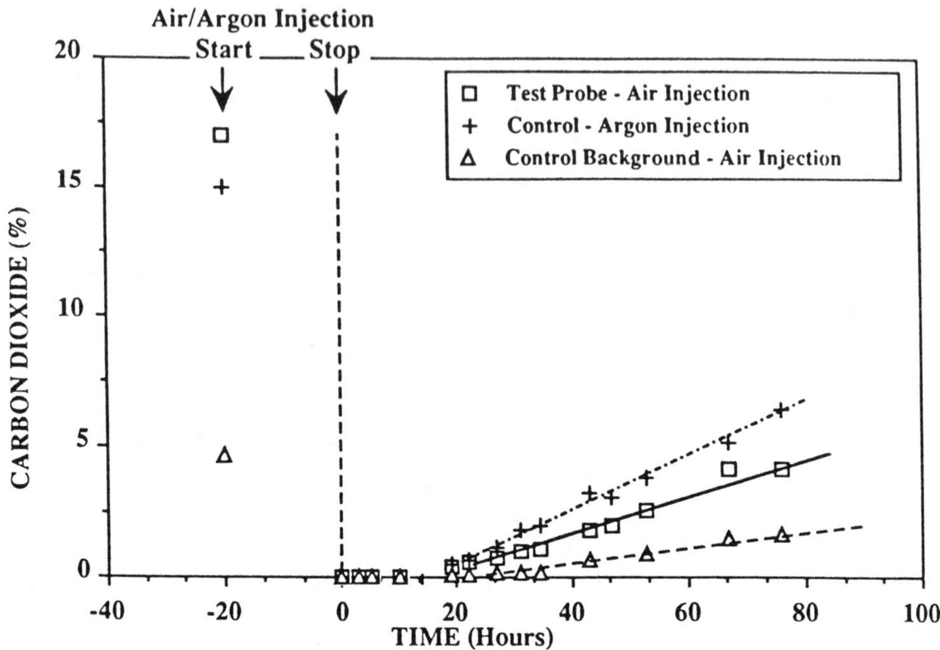

FIGURE 2. Change of carbon dioxide concentrations over time.

from this treatability test with other studies, hexane (C_6H_{14}) was selected as the respresentative hydrocarbon for the oxidation of jet fuel under aerobic conditions. The rate for biodegradation of JP-4 fuel for this site (after correction for background uptake of oxygen) is 2.9 ± 0.2 mg of hexane equivalent per kg of soil per day. The computed biodegradation rate assumes a porosity of 0.35 and a soil bulk density of 1,440 kg/m^3. This rate is of the same magnitude as the biodegradation rate (8 mg/kg/day) found for a diesel-contaminated soil using a mass balance approach (van Eyk & Vreeken 1989). Uptake of oxygen by inorganic sources such as oxidation of iron is assumed to be negligible, as seen by the low available iron (approximately 18 ppm) present in the soil. Also, the long aeration time (about 24 h) would be sufficient to oxidize any ferrous iron in the soil.

Figure 2 shows the change in carbon dioxide concentrations over time for the test and control probes. Biodegradation rate based on carbon dioxide production (after correcting for background concentrations) is 0.15 ± 0.04 mg of hexane equivalent per kg of soil per day. Carbon dioxide production was lower than expected for an equivalent change in oxygen concentration. A possible explanation is that carbon dioxide that evolved under respiration reacts with the carbonates to form bicarbonates in the highly alkaline (pH 9.1) soil at this site. As the uptake of carbon dioxide in the subsurface generally shows more sinks, biodegradation rates for this treatability test should be based on the change of oxygen concentration over time.

Conclusion

In situ biotreatment has been hailed as a potentially viable and economical approach to restoring contaminated sites. However, for this technology to work and to assist in remediation decision making, such as evaluating the biodegradability of the organic contaminants under aerobic conditions, data on biodegradation rates under field conditions are required. This type of data is currently lacking, and determination of these rates based on mass balances, for example, may be expensive. We have presented here a simple, rapid, and low-cost method for estimating the aerobic biodegradation rates of organic contaminants *in situ* without the use of expensive analytical instruments. This test method has been successfully tested at various sites.

REFERENCES

Hinchee, R. E.; Downey, D. C.; Dupont, R. R.; Aggarwal, P.; Miller, R. E. "Enhancing Biodegradation of Petroleum Hydrocarbon through Soil Venting." *J. Haz. Materials*, in press.

Miller, R. E.; Hinchee, R. E. "Enhanced Biodegradation through Soil Venting"; final report, Contract No. F08635-85-C-0122, Engineering and Service Laboratory, Air Force Engineering and Services Center, Tyndall Air Force Base, FL, 1990.

van Eyk, J.; Vreeken, C. "Venting-mediated Removal of Diesel Oil from Subsurface Soil Strata as a Result of Stimulated Evaporation and Enhanced Biodegradation." *Proceedings* of Envirotech, Vol. 2, Federal Environmental Agency, Vienna, 1989.

Subsurface Biogenic Gas Ratios Associated with Hydrocarbon Contamination

Donn L. Marrin[*]
InterPhase

Monitoring the *in situ* bioreclamation of organic chemicals in soil is usually accomplished by collecting samples from selected points during the remediation process. This technique requires the installation and sampling of soil borings and does not allow for continuous monitoring. The analysis of soil vapor overlying hydrocarbon-contaminated soil and groundwater has been used to detect the presence of nonaqueous phase liquids (NAPL) and to locate low-volatility hydrocarbons that are not directly detected by more conventional soil gas methods (Marrin 1989; Marrin & Kerfoot 1988). Such soil vapor sampling methods are adaptable to monitoring the *in situ* bioremediation of soil and groundwater contamination.

This paper focuses on the use of biogenic gas ratios in detecting the presence of crude oil and gasoline in the subsurface. Present soil gas surveying techniques are primarily applied to low boiling point petroleum distillates (*e.g.*, gasoline, stoddard solvent, JP-4), even though surveying for products with higher boiling points (*e.g.*, diesel and fuel oils) has been reported by Tillman *et al.* (1989). The use of biogenic gas analyses in locating nonvolatile, as well as volatile, hydrocarbons and monitoring their biodegradation could be of considerable value because these petroleum compounds are difficult to locate and are commonly remediated by *in situ* bioventing and other bioreclamation techniques (Marrin 1988; Molnaa & Grubbs 1989; Thomas & Ward 1989).

[*] InterPhase, 11558 Sorrento Balley Road, San Diego CA 92121

Site Description

Soil and groundwater contamination is present underlying two adjacent sites at the same bulk storage facility, which has been used for the transfer of both crude oil and refined petroleum products. Vadose zone soils underlying the storage facility consist of silty to fine sands from the ground surface to a depth of 3 m below ground surface (bgs), at which point the soils become more coarse-grained down to the depth of the water table (approximately 5.2 m bgs). Regional hydrogeological data suggests that the water table fluctuates seasonally and, at the time of the investigation, may have been at its lowest level in several years. A representative geological cross-section of the bulk storage facility is presented in Figure 1.

The ground surface at both sites is paved with a combination of asphalt and concrete; however, there are localized areas of bare soil and landscaping that could contribute minimally to on-site groundwater recharge. The lateral extent of subsurface contamination for both the crude oil and refined products (principally gasoline) was not delineated prior to the soil gas survey because conventional investigation procedures were hindered by limited site access. Primary objectives of the soil vapor survey included identifying sources and estimating the lateral extent of NAPL at the water table. Source identification was difficult because unidentified leaks from the extensive above-ground and underground piping systems have probably contributed substantially to the hydrocarbon contamination.

The vertical extent of gasoline contamination appears to be greatest in soils located just above the water table, although shallow soil contamination is present at locations where surface or near-surface leaks have been identified. Crude oil contamination is apparently more uniformly distributed throughout the vadose zone; however, black oily liquids were encountered in the capillary fringe zone. Maximum detected concentrations for total petroleum hydrocarbons (TPH) in soil, according to EPA Method 418.1, were approximately 11,000 mg/kg for gasoline and 87,000 mg/kg for crude oil; however, it is not known whether either of these samples were collected from the most contaminated areas. While no data are currently available regarding the physical properties of the soils (*e.g.*, porosity, water content, organic carbon content), observations recorded during the initial soil investigation suggest that moisture content generally increased with depth.

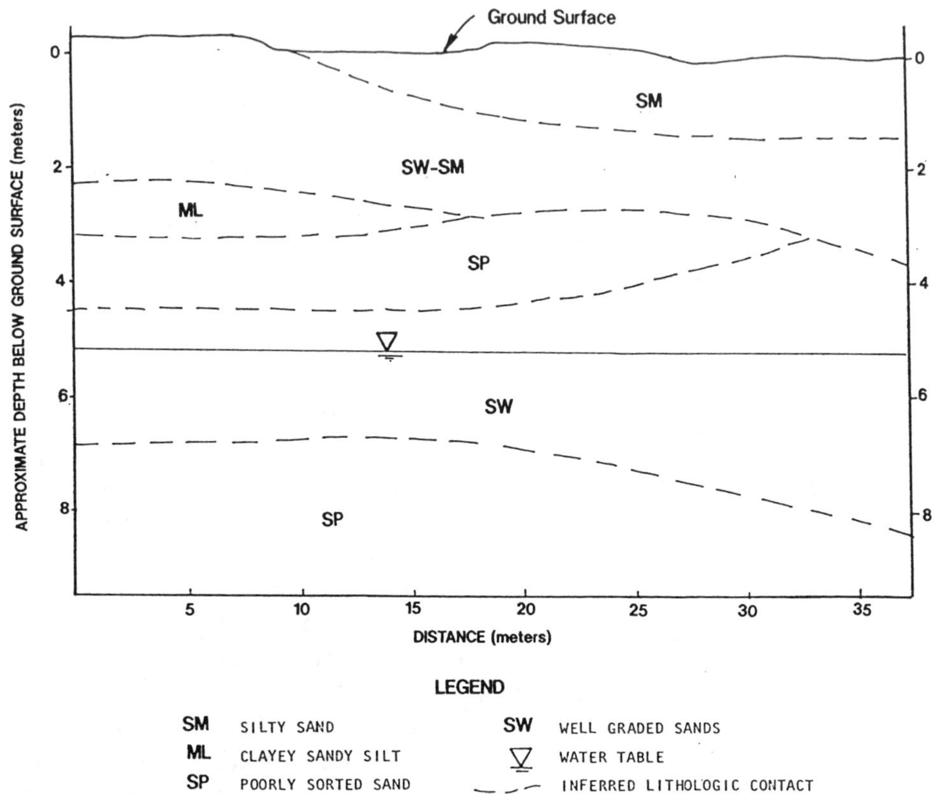

FIGURE 1. Geologic cross-sections for the gasoline and crude oil sites.

Sampling and Analysis Methods

Approximately 100 soil gas samples were collected at the two sites from depths of 1.5 to 4.5 m bgs. The target sampling interval at the facility was 3 to 4 m deep; however, soil gas was collected from multiple depths at about 10 percent of the locations in order to investigate the vertical distribution of fixed and biogenic gas concentrations. Sampling points were located (a) near confirmed and suspected leaks, (b) in background or uncontaminated areas of the facility, and (c) adjacent to piping or tanks where no contaminant information was available.

A predetermined volume of soil gas was pumped through steel sampling probes, which were driven to the sampling depth using a vehicle-mounted hydraulic hammer. Vapor samples were withdrawn using two-stage evacuation system that permitted the measurement of air flow rates over a specified vacuum drop. This sampling procedure was used to estimate the resistance to flow (a relative indicator of gas permeability) at each of the sampling locations and depths. Vapor samples were collected with a gas-tight syringe as soon as the evacuation system had returned to atmospheric pressure.

Gas samples were analyzed on site by gas chromatography using either a Varian model 3400 or a Hewlett-Packard model 5890 instrument, which was equipped with flame ionization (GC/FID) and thermal conductivity (GC/TCD) detectors. Separation of analytes was achieved by using chromatographic columns with a variety of stationary phases including carbowax, molecular seive, and chromosorb. Identification and quantification of the analytes were performed using external standards which were commercially-prepared in the appropriate carrier gas (i.e., nitrogen or helium). All samples were analyzed for oxygen (O_2), methane (CH_4), carbon dioxide (CO_2), and volatile fuel hydrocarbons (VFH) in the range of C_4 through C_{12}. As a result of their physiochemical properties (i.e., vapor pressure and octanol-water partition coefficient), C_{12}+ hydrocarbons present under normal soil conditions do not partition into the vapor phase at concentrations that are detectable in soil gas (Marrin 1988).

Results and Discussion

Because the vertical and lateral distribution of petroleum hydrocarbons were incompletely delineated by existing soil borings and groundwater monitoring wells, the fixed/biogenic gas data can only be compared to subsurface contamination on a qualitative basis. The data generated during these soil gas surveys are currently being used to select locations for confirmatory wells and borings, which should more accurately define the areal extent of contamination.

Vertical Distribution of Biogenic Gases. Depth profiles for methane, carbon dioxide, and oxygen are given on Figures 2 through 4. Figure 2 indicates that background concentrations these fixed and biogenic gases varied slightly within the 2- to 4.5-m depth interval, but were in the range of concentrations which have been described for uncontaminated soil conditions (Sposito 1989). Background conditions at this facility were characterized by oxygen levels greater than 15 percent, carbon dioxide levels less than 5 percent, and methane concentrations below the 1 ppmv detection limit. Slight variations in CO_2 and O_2 concentrations in background samples probably reflect localized variations in natural organic carbon or moisture content of the soils.

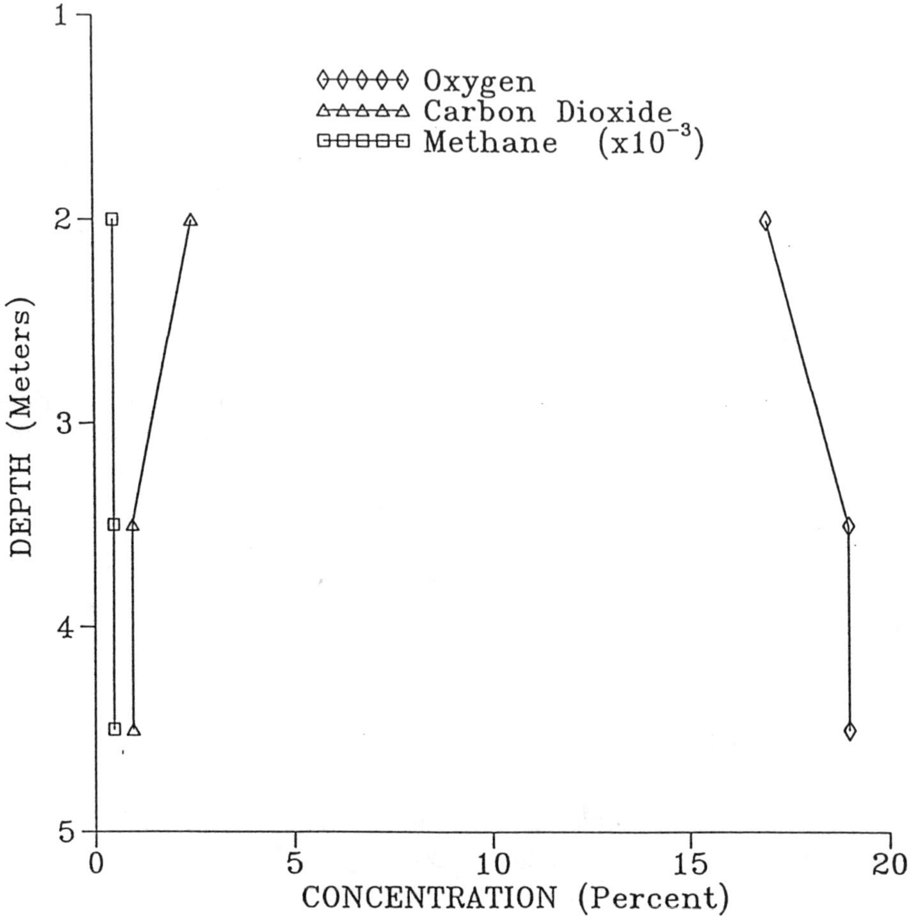

FIGURE 2. Depth versus concentration for fixed/biogenic gases at a background location.

Figure 3 indicates relatively high concentrations of methane and carbon dioxide in soil gas at the gasoline site, with methane reaching 40,000 ppmv at a depth of about 0.7 m above free product on the water table. Although the changes in oxygen and carbon dioxide levels with depth at this location are not as pronounced as those for methane, the highest CO_2 and lowest O_2 concentrations were analyzed at the 4.5 m depth. Oxygen levels less than 5 percent, carbon dioxide levels greater than 15 percent, and methane concentrations exceeding 10,000 ppmv characterized soil gas samples which were collected above free product layers on the groundwater. Maximum concentrations of methane were consistently encountered at sampling depths closest to the water table;

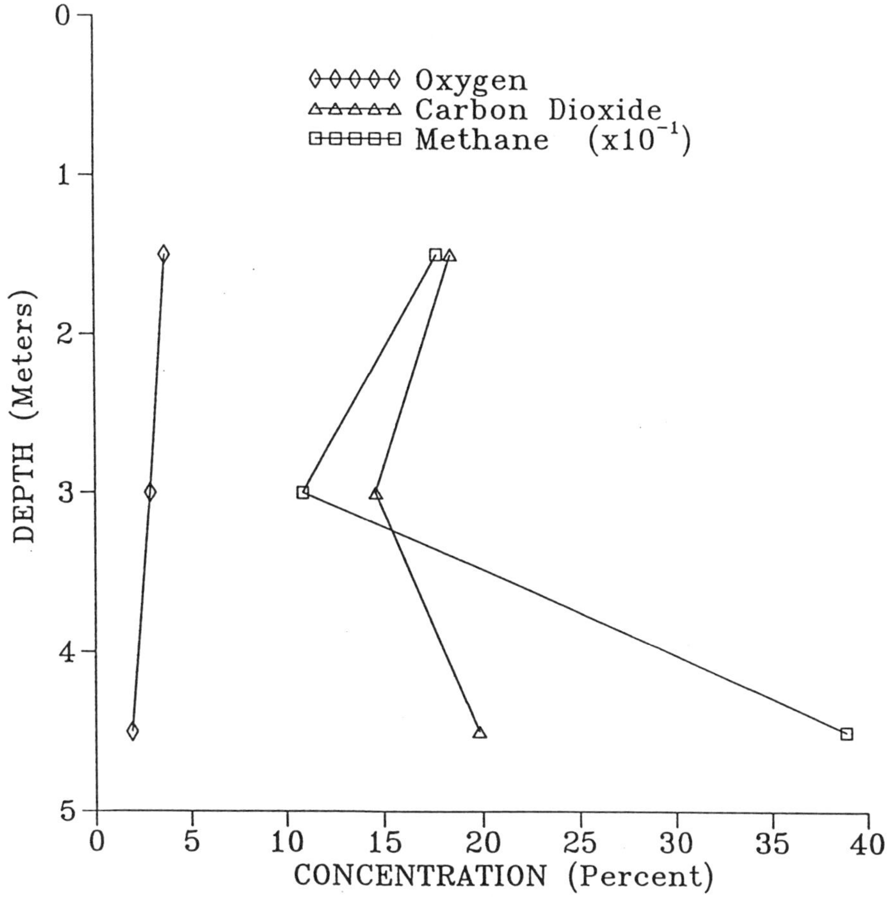

FIGURE 3. Depth versus concentration for fixed/biogenic gases at the gasoline site.

however, CO_2/O_2 ratios greater than one were present at all depths sampled above free product.

Figure 4 indicates a sharp increase in methane concentration (*e.g.*, <5 to 5,200 ppmv) from the 2.5- to 4.9-m depth in soils adjacent to a known crude oil leak. Although carbon dioxide showed an increase to 10 percent at the 4.9-m depth, oxygen concentrations dropped only as low as 13 percent. Oxygen concentrations greater than 3 percent have rarely been observed by the author at locations where methane is detected in soil gas, presumably because methanotrophic bacteria can readily oxidize CH_4 in the presence of oxygen (Wolin & Miller 1987). The presence of methane under seemingly aerobic conditions at the crude oil site suggests that CH_4 in soil gas may not be solely related to

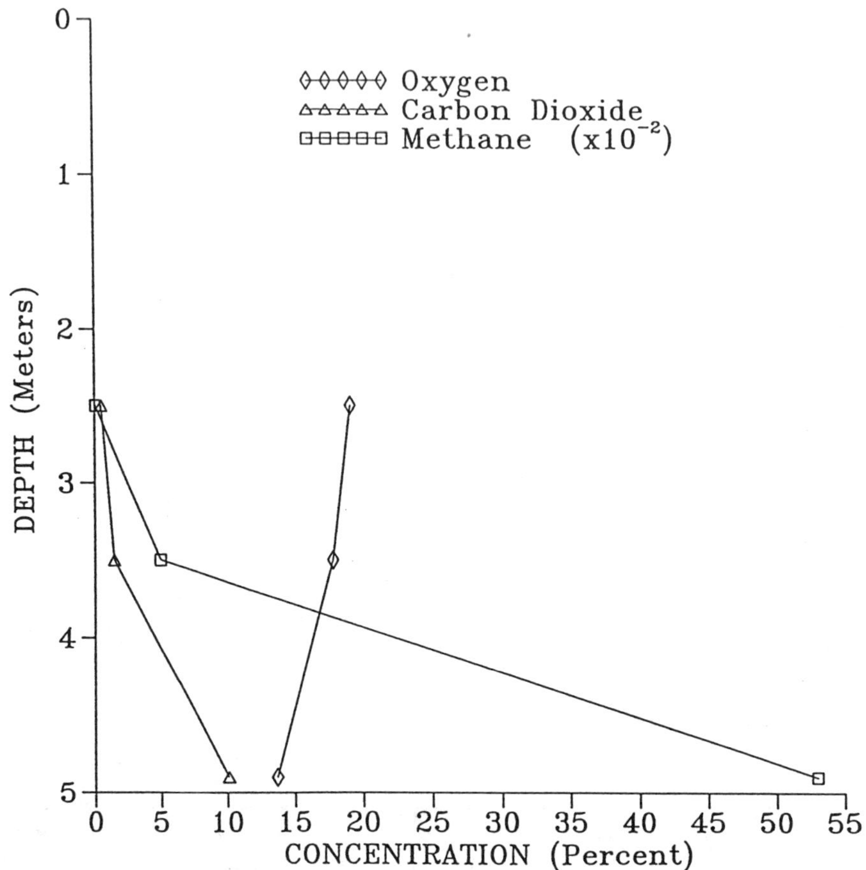

FIGURE 4. Depth versus concentration for fixed/biogenic gases at the crude oil site.

the anaerobic degradation of crude oil hydrocarbons. Although no alternative sources of methane (*e.g.*, underground sewer lines, natural gas reservoirs or pipelines) were anticipated at this site, the possibility of CH_4 introduction to the subsurface cannot be eliminated.

The relationship between biogenic gas concentration and depth is affected by a number of localized soil conditions such as moisture content, air permeability, depth and extent of hydrocarbon contamination, and degradation rates for naturally-occurring organic matter. Suchomel *et al.* (1990) identified plant root respiration, atmospheric venting of the upper soil, calcite dissolution, methane oxidation, and wetting fronts as some of the factors which influenced the vertical distribution of carbon dioxide in the vadose zone. In addition, their data suggested that the concentrations and vertical distribution of major fixed and biogenic gases were subject to minor changes on a seasonal basis.

The relationship between depth, gas flow rate, and CO_2/O_2 concentration ratios at the bulk storage facility is shown on Table 1. These data represent a composite of contaminated, background, and uncharacterized locations at both the gasoline and crude oil sites. Mean values for gas flow rate (used as a relative indicator of air permeability) suggest that shallow soils may be the most permeable; however, the very large standard deviation indicates that depth is only one variable which influences gas flow rates. CO_2/O_2 ratios also show a large standard deviation for each depth interval, suggesting that factors other than depth and permeability (*e.g.*, proximity to contamination, moisture conditions, soil structure, redox conditions) may have a substantial influence on the ratios. Observed flow rates varied from 0.3 to 40 L/min, whereas CO_2/O_2 concentration ratios ranged from <0.01 to 11. The large standard deviations calculated for both parameters are due to observed values varying by as much as two orders-of-magnitude within each depth interval.

TABLE 1. Mean flow rates and CO_2/O_2 concentration ratios within three depth intervals at the bulk storage facility.

Depth (m)	Gas Flow Rate (L/min)	CO_2/O_2 Ratio (%/%)
≤2.0 [n=12][(a)]	14 [±11][(b)]	2.2 [±3.9]
2.0–3.5 [n=34]	4.2 [±3.1]	4.1 [±2.7]
3.5–5.0 [n=48]	8.8 [±12]	3.8 [±3.9]

(a) Number of soil gas samples representing each depth interval.
(b) One standard deviation from the mean values.

CO_2/O_2 Ratios. Figure 5 shows the relationship between oxygen and carbon dioxide concentrations in soil gas obtained from the 3- to 4-m depth interval at both sites. Oxygen concentrations ranged from 1.5 to 20 percent, the higher value reported to be indicative of well-aerated surface soils (Sposito 1989). Carbon dioxide was present in soil gas at concentrations of <0.5 to 33 percent. Carbon dioxide levels in soil gas overlying uncontaminated soils have been reported to range

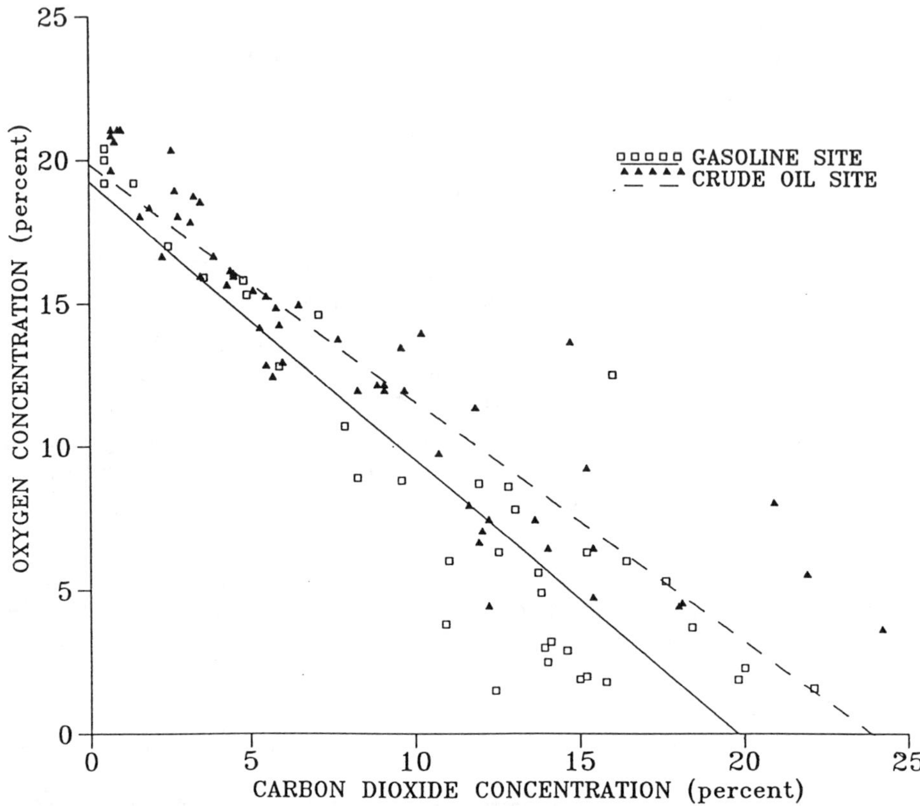

FIGURE 5. Carbon dioxide versus oxygen concentrations in soil gas.

from 0.3 to 3 percent (Sposito 1989); while CO_2 concentrations in soil gas above highly-contaminated ground water have been reported to be as high as 26 percent (Bishop *et al.* 1966).

Two general trends are evident from the CO_2 versus O_2 Plot shown on Figure 5. First, there is an inverse relationship between carbon dioxide and oxygen levels in soil gas, with an increase in scatter of the data points representing higher CO_2/O_2 ratios. Sampling locations represented in Figure 5 range from background or uncontaminated areas (i.e., where CO_2/O_2 ratios are less than one) to locations of known soil and/or groundwater contamination, where carbon dioxide exceed oxygen concentrations. Secondly, the inverse relationship between CO_2 and O_2 concentrations in soil gas is very similar between the gasoline and crude oil sites. According to mathematical analyses, the two populations are not statistically different and both linear regression curves have a correlation coefficient (r) greater than 0.9.

Oxygen is utilized as an electron acceptor during the aerobic oxidation of organic contaminants and is also consumed by microorganisms metabolizing organic substrates via more reducing pathways such as denitrification (Bouwer & Cobb 1987). Carbon dioxide is produced during the breakdown of petroleum hydrocarbons via the same chemical pathways specified for the utilization of oxygen. Oxygen concentrations in soil gas below 10 percent were observed only at locations where carbon dioxide concentrations exceeded about 8 percent. This inverse relationship between carbon dioxide and oxygen levels in soil gas has been noted by several authors (Kerfoot *et al.* 1988; Marrin 1989; Robbins *et al.* 1990) for petroleum-contaminated soils.

Carbon dioxide in soil gas is most often present as a result of the biodegradation of organic substrates (*e.g.*, contaminants and/or natural material such as humic and fulvic acids) or a shift in the inorganic carbonate system resulting from the dissolution of carbonate rocks (McMahon 1990). Background concentrations Of CO_2 and O_2 in soil gas within the target sampling interval at both sites were less than 5 percent and greater than 15 percent, respectively. Hence, it is unlikely that the high CO_2/O_2 ratios present in soil gas at this site were related to locally high levels of natural organic matter or to the weathering of soils. Despite the fact that carbon dioxide has a maximum aqueous solubility which is approximately 1,000-fold greater than that of oxygen, the inverse relationship between the two gases was observed over a range of soil conditions and contamination levels. This observation suggests an equilibrium between CO_2 in soil gas and soil moisture may exist; otherwise, vadose zone water would act as a major sink for carbon dioxide produced during hydrocarbon degradation.

A comparison of volatile fuel hydrocarbons (VFH) and CO_2/O_2 ratios in soil gas samples collected from the gasoline site are plotted on Figure 6. This relationship was investigated because current research has indicated that the partitioning of vapor-phase hydrocarbons into soil moisture can result in their aerobic biodegradation (Ostendorf & Kampbell 1991). VFH concentrations exceeded 10,000 ppmv at all sampling locations where CO_2/O_2 ratios were greater than 4; however, there was no statistical correlation between the two variables even within this subset of the data. These results suggest that while vapor-phase hydrocarbons may affect CO_2/O_2 ratios at very high concentrations (i.e., greater than 10,000 to 100,000 ppmv), these ratios are apparently not correlated with lower VFH concentrations in soil gas. The effects of vapor-phase VFHs on CO_2/O_2 ratios in soil gas may have been masked by the presence of sorbed and free phase hydrocarbons in soil and groundwater.

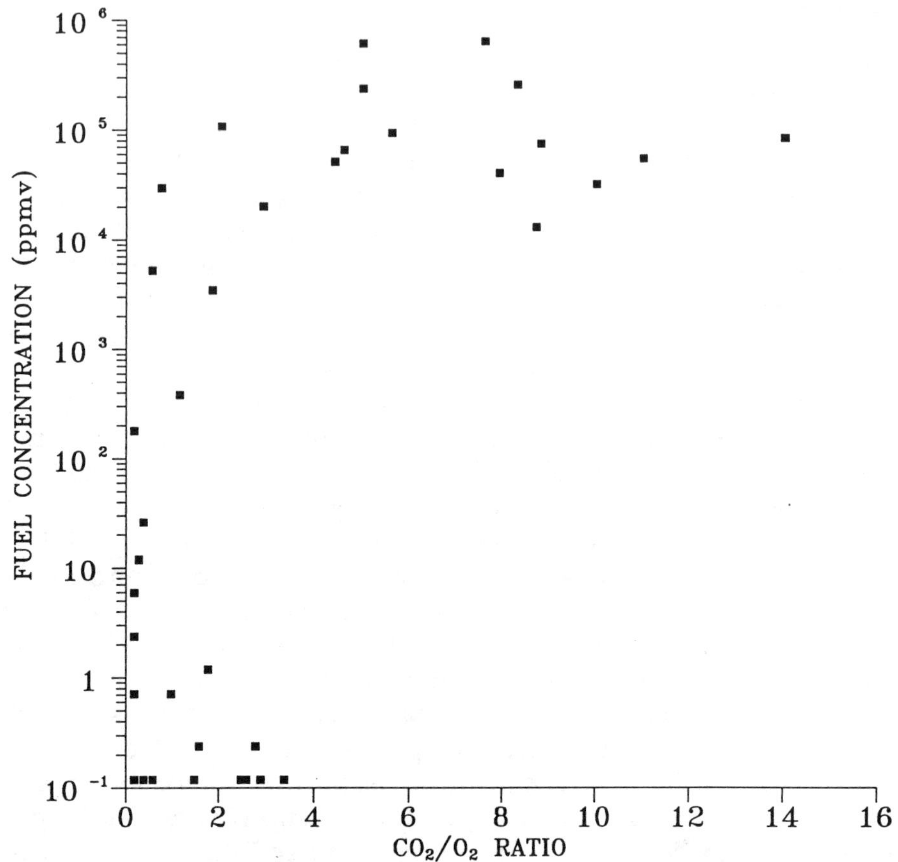

FIGURE 6. Volatile fuel hydrocarbons versus CO_2/O_2 ratios at the gasoline site.

CO_2/CH_4 Ratios. Concentrations of methane and carbon dioxide in soil gas are plotted for sample points within the target depth interval on Figures 7 and 8. Figure 7 indicates that methane concentrations at the gasoline site were above the 1 ppmv detection limit only at locations where carbon dioxide concentrations exceeded 10 percent. Methane concentrations ranged over four orders-of-magnitude and were highest adjacent to known leaks and a monitoring well which contained free gasoline product. The presence of methane in soil gas is probably related to highly reducing conditions in the subsurface, which may be created by free product on the water table. More oxidizing conditions generally result in biodegradation processes which release carbon dioxide, rather than methane, as the predominant biogenic gas.

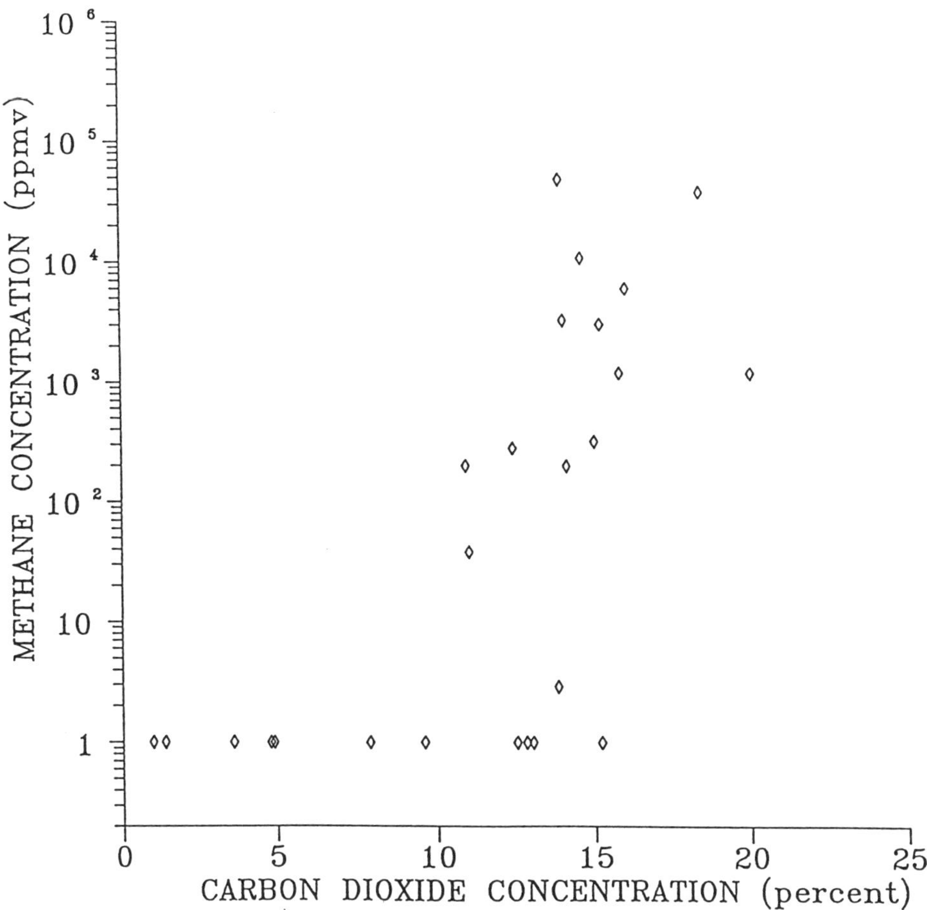

FIGURE 7. Methane versus carbon dioxide concentrations at the gasoline site.

There are two groups of data points which may be identified on Figure 7. The first group represent those samples which contain carbon dioxide at concentrations as high as 16 percent, but which contain no detectable methane. The second group consists of samples that contain methane, but only when corresponding CO_2 concentrations exceed 10 percent. Although the relationship between soil/groundwater contamination and biogenic gas levels could only be compared on a qualitative basis, 9 of the 12 samples which were collected from points adjacent to known locations of free product or soil contamination contained detectable methane. Although CO_2 acts as the electron acceptor in methanogenesis, both carbon dioxide and methane are produced under

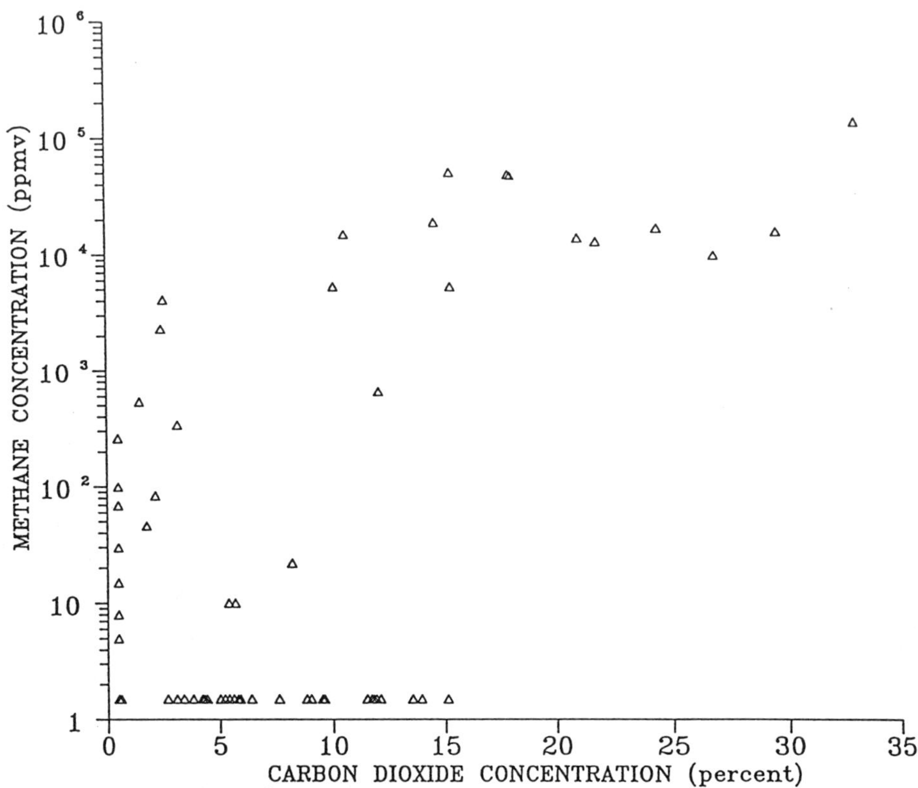

FIGURE 8. Methane versus carbon dioxide concentrations at the crude oil site.

anaerobic (reducing) conditions. Futhermore, O_2 concentrations, often associated with reducing conditions, correspond to high CO_2 levels in soil gas at this site.

As indicated on Figure 8, peak methane concentrations at the crude oil site exceeded 10 percent (100,000 ppmv). Similar to observations for the gasoline site, there was one group of samples that contained no detectable methane (even though CO_2 levels were as high as 15 percent), and another group of samples that contained methane only when corresponding CO_2 concentrations exceeded 10 percent. Unlike the data from Figure 7, there appears to be a third group of points on Figure 8 which indicate detectable concentrations of methane associated with carbon dioxide levels in the 0.5 to 10 percent range. This latter grouping of points suggest that methane is present even in the absence of elevated CO_2 concentrations. Differences in CO_2/CH_4 ratios at the two sites may be related to the vertical distribution of contamination or to the carbon range of compounds present in gasoline and crude oil. For example, the presence of methane in soil gas samples which have high CO_2 levels may indicate that a combination of methanogenic and nonmethanogenic processes are occurring in the same vicinity (*e.g.*, CH_4 production at the water table and CO_2 production in the vadose zone).

High concentrations of methane in soil gas are generally due to methanogenic bacteria, which are capable of metabolizing various organic substrates, as well as hydrogen and carbon dioxide gases (Wolin & Miller 1987). The presence of methane in soil gas at concentrations as high as 3,000 ppmv in soil gas at locations where carbon dioxide levels are less than 5 percent (and corresponding oxygen concentrations greater than 10 percent) could be a result of several processes. As discussed previously, methane could be introduced from a natural gas line, subsurface petroleum deposit, or other source not related to the degradation of hydrocarbon contaminants. Secondly, methane could persist if it were not chemically or biochemically oxidized (*e.g.*, via methanotrophic bacteria) in the vadose zone as rapidly as it were being produced by the methanogens. Finally, methane may appear persistent in the presence of relatively high oxygen and low CO_2 if atmospheric air were being introduced to the sample during collection or analysis. This air dilution hypothesis was tested and found not to be the cause of the observed biogenic gas ratios in soil vapor samples.

Summary

The analysis of fixed and biogenic gases in soil vapor samples collected from a petroleum-contaminated facility indicated that a correlation between these gases and general subsurface conditions may exist. Carbon dioxide and oxygen concentrations displayed a statistically significant inverse relationship and appeared to be a good indicator of background conditions at both the crude oil and gasoline sites. Although a quantitative comparison of soil gas and soil/groundwater data was not possible, all background (uncontaminated) locations were characterized by CO_2/O_2 ratios less than one. Carbon dioxide to methane ratios were lowest in soil vapor samples overlying free gasoline product and may have been indicative of highly reducing conditions caused by the degradation of hydrocarbons at the groundwater table, where O_2 replenishment is limited. At the crude oil site, methane was encountered at high concentrations in soil gas samples which contained background CO_2/O_2 ratios, perhaps indicating an unknown source of CH_4.

The vertical distribution of oxygen and carbon dioxide was investigated at several locations within the 1.5- to 4.8-m depth interval. Although the highest CO_2 and CH_4 concentrations and lowest O_2 levels were encountered in the deepest samples (closest to the contaminated capillary fringe zone), the shallower samples also contained fixed and biogenic gas levels that were indicative of hydrocarbon contamination. Although depth and air permeability may have influenced the CO_2/O_2 ratios at this site, statistical analyses suggest that they were not the major factors influencing these ratios. It appears that ratios of fixed and biogenic gases in the subsurface may be an indirect indicator of hydrocarbon degradation; however, considerable study is still required to quantify the effects of site-specific parameters (*e.g.*, soil moisture, contaminant type, proximity to source, redox potential) on the relative concentrations of these gases.

REFERENCES

Bishop, W. D.; Carter, R.; Ludwig, H. In *Proceedings,* Third International Conference on Water Pollution Research; Water Poll. Contr. Fed.; Washington, DC, 1966; 19 pp.

Bouwer, E. J.; Cobb, G. D. *Water Sci. Technol.* **1987,** *19,* 769–779.

Kerfoot, H. B.; Meyer, C. L.; Durgin, P. B.; D'Lugosz, J. J. *Ground Water Monitor. Rev.* **1988,** *8,* 67–71.

Marrin, D. L. *Ground Water Monitor. Rev.* **1988,** *8,* 51–54.

Marrin, D. L. In *Proceedings,* Conference on Petroleum Hydrocarbons and Organic Chemicals in Ground Water; Nat. Water Well Asso.: Dublin, OH, 1989; pp 357–367.

Marrin, D. L.; Kerfoot, H. B. *Environ. Sci. & Technol.* **1988,** *22,* 740–745.

McMahon, P. B.; Williams, D.; Morris, J. *Ground Water* **1990,** *28,* 693–702.

Molnaa, B. A.; Grubbs, R. In *Petroleum Contaminated Soils,* Calabrese, E.; Kostecki, P., Eds.; Lewis Publ.: Chelsea, MI, 1989; Vol. 11, Chapter 19.

Ostendorf, D. W.; Kampbell, D. *Water Resourc. Res.* **1991,** in press.

Robbins, G. A.; Deyo. B.; Temple, M.; Stuart, J.; Lacy, M. *Ground Water Monitoring Review* **1990,** *10,* 110–117.

Sposito, G. *The Chemistry of Soils;* Oxford Press: New York, 1989.

Suchomel, K.; Kreamer, D.; Long, A. *Environ. Sci. & Tech.* **1990,** *24,* 1824–1831.

Thomas, J. M.; Ward, C. *Environ. Sci. & Tech.* **1989,** *23,* 760–766.

Tillman, N.; Ranlet, K.; Meyer, T. *Pollution Engr.* **1989,** *7,* 86–89.

Wolin, M. J.; Miller, T. L. *Geomicrobiol. Jour.* **1987,** *5,* 239–259.

In Situ Bioremediation of JP-5 Jet Fuel

M. P. Eisman[*], *E. Dorwin*
Computer Sciences Corporation
D. Barnes, *B. Nelson*
Naval Civil Engineering Laboratory

Fuel leaks and spills of the jet fuel JP-5 at various Naval installations are required by law to be remediated. Use of microorganisms for fuel spill remediation is the focus of this investigation, which examines biodegradation of JP-5 by means of CO_2 evolution in batch cultures. In particular, the aerobic biodegradation of fresh and weathered JP-5, along with a representative fuel mix of three pure compounds, was examined. Since microorganisms exist in aqueous environments, the solubility in water of fuels and fuel components was also examined. These laboratory studies oversimplify the field situation, where solubility of fuel in water is affected by many factors, including humic content, amount of water, adsorption onto clay and silica colloids, the selective partitioning of the more hydrophilic components out of the fuel, different solubilities of fuel mixes compared with pure compounds, and even the presense of the microorganisms themselves.

Other chemical properties of the complex mixture of hydrocarbons in JP-5 may affect bioavailability, such as polarizability (ability to separate positive and negative charge) of fuel molecules, or carbon-hydrogen bond strength in the enzyme-catalyzed insertion of oxygen molecules into hydrocarbon molecules, or the type of hydrocarbon present, or volatility. This investigation will attempt to relate biodegradation to these properties, particularly water solubility and type of hydrocarbon. Water solubility, as mentioned, may relate to the ability of the microorganism to uptake the fuel or fuel component.

[*] Naval Civil Engineering Laboratory, Code L71, Port Hueneme CA 93043

Materials and Methods

Fresh JP-5 fuel was fractionated into rough boiling point fractions using a vacuum, high-speed spinning band column (Nester/Faust Manufacturing Corp., Newark DE) at a pressure of 10 mm Hg using the method developed by Nester (1956). The sample (132 g of JP-5 from a nearby Naval fuel tank farm) was maintained at a constant reflux ratio of 2:1. After 30 min, a steady rate of condensate was observed at the head of the column, and samples were collected (Table 1). The relationship between vapor pressure of water solubility has been discussed by McAuliffe (1965). Measuring the water solubility of these fractions, separated into rough boiling point distributions according to Raoult's Law, will aid this investigation.

TABLE 1. Results of vacuum distillation of JP-5 jet fuel.

Fraction	Distillation Pot Temp., C	Temp. at Top of Column, C	Weight Collected, g
1	124	41	5.1
2	145	61	9.0
3	144	95	8.5
4	141	85	11.4
5	143	96	14.4
6	165	119	18.0
7	175	125	16.2
		Total wt	82.6
			(62% of total)

The water solubility of pure hydrocarbon samples was also determined. The aqueous phase, which had been vigorously mixed with hydrocarbon and allowed to equilibrate for 5 days, was extracted with hexane followed by GC analysis of the hexane layer. This method was based upon a published procedure (Klein & Jenkins 1981). Pure hydrocarbons, decane, dodecane, and para-cymene (Aldrich Chemical Company, Milwaukee WI 99+ percent grade) were analyzed. The GC detector (FID or MS) was calibrated in the concentration range of hydrocarbon solubility in water in order to make a quantitative measurement of solubility.

The JP 5 boiling point fractions, diluted 1:500 with hexane prior to on-column injection, were analyzed using a Hewlett-Packard Model 5890 gas chromatograph (Hewlett Packard Co., Avondale PA), a DB-5 column (J & W Scientific, Folsom CA), flame ionization detection (FID), and hydrogen as the carrier gas at 31 cm/sec. After injection at 30 C, the oven temperature was held at 30 C for 4 min, ramped to 100 C at a rate of 2.5 C/min, then to 300 C at 10 C/min. Data were analyzed using HP 5895A ChemStation software (Hewlett Packard Co., Avondale PA).

Gas chromatography with mass spectrometry (GC/MS) analysis on the boiling point fractions used a 1/50 split ratio for injection (Varian Model 1075 split/splitless injector) into a Varian 3400 GC interfaced to an Extrel 400 quadrupole mass spectrometer. The oven program was identical to the one described above, with the injector at 200 C and GC/MS interface at 250 C. The carrier gas was helium at a velocity of 45 cm/sec.

A retention time library was built up that contained 41 standard compounds from the four classes of hydrocarbons present in JP-5. Using Extrel 2000 software Version 8.1 (Extrel Corporation, Pittsburgh PA), it was possible to do on-line computer searching of the GC peaks observed in the JP-5 sample analysis. Quantitation was carried out by spiking the JP-5 fractions with deuterated internal standards, making reference to calibration curves prepared with the Extrel quantitation programs. Data analysis used the Extrel 2000 software run on a Digital Equipment Corporation PDP 11/73 microcomputer (Extrel Corporation, Pittsburg PA).

Soil from a JP-5 contaminated site at a Naval Air Station had been excavated for a laboratory study. Approximately 5 kg of this soil was obtained and stored at 4 C. The original soil samples had been stored in doubled plastic bags and stacked in a fume hood for several months. Hydrocarbon permeation or adsorption to the bags was not examined, as the soil was of interest for the microorganisms able to degrade JP-5. For biodegradation experiments, soil from the 5 kg sample was weighed into 1-g portions, each of which was placed in a sterile 60-mL borosilicate glass vial with a sterile Teflon® insert and an aluminum crimp top. These 1-g portions served as inocula of both weathered JP-5 and adapted microorganisms.

During a biodegradation experiment, at least two replicates of these microcosms were made for each treatment for each time point. Treatments consisted of media alone, media with fresh JP-5, media with soil, media with soil and fresh JP-5, and media with soil and a representative fuel mix of dodecane, para-cymene and tert-butyl cyclohexane.

The JP-5 was stored in glass at 4 C. When required, the fresh JP-5 was sterilized by filtering through sterilized Gelman filters (Acrodisc CR PTFE 0.2 µm) using sterile syringes.

HCMM2 media was a minimal salts media consisting of minerals, salts and trace minerals as shown in Table 2 (Ridgway *et al.* 1988).

TABLE 2. Composition of HCMM2 mineral salts medium.[(a)]

Component	Amount	Element	Molarity in Final Dilution
Phosphate buffer (50 x)[(b)]			
KH_2PO_4	1.36 g/L	–	10 mM
Na_2HPO_4	1.42 g/L	–	10 mM
Major ions (100 x)			
KNO_3	0.50 g/L	–	4.99 mM
$(NH_4)_2SO_4$	2.38 g/L	–	18 mM
$MgSO_4 \cdot 7H_2O$	0.05 g/L	–	200 µM
$CaCl_2$	0.01 g/L	–	68 µM
Trace elements (1,000 x)[(c)]			
H_3PO_4	2.862 mg/L	B	46.3 µM
$MnSO_4 \cdot H_2O$	1.538 mg/L	Mn	9.1 µM
$Fe(NH_4)_2(SO_4)_2 \cdot 6H_2O$	3.529 mg/L	Fe	9.0 µM
$CuSO_4 \cdot 5H_2O$	0.0392 mg/L	Cu	0.157 µM
$ZnCl_2$	0.0209 mg/L	Zn	0.153 µM
$CoCl_2 \cdot 6H_2O$	0.0405 mg/L	Co	0.170 µM
$Na_2MoO_4 \cdot 2H_2O$	0.0252 mg/L	Mo	0.104 µM

(a) For petri plates or slants, medium is solidified with 0.8 percent w/v purified agar (Difco). Nonpurified agar gives false positives: hydrocarbon degraders are able to utilize something in the nonpurified agar as a carbon source.

(b) Buffer is pH adjusted to 7.2 with 10 M NaOH or NaOH pellets, as it will be acidic.

(c) Trace elements were stored in an amber bottle at 4 C to help slow precipitation, and were filter sterilized since iron precipitates during autoclaving.

The CO_2 analysis, which determines complete mineralization, was done by liberating CO_2 inside a small reaction chamber. Figure 1 shows a diagram of the custom-built system modelled after that developed by Ridgway *et al.* (1988). The gas concentration was then recorded on a computerized nondispersive infrared CO_2 gas analyzer (CNDIR; Automated Custom Systems, Inc., Orange CA, Model 3300) with 125 mm cell. When the CO_2 reading was at maximum, the data were transferred through an analog-to-digital converter (parts from Dow Electronics, Oxnard CA and American Analog Co., Inc., Costa Mesa CA) and then to a computer.

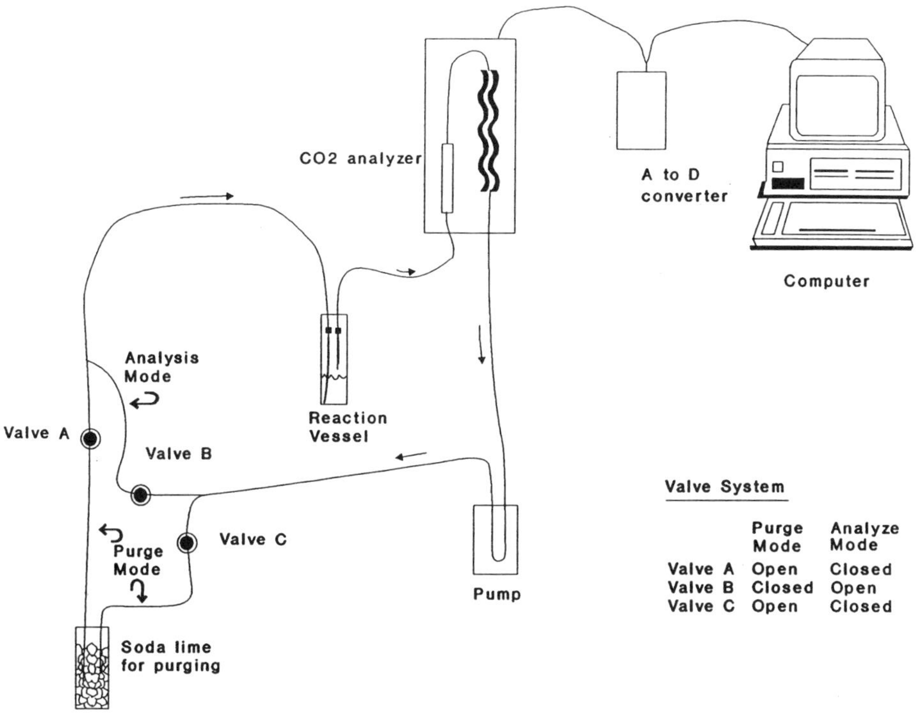

FIGURE 1. Carbon dioxide analysis layout.

The apparatus was configured with Teflon tubing and Teflon valves such that two modes were possible: one for purging the tubing of atmospheric CO_2 by moving the air through a vial filled with soda lime, and the other mode for analyzing. Gasses were moved through the tubing with a small Pipette-Aid pump adapted for this use.

Each microcosm vial was sacrificed by injecting 2.5 N NaOH into the vial to bring the liquid to a final concentration of 1 percent

NaOH. The vial was then returned to the shaker for at least 10 min, stopping the biodegradation reaction and causing gaseous CO_2 to become carbonate and enter the liquid phase.

Standards were made fresh each time an experimental run was done. Helium-purged, double-deionized water was used for standards. Kill controls were also done to determine what, if any, CO_2 was evolved from the soil. Kill controls had 5 percent $HgCl_2$ added to a final concentration of 0.1 percent, or Amphyl disinfectant (National Laboratories, Montvale NJ) at 1 percent.

Results and Discussion

It was assumed that the JP-5 in the soil samples was weathered, but to determine the extent of weathering, an extract of the soil was made by washing 5 g of soil with two 15-mL aliquots of hexane. GC/FID analysis of the extract showed a distribution of JP-5 components similar to those contained in the residue left after vacuum distillation of JP-5 (Figure 2). The trace from the vacuum distillation residue had distinct peaks; these are alkanes C13 through C18, confirmed by retention index standards. The trace from the hexane extraction of soil elutes at the same general retention time. Note that this trace has no distinct alkane peaks, which may result from interaction with the column of polar paraffin oxidation products, or the sample may have been too concentrated. There is some indication that the hump under the alkane peaks in the upper trace from the distillation residue may represent additives to the fuel such as corrosion and icing inhibitors.

Thus far, solubility in water has been measured on pure compounds. Using the extraction method described above, the solubility of decane has been measured at 0.057 mg/L and dodecane at 0.0024 mg/L. Both of these values are in good agreement with those reported by Yaws *et al.* (1990). The solubility of the substituted benzene, para-cymene $C_{10}H_{14}$, was measured at 44 mg/L. After 5 days of equilibration with water, several peaks were observed in the GC/FID trace and one was identified to be $C_{10}H_{18}O$ by GC/MS. These peaks were not present when the para-cymene was analyzed prior to equilibration. It is possible that this product and perhaps the other unidentified peaks arose from biotransformation of para-cymene in the nonsterile water used for the solubility study.

Mineralization was measured in the microcosms by measuring CO_2 evolution. The concentration of CO_2 was at a low and consistent level for the controls, rose to a slightly higher level for the weathered

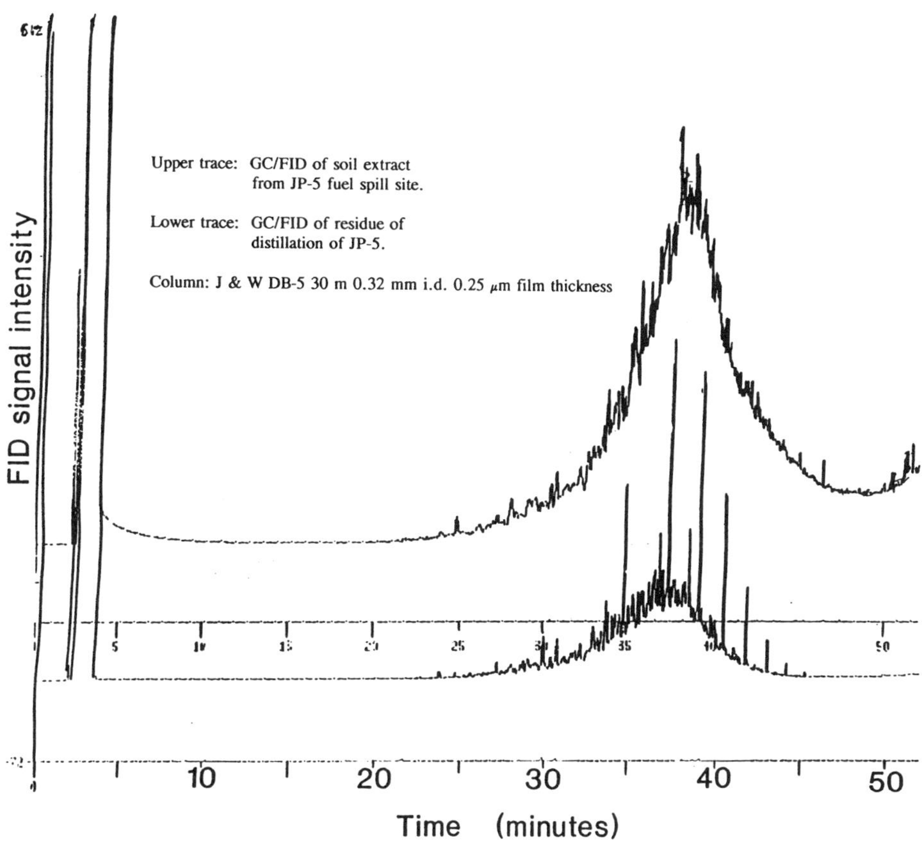

FIGURE 2. GC/FID traces of artificially and naturally weathered JP-5.

JP-5 soil, and to a greater level when fresh, sterile JP-5 was added to the weathered JP-5. The representative fuel mix mineralization level was approximately the same as the fresh JP-5. Results are shown in Figures 3 and 4. Note that the drop in the CO_2 levels after day 6 was expected, as the amount of oxygen in the headspace of the vials was limited.

It is possible that the higher level of CO_2 produced when fresh JP-5 was added to the microcosms resulted from the presense of more volatile aromatic compounds that were not present in the weathered sample. Comparison of results in Figure 2 with GC traces of the early fractions of fresh JP-5 (data not shown) demonstrate that the volatile compounds are not present in the weathered JP-5. These data suggest that cell metabolism is stimulated by the volatile components that are more water soluble and thus more bioavailable.

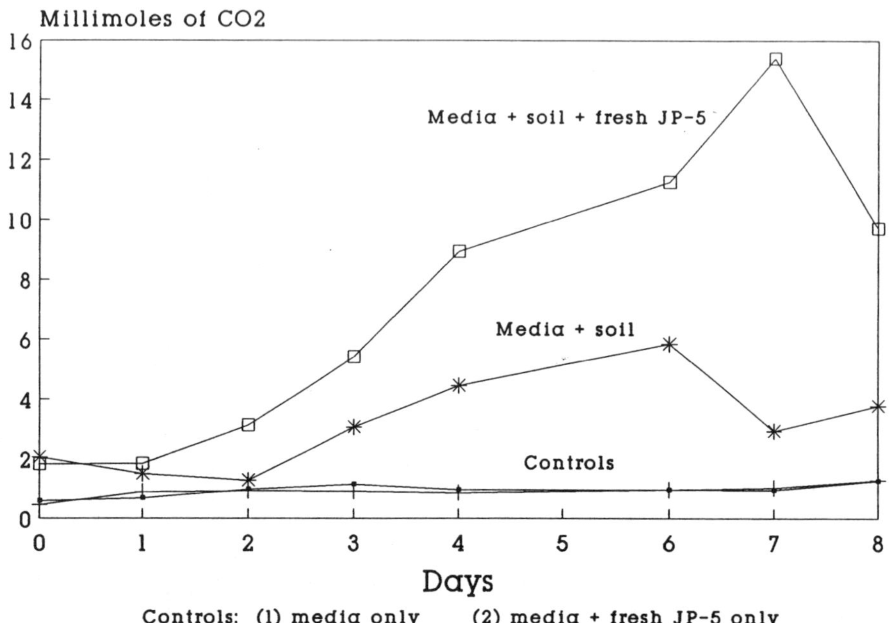

FIGURE 3. Growth curves in aerobic batch cultures; millimoles of CO_2 generated—experiment 2.

Since the representative fuel mix showed CO_2 evolution similar to fresh JP-5, the next step will be to determine which of the three pure compounds in the mix gave rise to this result. Also, the boiling point fractions will be used in subsequent solubility and biodegradation studies. GC/FID and GC/MS analysis will be conducted to detect intermediate metabolites. Biodegradation of the chemical groups of JP-5 will be done after separating the fuel into chemical groups by high-performance liquid chromatography.

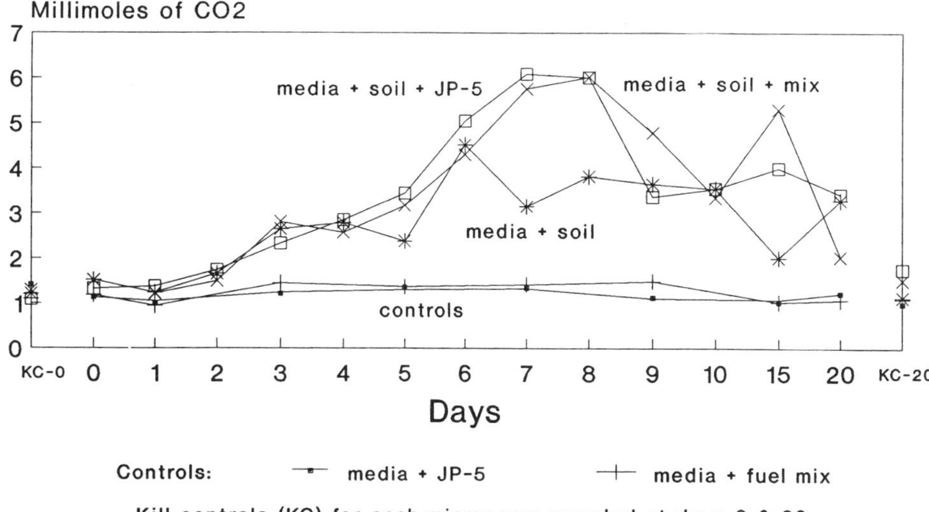

FIGURE 4. Growth curves in aerobic batch cultures; millimoles of CO_2 generated—experiment 2.

REFERENCES

Klein, S. A.; Jenkins, D. *Water Res.* **1981,** *15,* 75–82.

McAuliffe, C. J. *Physical Chem.* **1965,** *70,* 1267.

Nester, R. G. "Spinning Band Still for Vacuum Operation." *Analytical Chemistry* **1956,** *28,* 278.

Ridgway, H. F.; Phipps, D.; Safarik, J.; Haag, F.; Reinhard, M.; Ball, H.; McCarty, P. L. "Investigation of the Transport and Fate of Gasoline Hydrocarbon Pollutants in Groundwater"; a final project report to the United States Geological Survey on Grant #14-08-0001-G1126; Reston, VA, 1988.

Yaws, C. L.; Yang, H.; Hopper, J.; Hansen, K. C. "Hydrocarbons: Water Solubility Data." *Chemical Engineering April 1990,* p 177.

Hydrocarbon Degradation Potential in Reference Soils and Soils Contaminated with Jet Fuel

*Richard F. Lee**
Skidaway Institute of Oceanography
Ronald Hoeppel
Naval Civil Engineering Laboratory

A series of test wells were drilled adjacent to a fuel farm and a JP-5 jet fuel pump station located at a naval air station in Maryland. At least 5 ha of subsurface soil (to an average depth of 4 m) above a local aquifer were found to contain high concentrations of petroleum compounds, including such volatile aromatics as benzene, toluene, ethylbenzene, and xylenes. Horizontal transport has resulted in slow seepage from banks into streams of the affected area. The source of the petroleum is due to various spills over the past 10 years and possibly continuous leakage from the tanks.

There is a large body of literature describing the microbial metabolism of polycyclic aromatic hydrocarbons in aerobic solid-water systems (Arvin *et al.* 1988; Atlas 1981; Bauer & Capone 1985; Mihelcic & Luthy 1988; Swindoll *et al.* 1988; Van der Hoek *et al.* 1989). Petroleum degradation in surface and subsurface soils is affected by such factors as moisture content, pH, soil type, soil organics, temperature, and oxygen concentrations. We determined the degradation rates of ^{14}C-labeled hydrocarbons added to soils collected from a contaminated surface site (Site D), contaminated subsurface sites (Wells A and B), and a clean reference site (Well C). The radiolabeled hydrocarbons used include benzene, toluene, naphthalene, 1-methynaphthalene, phenanthrene, fluorene, anthracene, chrysene, and hexadecane.

Microbial degradation rates were based on determination of mineralization rates (production of $^{14}CO_2$) of hydrocarbons that were

* Skidaway Institute of Oceanography, P.O. Box 13687, Savannah GA 31416

added to soil samples. This technique, often referred to as hydrocarbon degradation potential, has been used to evaluate hydrocarbon degradation in natural waters and soils (Atlas 1979; Lee & Ryan 1983; Scheunert *et al.* 1987). Since water was added and oxygen was not limiting, the hydrocarbon rates determined are likely to be higher than those occurring *in situ*. Using radiolabeled hydrocarbons, information can be provided on differences in the degradation rates of various petroleum compounds in different types of soils at a site, on possible production of petroleum metabolites in the soil, and on the importance of anaerobic petroleum degradation and the effects of nutrient, water, and surfactant addition on biodegradation rates.

Materials and Methods

Test wells (A, B, and C) were augured to a clay confining layer. We used core samples from above the observed water table. A few samples from the cores were collected by aseptic techniques (i.e., undisturbed, sealed samples) to compare with soils not aseptically collected. Gas samples were withdrawn from a depth of 1 m from the top of the casing prior to water sample retrieval. Volatile hydrocarbons were determined using a field photoionization detection-gas chromatograph. Well waters were also analyzed using a gas chromatograph/mass spectrometer. These analyses were conducted by International Technology Corporation under contract to the Naval Civil Engineering Laboratory (Port Hueneme CA). Standards used included benzene, toluene, xylenes, and ethylbenzene. ^{14}C-labeled hydrocarbons were added to soil mixed with water (5 g soil with 10 mL water) in 250 mL flasks capped with silicon stoppers. After incubation at room temperature (20 C), the respired $^{14}CO_2$ was collected by trapping on phenethylamine paper and counted in a liquid scintillation counter (for details of this procedure see Lee & Ryan 1983). Controls were soil samples containing 10 percent formalin. All samples were in triplicate for each time interval.

Degradation can often be expressed by the first-order equation

$$dc/dt = kc \tag{1}$$

where k is the rate constant and c is the concentration of the hydrocarbon at time t. Half-lives were calculated by the equation

$$t_{1/2} = \frac{0.693}{k} \quad (2)$$

Radiolabeled hydrocarbon used included 2-[8-^{14}C] methylnaphthalene (136.9 MB q/mM); 1-[^{14}C]-naphthalene (135.8 MB q/mM); 9-^{14}C-fluorene (95.1 MB q/mM); ^{14}C-methylbenzene (647 MB q/mM); 5,6 (11,12-^{14}C) chrysene (2.33.1 MB q/mM); 9-anthracene (76.6 MB q/mM); UL-^{14}C-benzene (370 MB q/mM); and ^{14}C-hexadecane (2268 MB q/mM).

Results

Soils at the various sites are primarily sand, but Site D has surface seeps and includes a layer of oil-saturated peat. The concentrations of some selected volatile hydrocarbons at the study sites is given in Table 1. Soils from Wells A and B were contaminated with benzene but no toluene was detected. These wells are adjacent to a previously leaking tank and JP-5 usually has low toluene concentrations. Site D and a nearby seep are adjacent to a fuel farm and tanks here have held JP-4, JP-5, fuel oil #2, and AVGAS. Fuel oil, JP-4, and AVGAS all contain toluene and benzene and thus both toluene and benzene were present in the seep soils. Site D, where the soil was collected near the surface, appears to be contaminated by an old spill since benzene was absent and toluene concentration was very low, and mostly high molecular weight alkanes were found. Soil taken from cores at depths of 2 to 3 m from Wells A, B, and C had petroleum hydrocarbon concentrations of 570, 530, and less than 20 µg/g soil, respectively.

Soil from Wells A and B rapidly degraded benzene, toluene, naphthalene, 1-methylnaphthalene, and phenanthrene with half-lives ranging from 1 to 3 days (Figures 1-4). Degradation of these compounds in "clean" soil from reference Well C was very low with half-lives for these compounds ranging from 20 to 125 days (Figures 1-4). No differences in degradation potential were found between soils collected by aseptic or non-aseptic techniques. At Site D, hydrocarbon degradation rates were lower in oiled peat compared with oiled sand with naphthalene half-lives ranging from 0.7 to 2.5 days, (Figures 1, 2, and 3). Toluene was more slowly degraded in soil from Well B than in soil from Site D (Figure 1). This may be due to the absence of toluene

TABLE 1. Volatile hydrocarbons found in soil cores or surface soils at Naval Air Station on Patuxent River MD.[(a)]

Compounds (μg/L)	Site D	Well B (μg/L)	Well A	Well C	Seep[(b)]
Volatile Alkanes	—	2,535	2,487	2	11,000
Benzene	N.D.[(c)]	850	1,020	N.D.	1,400
Ethylbenzene	26	430	23	2	7,000
Toluene	6	N.D.	N.D.	N.D.	5,600
m, p, o-xylenes	186	240	1,260	N.D.	N.D.

(a) Site D and nearby seep are adjacent to fuel farm containing JP-4, JP-5, fuel oil #2 and AVGAS; Wells A and B are adjacent to previously leaking JP-5 tank; Well C is from a reference site where there was no evidence of contamination from the fuel tanks.
(b) μg/g soil.
(c) Not detected at <1 μg/L.

from soil at Well B so that the microflora at this site was not adapted to toluene, even though other aromatics were rapidly degraded including benzene, naphthalene, and methylnaphthalene (Figures 2 and 3). Fluorene, chrysene, and anthracene were only slowly degraded in both oiled soils and reference soils (data not shown). These compounds are in low concentrations in JP-4, JP-5, and AVGAS. Hexadecane, a long chain alkane which can be produced by plants and is normally found in soil, was degraded at a much higher rate in reference soil compared with oiled soils. This hydrocarbon is in low concentrations in jet fuels. In addition to the type of soil (i.e., peat or sand), the amount of water in the soil was shown to be important. Soil with a moisture content of 20 percent degraded naphthalene at a lower rate than contaminated sand with a moisture content of 66 percent (Figure 5).

Discussion

Soils contaminated with petroleum are characterized by high concentrations of hydrocarbon-degrading bacteria (Dragum 1988; Raymond *et al.* 1976). Oil was added to one study site, and after 8 months, the soil showed a tenfold increase in the concentration of oil-degrading bacteria

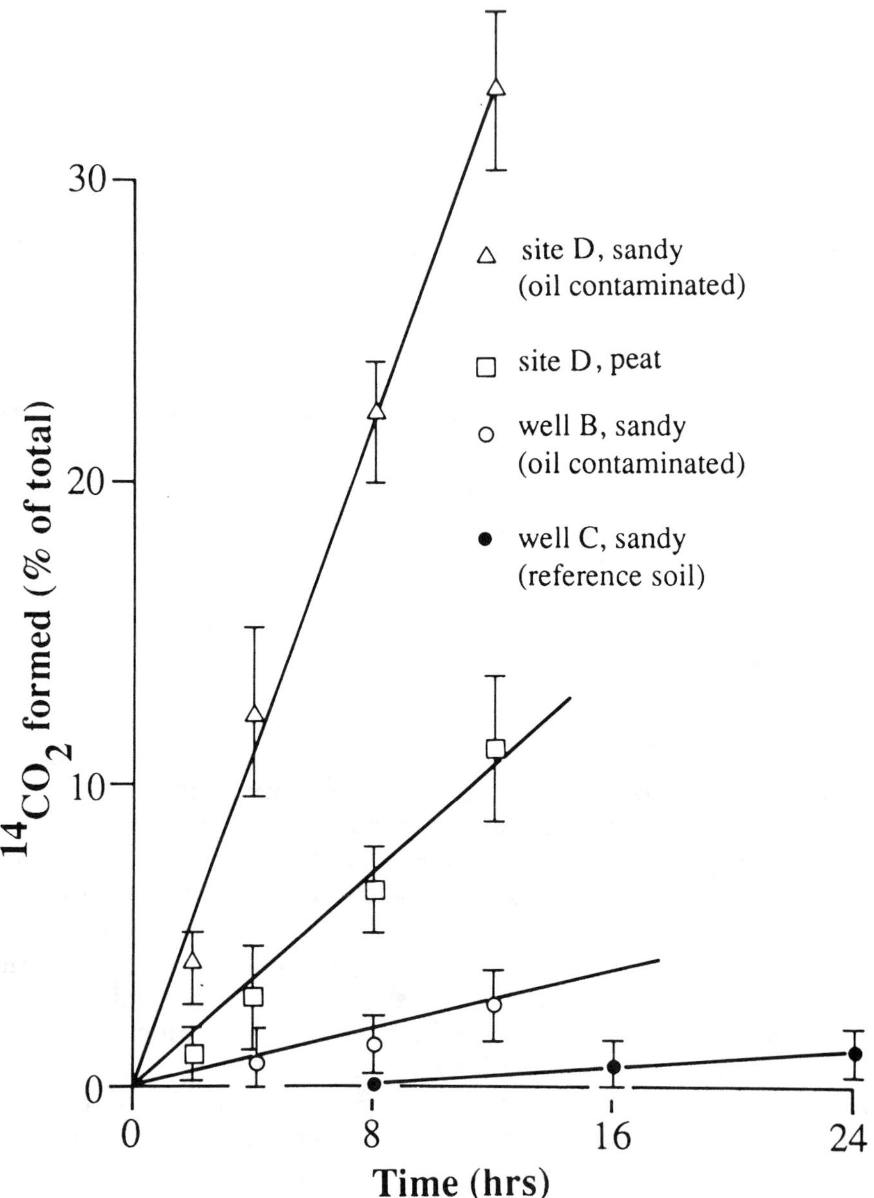

FIGURE 1. **^{14}C-toluene added to soil (5 µg/5 g soil) from Site D and Wells B and C. Error bars are standard deviation (*n* = 3). Calculated half-lives for toluene in Wells B and C were 7 and 50 days, respectively. Calculated half-lives for toluene in sand and peat of Site D were 0.6 and 1.9 days, respectively.**

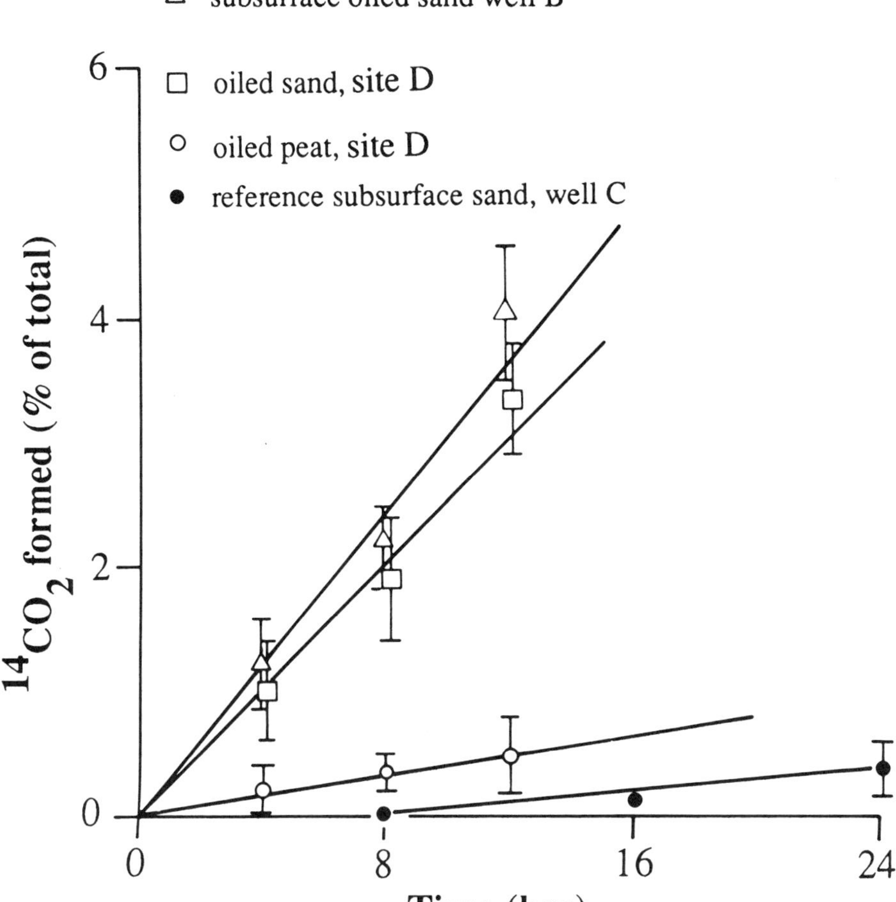

FIGURE 2. **^{14}C-benzene added to soil (5 µg/5 g soil) from Site D and Wells B and C. Error bars are standard deviation (*n* = 3). The calculated half-lives of benzene in Wells B and C were 5 and 125 days, respectively. The calculated half-lives of benzene in sand and peat of Site D were 6 and 41 days, respectively.**

(Pinholt *et al.* 1979). Thus, soils previously exposed to various foreign compounds, including pesticides or petroleum, show an enhanced ability to degrade such compounds or mixtures (Aurelius & Brown 1987; Chapman *et al.* 1986; Harris *et al.* 1988; Heitkamp *et al.* 1987; Hendry & Richardson 1988; Lee & Ryan 1983; Lee *et al.* 1988). The lag period found in "clean" reference soils before hydrocarbon degradation begins

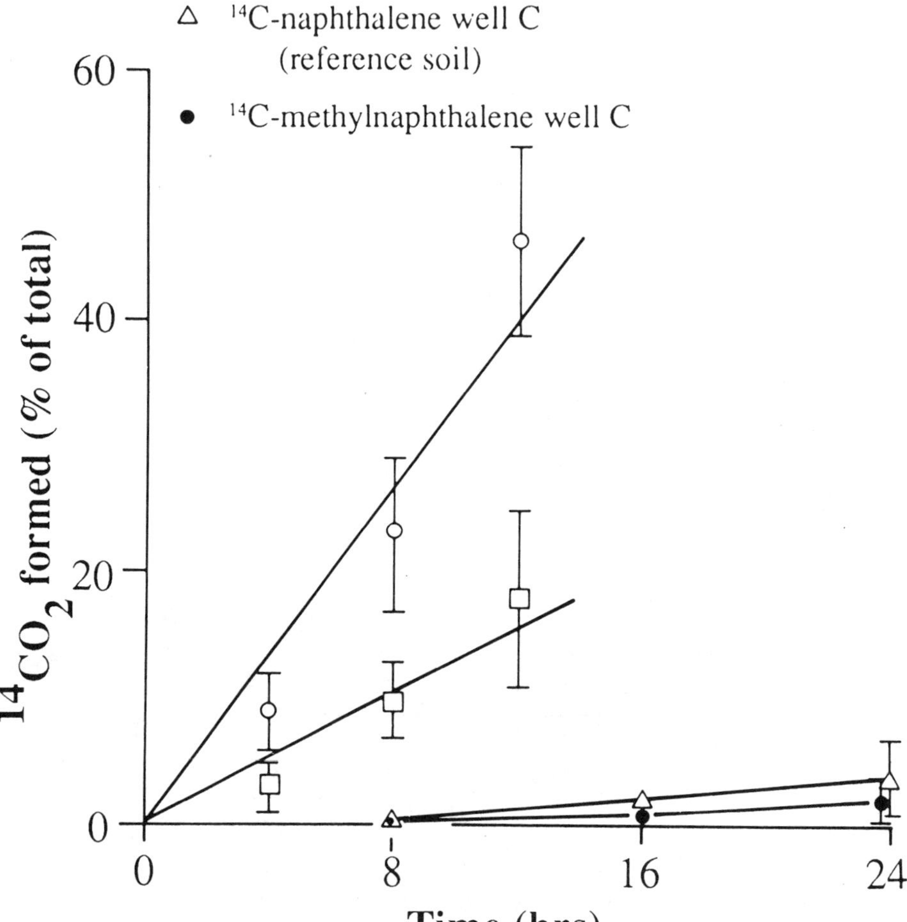

FIGURE 3. **^{14}C-naphthalene and ^{14}C-methylnaphthalene added to subsurface soils (5 µg/5 g soil) from Wells B and C. Error bars are standard deviation (*n* = 3). Calculated half-lives for naphthalene in Wells B and C were 0.7 and 20 days, respectively. Calculated half-lives for methyl-naphthalene in Wells B and C were 1.2 and 25 days, respectively.**

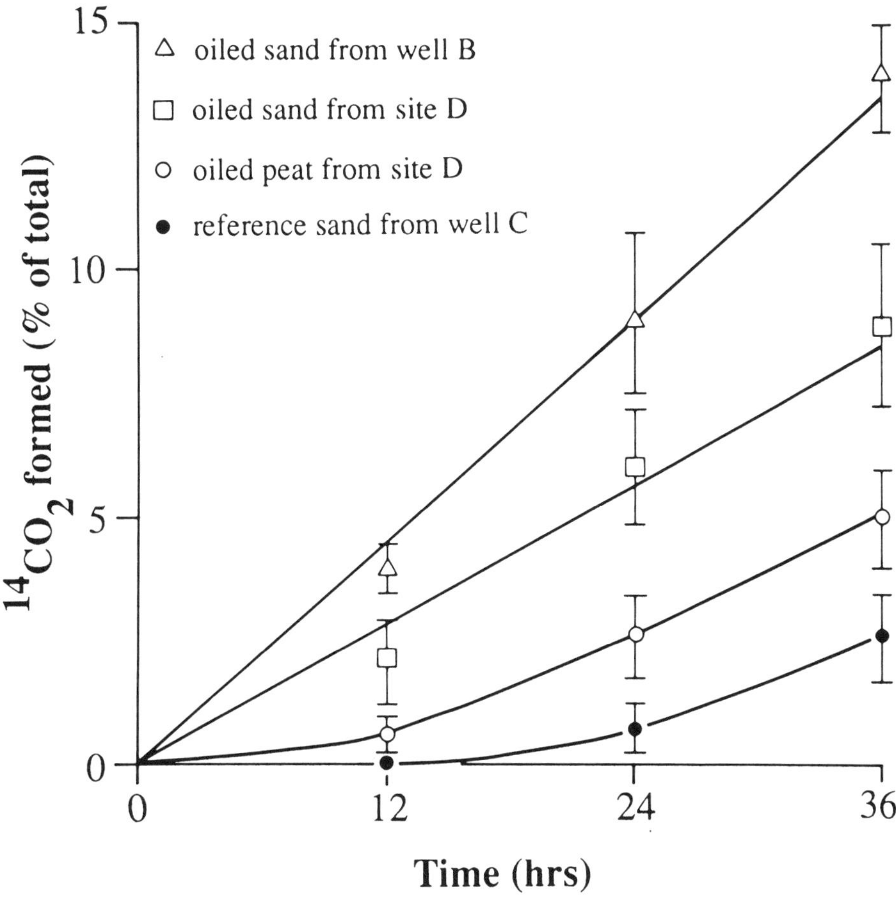

FIGURE 4. ^{14}C-phenanthrene added to soil (5 µg/5 g soil) from Site D and Wells B and C. Error bars are standard deviation (n = 3). The calculated half-lives for phenanthrene in soils of Site D, Wells B and C were 10, 6, and 21 days, respectively.

is assumed to be due to the time needed for the microbial community to adapt to the added hydrocarbons (Cripe *et al.* 1987). Such adaptation has been defined as a change in the microbial community that increases the rate of transformation of a compound as a result of prior exposure to the compound (Spain & Van Veld 1983). The time needed for this adaptation depends on the compounds and can vary from days to weeks. Thus, because of such adaptation at our study site, we observed

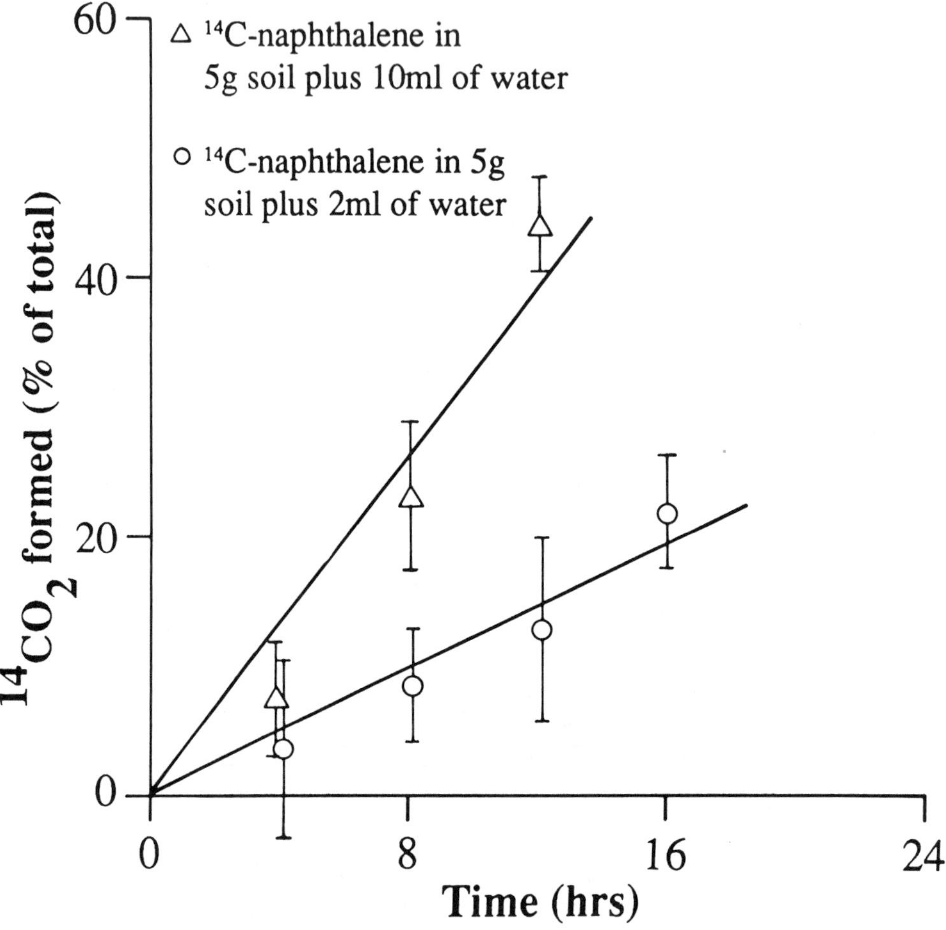

FIGURE 5. Effect of moisture content on ^{14}C-naphthalene degradation. ^{14}C-naphthalene added to subsurface soil (5 µg/5 g soil) from Well B (oiled). Error bars are standard deviation (n = 3).

rapid degradation of toluene, benzene, naphthalene, methylnaphthalene, and phenanthrene.

Work with pure cultures has shown that oil degraders only slowly degrade certain xylene (dimethybenzene) isomers and highly branched alkanes because of steric hindrances inhibiting the ability of the bacterial oxygenases to attack the compounds (Bailey *et al.* 1973; Hopper 1978). The *p*- and *m*-xylenes were readily degraded by pure cultures but *o*-xylene was only slowly degraded. Therefore, even after extensive degradation of petroleum, certain highly branched aromatics and alkanes can remain in the soil. Analysis of the petroleum-contaminated soils at the study sites indicated that the major compounds present were branched-chain alkanes (Hoeppel 1988). The degradation we report here are potential rates and are probably not the rates occurring in the subsurface soils, but indicate a degradation rate that can be obtained when such factors as moisture and oxygen are optimal. Naphthalene added to sediments from a stream heavily contaminated with petroleum had half-lives of less than 1 day (Herbes & Schwall 1978), while addition to less contaminated sediments resulted in a 17 day half-life (Heitkamp *et al.* 1987).

An interesting observation was the very slow degradation of certain hydrocarbons that were present in very low or nondetectable levels in the contaminated soils. It appeared that the microbes were adapted to degrade only the compounds in the petroleum products at the site. Thus, toluene was only slowly degraded in a site contaminated with JP-5 (very low in toluene content), while benzene, naphthalene, and methylnaphthalene were rapidly degraded in soil from this site.

The data indicated that under the conditions used in our incubations, microflora of the contaminated sites had the potential to rapidly degrade many of the most toxic components found in jet fuels contaminating the area. From a practical standpoint, it appears that it is unnecessary to add hydrocarbon degrading microbes to such contaminated sites, but rather a need to optimize biological degratation by monitoring and modifying oxygen, nutrients, and moisture. (For summaries of work on stimulating biodegradion of petroleum in soils, see Lee *et al.* 1988; Morgan & Watkinson, 1990.)

Acknowledgments. The work reported here was supported by Contract No. N66-001-87-C-0377 from the Naval Ocean Systems Center (San Diego CA) and Naval Civil Engineering Laboratory (Point Hueneme CA). We thank R. K. Johnston at the Naval Ocean Systems Center for his advice and encouragement.

REFERENCES

Arvin, E.; Jensen, B.; Aamand, J.; Jorgensen, C. *Water Science and Technology* **1988**, *20*(3), 109–118.

Atlas, R. M. *American Society for Testing and Materials* **1979**, *695*, 196–206.

Atlas, R. M. *Microbiology Reviews* **1981**, *45*, 180–209.

Aurelius, M. W.; Brown, K. W. *Water, Air and Soil Pollution* **1987**, *36*, 23–31.

Bailey, N.J.L.; Jobson, A. M.; Rogers, M. A. *Chemical Geology* **1973**, *11*, 203–221.

Bauer, J. E.; Capone, D. G. *Applied and Environmental Microbiology* **1985**, *50*, 81–90.
Chapman, R. A.; Harris, C. R.; Harris, C. *Journal of Environmental Science and Health* **1986**, *B21*, 125–141.
Cripe, C. R.; Walker, W. W.; Pritchard, P. H.; Bourquin, A. W. *Ecotoxicology and Environmental Safety* **1987**, *14*, 239–251.
Dragum, J. "Microbial Degradation of Petroleum Products in Soil." In *Soils Contaminated by Petroleum*; Calabrese, E. J.; Kostecki, P. T.; Fleischer, E. J., Eds.; John-Wiley and Sons: New York, 1988; pp 289–300.
Harris, C. R.; Chapman, R. A.; Morris, R. F.; Stevenson, A. B. *Journal of Environmental Science and Health* **1988**, *B23*, 301–316.
Heitkamp, M. A.; Freeman, J. P.; Cerniglia, C. E. *Applied and Environmental Microbiology* **1987**, *53*, 129–136.
Hendry, K. M.; Richardson, C. J. *Environmental and Toxicological Chemistry* **1988**, *7*, 763–774.
Herbes, S. E.; Schwall, T. R. *Applied and Environmental Microbiology* **1978**, *35*, 306–326.
Hoeppel, R. "Status of the NAS Patuxent River Fuel Farm Bioremediation Studies"; progress report to Naval Civil Engineering Laboratory, Port Hueneme, CA (Contract No. N62474-86-C07280/00004); Port Hueneme, Oct.–Dec. 1988.
Hopper, D. J. *Applied Science* **1978**, 85–112.
Lee, M. D.; Thomas, J. M.; Borden, R. C.; Bedient, P. B., Ward, C. H.; Wilson, J. T. *CRC Critical Reviews in Environmental Control* **1988**, *18*, 29–89.
Lee, R. F.; Ryan, C. *Canadian Journal of Aquatic Science* **1983**, *S2*, 86–94.
Mihelcic, L.J.R.; Luthy, R. G. *Applied and Environmental Technology* **1988**, *54*, 1182–1187.
Morgan, P.; Watkinson, R. J. *Water Science and Technology* **1990**, 22, 63–68.
Pinholt, Y.; Strowe, S.; Kjoller, A.; *Holartic Ecology* **1979**, 2, 195–200.
Raymond, R. L.; Hudson, J. O.; Jamison, V. W. *Applied and Environmental Microbiology* **1976**, *31*, 522–535.
Scheunert, I.; Vockel, D.; Schmitzer, J.; Korte, F. *Chemosphere* **1987**, *16*, 1031–1041.
Spain, J. C.; Van Veld, P. A. *Applied and Environmental Microbiology* **1983**, *45*, 428–435.
Swindoll, C. M.; Aelion, C. M.; Dobbins, D. C.; Jiang, O.; Long, S. C.; Phaender, F. K. *Environmental and Toxicological Chemistry* **1988**, *7*, 291–299.
Van der Hoek, J. P.; Urlings, L. G.; Grobben, C. M. *Environmental Technology Letters* **1989**, *10*, 185–194.

Bioremediation of Soil and Groundwater Contaminated with Stoddard Solvent and Mop Oil Using the PetroClean® Bioremediation System

E. K. Schmitt, M. T. Lieberman, J. A. Caplan*
ESE Biosciences, Inc.
D. Blaes, P. Keating, W. Richards
Environmental Science & Engineering, Inc.

Environmental Science & Engineering, Inc. (ESE) was contracted by a confidential industrial client to perform a three-phased project. This project was conducted jointly by ESE personnel from the Fountain Valley CA office and ESE Biosciences, Inc. (EBIO) of Raleigh NC. Phase I involved characterizing the site and delineating the extent of subsurface contamination. Phase II included biofeasibility and pilot-scale evaluations, determining remedial requirements, and designing the full-scale treatment system. Phase III involved implementing and operating the designed *in situ* bioremediation system (i.e., PetroClean® 4000) to achieve site closure.

Site Description and History. Exploratory borings at the site indicated that the shallow soil, from the ground surface to a depth of approximately 1.5 m, consisted of sandy clay. Underlying the sandy clay was decomposed granite. Groundwater was at approximately 12.2 m below ground surface (BGS).

This facility was constructed in the early 1960s to distribute, rent, and clean uniforms, wiping towels, and related cloth products for a variety of commercial, industrial, and governmental firms and agencies. In 1961, five underground storage tanks (UST) were installed to store Chevron 325 solvent (commonly known as Stoddard solvent)

* ESE Biosciences, Inc., 3208 Spring Forest Road, Raleigh NC 27604

to be used in the dry cleaning process. The chemical composition of this Stoddard solvent is 98 percent paraffins (including naphthenes, which are cycloaliphatic hydrocarbons), 2 percent C_8 and larger aromatic hydrocarbons, and less than 0.1 percent benzene. Two USTs were also installed to store Farbest Corporation's L-100 oil (commonly known as mop oil) to be used in the facility's cleaning and dust control processes. The USTs containing Stoddard solvent included one 37,850-L tank storing new solvent, and four 7,570-L tanks storing used and reprocessed Stoddard solvent. The other two USTs consisted of a 1,136-L tank storing new mop oil and a 4,542-L tank storing used mop oil. The Stoddard solvent and mop oil systems had been in operation since 1961 at this facility. In 1987, it was decided to convert the cleaning system to an aboveground chemical treatment system. As a result, the decision was made to remove the seven underground storage tanks.

In December 1987, the seven USTs were removed and soil samples were collected from the excavations. These soil samples were analyzed using EPA methodologies as specified in U.S. EPA SW-846 (1986). Stoddard solvent and mop oil were quantified using EPA Method 8015. Total petroleum hydrocarbons (TPHC) were quantified using EPA Method 418.1. The results of the soil sample analysis indicated the soil contained elevated concentrations of Stoddard solvent and mop oil from the former UST systems.

Phase I—Site Characterization and Assessment

Following the discovery of Stoddard solvent and mop oil in the soil beneath the USTs, a preliminary subsurface investigation was performed to assess the vertical extent of subsurface contamination within the dense, decomposed granite underlying the former tank farm area. Three soil borings (i.e., borings B-1, B-2, and B-3) were installed to a depth of approximately 9.0 m BGS. All three borings were located within the former tank farm area as shown in Figure 1. As a result of the UST excavation, undisturbed, native soil was not encountered until a depth of 3.5 to 4 m was reached in each boring. Therefore, samples were collected at depths of approximately 4.5, 6.0, 7.5, and 9.0 m BGS. Soils collected from borings B-1 and B-2 were analyzed for Stoddard solvent, while soils from B-3 were subjected to both Stoddard and TPHC analysis. Results of the chemical analyses of the soil samples are shown in Table 1.

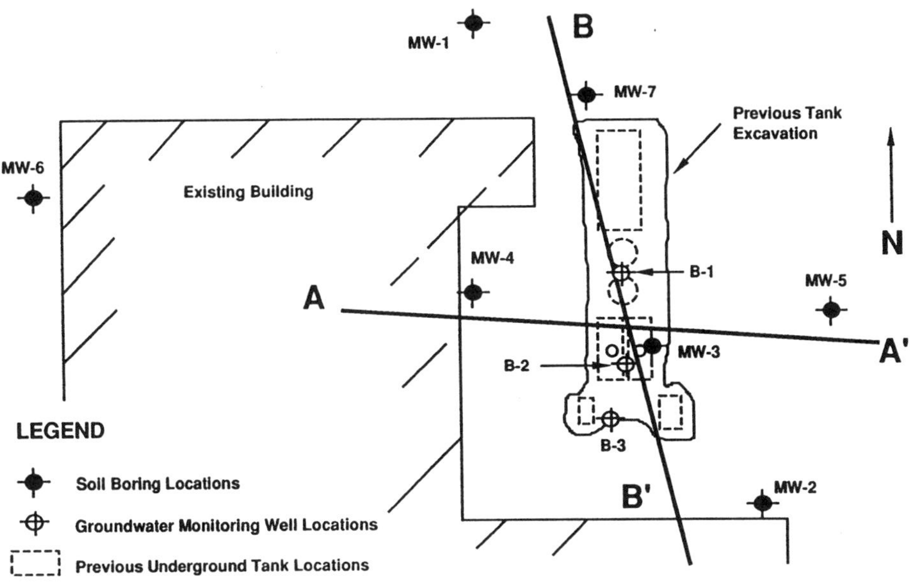

FIGURE 1. Original site map showing building prior to expansion, the locations of the seven USTs prior to removal, and the locations of the soil borings (B-1, b-2, B-3) and monitor wells (MW-1 through MW-7) installed during the UST closure and site assessment phases.

The results of this preliminary subsurface investigation indicated the soil below the former USTs was contaminated to a depth of at least 9.0 m BGS. An additional investigation was recommended and performed to determine the lateral and vertical extent of subsurface contamination. This involved drilling seven soil borings (MW-1 through MW-7), collecting soil samples for chemical analysis, installing groundwater monitoring wells within each boring (Figure 1), collecting groundwater samples for chemical analysis to assess the extent of the contaminant plume, and collecting groundwater elevation measurements to evaluate groundwater flow direction.

Depth to groundwater was found to be approximately 12.2 m BGS. Groundwater flow was generally in a westerly direction at this site. Results of the laboratory analyses performed on the soil samples are presented in Table 2. The results of chemical analyses for the presence of TPHCs, benzene, toluene, ethylbenzene, and xylenes (BTEX), and Stoddard solvent in groundwater samples collected from each monitor well are presented in Table 3. Figure 2 illustrates the

TABLE 1. Results of preliminary investigation soil sample analyses.

Boring Number	Sample Depth (m)	Stoddard Solvent (mg/kg)	Total Petroleum Hydrocarbons (mg/kg)
B-1	4.6	340	na[(a)]
	6.1	1,000	na
	9.1	1,100	na
B-2	4.6	1,600	na
	6.1	780	na
	9.1	930	na
B-3	4.6	420	7,000
	6.1	410	4,300
	9.1	50	1,700

(a) na = not analyzed.

approximated lateral extent of dissolved Stoddard solvent contamination, as well as the estimated extent of phase-separated Stoddard solvent.

The results of the site assessment indicated that soil beneath the former USTs, down to the groundwater surface, contained Stoddard solvent and mop oil. However, the lateral extent of soil contamination was limited to the immediate tank area. Three areas of high Stoddard solvent concentrations (in excess of 100 mg/kg) were detected in the soil. The first area was in boring MW-4, at a depth of 3.0 to 4.5 m. This boring was located in the area of the former dry cleaning room where underground pipes delivered Stoddard solvent from the USTs to the dry cleaning equipment. Contamination in this area most likely originated from pipe leakage. Stoddard contamination in this area was found to be localized, and attenuated rapidly with depth. At a depth of approximately 9.0 m, Stoddard solvent concentrations decreased to below the detection limit of 1 mg/L.

The second area was detected at a depth of 4.5 to 6.0 m BGS in boring MW-3 and at a depth of 4.5 m in boring B-2. These two borings were located in the area where the two 7,600-L (2,000-gal) Stoddard solvent tanks had been positioned. The highest concentration of Stoddard solvent found in the soil (3,500 mg/kg) was detected at a depth of 6.0 m in MW-3. This contamination is possibly the result of periodic overflowing of solvent from the Stoddard USTs.

TABLE 2. Results of site assessment and soil sample analyses.

Boring Number	Sample Depth (m)	Stoddard Solvent (mg/kg)	Total Petroleum Hydrocarbons (mg/kg)
MW-1	1.5, 3.0, 4.6, 6.1 (composite)	nd (1)[a]	nd (1)
	9.1	nd (1)	nd (1)
	12.2	nd (1)	nd (1)
	13.7	nd (1)	nd (1)
MW-2	1.5, 3.0, 4.6, 6.1 (composite)	nd (1)	5.0
	9.1	nd (1)	5.0
	12.2	nd (1)	6.0
	13.7	nd (1)	4.0
MW-3	4.6	1,100.0	2,600.0
	6.1	3,500.0	1,800.0
	9.1	900.0	2,000.0
	12.2	2,300.0	9,700.0
	13.7	1,600.0	4,500.0
MW-4	1.5	nd (1)	7.0
	3.0	2,200.0	3,800.0
	4.6	1,200.0	400.0
	6.1	300.0	400.0
	9.1	nd (1)	6.0
	12.2	nd (1)	5.0
	13.7	nd (1)	nd (1)
MW-5	1.5, 3.0, 4.6 (composite)	55.0	nd (1)
	9.1	nd (1)	1.0
	12.2	20.0	1.0
MW-6	1.5, 3.0, 4.6 (composite)	nd (1)	17.0
	9.1	nd (1)	16.0
	12.2	nd (1)	11.0
	13.7	nd (1)	2.0
MW-7	1.5, 3.0, 4.6, 6.1 (composite)	nd (1)	1.0
	9.1	nd (1)	3.0
	12.2	450.0	200.0
	13.7	280.0	37.0

(a) nd = not detected (detection limit).

TABLE 3. Results of groundwater sample analyses.

Well Number	TPHC (mg/L)	Stoddard Solvent (mg/L)	Benzene (mg/L)	Toluene (mg/L)	EB (mg/L)	Xylenes (mg/L)
MW-1	3.0	nd (0.1)[a]	nd (0.007)	nd (0.001)	nd (0.001)	nd (0.001)
MW-2	3.0	nd (0.1)	nd (0.007)	nd (0.001)	nd (0.001)	nd (0.001)
MW-3	3,600.0	500.0	nd (0.003)	0.37	62.0	140.0
MW-4	7.0	nd (0.1)	nd (0.007)	nd (0.001)	nd (0.001)	2.0
MW-5	9.0	nd (0.1)	nd (0.007)	nd (0.001)	nd (0.001)	2.0
MW-6	nd (0.5)	nd (0.1)	nd (0.007)	nd (0.001)	nd (0.001)	nd (0.001)
MW-7	3.0	0.1	nd (0.007)	nd (0.002)	nd (0.002)	nd (0.002)

(a) nd = not detected (detection limit).

The third area of high Stoddard solvent soil contamination was found below the same two Stoddard USTs, but at the groundwater surface. Soil samples collected from boring MW-3 at depths of 12.2 to 13.7 m exhibited high concentrations of Stoddard solvent (1,600 mg/kg).

High concentrations of TPHCs in the soil generally occurred in the same areas as the high concentrations of Stoddard solvent. The soil directly beneath the two mop oil USTs contained the highest concentrations of contaminants. As with the Stoddard solvent, TPHC soil contamination was localized and decreased with depth. The highest concentration of TPHCs (9,700 mg/kg) at a depth of 12.2 m in boring MW-3 was likely the result of accumulation of contaminants at the groundwater surface. Based on these soil investigation results, it was estimated that approximately 4,600 m^3 (6,000 yd^3) of soil had been contaminated by the released Stoddard solvent and mop oil.

Groundwater below the former USTs had been impacted by both Stoddard solvent and mop oil. However, as with the soil contamination, groundwater contamination was localized in the immediate former UST area and had not migrated off-site. Phase-separated Stoddard solvent (i.e., 0.75 m of product) was found in MW-3. Phase-separated Stoddard solvent was not found in any other monitor well. The only other monitor well in which Stoddard solvent was detected was MW-7. A Stoddard solvent concentration of 0.1 mg/L was detected at this location.

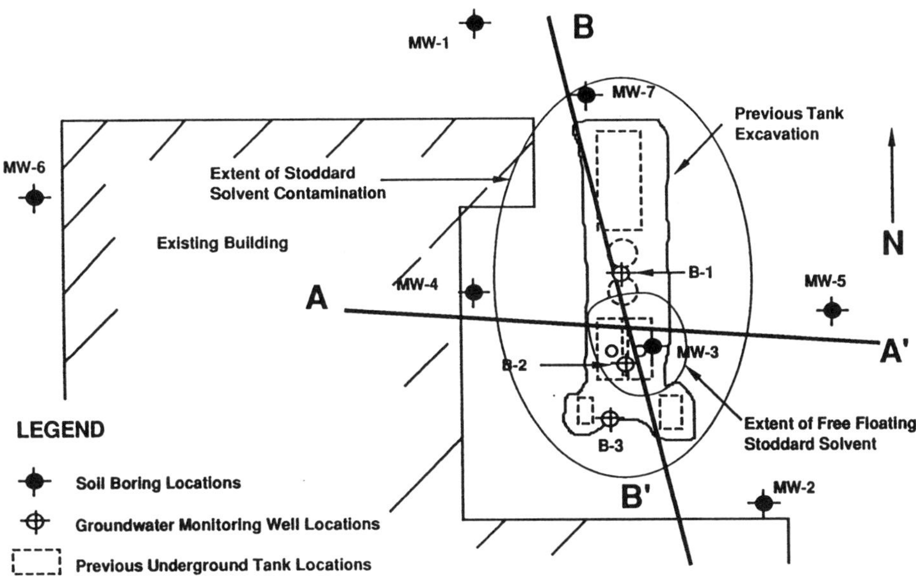

FIGURE 2. Site map depicting the extent of Stoddard solvent contamination in groundwater.

Low mg/L concentrations of TPHCs in the groundwater were found at all locations sampled, with the exception of MW-3 which contained 3,600 mg/L TPHCs and MW-6 which contained less than 0.5 mg/L. These generally low TPHC concentrations typically result from migration of dissolved petroleum hydrocarbons in the groundwater. The high TPHC concentration detected in MW-3 was anticipated since phase-separated Stoddard solvent was known to be present in this location.

Phase II—Biofeasibility Analysis and Pilot-Scale Evaluation

Based on the results of the site assessment, *in situ* biological treatment was considered to be a viable remedial technology for this site. Because both soil and groundwater had been impacted as a result of leaking USTs and/or associated piping, it was imperative that the remedial technology employed at this site be capable of effectively treating both matrices concurrently to achieve the most rapid restoration. An additional factor influencing the selection of the remedial technology was the fact that a building expansion was in progress. As a result of this expansion, the area of contamination would be underneath a new building. The selected remedial technology, therefore, had to be unobtrusive and capable of achieving site restoration without disturbing the building structure or facility operations.

In situ biological treatment is fully capable of achieving these goals. It has been demonstrated that biodegradation is an effective means of eliminating petroleum hydrocarbon contaminants from both soil and groundwater (Alexander 1977; Thomas & Ward 1989). Schmitt and Caplan (1987) successfully used a combination of biological land farming and *in situ* technology at a decommissioned oil refinery site in Michigan and achieved biologically mediated reductions in TPHC concentrations of 75 to 85 percent in just 3 months of treatment. Lieberman *et al.* (1989) utilized *in situ* biorestoration to remediate diesel fuel contaminated soil and groundwater at a site in northern Michigan. Within 13 months of operation, clean-up criteria for both soil and groundwater had been attained.

To investigate the applicability of *in situ* bioremediation for the treatment of Stoddard solvent in this particular site matrix, a complete biofeasibility analysis was performed. Representative samples of contaminated site matrix soil and groundwater were collected and used for this biofeasibility analysis. No toxicity was demonstrated by either site matrix soil or groundwater. Site matrix soil toxicity was evaluated by measuring the oxygen uptake of an active, nonacclimated microbial inoculum in the presence of increasing concentrations of site matrix soil. Toxicity assay broth, which provides all the necessary components for microbial respiration, was added to replicate vessels containing increasing percentages of site matrix soil. Replicate control vessels were also prepared, one containing 100 percent site matrix soil and the other containing 100 percent clean sand. Instead of adding toxicity assay broth, these two controls received only phosphate buffer solution. The control containing 100 percent site matrix soil provided a means of quantitating respiration due to indigenous nutrients, while the control containing 100 percent clean sand was used to measure endogenous respiration due to the microbial inoculum. Any inhibitory or toxic effects due to site matrix soil were quantitated by reductions in the rate of oxygen uptake or reductions in the total oxygen uptake as the concentration of site matrix soil increased (Figure 3).

Site groundwater toxicity was evaluated by measuring the optical density of a microbial culture after 24 h incubation in the presence of increasing percentages of site groundwater. Nutrient broth and inorganic salts solutions were prepared in replicate vessels containing increasing concentrations of site groundwater. Each vessel was

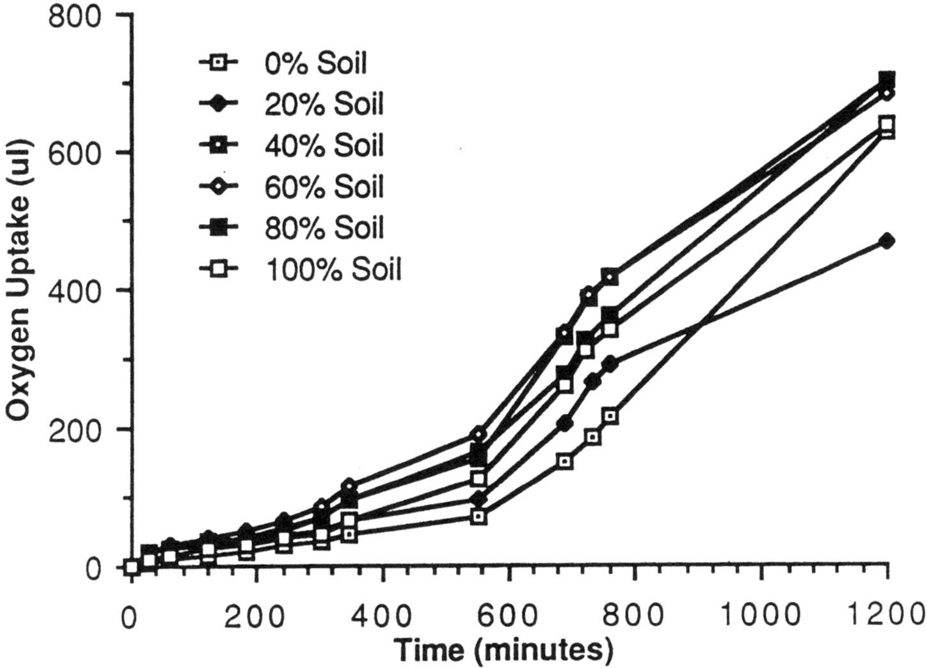

FIGURE 3. Toxicity evaluation of site matrix soil. Cumulative oxygen uptake by a nonacclimated microbial inoculum in the presence of increasing concentration of site matrix soil was measured over time.

inoculated with nonacclimated microbial culture. The vessels were incubated for 24 h and the change in optical density was recorded and compared to a zero percent site groundwater control. Any inhibitory or toxic effects of site groundwater were indicated by less growth (i.e., percent change in optical density) attained during incubation as the concentration of site groundwater increased (Figure 4).

The pilot-scale, closed-loop PetroClean Bioremediation System was employed to simulate the anticipated flushing action, groundwater recovery, aboveground biotreatment, and aquifer recharge of a full-scale, *in situ* biological treatment system. A scale-model of the system was used for this bench-scale evaluation. This model incorporated a soil vadose zone/aquifer treatment cell and a fixed-film bioreactor. Site matrix groundwater was used to saturate a portion of the soil (i.e., the aquifer) which was overlain by contaminated site matrix soil. The model bioreactor was filled with site groundwater, and a closed-loop groundwater recovery and recharge system was initiated. The treated effluent from the aboveground bioreactor was reapplied to the surface

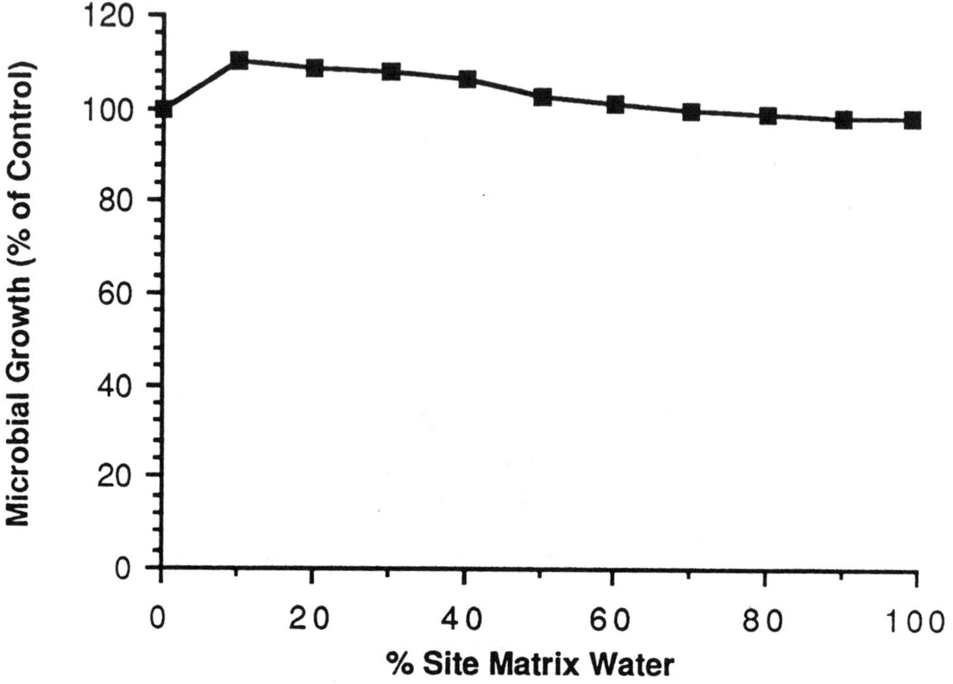

FIGURE 4. Toxicity evaluation of site groundwater. A comparison of microbial growth during 24-h incubation in the presence of increasing concentrations of site matrix groundwater. Growth, as measured by optical density, was compared with a control containing no site matrix groundwater.

of the contaminated soil. This recharge water percolated through the soil and was captured by the recovery well. The recovery rate was adjusted to maintain a bioreactor hydraulic retention time (HRT) of approximately 6.7 h. A proprietary blend of nutrients was added to the closed-loop system to satisfy the biological requirements for efficient microbial degradation of the Stoddard solvent.

A 1,000-fold increase in bacterial population density and a concommitant reduction in pH and dissolved oxygen concentration in the model system were evidence of biological activity. This evidence of biological activity was substantiated by examining the Stoddard solvent concentration in the soil during the 7-day treatment period (Figure 5). The initial Stoddard solvent concentration in the soil was 2,778 mg/kg. After 7 days of closed-loop, enhanced biological treatment, the Stoddard

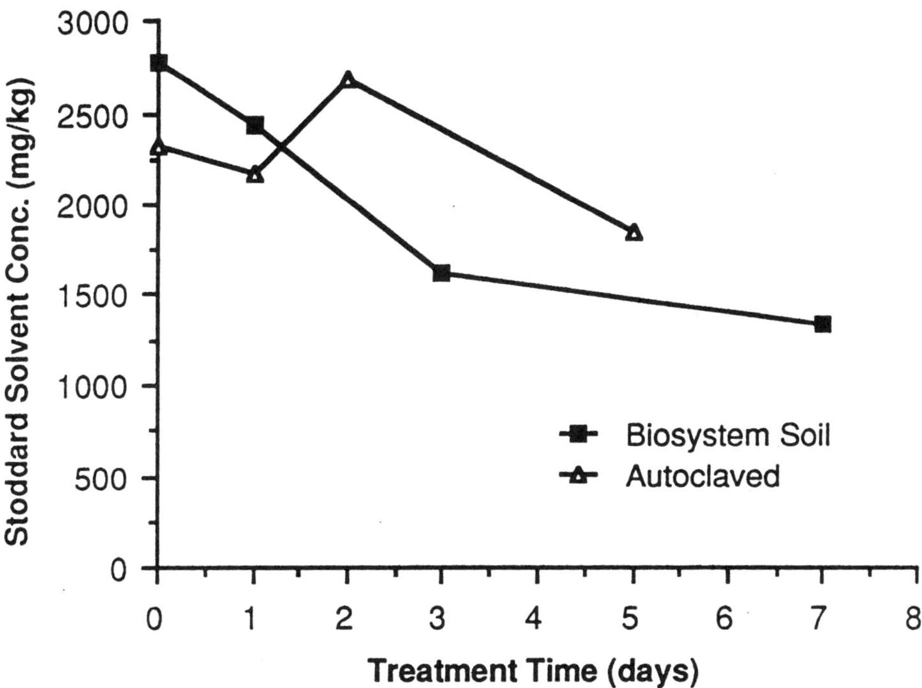

FIGURE 5. Stoddard solvent removal from site matrix soil under conditions of enhanced bioremediation using a pilot-scale PetroClean system versus an autoclaved control.

solvent concentration in the soil had been reduced to 1,335 mg/kg. This was a reduction of approximately 50 percent.

The results of the biofeasibility analysis and the pilot-scale bioremediation system demonstrated the feasibility of implementing a closed-loop, *in situ* biotreatment design for the removal of Stoddard solvent from both subsurface soil and groundwater at this site. Based on the results of the pilot-scale evaluation, it was estimated that full-scale bioremediation of this site would require approximately 12 to 18 months, employing the modelled, *in situ* design.

Full-Scale Bioremediation

Based on the results of the site assessment, the biofeasibility/pilot-scale evaluation, and the client's plans to construct a building directly on top of the contaminated area, an *in situ* biological treatment system was selected as the most appropriate remedial technology for this site. A

15,000-L (4,000-gal) PetroClean Bioremediation System was designed to allow building expansion activities to proceed, while concurrently achieving the clean-up goals dictated by the California Regional Water Quality Control Board. The patented system (Caplan *et al.* 1991) employs both aboveground and subsurface components to enhance biological degradation of contaminants in recovered groundwater, as well as in the *in situ* soil and groundwater (Figure 6).

Prior to constructing the foundation for the building expansion, ESE excavated the upper 1.5 m of sandy clay overlying the contaminated decomposed granite. Four groundwater extraction wells (EW) were installed to a depth of 18.3 m BGS. Four recharge wells (RW) were installed in borings drilled to a depth of 7.6 m BGS. The final components of the subsurface recharge system, three infiltration galleries, were then installed. All piping for the infiltration galleries, extraction and recharge wells, as well as electrical conduits, were run to a location that would be outside the exterior wall of the new building once construction was complete. The infiltration galleries were then covered with 0.3 m of pea gravel, and the remainder of the excavation was backfilled and compacted to its original elevation using the sandy clay stockpiled on site.

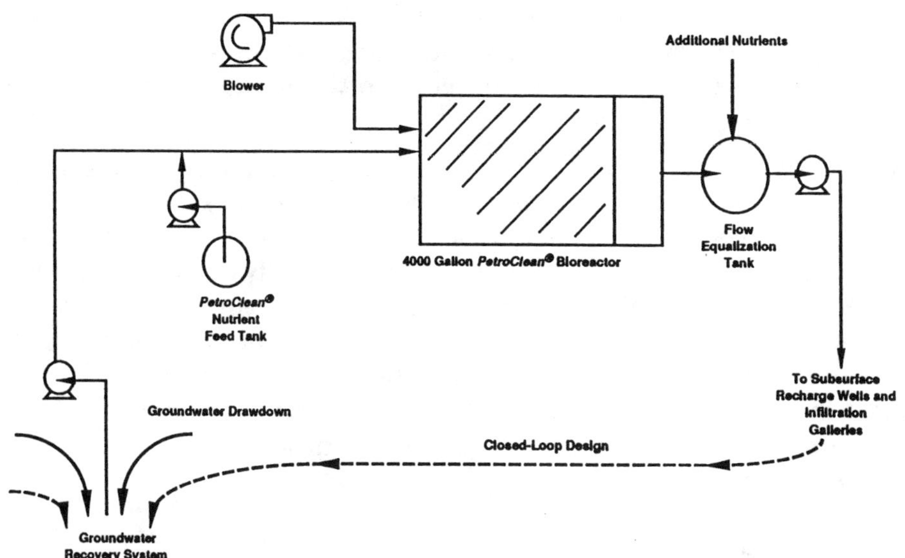

FIGURE 6. Schematic representation of the closed-loop, *in situ* PetroClean Bioremediation System used for the treatment of Stoddard solvent contaminated soil and groundwater.

Once the subsurface recharge system was installed, the client began constructing the new building. When construction was completed, the bioreactor and associated equipment were mobilized and installed. The bioreactor was set in place outside the new building. Electrical connections were made and the bioreactor system was plumbed into the previously installed groundwater extraction and subsurface recharge systems. The bioreactor was then filled with recovered groundwater and inoculated with a specifically adapted, Stoddard-degrading microbial culture. This microbial culture was generated in the EBIO bacteriology laboratory using microorganisms isolated from the site during the biofeasibility/pilot-scale evaluation phase. Continuous flow of recovered groundwater through the bioreactor was then initiated.

Nutrients were continuously fed to both the aboveground bioreactor and the subsurface soil environment, via the bioreactor effluent, to balance the contaminant concentrations. The bioreactor effluent served as the vehicle for providing increased concentrations of essential nutrients, dissolved oxygen, and a continual source of Stoddard solvent- and petroleum hydrocarbon-degrading microorganisms to the subsurface zone of soil contamination. These additions to the subsurface soils created the proper environmental conditions for enhanced biodegradation.

To monitor contaminant removal efficiencies of the bioreactor, influent and effluent samples were collected twice per month and analyzed for TPHC, Stoddard solvent, and mop oil concentrations, as well as biological parameters. During the course of the *in situ* treatment, the bioreactor averaged 99 percent removal of Stoddard solvent (Figure 7), 96 percent removal of mop oil (Figure 8), and 95 percent removal of TPHCs (Figure 9). The bioreactor operated at a continuous flow rate of 11 to 19 L/min (3 to 5 gpm). Based on the bioreactor volume of 15,000 L (4,000 gal), the HRT varied between 13.3 h and 22.2 h, depending on the recovery rate at any given time.

Typically, when a closed-loop *in situ* treatment system is employed at a site where the affected soils have a good permeability and the contaminants have been fairly uniformly dispersed throughout a portion of the vadose zone and at the groundwater interface, the recovered groundwater displays a sharp increase in contaminant concentrations as the *in situ* loop is closed. After sufficient soil flushing and biodegradation have occurred, the bioreactor influent contaminant concentrations will steadily decline. Therefore, the influent contaminant concentrations can be used as indicators of remedial progress. However, at this particular site the soils were not highly permeable,

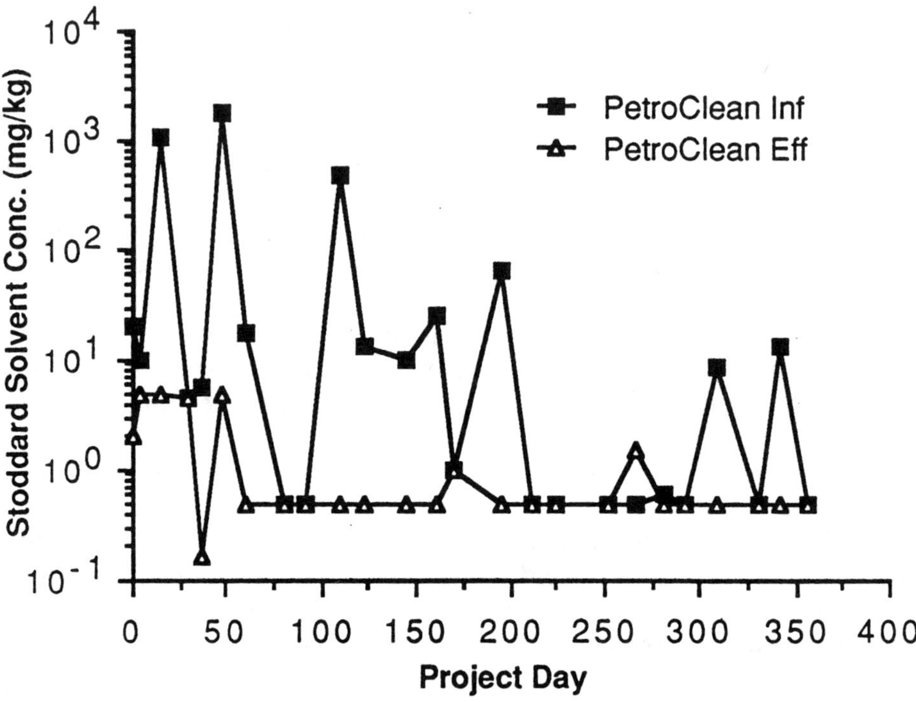

FIGURE 7. Efficiency of Stoddard solvent removal.

and the contaminants were apparently localized in discrete zones at various depths above the groundwater, making it more difficult to distinguish a steady decline in influent contaminant concentrations. As a result of these factors, the influent contaminant concentrations periodically increased sharply as these pockets of contaminants were successfully moved toward the recovery wells and pumped up to the bioreactor for treatment. During the first 2 months of treatment, the bioreactor influent contained significant concentrations of the three contaminants, ranging as high as 1,800 mg/L Stoddard solvent, 880 mg/L mop oil, and 8,800 mg/L TPHCs. Although the influent contaminant concentrations showed some dramatic fluctuations throughout the remediation, the overall trend after the second month of treatment was decreasing. In fact, after the fourth month of treatment, the majority of influent samples were below the detection limit (BDL) for both Stoddard solvent and mop oil. After the fourth month of treatment, the influent concentrations of both Stoddard solvent and mop oil peaked once more, during the sixth month of operation. The concentrations detected were 64 mg/L for Stoddard solvent and 53 mg/L for mop oil. During the 12 sampling events after this peak,

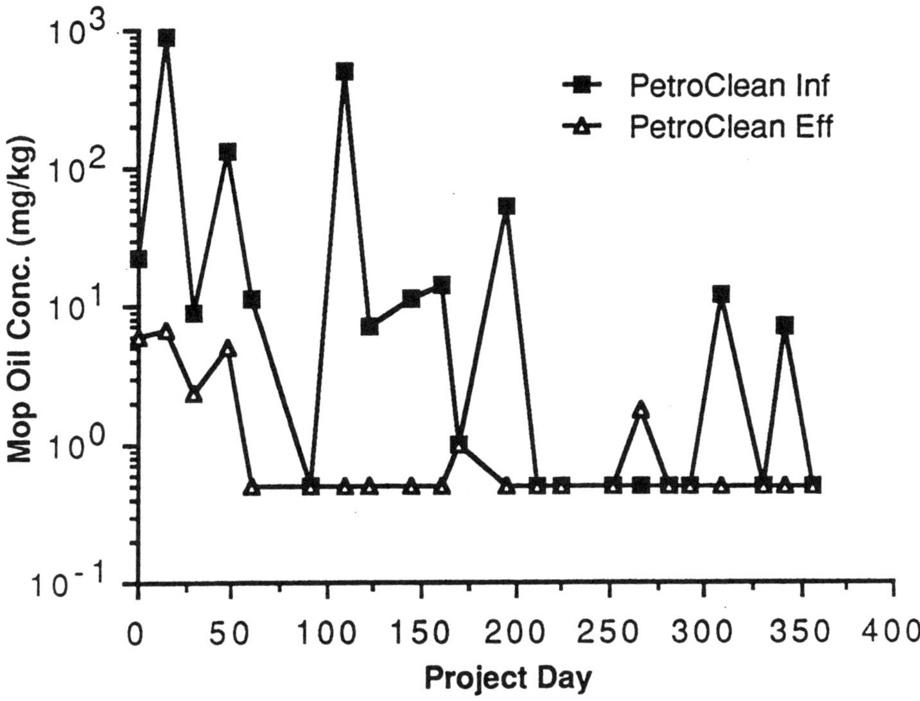

FIGURE 8. Efficiency of mop oil removal.

only 4 influent samples contained concentrations of Stoddard solvent above the limit of detection, and only 3 contained detectable mop oil concentrations. The influent Stoddard solvent concentrations ranged from 13 mg/L to nondetectable. The influent mop oil concentrations ranged from 14 mg/L to nondetectable. In all cases, the corresponding effluent samples contained no detectable concentrations of either contaminant. During this same time period, only low mg/L concentrations of TPHCs were detected in the bioreactor influent. Concentrations of TPHCs after the fourth month of operation ranged from 58 mg/L (during the sixth month) to nondetectable.

Aggressive treatment of the contaminated subsurface soils is the distinguishing advantage of *in situ* bioremediation over other remedial technologies. By creating the proper environmental conditions in the subsurface soils, natural biological activity can be enhanced to achieve, in a matter of months, what nature may require decades to accomplish. To monitor and evaluate overall treatment progress, split-spoon soil samples were collected prior to initiating biotreatment, and then on a quarterly basis during treatment system operation. Three locations were selected within the area of initial Stoddard solvent soil

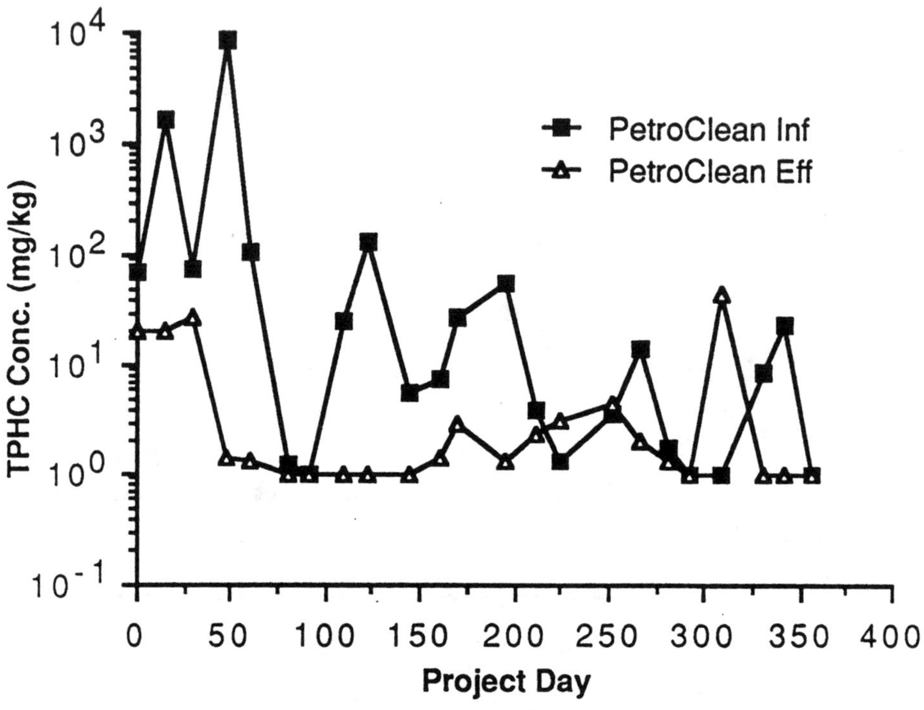

FIGURE 9. Efficiency of TPHC removal.

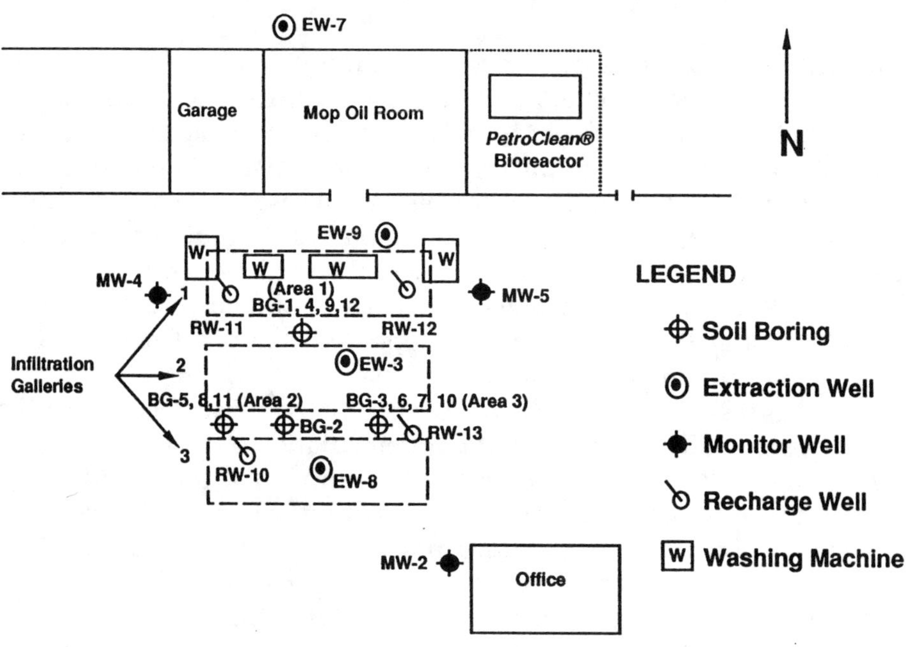

FIGURE 10. *In situ* biotreatment system design.

contamination. Area 1 (i.e., BG-1, BG-4, BG-9, and BG-12) was located downgradient of extraction well EW-3. The location of Area 2 (i.e., BG-2, BG-5, BG-8, and BG-11) was near recharge well RW-10. Area 3 (i.e., BG-3, BG-6, BG-7, and BG-10) was located upgradient of EW-3, near recharge well RW-13 (Figure 10). These three locations were sampled during the quarterly sampling events, and the results of the soil analyses are shown in Table 4.

TABLE 4. Bioremediation progress in subsurface soils.

Boring	Sampling Interval	Stoddard Solvent [a] (mg/kg)	Mop Oil[a] (mg/kg)	TPHCs[b] (mg/kg)
Area 1:				
BG-1	Pre-startup (11/88)	410	1,200	1,500
BG-4	2nd Quarter (7/89)	9	17	35
BG-9	3rd Quarter (11/89)	1,800	6,300	6,200
BG-12	4th Quarter (3/90)	nd (5)	10	50
Area 2:				
BG-2	Pre-startup (11/88)	930	2,200	2,700
BG-5	2nd Quarter (7/89)	nd (1)[c]	20	25
BG-8	3rd Quarter (11/89)	nd (1)	nd (1)	nd (1)
BG-11	4th Quarter (3/90)	nd (5)	nd (5)	nd (5)
Area 3:				
BG-3	Pre-startup (11/88)	410	2,200	2,300
BG-6	1st Quarter (5/89)	230	320	4,200
BG-7	3rd Quarter (11/89)	200	1,500	320
BG-10	4th Quarter (3/90)	nd (5)	nd (5)	nd (5)

(a) Analysis by GC-FID using EPA Method 8015.
(b) Analysis by IR using EPA Method 418.1.
(c) nd = not detected (detection limit).

The results of chemical analysis of the soil samples indicate that enhanced *in situ* bioremediation was very successful in removing the Stoddard solvent and mop oil contaminants. The contaminants in Areas 2 and 3 were reduced to below the limit of detection within 12 months. Area 1, which was in the area containing the highest contamination during the initial site investigations (i.e., in the area around EW-3), also showed significant reductions in the contaminant concentrations during the first 12 months of *in situ* treatment. Soils in this area were found to contain as much as 3,500 mg/kg Stoddard solvent and 9,700 mg/kg TPHCs during the initial site assessments. In Area 1, after 12 months of enhanced *in situ* biological treatment using the PetroClean system, Stoddard solvent was reduced to below the limit of detection. Mop oil concentrations were reduced to 10 mg/kg, while TPHC concentrations were reduced to 50 mg/kg.

The *in situ* bioremediation system remained in operation for an additional 4 months while closure of this site was pursued with the California Regional Water Quality Control Board. Approval for site closure was received in July 1990. The aboveground system components were disconnected from the recovery and recharge components and shipped back to the EBIO facilities in Raleigh NC. A quarterly groundwater monitoring program has been implemented for a period of 1 year.

Conclusions

An *in situ* bioremediation system was designed and implemented at a site which had been contaminated by leaking underground storage tanks. Based on the results of the site investigation and pilot-scale bioremediation evaluation, a closed-loop, PetroClean Bioremediation System was installed to achieve *in situ* bioremediation after a new building had been constructed over the contaminated area. This patented system employs both aboveground and subsurface components to enhance biological degradation of contaminants in recovered groundwater, as well as in the *in situ* soil and groundwater.

Contaminant removal efficiencies of the bioreactor averaged 95 percent for TPHCs, 96 percent for mop oil, and 99 percent for Stoddard solvent. During the first 2 months of treatment, the bioreactor influent contaminant concentrations ranged as high as 1,800 mg/L Stoddard solvent, 880 mg/L mop oil, and 8,800 mg/L TPHCs. After the fourth month of treatment, the majority of bioreactor influent contaminant concentrations were BDL for both Stoddard solvent and mop oil. During this same time period only low mg/L concentrations of TPHCs were detected in the bioreactor influent.

The three areas sampled quarterly for soil analysis showed significant improvement as a result of enhanced, *in situ* bioremediation. Contaminants in two of the three soil sampling areas were reduced to BDL within 12 months of implementing the *in situ* biotreatment system. The third area, which originally contained the highest concentrations of soil contaminants (i.e., the area around MW-3), showed contaminant reductions from the original concentrations of 3,500 mg/kg Stoddard solvent and 9,700 mg/kg TPHCs to BDL for Stoddard solvent, 10 mg/kg for mop oil, and 50 mg/kg for TPHCs. Approximately 4,600 m^3 (6,000 yd^3) of contaminated soil were remediated at a cost of approximately \$59/m^3 (\$45/yd^3). In addition, more than 5.7 million L (1.5 million gal) of contaminated groundwater were recovered and treated at no additional cost. The biological restoration of this site was accomplished in 16 months. This remediation technology was completely compatible with the client's desire to expand production capabilities and the California Regional Water Quality Control Board's desire to achieve site clean-up. The PetroClean Bioremediation System was capable of achieving these goals without disrupting the client's daily operations.

REFERENCES

Alexander, M. *Introduction to Soil Microbiology,* 2nd ed. John Wiley & Sons; New York, 1977.

Caplan, J. A.; Schmitt, E. K.; Malone, D. R. U.S. Patent 4 992 174, 1991.

Lieberman, M. T.; Schmitt, E. K.; Caplan, J. A.; Quince, J. R.; McDermott, M. P. "Biorestoration of Diesel Fuel Contaminated Soil and Groundwater at Camp Grayling Airfield Using the PetroClean® Bioremediation System." *Proceedings,* Conference on Petroleum Hydrocarbons and Organic Chemicals in Groundwater: Prevention, Detection and Restoration, Houston TX, 1989.

Schmitt, E. K.; Caplan, J. A. "*In situ* Biological Cleanup of Petroleum Hydrocarbons in Soil and Groundwater." *Proceedings,* Fifth Annual Hazardous Materials Management Conference, Atlantic City NJ, 1987.

Test Methods for Evaluating Solid Waste; U.S. Environmental Protection Agency Manual SW-846, Office of Solid Waste and Emergency Response: Washington, DC, 1986.

Thomas, J. M.; Ward, Jr., C. H. "*In Situ* Biorestoration of Organic Contaminants in the Subsurface." *Environmental Science and Technology* **1989,** *23,* 760–765.

In Situ Decontamination of Heavy Metal Polluted Soils Using Crops of Metal-Accumulating Plants—A Feasibility Study

A.J.M. Baker[*]
University of Sheffield,
R. D. Reeves
Massey University
S. P. McGrath
A.F.R.C. Rothamsted Experimental Station

INTRODUCTION

Decontamination of soils and wastes polluted with heavy metals remains one of the most intractable problems of clean-up technology. Techniques currently in use are based on either immobilization *in situ* or extraction by physicochemical methods (U.S. Army 1987). They require sophisticated equipment and are accordingly expensive and appropriate only for small areas where rapid, complete decontamination is required. Furthermore, at the end of the decontamination process, virtually all biological activity has been lost from the soil medium. We are developing a low-technology "green," *in situ* approach to achieve both partial decontamination and site restoration for sites where a more rigorous, rapid, and costly treatment is not appropriate. If successful, it could represent an early stage of site treatment prior to more conventional revegetation procedures.

Some native plant species are able to accumulate unusually high concentrations of potentially phytotoxic elements such as Cd, Co, Cu, Pb, Ni, and Zn from metalliferous soils (Baker & Brooks 1989). Over the last two decades, comprehensive analytical data have been collected

[*] Department of Animal and Plant Sciences, University of Sheffield, P.O. Box 601, Sheffield, South Yorkshire, S10 2UQ, A.E.R.C. Ltd, Colchester, United Kingdom

from plants growing over metal-enriched soils to identify hyperaccumulators. Such plants in Europe are species endemic to areas of natural mineralization and mine spoils resulting from ore extraction. They include species of *Alyssum* and *Thlaspi* (Brassicaceae) from serpentine soils, which can take up Ni to concentrations of more than 2 percent (dry weight), and species of *Thlaspi* from calamine soils, which can accumulate >3 percent Zn, 0.5 percent Pb, and 0.1 percent Cd in their shoots. The use of such plants for specific extraction of metals from mine wastes seems impractical because of the depth of materials, but they may prove an effective and practical means of metal removal from superficially contaminated soils such as those produced by the land disposal of industrially polluted sewage sludges or other metalliferous wastes. In view of the current concern about past and present land application of such sludges, the feasibility of using metal-accumulating plants to decontaminate polluted soils is being assessed.

EXPERIMENT

The Woburn Market Garden Experiment at Rothamsted Experimental Station was originally set up in 1942 to examine the usefulness of organic wastes as nutrient sources for horticultural crops. One treatment was an application of sewage sludge that had been inadvertently obtained from a works receiving industrial effluents mixed with domestic sewage. Twenty years of application of sludge from this works (up until 1961 when treatments were discontinued) and mixing to plough depth have resulted in typical total soil concentrations (mg/kg dry soil) of Zn-381 and Cd-11, in excess of the Commission of the European Communities Directive (C.E.C. 1986), and of Ni-42, Cu-126, Pb-141, and Cr-147 approaching maximums. Co is slightly enriched at 6 mg/kg. The organic carbon content is <2 percent. Direct toxic effects of this soil on crop growth and indirect effects on microbial activity have been detected (Giller & McGrath 1989). Other plots have been treated at similar rates of application with uncontaminated farmyard manure. Typical metal loadings (mg/kg) here are Zn-104, Cd-2, Ni-17, Cu-30, Pb-53, Cr-42, and Co-5. In 1988, bulk soil samples were collected from the center of plot numbers 39, a heavily sludged soil (SL), and 2, a farmyard manured plot (FY); an intermediate soil (MX) was created by mixing soils SL and FY in equal proportions.

A pot trial was set up with these three soils to assess the growth and metal uptake of six species of Brassicaceae under glasshouse conditions. The species were: *Brassica oleracea* cv. Greyhound, cabbage;

Raphanus sativus cv. French Breakfast, radish; *Thlaspi caerulescens*, a strong Zn accumulator from N.W. Europe; *Alyssum lesbiacum* and *A. murale*, both Ni hyperaccumulators from serpentine soils in Greece; and *Arabidopsis thaliana*, a widespread weedy species. The first two species were included to compare metal uptake by potential accumulator plants with leaf and root brassica crops. Plant/soil combinations were set up in a randomized block design [Species(6) × Soils(3) × Blocks(6)]. After 5 weeks, shoot material was harvested from five uniformly sized plants per pot, dried, and heavy metal contents determined by I.C.P. spectrometry of tissue digests.

RESULTS

The summarized plant analyses for Zn, Ni, Cd, Cr, Cu, Co, and Pb for the six species are shown in Figure 1. Data for the two crop species show increasing uptake of most metals in the soil series FY-MX-SL. *Thlaspi caerulescens* accumulated similar high concentrations of Zn from all soils and was about 10 times more efficient than cabbage and radish; it also showed enhanced uptake of Cd, Cu, and Ni. The two *Alyssum* species showed the predicted extreme accumulation of Ni, but also unexpectedly high Zn and to a certain extent Cd, Cr, and Cu. *Arabidopsis thaliana* showed some Zn accumulation, but in all other respects mirrored the two crop species.

DISCUSSION

The pot trial reported here detected enhanced levels of most of the heavy metals in the edible parts of the two crop species employed. However, although uptake increased in the soil series FY-MX-SL, most metals were present at trace concentrations only. This is both good and bad: the levels accumulated at present do not pose a significant health risk (Cd excepted), but the findings also suggest that it will take centuries to millenia before normal cropping can reduce the soil concentrations of these elements to background values. The experiment also showed that notable accumulator species for Zn and Ni are capable of accumulating these metals from low as well as high background concentrations (Morrison *et al.* 1980). A further important discovery was that the accumulation was not as metal-specific as predicted, suggesting

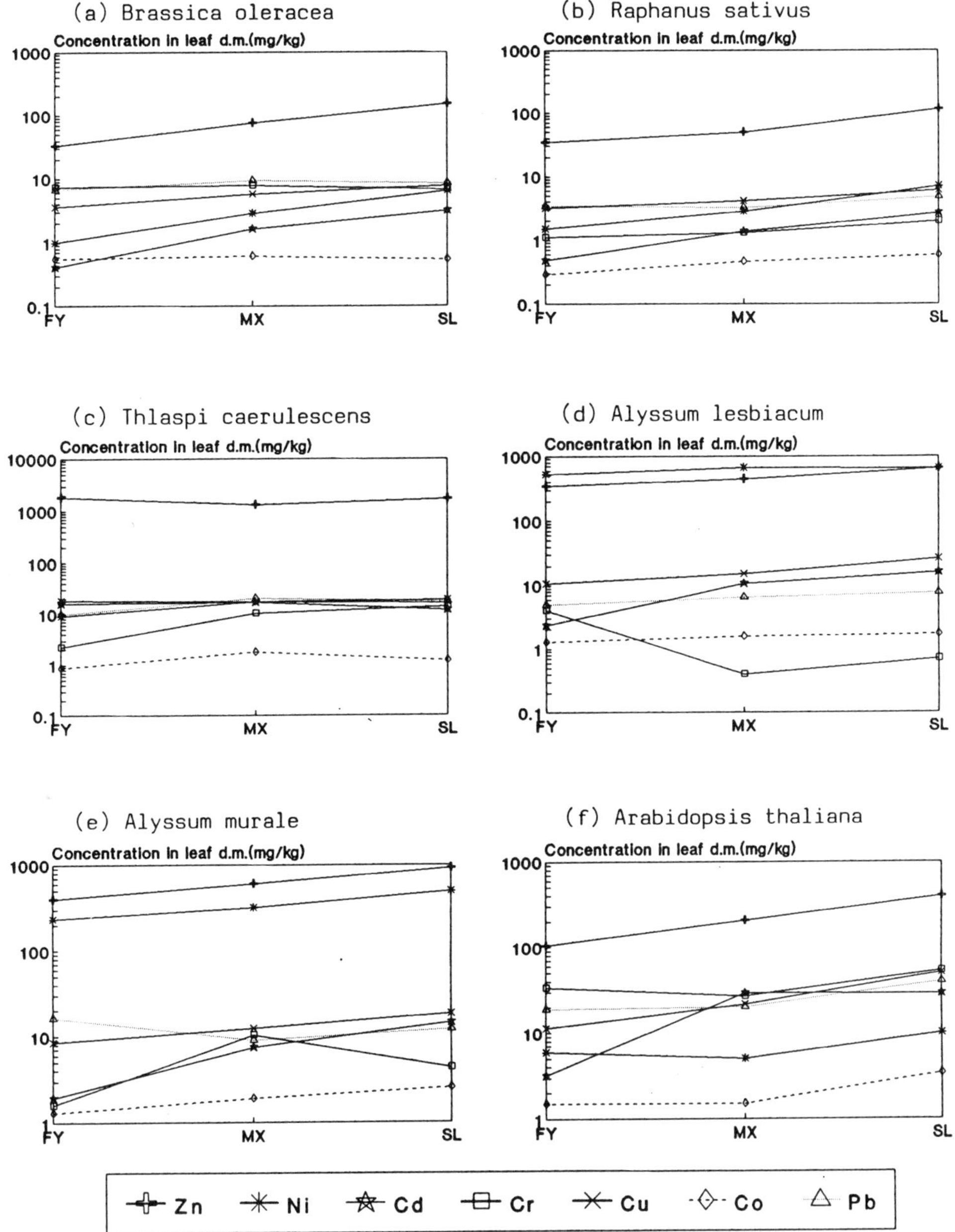

FIGURE 1. **Mean concentrations of Cd, Co, Cr, Cu, Ni, Pb, and Zn in the foliar dry matter of six species of Brassicaceae, grown for 5 weeks on soils from the Woburn Market Garden Experiment.**

that hyperaccumulators could be effective in situations of multiple metal contamination.

The potential exploitation of metal uptake into plant biomass as a means of soil decontamination is clearly limited by plant productivity and the concentrations of metals achieved. This problem was discussed by Ernst (1988), who concluded that the option was not a viable one for mine tailings, spoil heaps, and smelter wastes where regular cropping with metal-tolerant plants can result in a decontamination percentage of <1 percent in a century. However, for crops of hyperaccumulator plants, a trade-off between high metal concentration and slow growth rate may produce a more favourable scenario (Chaney 1983). The plants raised in this pot trial achieved only a very small fraction of their potential biomass (a few mg) over the limited duration of the experiment. For example, plants of the *Thlaspi* have been collected from the field with a dry weight of >20 g, which if scaled up under intensive cropping conditions (100 plants/sq m) could produce a yield of >20 t/ha above-ground dry matter. Field observations also suggest that several of the biennial and short-lived perennial *Alyssum* species can be even more productive.

Assuming that the same metal concentrations can be maintained in mature plants as in the 5-week seedlings in this trial (other evidence suggests that they may actually increase), it is possible to extrapolate to the crop situation. On this basis, a crop of *Thlaspi* could take up 34.3 kg/ha of Zn, 0.16 of Cd, 0.25 of Ni, 0.22 of Pb, 0.40 of Cu, 0.27 of Cr, and 0.02 of Co. The efficiency of the uptake can be estimated by comparing this with the extra metal loadings on soil SL above soil FY. Assuming that metal concentrations are uniform to a depth of 20 cm and a soil bulk density of 1.3 g/cu cm, the percentage of metals removed in one crop could reach 4.8 Zn, 0.7 Cd, 0.4 Ni, 0.1 Pb, 0.2 Cu, 0.1 Cr, and 0.9 Co. In order to polish soil SL down to current U.K. D.o.E. (1989) limits (and C.E.C. limits), a minimum of six croppings with *Thlaspi* would be needed for Zn and about 130 for Cd. Similar calculations for *Alyssum* suggest it could also be equally useful in both Ni and Zn decontamination. Cropping with mixtures of hyperaccumulator species rather than using monocultures may present an attractive option for multiple metal accumulation from soils (such as SL) contaminated with a mixture of heavy metals. The scaling up from pot trials to the field situation is clearly fraught with practical problems, but the calculations from this and other similar trials have encouraged us to extend the study to a field scale. A project supported under the C.E.C.'s A.C.E. Programme is now under way.

The technique proposed here presents an attractive, low-technology biological solution to an otherwise intractable problem. However, the cropping of contaminated land with hyperaccumulator plants presents its own problem—the disposal of a potentially hazardous biomass. It is suggested that one option could be the controlled ashing of harvested material to yield a residue in which metals such as Zn and Ni may be concentrated at >10 percent. Economic recovery could be achieved by adding the ash to an appropriate smelter feedstock. A less attractive alternative would be the disposal of the ash to landfill in the normal way for hazardous materials. Bioaccumulation may also result in the recovery of economic quantities of other rarer metals not discussed here.

REFERENCES

Baker, A.J.M.; Brooks, R. R. *Biorecovery* **1989,** *1,* 81–126.

C.E.C. (Commission of the European Community). *Official Journal of the European Communities* **1986,** *L 181,* 6–12.

Chaney, R. L. In *Land Treatment of Hazardous Wastes*; Parr, J. F.; Marsh, P. B.; Kia, J. M., Eds.; Noyes Data Corporation: Park Ridge, NJ, 1983, pp 50–76.

Ernst, W.H.O. In *Proceedings of the 3rd International Conference on Environmental Contamination, Venice*; Orio, A. A., Ed.; C.E.P. Consultants: Edinburgh, 1988; pp 305–310.

Giller, K.; McGrath, S. P. *New Scientist* **4 November 1989,** 31–32.

Morrison, R. S.; Brooks, R. R.; Reeves, R. D. *Plant Science Letters* **1980,** *17,* 451–457.

U.K. Department of the Environment. *Code of Practice for the Agricultural Use of Sewage Sludge*; HMSO: London, 1989.

U.S. Army (U.S. Army Toxic and Hazardous Materials Agency). "Heavy Metal Contaminated Soil Treatment"; Interim Technical Report AMXTH-TE-CR-86101; Roy F. Weston Inc.: West Chester, PA, 1987.

List of Contributors

* Contributions of authors whose names appear in *italics* on this list can be found in the companion volume, *On-Site Bioreclamation*.

Index

[*] Page numbers in *italics* can be found in the companion volume, *On-Site Bioreclamation*.

F

G

H

O

P

T